Water, Energy & Food Sustainability in the Middle East

Adnan Badran
Chief Editor

Sohail Murad • Elias Baydoun • Nuhad Daghir
Editors

Water, Energy & Food Sustainability in the Middle East

The Sustainability Triangle

Springer

جامعة البترا

Chief Editor
Adnan Badran
University of Petra
Amman, Jordan

Editors
Sohail Murad
Chemical and Biological Engineering
Illinois Institute of Technology
Chicago, IL, USA

Elias Baydoun
Department of Biology
American University of Beirut
Beirut, Lebanon

Nuhad Daghir
Faculty of Agricultural and Food Sciences
American University of Beirut
Beirut, Lebanon

ISBN 978-3-319-48919-3 ISBN 978-3-319-48920-9 (eBook)
DOI 10.1007/978-3-319-48920-9

Library of Congress Control Number: 2017934068

Printed on acid-free paper

This Springer imprint is published by Springer Nature
The registered company is Springer International Publishing AG
The registered company address is: Gewerbestrasse 11, 6330 Cham, Switzerland

Introduction and Scope

Adnan Badran, Elias Baydoun, and Sohail Murad

The importance of climate change and its impact on society in general is clear from the proceedings of the recent 21st Conference of the Parties of the UNFCCC (United Nations Framework Convention on Climate Change) in Paris and adopted by consensus on 12 December 2015.[1] The aim of the convention was as follows[2]:

(a) Holding the increase in the global average temperature to well below 2 °C above preindustrial levels and to pursue efforts to limit the temperature increase to 1.5 °C above preindustrial levels, recognizing that this would significantly reduce the risks and impacts of climate change
(b) Increasing the ability to adapt to the adverse impacts of climate change and foster climate resilience and low greenhouse gas emissions development, in a manner that does not threaten food production
(c) Making finance flows consistent with a pathway toward low greenhouse gas emissions and climate-resilient development

Countries furthermore aim to reach "global peaking of greenhouse gas emissions as soon as possible."

While climate change is a problem that will challenge all countries, countries in the Middle East are especially vulnerable, because they are in a weak position to politically and economically cope with the expected damage to the environment

[1]"Framework Convention on Climate Change" (PDF). United Nations FCCC Int. United Nations. 12 December 2015. Retrieved 12 December 2015.

[2]"FCCC/CP/2015/L.9/Rev.1" (PDF). UNFCCC secretariat. Retrieved 12 December 2015.

Adnan Badran
University of Petra, Amman, Jordan

Elias Baydoun
Department of Biology, American University of Beirut, Beirut, Lebanon

Sohail Murad
Chemical and Biological Engineering, Illinois Institute of Technology, Chicago, IL, USA

and the economic well-being of the region. This is primarily because many governments in the region are dysfunctional and are unlikely to undertake the long-term planning and complicated policies needed to cope with a changing world climate. Climate change has also resulted in major political upheavals in the area. In Syria, for example, the 5-year drought that began in 2006 resulted in drastically impoverishing farmers which led to their movement to urban areas, where no government support services were provided. This exodus led to unrest, social discontent, and tearing down the entire social and political fabric of the country – the consequences of this unrest are being felt all over the world, especially in neighboring countries and Europe. While climate change alone is not clearly responsible for the entire problem in, for example, Syria, it would be equally wrong to discount the role of climate change completely.

Most of the countries in the Middle East are essentially arid and receive very little rainfall (250–400 mm annually), and the expected drop from such limited amounts can be especially disastrous to domestic, agricultural, and industrial diversity.

Another problem facing countries in the region is the rise of sea level resulting from polar ice melts; for example, it is estimated that up to 30% of coastal areas could be submerged under water over the course of the twenty-first century in Middle East regions, such as the Nile Delta, the food basket of Egypt, where the sea level is expected to rise by half meter by 2025. The fertile Nile Delta provides around a third of the crops for Egypt's population of 80 million. "As a result, over half a million inhabitants may be displaced and approximately 70,000 jobs could be lost." Other estimates include serious threats to Alexandria, Egypt, of 4.1 million people and 40% of the country's industry. In addition, the Middle East is unlike other regions that are very susceptible to climate change in Asian and North American countries, where it imports almost half its food supplies. If food production in these areas is threatened, it could have additional impact on the Middle East.

According to the latest Intergovernmental Panel on Climate Change (IPCC),[3] higher temperatures (1–3 °C) are a likely outcome of climate change in the MENA region, and reduced precipitation will result in an addition of 80–100 million people in water-stressed areas. This will result in a decrease in agricultural yields, in rain-fed areas, coastal flooding, heat waves, and lower air quality, all further lowering the quality of life in the region.

There are, however, some encouraging signs. There continues to be increased awareness among decision makers in the Middle East region on the possible disastrous consequences of climate change, which likely is caused by the global increase in such awareness especially among the more enlightened leaders of the region. The awareness is also a result of more frequent droughts in the region and the worsening water supply shortage.

During the annual meeting of the Arab Academy of Sciences in Beirut in December 2015, as part of the final discussions, the urgent need of a book on the

[3]https://www.ipcc.ch/index.htm

consequences of climate change on the water, energy, and food triangle was discussed, as well as a comprehensive book that first describes the problem and then also provides possible solutions. This book is the end result of those discussions. The book is divided into three sections (although, as is often in science, the sections overlap). The first section is concerned with water. Water sustains life, environment, and development. Water crisis in terms of quantity and quality is a man-made disaster linked to the degradation of the life support ecosystem. The section on water includes chapters on the politics of water and the concept of virtual water, as well as water treatment and conservation strategies that are particularly applicable to Middle Eastern countries and are relevant for efficient agriculture. By 2050, it is estimated that half of the people on our planet will be living in water scarcity. Sustainable management of an integrated approach of water, land, and people for a sound ecosystem is needed. Sixty percent of renewable water resources in Arab countries originate from sources outside the region. Shared management of downstream with upstream riparian countries is imperative. The total Arab water resources is 371.8 billion m^3 distributed as follows: 41% Mashreq Arab States, 23.4% Maghreb Arab States, 31% Nile Arab States, and 4.6% Arab Peninsula (Gulf States).

Already one-third of the world population is living in water-scarce or water-short areas. Climate change will accelerate the figure to one-half; there are 13 Arab countries among the world's most water-scarce nations, and water availability in 8 Arab countries is less than 200 m^3/capita/year, less than half of the UN-designated water-severe country (UN severe water scarcity below 500 m^3/capita/year).

The Arab region houses 5% of the world population and occupies 10% of Earth space, with only 1% of world water resources. This is why the Arab region shares 50% of the world desalination capacity and is expected to double by 2020. Agriculture consumes 87% of available water resources, (the highest) as compared to 70% of the world average. The industrial sector consumes 7% and domestic use 6%.

It is a crisis of water management, fragmented institutions, inadequate policies and legal systems, lack of political will, and a widening gap between science and policy making at the national, regional, and global levels. Twelve percent of the world's population uses 85% of its fresh water. And water supply resources are being stretched to their limits. By 2050, an additional 3 billion people will be born mostly in countries already suffering from water shortage.

The second section is concerned with energy. Energy according to the IPCC Nov 2014 report; the world's electricity must be produced from zero carbon sources by 2050; otherwise, our planet faces irreversible damage. The report says renewables have to grow from 30% share of the power sector to 80% and all fossil fuel generation without carbon capture and storage (CCS) has to be phased out by 2100. Strategies to harvest it from abundant solar sources are suited to the Middle East, as it is situated in the desert sunbelt where sun is available over 90% of the year over vast areas in most countries of the region, as well as steps that can be undertaken for its efficient use as well as conservation. It includes producing hydrocarbon fuels from lingo-cellulose (agricultural waste is one such example). This is especially suited for many already oil-producing countries in the area since it eliminates the need for new infrastructure for ethanol-based fuels, for example. Two chapters focus on two important alternate energy sources that are especially important for the Middle East, viz., solar energy and wind energy. Morocco is an advanced country in utilizing renewable energy, sun and wind, in their energy mix. The Desert Tech Solar Project is aiming to export electricity by feeding it to the grid of Europe from the North Africa MENA region. Jordan although is moving slowly because of its governmental bureaucracy and insufficient grid capacity to accommodate renewables and is targeting to achieve 10–20% renewables in the energy mix by 2020. Emirates and Saudi Arabia have ambitious plans, in advancing renewables to reduce dependence on fossil fuel energy. Experts say that renewables have to grow to 80% share of the power sector by 2050.

Two final chapters focus on energy conservation strategies. The first focusses on the use of phase change materials (PCM) for both energy storage during non-peak and peak hours and other simple steps that can be taken to reduce energy usage. The second describes NetZero energy building that can either be designed or retrofitted to make them more efficient during renovations. Finally, to demonstrate the close connection between water and energy, a chapter is included on strategies to minimize energy use for water treatment technologies. This chapter serves as the link between the sections on water and energy. This chapter then leads to the last section on food.

This last section of the book addresses issues related to food. Included are chapters on food safety and security as they pertain to specific conditions in the Middle East. Because of the abundance of sea water in many Middle Eastern countries, one chapter focuses on agriculture based on saline (brackish) water as well as improving fresh water use efficiency. In addition, there is also a chapter on the diminishing arable land in the Middle East and how erosion can be minimized or reversed. There is also a chapter on the impact of food losses and waste on food security since losses constitute over 35% of food produced in the region. Finally, a chapter is included which discusses the technologies available for agriculture in water-challenged regions, especially those areas dependent largely on rain water for agriculture.

With climate change and managing scarce resources of water, food and water are inextricably linked. Therefore, food security for self-sufficiency could be achieved through right policies, improved agricultural and irrigation technologies with high-yield cultivars suitable for semiarid zones, and conservation of water through protected agricultural innovative practices to bring down the high use of water of 87% to the world average of 70%.

There is no doubt that the Climate change and Water – Energy – Food security Nexus in the Arab Middle East, is becoming more complex due to rapid population growth and growing demands by industrial and agricultural developments. Therefore, science becomes crucial in providing the basis for sound governance and a holistic approach enlightened policy linked to energy and water management for sound food security.

Contents

Part I
Water

Elias Baydoun

Climate Change and Water Science Policy in Management

Adnan Badran

Abstract Water sustains life, the environment and development. Human rights to water, as water is becoming a commodity threatens the poor. Global water crisis in term of quantity and quality is a man-made disaster linked to environmental imbalance and degradation of the life-support ecosystem. It is a crisis of water management, fragmented institutions, inadequate policies and legal systems, lack of political will, and a widening gap between science and policy making at the national, regional and global levels. Already one third of the world population is living in water-scarce or water-short areas. Climate change will accelerate the figure to one-half. 12% of the world's population uses 85% of its fresh water. And water supply resources are being stretched to their limits. By 2050 an additional three billion people will be born mostly in countries already suffering from water shortage.

According to the IPCC-Nov 2014 report, the world's electricity must be produced from zero carbon sources by 2050; otherwise, our planet faces irreversible damage. The report says renewables have to grow from 30% share of the power sector to 80%. And all fossil fuel generation without carbon capture and storage (CCS) has to be phased out by 2100.

Global warming is unequivocally linked to human interference in the ecosystem, causing glaciers to melt on the polar ice caps resulting in the rise of sea level flooding of agricultural coastal areas. The Nile delta, which is the food basket of Egypt housing 46 million people, may disappear. Coastal fresh water aquifers may be flooded by seeping seawater threatening food security of many large regions of the world.

With the advent of climate change, most of water stressed areas particularly in arid and semi-arid zones (Middle East and MENA regions) will face a rainfall decline of 20% and a temperature rise of 2–3 °C that would result in large losses of water resources, basic food, basic needs, and increased poverty.

Water science is a "must" in developing a unique water management scheme. It contributes to well-defined policies for efficiency, sound strategy and sustainable plans of action. There is unlimited potential with what science can do on our planet,

A. Badran (✉)
University of Petra, Amman, Jordan
e-mail: abadran@uop.edu.jo

S. Murad et al. (eds.), *Water, Energy & Food Sustainability in the Middle East*,
DOI 10.1007/978-3-319-48920-9_1

where salt water and fresh water comprise 97.5% and 2.5% of planet waters, respectively. With 70% of waters tied in polar caps, only 30% is left in rivers, lakes and ground waters to humanity.

Food production accounts for 70% of water used in the world. Through the efficient use of water by renewables for desalination, recycling water for agriculture, using new cultivars under stress of low level of waters and brackish waters and genomes resistant to pests and droughts etc., have to intrigue scientists in our part of the world to find a lasting solution to the challenging problem.

Fundamental change in water policies and engaging science to develop a unique water management scheme is imperative. Currently, water policies are divorced from sound science. Demands should be managed by a new culture of efficiency, cutting losses, and protecting water from overuse and pollution.

There is no doubt that the Climate change and Water – Energy – Food security Nexus in the Arab Middle East, is becoming more complex due to rapid population growth and growing demands by industrial and agricultural developments. Therefore, science becomes crucial in providing the basis for sound governance and a holistic approach enlightened policy linked to energy and water management for sound food security. The potential of what modern science R&D can do is without limits.

Keywords Climate change • Water management • Global warming

Introduction

The improvement of water management techniques and technologies needed to cope with the projected increase in water scarcity will require new water science as well as extensive use of existing water science. Future water policies will have to be well informed by science if they are to be effective. Many existing water policies are not based on sound science and are aimed at goals other than ensuring that water is used efficiently, protected from qualitative degradation and maintained for future generations. The potential of science to contribute to the resolution of current and foreseeable water problems is virtually unlimited. There are numerous examples.

At the global level, the developments from nano-science can help in a variety of ways. Development of more effective ways of cloud seeding; development of nanomembranes for cleaning polluted water and improvements in diffusion technology which will lower the costs of desalination are important examples. Development of small-scale solar technology can improve energy generation and thus lower the costs of desalination. The importance of such a development can be illustrated by reference to the MENA region where solar energy falling on one square meter of surface annually is the BTU equivalent to one barrel of oil. Currently, the Arab region, with 5% of world population, produces 50% of desalinated water of the world (AFED 2010). Technology can help to extend the application of desalination and other water cleansing techniques to other areas throughout the world.

At the regional level, scientifically based management of shared water resources, whether surface or ground water, should be placed high on the agenda of countries with shared water basins. Effective bilateral and/or multilateral agreements should lead to stronger economic and political ties among countries with shared water-basins obviating the potential for conflict. The importance of dealing effectively with shared water is almost self-evident. This is particularly true in the MENA region where, "Of all renewable water resources…., two thirds originate from sources outside the region". (AFED 2010; El-Quosy 2009)

At the national level, science can contribute to the acquisition of knowledge about possible new water sources and about the application of techniques for using existing sources more efficiently. Thus, for example, agriculture accounts for 85% of water use in the Arab region as compared with a world average of 70%. On-farm irrigation efficiency remains at 35% so there is clearly room for improvement at the farm level (AFED 2010). Science can also contribute to the development of new crop strains that tolerate better both aridity and salinity. Rain-harvesting systems and efficiency improvements in science-based agricultural practices to achieve water savings should be emphasized. Other policy reforms leading to a new political economy of water management could focus on the acquisition of water "virtually" through imports of crops "from water-rich countries, while allocating scarce water resources to low-water consuming, high value crops that can generate foreign exchange". (AFED 2010, page 61).

In this way, food security may be achieved through a set of well-balanced trade and water management policies.

One potential new source of water is recycled wastewater. Wastewater generated by domestic and industrial sectors in the Arab region totals 10 km^3/year, of which 5.7 km^3 undergoes treatment. Of this volume of treated wastewater, only one third is reused. However, wastewater treatment plants currently handle waste loads that exceed their capacity limits. The untapped potential of wastewater should be the focus of appropriate policy interventions including national water management strategies for water reuse.

The focus of this chapter is on the importance of science in fashioning enlightened water policies to manage the intensifying global water scarcity. In the following section, the importance of water in sustaining life, the environment and economic development is discussed. Subsequently the decline of available water resources is characterized. The following sections focus on the importance of water science to fashioning solutions to the global water crisis; on the needs to build scientific competence and capacity and on issues related to making science based water policy. An important underlying theme that runs throughout the chapter is that the existing water scarcity in the arid and semi-arid Arab countries lies at the extreme edge of the global water scarcity picture. Moreover, it offers to other parts of the world, particularly those that are arid and semi arid, a picture of the future water situation likely to be visited upon them if the current situation is neglected.

Water Sustains Life, the Environment and Economic *Development*

Fresh water is tiny proportion of the water resources on earth, with salt water accounting for 97.5% of planetary waters and fresh water for only 2.5%. Seventy percent of the fresh water is tied up in polar caps, glacial ice and groundwater at inaccessible depths. This means that 30% of available freshwater or only 0.75% of total water supplies are available to humans for various uses (Shiklomanov 1997). Human water endowments, which are found in lakes, rivers and accessible ground, are but a tiny proportion of the total planetary water endowment. As documented by Vorosmarty et al. (2010) and others, water endowment is distributed unevenly around the globe in both spatial and temporal terms. This means that there are times and places where water is especially scarce as well as times and places where it is reasonably plentiful. This is shown in Fig. 1 where it can be seen that renewable fresh water is relatively scarce in the MENA and South Asia regions and relatively plentiful in the Americas, Australia and New Zealand. It is also important to recognize that there is significant variability within each region exhibited by water sparse and water rich locales.

The importance of water in sustaining life, the environment and development has been acknowledged in the Dublin-Rio water principles (Assaf 2010). In addition, Article 25 of United Nations Declaration of human rights 1948 (UN-Human-Development report 2006), that:

> Everyone has the right to a standard of living adequate for the health and well-being of himself and his family, including food, clothing, housing and medical care and necessary social services and the right to security.

Although water is not acknowledged explicitly, it is a crucially important part of the daily human diet and sustains life. It should be recognized that some efforts to manage water sustainably have unintended side effects that could have been predicted. A case in point is the treatment of water as a commodity, a practice that threatens the poor. Privatization of water resources has reduced the availability of fresh sanitary water. Two in three people survive on less that $2 a day and are simply unable to pay for water for simple washing, cooking and sanitation needs. One proposal for dealing with the problem is to create an escalating price system based on the quantities of water used. Under this system, costs to the poor are minimal since they use small amounts. This proved to be an effective social package policy related to poverty.

As resources are decreasing in quality and quantity, water policies promoted by developmental agencies with governments have concentrated on comprehensive integrated ecosystem of water management. Expanding demands for domestic, agricultural, and industrial water uses have made water a scarce resource in some countries in the Middle East where total water withdrawals exceed renewable water resources. In fact, most Arab countries are already below the water scarcity level (Plan Blue UNDP Database 2005–2009).

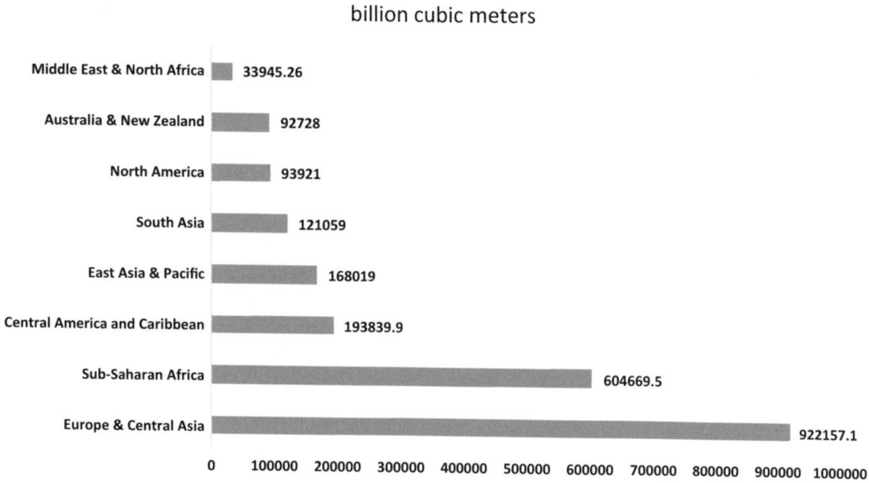

Fig. 1 Renewable internal freshwater resources per capita, by region 2014 (FAO 2016)

Therefore, science based water policies, strategies and management regimes are crucial if supplies, demands and allocations among stakeholders are provided in a balanced fashion that incorporates fairness and efficiency.

Fresh Water Resources Are Becoming Less Available

The UN & UNESCO classify rich-water countries as those who secure 8000 m^3 per capita per year. The World average is estimated to be 6000 m^3 per capita per year. Water scarce countries are defined as those with annual allocations below 1000 m^3 per capita while allocations of below 500 m^3 per capita per year constitute severe water scarcity. Annual per capita endowments of renewable water resources are shown for the 25 most populous countries in the world in Fig. 2. Global per capita renewable fresh water resources are declining at significant rates. Rayne and Forest (2013) reported "substantial reductions of global per capita stock of 54% between 1962 and 2011. There was a decrease of 75% in sub-Saharan Africa, 71% in the Middle East & North Africa (MENA), 64% in South Asia, 61% in Latin America and the Caribbean, 52% in East Asia & the Pacific, and 41% in North America". At current rates of depletion, global per capita renewable internal fresh water resources are projected to decline from levels observed for 1962 by 65% by 2020. Thirteen Arab countries are among the 19 most water scarce nations in the world (Jagannathan et al. 2009). Per capita water availability of eight of those countries is below 200 m^3, less than half of the level that the UN defines as severe scarcity. Per capita annual renewable freshwater resources for the MENA region are expected to decline from 1962 values by 80% in the year 2020.

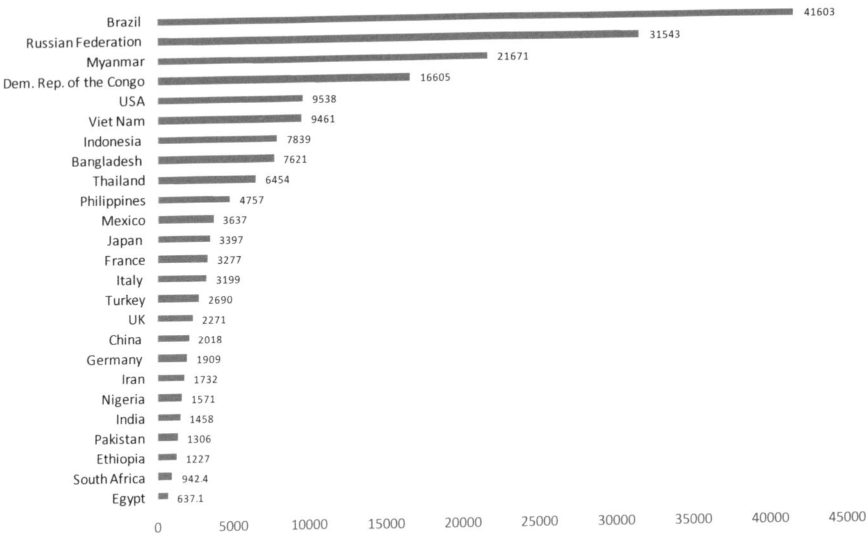

Fig. 2 Total renewable water resources per capita for the 25 most populous countries in 2014 (in 1000 m^3 per person per year) (FAO 2016)

The reasons for this decline are many and the importance of each varies by region. Population growth, which has occurred in all regions, is an obvious reason. Declines in the availability of the water resources also account for diminishing per capita availability. The world-wide trend of declining water quality means that there is less water available for consumptive uses. Declining water quality reduces available supplies just as surely as drought. Lower per capita endowments also result over time when non-renewable resources of water are persistently utilized as long-term supplies. Fossil ground water and quantities of water that are over drafted from renewable aquifers are the most obvious examples of non-renewable supplies. Persistent withdrawal of such supplies depletes them to the point where demand pressures (that were previously supplied by non-renewable sources) fall on renewable sources that are physically substitutable. A final explanation is climate change, which has occurred in the past and is expected to occur in the future. This means that for some regions water is less available than it was historically.

The picture that emerges, then, is one of intensifying scarcity. The fundamental cause of the intensifying scarcity is bound up in the fact that demands for water are growing at the same time that available supplies of water of appropriate quality are shrinking. Some of that scarcity is self-inflicted owing to the absence of effective water policies and management regimes. Some of that scarcity can be avoided by employing existing science in the making of policy and in the fashioning of improved techniques and technologies, which will permit water to be used both more efficiently and more extensively than it has been in the past. Commitments to programs of research and development will also be required since science is needed as a basis for the public policies and innovative technologies necessary to confront and manage the emerging global water crisis.

The Need for Science and More

There is little argument that science will need to be at the foundation of the policies needed to address the intensifying water scarcity. Nevertheless, while existing science and science to be developed in the future will be crucial, science by itself will not be sufficient to resolve global and regional water problems. Other needed elements will include the process of adapting scientific findings for use in managing water resources and building the necessary institutional linkages to facilitate the use of science in the making of policy.

Science As shown in Fig. 3 world-wide spending on research and development has grown from $522.5 million in 1996 to $1.275 trillion in 2009. Most of the growth was accounted for by OECD countries. Additionally, total spending on R&D as a percent of Gross Domestic Product ranged from 1.2% (Spain) to 3.37% (Sweden) as shown in Fig. 4. That figure also shows that the private sector contributes more than the public sector for the countries listed.

These figures mask several important facts about the investment in scientific research related to water. First, water is not large or even constant percentage of the research budgets of any of the nations considered. Moreover, it is not unreasonable to assert that water research budgets have not grown in parallel to the total R&D budgets over the period in question. Thus, for example, in the United States public spending on water research in real terms (adjusted for inflation) was at the same level in 2000 as it had been in the late 1970s and did not grow in parallel with the substantial growth of general R&D over the same period (National Research Council 2004).

A second important fact about investment in R&D generally and water research specifically is that there is a great deal of variation between countries and regions. Badran (2005) provided a comprehensive review on the state of science in the Arab region. He concluded that the region exhibited poor performance in science and technology. This was attributed to political turmoil, low quality education, and an inadequate R&D infrastructure. In short, the region has failed to deliver high quality science and failed to build capacity in R&D (Badran 2013). This has resulted in low rates of innovation and a below average evolution to a knowledge based economy compared to the rest of the world. The results are summarized in Figs. 5 and 6, which indicate that the scientific research personnel per million inhabitants varies but is generally low in the Arab States. Only Africa has fewer scientists per million inhabitants. Figure 7 shows that in the Arab region only 0.2% of GDP is directed to R&D and most of that is public sector or governments.

This measure of research capacity contrasts with the general levels of water availability and the effectiveness of water management in the region. With the

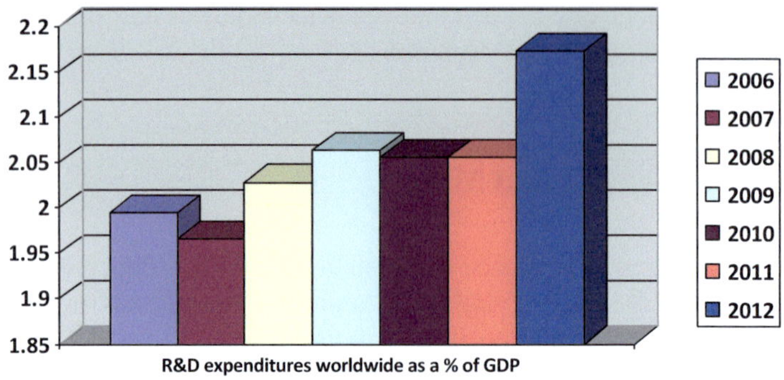

Fig. 3 R&D expenditures worldwide 2006–2012 as % of GDP (*United Nations Educational, Scientific, and Cultural Organization (UNESCO) Institute for Statistics.* ***License:Open***)

Country	Spending on R&D(% GDP)	The contribution of private sector (% GDP)	The contribution of public sector (% GDP)
Sweden	3.37	2.79	0.94
Japan	3.39	2.62	0.77
Finland	3.37	2.46	0.91
USA	2.61	1.84	0.77
Germany	2.53	1.77	0.76
France	2.09	1.34	0.75
Eu (27 countries)	1.84	1.11	0.73
China	1.42	1.01	0.41
Italy	1.9	0.54	0.55
Spain	1.20	0.67	0.53

Fig. 4 Expenditure on R&D: The contribution of private sectors vs. public sector (*United Nations Educational, Scientific, and Cultural Organization (UNESCO) Institute for Statistics.* ***License: Open***)

exception of Sudan and Iraq, all Arab countries are water-poor. In agriculture, there is an effort to utilize technology for saving water in irrigation but the effort needs expansion. Also, the bio saline center for agriculture in Abu Dhabi is developing sustainable crop production by using saline water (Badran and Zoubi 2010) and this effort also needs to be built upon. In another area, research-based universities in the MENA region have started to give priority to research focused at the nexus of water and energy for a sustainable model of technical knowledge and a system linking technology with policy. Badran (2011) has found a strong connection between human rights, levels of R&D and rates of innovation. Indicators have shown a strong correlation between human rights as a contributor to effective science and technology among Islamic (OIC) and Arab countries. Globally, the World Bank

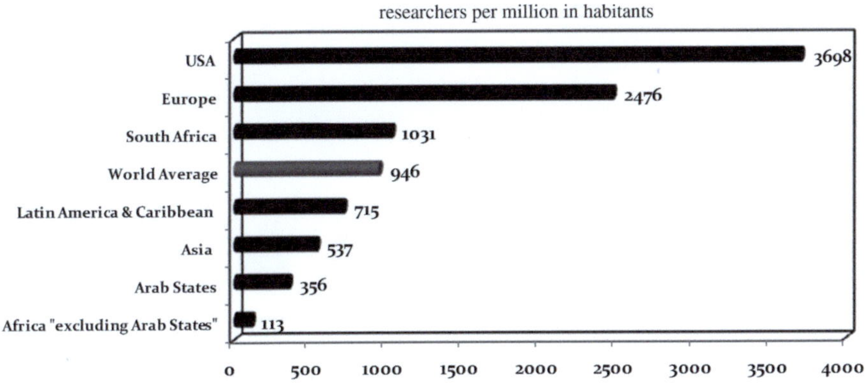

Fig. 5 Researchers in the Arab region as compared to other regions of the world (*Source: UNESCO Institute for Statistics 2012.* ***License: Open***)

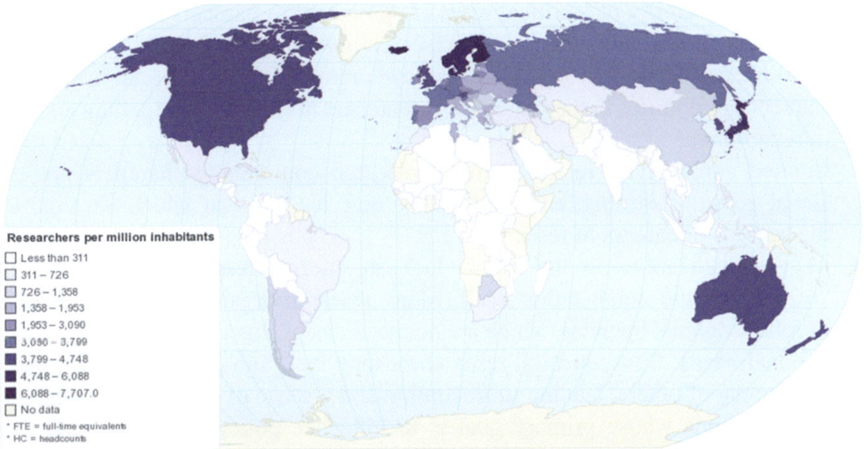

Fig. 6 Researchers Per Million Inhabitants, 2010 (*Source: UNESCO Institute for Statistics 2012.* ***License: Open***)

found effective performance of water research in various sectors such as agriculture, but concluded that very little of this was being conducted in the Arab region (World Bank 2008). In another report, the Bank concluded that when compared with other regions water science was at a low level in the Arab region (World Bank 2007). Demand for water research is not yet an integral part of water policy in many countries in the region. There are, however, a few bright and promising spots such as the Masdar Institute in Abu Dhabi, Kaust in Saudi Arabia and the Science and Technology Foundation in Qatar.

Taylor et al. (2008) identified constraints on the ability of science to influence policy in water management in the MENA region as follows:

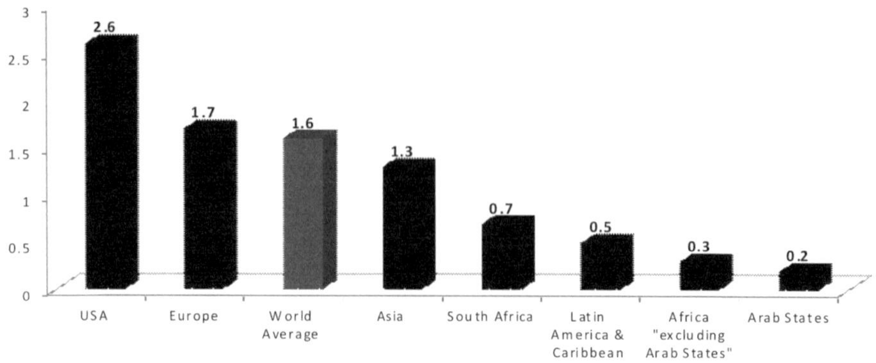

Fig. 7 Gross expenditure on research and development (GERD) as % of Gross Domestic Product (GDP) (*Source: UNESCO Institute for Statistics 2012.* ***License: Open***)

- "The unavailability of a critical mass of competent researchers in the region.
- The management and leadership of research organizations are ineffective.
- The linkages between research and policy communities are not established.
- Career opportunities in the region for researchers may not be compelling enough to retain them.
- Limited connectivity to international research communities hampers professional growth, learning, and exposure to new ideas, all of which are vital to the success of careers in research.
- Many organizations in the region lack an internal research agenda that is "owned" by the organization itself. Many research organizations feel obliged to follow donors' agendas, which are not necessarily aligned with community or national needs. Consequently, researchers may feel "sub-contracted" to pursue the agenda of others, leading to frustration and a sense of disempowerment.
- Organizations whose primary goal is to influence policy will often resort to recruiting well-connected and reputable researchers in order to increase policy makers' confidence in their research. However, reputation in the Arab region seems to be closely associated with seniority rather than performance in terms of relevant, high quality research. The importance of seniority appears to make it difficult for young researchers to attract funding or support for their own research ideas."

Building Scientific Capacity and Infrastructure

In regions where science needs to be more available to support water policy-making, the scientific community needs to identify research priorities. The research itself should be conducted in an integrated and interdisciplinary fashion, which will allow scientists to find solutions to complex problems arising from an increasingly dynamic environment. The international water research community should be involved in strong partnerships that are appropriate and feasible. Simultaneously efforts should be made to improve and expand the educational opportunities for young scientists with interests in hydrology and related water disciplines. An interdisciplinary approach should be part of the educational and training effort. It no longer suffices for water research to be done exclusively by agricultural schools or civil engineering departments, as has been the case in the past. Rather water science needs to be viewed for both educational and research purposes as an integrated and interdisciplinary field of endeavor.

In Germany, there are 500 institutes conducting research on water and related fields. The resulting fragmentation of the research structure can be attributed to the heterogeneous funding system for universities and research institutions. To combat this, a "water science alliance" was formed in 2009 with the aim of joining and strengthening existing competences in water research and creating a framework for complex research in water sciences over the foreseeable future. The alliance is a tool for bringing together and interlinking leading groups and institutions in "thematic clusters" to conduct research leading to concrete solutions to water problems. The alliance will bring synergy and added-value by integrating different disciplines (Teutsch and Krueger 2010). This is one model that has great promise in reducing fragmentation and creating integrated programs of water research.

In Australia, South East Queensland (SEQ) has faced intensive pressures on its water resources, which may be compounded by climate change. An alliance for scientific research on water was established as a partnership between the Queensland Provincial government, the Commonwealth Scientific and Industrial Research Organization (CSIRO), and the University of Queensland and Griffith University. The resulting partnership was to tackle problems of uncertainty and development of a strategic plan for managing the water resources of SEQ. It was supported with a $50 million appropriation over a period of five years (Clayden et al. 2010). Specific elements of the task include:

- Ensuring the reliability and safety of recycled water.
- Identifying needed infrastructure and developing needed technology for recycling waste water and storm water.
- Building scientific knowledge into the procedures for planning and management of water supply systems.
- Developing methods for increasing public confidence in water supplies of the future.

Integrated water management analyses have shown that coordinated development of water, land and related resources cannot be solved by structural measures alone, but require linkages of knowledge with action for sustainable development. Institutional and organization structures that effectively link scientific knowledge to decision-making contribute to problem solving and innovation for integrated urban water systems.

There is also a need for business-based models for technology transfer. These can be established by research managers in the form of incubators and science business parks. Funding mechanisms that are mostly governmental tend to be inadequate and not sustainable. Although external funding has contributed importantly to meet water research challenges, the emerging research agendas were not based on national needs, but rather on the donors' agenda. High quality research requires a national science and research agenda, political endorsement, outstanding research managers, as well as sustainable funding and linkage between research and policy (Laarmani and Salih 2010). Ultimately, water decision makers must employ the results of carefully targeted research and development in establishing water policy or the most important water challenges are unlikely to be addressed.

Bridging Science and Policy

Interactive knowledge sharing in the development of policy for the sustainable management of water resources is sometimes hampered by stakeholders who oppose certain policies on political or ideological grounds (Howari 2008). Solid scientific knowledge can provide the basis for a credible common ground among stakeholders and lead to effective science-based water policies. The key element in linking science to policy is governmental requirement that research be used in the formulation of policy (Carden 2009). Taylor et al. (2008) found in a survey that building institutional relationships between independent or private research organizations and policy-making bodies is difficult to sustain.

A UNESCO International conference held in 1977 on multiple uses of water and integrated water management is seen by many as the genesis of integrated water management. Many countries including the United States, South Africa, Australia and the United Kingdom have adopted integrated watershed management as the fundamental approach to water policy. Powerful environmental movements in North America and Europe in 1980 were confronted with existing governmental policies that have concentrated exclusively on economic growth, with no equity and sustainability.

Global efforts that resulted in the development of the Dublin-Rio water principles for a holistic approach for integrated water management (Assaf 2010) gave further voice to the notion of integrated watershed management.

- 1st principle: Fresh water is both of finite quantity and essential to sustain life, development and environment. Fresh water is needed to maintain all forms of life and for socio-economic human development.
- 2nd principle: Management should include participation of stakeholders, users and policy-makers.
- 3rd principle: Women should be purposefully involved in the making of water policy and in water management. Women in rural areas spend most of their time looking for water and carrying it over long distances.
- 4th principle: Water has an economic value and should be considered as a commodity.

The last principle, which leads to the pricing of water, is not popular unless the poor are secured and consumers understand the value of water, so as not to pollute it or waste it. Still in many countries, water is perceived as a public good. After all, research for water policy should be conceived of within the framework provided by the notion of integrated watershed management.

The concepts of green water, blue water and virtual water also need to inform the research-policy interface. Green water is the soil moisture within two meters depth which is made available for absorption by the root system of the plants. Additionally, it contributes to water vapor in the atmosphere through direct evaporation or transpiration from plants. To conserve green water, soil ploughing particularly for rain fed summer crops and fallowing is practiced. Also, the use of plastic culture to conserve soil moisture has been found effective as a water-management technique. Engineered crops that utilize less water and sometimes are subject to moisture stress, is another way to conserve green water. The management and manipulation of green water is a highly promising area for water research and development.

Blue water is found in rivers, lakes, aquifers. It includes transboundary waters, whether surface waters or aquifer waters. Globally, blue water is becoming fully appropriated. This means that research should be directed at means for economizing on blue water but also at economical ways of using green water. Protecting blue water from pollution and from diversions that are not renewable are other ways to conserve blue water.

Virtual water is water embodied in foodstuffs and other commodities which can be imported. It permits water short countries to acquire water from water-rich countries and utilize internal supplies to grow high valued crops that will generate foreign exchange. The figures below show the quantities of virtual water that are embodied in wheat, rice and red meat.

Jordan's Water Situation: A Case Study

In Jordan, there is a strict policy on national (as opposed to transboundary) water aquifers. Hydrological studies and geological surveys have been done to provide the basic data needed to manage ground water. No new licenses are offered for drilling water wells, with few exceptions for university campuses and hospitals. All wells are controlled and supervised by the water national authority, where meters are installed and monitored very closely. The amount of water pumped out from the underground is measured and actions are taken accordingly.

The landscape of water resources in Jordan is as follows:

1. **Rainfall distribution of water in Jordan**

   - 8200 million m^3 annual rainfall, 80% loss to evaporation.
   - 1640 million m^3 annually left: (Haddadin 2011)

     • 510 million m^3 blue water – surface.
     • 200 million-m^3 ground water blue water - aquifers.
     • 860 million-m^3 soil moisture green water-soil moisture.
     • 70 million m^3 reclaimed water recycled water.

2. Jordanian share from transboundary annually (Blue water):
   - 80 million m^3 Yarmouk transboundary basin (original 296 million m^3) according to Johnston plan of distracting between Syria. Israel and Jordan.
   - 60 million m^3 Tiberias.
   - 68 million m^3 Syrian-Jordan underground basins.
   - 100 million m^3 Saudi –Jordan basin underground (Disi).

   Total: 1948 million m^3 annually (314 m^3\capita\year), which puts Jordan according to UN classification as a severely water scarce country. Virtual water is hard to calculate since there are both imports of meat, grains, fruits and vegetables as well as exports of fresh produce.

1 kg wheat	**Needs 1000 liters of water**
1 kg rice	**Needs 1400 liters of water**
1 kg red meat	**Needs 13000 liters of water**

Research aimed at developing technologies and techniques for exploiting green water, conserving blue water and identifying economical opportunities to acquire virtual water is needed.

Finally, demand management policies, which emphasize on rationing or economizing on water have not been fully utilized though many of them have a strong basis in science. Any research that can increase the public acceptability of demand management policies will be helpful.

Scientists and Policy-Making Scientists in the labs and the field believe that their mandate is to create knowledge and disseminate it through publication in peer-reviewed journals – that is it. Other professionals should take the task of bridging the scientific findings with policy and decision-making. Therefore, the missing link in politics is to bridge science output with policy. Carden (2009) suggested the creation of knowledge brokers, such as Intergovernmental Organizations (IGO's) or Nongovernmental Organizations (NGO's). The National Water Research Center of Egypt is a consortium and is an ideal mechanism. The Royal Water Commission in Jordan and the Higher Council for Water and Climate in Morocco are other examples.

Policy-science interaction (PSI) in the water sector aims to bridge the science-policy gap with different types of knowledge brokering instruments (KBI) at national and regional levels. KBI aims to increase the quality of science-policy interactions by positioning the public to learn about the complexity of the issue, and by understanding the impact of driving forces affecting their future.

The major challenge to sustainability is how to use science to overcome uncertainty in basic issues of agenda **21** related to environment and development (Earth summit Rio 1992). These are complex issues and cannot be addressed except through alliances and interdisciplinary, holistic approaches of physical, life sciences (hydrology, ecology, agriculture-human food and health… etc) and social sciences (policy, social sciences, economics, human development etc.). An approach of this type will lead to a physical-social earth agenda for the use of natural resources linked with managerial skills responsive to stakeholders.

Escalating pressure on less than 1% of the world's total supply of water is made more difficult to manage because of population increase and is exacerbated by climate change and degradation of water quality. In addition, research activity is fragmented and poorly linked to policy and management needs (jimwallace@easynet.co.uk). The UNESCO and WMO set up the HELP (hydrology for the environment, life and policy) initiative to deliver social, economic and environmental benefits through the sustainable use of water by deploying hydrological science, which is an interdisciplinary science to achieve integrated catchment area. The objective here is to form a global network to bring together hydrologists, water resource managers, and policy and legal experts to address water issues defined by local stakeholders (www.unesco.org\water\ihp\help). Twenty-five basins were established from different climatic, social and economic regions around the world. These basins will serve as "outdoor labs". The main outcome is to integrate hydrological, social-economic and legal research responsive to related water-policy. HELP has created a platform for dialogue between physical and social scientists, water resource managers and policy makers.

Conclusion

Analyses of science-links-policy in water management in the MENA region show that water research is not part of the water policy-making. There is an absence of cutting-edge scientific research, and linkage of knowledge to policy is not well developed. There is a lack of national science and technology policy and coordination. Agendas may be dictated by donor agencies and water policy is influenced by politics of interest groups more than by science-based discourse. Capacity building in training scientists to excel in water research is needed. This requires a national agenda that includes water research priorities, political commitment, sound research management, sustainable funding, manpower plan to attract outstanding scientists and to send outstanding graduates abroad on scholarship in distinguished institutions for Ph.D and postgraduate studies to develop the critical mass in water research nationally.

Conferences and other interactions between scientists, planners, communicators, managers and public officers, should bring water science to the policy making process for water management.

Stakeholders should be engaged in water policy-making and they should utilize knowledge in overcoming differences. Universities and research centers should develop water science research groups to tackle priorities of the water sector. This will develop an interdisciplinary approach that will pool resources and develop the critical mass of the know-how and attract funding for joint proposals related to the water national agenda.

National water agendas should not be subjected to the whims of outsiders but international scientific cooperation with world-class research institutions should be encouraged and maintained. Scholars should produce tangible scientific results that are perceived by the government as credible. Research must be placed at the center of water policy and governments should implement sustainable water policies to rationalize demand to ensure efficient use. The government's role should be shifted from being an exclusive provider to being an effective regulator and planner.

A report from the economist intelligence unit published recently on challenges to meet water supply in 2030, emphasized that shortage and stress will yield scientific innovations. The report (2012) cited a few promising technologies:

- Lower-cost water desalination: carbon nano-tubes with membranes to radial deionization for removing salts from water.
- Better wastewater reuse: R&D is cutting energy now by 30–45%.
- Managing aquifers recharge: storing surplus water

Finally, with the right political will and consumer backing, water stress will force technology development and innovation in all phases of the water cycle.

References

AFED (Arab Forum for Environmental Development). (2010). Arab environment: Water, sustainable management of a scarce resource, Beirut, Lebanon. www.afedonline.org.

Assaf, H. (2010). *Integrated water resources management*, 2010 AFED report, chapter 6 (pp. 92–105). Beirut, Lebanon.

Badran, A. (2005). *S&T in the Arab region*. In UNESCO Science report 2005. Paris: UNESCO.

Badran, A. (2011, October 22). *The science & technology & human rights Nexus in the Arab World*. Paper presented at 18th IAS conference, Doha.

Badran, A. (2013, May 6–9). *Excellence in higher education for building S&T proficiency: A global perspective*. Paper presented at the 19th IAS conference, Dhaka, Bangladesh.

Badran, A., & Zoubi, M. (2010). *Arab States: UNESCO World Science Report 2010*. Chapter 13 (pp. 251–274).

Carden, F. (2009). *Knowledge to policy: Making the most of development research*. Ottawa: Publications/International Development Research Center (IDRC).

Clayden, G. et al. (2010). *Action systems for integrated water management: Implications for S.E. Queensland-Australia*.

Economist Intelligence Unit. (2012). *Water for all? A study of water utilities preparedness to meet supply challenges to 2030*.

El-Quosy, D. E. (2009). Fresh water. In M. K. Tolba & N. W. Saab (Eds.), *Impact of climate change on Arab countries*. Beirut: Arab Forum for Environment and Development.

FAO. (2016). AQUASTAT main database – Food and Agriculture Organization of the United Nations (FAO). Website accessed on 13 July 2016 8:45.

Global Issues. (2010). *Water & Development*. www.globalisones.org

Haddadin, M. (2011). Water: Triggering cooperation between former enemies. *Water International*, *36*(2), 178–185.

Howari, F. (Ed.). (2008, April). *Bridging the gap between water science & policy making*. Water policy in GCC countries conference, Kuwait.

Jagannathan, N. V. et al. (2009). Unbundling water management in MENA countries. *In water in the Arab world, AFED Report 2010*.

Laarmani, H., & Salih, A. (2010). *Linking water research & policy*. AFED report chapter (12) (pp. 190–199).

National Research Council. (2004). *Confronting the Nation's water problems: The role of research*. Washington, DC: National Academy Press.

Rayne, S., & Forest, K. (2013). *The decline of global per capita renewable internal freshwater resources*. Sask: Saskatchewan Institute of Applied Sciences.

Rio Declaration on Environment and Development. (1992). Earth Summit, Brazil. www.unep.org

Saleh, W. et al. (2010). *State of freshwater ecosystems*. AFED Report 2010, Beirut, Lebanon. www.afedonline.org

Shiklomanov, I. (1997). *Assessment of water resources and water availability in the World*. Geneva: World Meteorologial Organization.

Taylor, P., et al. (2008). *Researcher perceptions of capacity for adaptiveness & management of research organizations in the MENA region*. Cairo: IDRC.

Teutsh, G., & E. Krneger (2010). *Water Science. Alliance – White paper: Priority research fields*. HELMHOLTZ, Germany.

United Nations Human Development Report (2006). Universal Declaration of Human Rights, UN, December 10, 1948.

Vorosmarty, C. J., McIntyre, P. B., Gessner, M. O., Dudgeon, D., Prusevich, A., Green, P., Glidden, S., Bunn, S. E., Sullivan, C. A., Reidy Liermann, C., & Davis, P. M. 2010. Global threat to human water security and river biodiversity. *Nature, 467*(7315), 555–561.

World Bank. (2007). *Making the most of scarcity: Accountability for better water management in the Middle East and North Africa*. Washington, DC: World Bank.

World Bank. (2008). *Agriculture for development*. World Development Report. Washington, DC: The World Bank.

The Triangle: Energy, Water & Food Nexus for Sustainable Security in the Arab Middle East

Peter Rogers

Abstract This chapter examines water security in the broader relationships governing the Food-Water-Energy-Climate Nexus. It particularly stresses the role of the great global transmissions of the nineteenth and twentieth centuries in presenting intractable barriers to returning to less complicated eras of resource conflicts. These transitions are manifest in total and urban populations' growth and shift to urbanization; radical shifts in the nutrition demanded by the new economic and social developments; the radical changes in land use and chemicals in agriculture; a rapid shift in emphasis on renewable energy resources and reduced reliance on fossil fuels; and finally the great challenge of climate change. All of these transitions have major implications for water security both globally, and regionally. Globally this is well articulated by DuBois (The case for "energy-smart food for people and climate". Food and Agriculture Organization of the United Nations, World Food Day-Oct 16, 2015):

> Our agrifood systems currently consume 30 percent of the world's available energy—with more than 70 percent occurring beyond the farm gate, and produce about 20 percent of the world's greenhouse gas emissions. More than one third of the food we produce is lost or wasted, and with it about 38 percent of the energy consumed in the agrifood chain.

To this we can add that the greatest loss of water in the overall national water balances is that of the water used to grow the food that is wasted.

While the water security situation for the Arab Middle East Region is generally considered bleak, the paper is fairly optimistic that, at least water resource use, until 2050 will be still manageable if the eleven "technical fixes," outlined in the paper are pursued. These technical fixes are not to be construed as purely engineering the water supply, but fixes to many of the economic and social barriers to a more secure water future. They cover major national policy choices such as international trade in virtual water, traditional water engineering of traditional and non-traditional sources, improving efficiency in use via agronomic research, improvement of post harvest food and value chains, and softer options

P. Rogers (✉)
Gordon McKay Research Professor of Environmental Engineering, John A. Paulson School of Engineering and Applied Sciences, Harvard University, Cambridge, MA 02138, USA
e-mail: rogers@seas.harvard.edu

S. Murad et al. (eds.), *Water, Energy & Food Sustainability in the Middle East*,
DOI 10.1007/978-3-319-48920-9_2

such as trading among users, pricing, rationalizing property rights and legal protection for third parties.

Keywords Agrifood • Water security • Food security

Origins of the Concept of an Energy, Water, and Food Nexus

While the concept of the energy, water and food Nexus has been discussed for about a decade, only recently has there been a plethora of reports by major agencies, institutions, and academics. So much is now available that it is impossible to analyze all of them and arrive at a coherent picture of the whole. For example, Allouche et al. (2014) are skeptical that there is anything new in the reframing of water security and climate change in this way, and that it may not provide new and innovative solutions to water, food, and energy problems. Even the title of their report (Nexus Nirvana or Nexus Nullity) expresses a lack of confidence in the usefulness of the concept. However, Unver and Pluschke (2015) perhaps reflecting FAO's institutional biases, see the Nexus as "not only relevant, but also potentially preferable to the more traditional approaches." This is probably the most widely held view of the concept.

The World Bank has been a major contributor to the literature of the Nexus generally, but also of its analysis of the Middle East and North African (MENA) region (Hallegatte et al. (2012), Ray et al. (2015) for examination of the concepts, and Verner and Breisinger (2013) for specific country analyses). The Arab Forum for Environment and Development (AFED) was an early contributor to the debate over the Nexus (see Tolba and Saab 2009) as was the Arab Organization for Agriculture and Development (AOAD 2012). The FAO has also been a major participant in the discussion, as for example, through its Swedish FAO Committee (2014). The various bilateral agencies have been very active in promoting the concepts and potential remedies. For example the USDA's contribution stressing the Nexus, was released as late as December 1, 2015 to the Paris COP 21 (USDA 2015). The International Food Policy Institute (IFPRI) has made several country specific studies in the MENA region (Breisinger et al. 2010, 2012).

Academicians have weighed in recently with many journal articles; as for example, in the September 2015 special issue, volume 31 of the *International Journal of Water Resources Development (IJWRD)* devoted to the Water-Food-Energy-climate nexus (see Allan et al., 2015). This issue should be of great interest, and required reading, for regional resources managers and policymakers. The journal *Water* has also published several papers this year (see Endo et al., 2015).

There are two lengthy political commentaries Waterbury (2013) and Greenwood (2014) focusing on climate change, and political economy and governance of the Arab nations. Each focuses on the role of the components of the Nexus, not the Nexus itself. Both conclude that the water, food, and energy problems are extremely serious, but with Waterbury saying more depressingly "Despite the apparent urgency of the challenges facing the Arab region as a result of climate change, inaction is a viable political strategy and, in ways that I attempt to specify, the most likely one." Greenwood is equally pessimistic concluding that "...there is the possibility that growing water scarcity and the effects of climate change could exacerbate existing political tensions." In this chapter we are less pessimistic and envisage a series of strategic options for managing the resources that could soften the Waterbury-Greenwood pessimism and ensuring water, food, and energy security at least until 2050.

Concept of Water Security

The concept of water security seems at first glance relatively simple and straightforward, water demand is estimated to lie between 500 and 1000 cubic meters per capita per year for all uses including food and industrial production in a nation or region (Falkenmark and Lundqvist 1998). For direct domestic use the figure is only a fraction of this amount, probably about 20 cubic meters per year. However when applied to many different uses of water it, becomes increasingly difficult to tie down the precise meaning of water needs. Water security is seen as a part of a complex web of food, energy, and social security. Maslow (1943) proposed a six-fold hierarchy of the most important of humanity's needs:

Physiological (food, shelter, water, sleep, and clean breathable air),
Safety (security, stability, order, physical safety),
Love and Belonging (affection, identification, companionship),
Esteem and Recognition (self-esteem, self-respect, prestige, success, esteem of others),
Self-actualization (self-fulfillment, achieving one's capabilities),
Aesthetic (beauty, harmony, spiritual).

Food-water-energy security is clearly amongst the most important of Maslow's needs and transcends physical and social safety and security, but when we talk about national water security we need to conjoin the physiological needs with the social needs for safety.

In addition to Falkenmark and Lunqvist's (1998) widely used definition of water security, other authors have suggested the following definitions:

- The availability of an acceptable quantity and quality of water for health, livelihoods, ecosystems and production coupled with an acceptable level of water-related risks to people, environments and economics (Grey and Sadoff 2007)
- Linked to state security, to avoid water wars (Gleick 1993)
- As competition over uses between competing sectoral demands
- As a right to individual access to water (freedom from fear, freedom to live with dignity (Gutierrez 1999).

The thrust of this book is the nexus of energy, water, and food; hence we need to go beyond defining water security as an isolated concept, to defining it as connected to energy and food security.

The FAO has defined food security as "existing when all people at all times have access to sufficient, safe, nutritious food to maintain a healthy and active life."

Breisinger et al. (2012) suggest a measure called the "risk of food insecurity," which is based upon two factors; expressed as the percentage of food imports divided by the value of all exports, and the prevalence of child stunting as a percentage of all children.

Patel (2009) suggested that food security hinges on "food sovereignty" which is "the right of each nation to maintain and develop its own capacity to produce its basic foods respecting cultural and productive diversity" This concept is still hotly contested by many economists, agricultural planners and the NGO community.

Equally vague definitions are applied to energy security, such as by Hoff (2011) "access to clean, reliable and affordable energy services for cooking and heating, lighting, communications, and productive uses; uninterrupted physical availability of energy at a price which is affordable, while respecting environmental concerns."

Nevertheless, the water security definitions, while still vague, appear to be rigorous, pursuable objectives when compared with the food and energy security definitions. The definitions indicate that water, food, and energy security are social constructs rather than an absolute physical quantity per capita. Equivalent absolute measures; for food would deal with starvation diets (less than 1300 kcal per day); for water extreme thirst (less than 5 l per day); and for energy limited to energy for cooking, and transport etc. As social and political situations differ widely in the Arab Middle East it is important to treat each country separately to determine which ones are insecure with regard to the nexus resources. The definitions are so broad and contentious that they can lead to each author choosing to focus on some particular facet of security to promote their own agendas. We believe that the best approach is to stay with the broad definitions, but flag the issues of conflict between and among the various goals, users, and national policy. We recognize that all of these security definitions are held hostage to domestic political processes given the relative importance of agricultural, industrial and energy interest groups in controlling the political dialog. In addition, the uncertainty induced by natural and human enhanced impacts on the physical environment differs among the countries of the Region. Even when we are willing to accept the operational definitions of the Nexus, we still have to embed our analysis in the great social and economic transformations inherited from the twentieth century.

Inherited Global Transitions

The world is undergoing major transitions inherited from earlier periods when populations roughly *matched the available resources and is now moving towards new global conditions which require* new thinking about the nexus of food, water, energy. The transitions are happening so fast that the training and mindset of most senior planners and managers have long since been overtaken by these equilibrium shifts; and well-tried solutions to food, water, and energy management of the past are no longer viable.

There are five major global transitions that make historically-based thinking obsolete (Rogers and Daines 2014).

The first is the "urban population transition;" the majority of the global population now resides in cities and has increased purchasing power.
The second is the "nutrition transition;" demands for a new basket of foodstuffs with greatly increased consumption of animal products and other high-value foods.
The third is the "agricultural transition;" bringing almost all of the arable land into agricultural food and fiber production.
The fourth is the "energy transition;" from cheap fossil fuels to renewable energy resources; and finally.
The fifth is the "climate transition;" increasing temperatures and increasing variability in water supplies and growing conditions for plants.

All of these transitions are exacerbated by a growing world population and the deterioration of the quality of water and land. Coping with any one of these alone would be a major problem, but the transitions are actually happening simultaneously with differing rates of change in different countries and regions. New political leaders and resource managers need to be cognizant of what is going on globally, and on which flexible approaches will enable us to sustain our livelihoods during the transitions.

Unfortunately, the effects of these transitions are not simply additive: each domain affects the others and causes a cascade of effects which then cause non-linear impacts. Because of the non-linear synergistic interactions among water, energy, climate, and other inputs into food production, it can be quite misleading if we treat them as separate inputs to the production process. For example, since almost all of the cultivable land is currently in use, producing more food requires intensification in yields. This will require intensified use of chemical fertilizers, or extensive double cropping, which will mean large additional quantities of irrigation water. However, we are increasingly approaching limits to water available for conventional agriculture. Further since 1962, the use of chemical fossil fuel-based fertilizers began to surpass the amount of globally available natural organic fertilizers (IFA 2013). Today they exceed organic fertilizers by a factor of two. The production of chemical fertilizers, however, uses large amounts of fossil fuels, which exacerbates green house gas production. Current estimates for the footprint of agriculture, including its fertilizer production and use, lies between 17 and 32% of the total greenhouse gases (Bellarby et al. 2008). Additionally, the development of bio-energy from crops such as corn (about 20% of the US corn crop is now diverted to bio-fuel production) also creates a conflict between food and energy supplies.

In 2012, the US Department of Agriculture estimated that the global number of food-insecure people (the number of people consuming less than a nutritional target of 2100 calories per day) to be about 800 million. By 2022, it projected an increase of 4.6%, which means that the total is inching up toward one billion hungry people by 2050!

Despite globally declining human fertility, by 2050 an extra 2 billion people will need to be fed; moreover one half of the total expected 9 billion population will be urban, and wealthier, with demands for foods which require much more water and other inputs than the traditional grains, and more grains to feed the bourgeoning livestock. A doubling of the global food production will be required to meet these additional needs. The corresponding projections for the Arab Middle East are for a 2.0% annual increase of the total 2010 population of 372.6 million to more than double, with possibly a tripling of the urban populations by 2050 (World Bank 2014a, b).

Population size by itself is not the problem. The current total world food production is capable of providing everyone with an adequate diet. The real problems are the distribution of food and meeting the demands for higher quality diets. Moreover in many countries fully 50% of the food production is lost in the post-harvest system from "farm to fork." Reducing these food losses and supply fluctuations are major problems requiring modern management of food chains relying on more efficient agricultural inputs, storage, transport, and marketing. For the Arab Middle East the distribution of access to adequate diets is similar to the global figures depending upon the economic and social conditions prevailing in the individual states. In the MENA region using Breisinger et al.'s (2012) food insecurity index, all the six nations of the Gulf Cooperation Council GCC are highly rated, and all of the seven oil importing countries in the GCC were rated as seriously insecure or worse. Two countries, Djibouti and Yemen being rated as extremely alarming, and Syria was not rated. Even countries rated moderate scores on the index still had significant levels of prevalence of child stunting. Given the current turmoil in the region it is likely that food insecurity will be of major concern in the region over the next decades.

Urban Demographics in the MENA Region

The good demographic news coming from the MENA region is the decline in population growth rates since 2010 (MENA Data Book, World Bank 2014a, b) with significant declines until 2025. Despite the fact that more than 50% of the world's population now lives in cities, the MENA region is typically less than 50% urbanized, but is urbanizing rapidly and will be more than 50% by 2025. According to Tolba and Saab (2009), unlike the large cities in Asia, no MENA city, apart from Cairo, will exceed the mega-city definition of 10 million plus by 2025. Even so, rapid urbanization is one major component of the Nexus that may cause serious social and water and energy resource conflicts. As the urban population increases, disposable income rises and food habits and water use changes leading to large increases of water demanded. The increased demand for animal products raises the amount of water used for feedstock and watering of the animals. For large Asian cities, Rogers (2009) calculated an almost 10-fold increase in total (rural, agricultural, and urban) water demanded over the next 25 years. For the MENA region,

Breisinger et al. (2012) show urban population increases from 2010 to 2025 in Egypt of 15 millions, Morocco 6.4 millions, and Jordan almost doubling its urban population by adding a further 5.6 millions. Each of these countries already has major difficulties in servicing their current urban water and energy systems.

Resource managers must cope with the resource constraints and global transitions in a way that is cost effective and sustainable. Unfortunately, these global transitions are experienced more harshly in some parts of the globe than in others. For example, South, Southeast and East Asia will be faced with a monumental increase in urban populations, more than doubling by 2050. When coupled with the large increases in demand for higher quality foods, this could lead to a more than doubling in the demanded levels of agricultural inputs of water, land, energy, and agricultural chemicals.

With regard to water and expected climate change, the Arab Middle East is likely to be one of the worst hit regions. All of the countries in the region, with the exception of Iraq, Iran, and Lebanon are already facing serious water shortages (Breisinger et al. 2012). The richer countries have been so far able to buy their way out of this problem by investing heavily in desalination for domestic and industrial supplies and relying on virtual water imports in lieu of developing agriculture. This is not an option for poor countries with large agricultural sectors, such as Morocco. Over the coming decades dealing with water supplies will be a major problem for all of the countries, particularly when faced with the uncertain consequences of climate change.

Climate Change: Dealing with an Uncertain Future

Any decision concerning changing the global climate in ways that make it more unfavorable habitat for the human species raises the possibility of "ruin" in a global sense: there is a remote chance that we will cause changes in the ecosystem due to cascading (non-linear) effects from the climate change so that we could "lose" the whole planet. Here we are faced with a classical problem—the extremely rare event with huge consequences. In statistical terms, we can assume that total collapse of earth's agricultural system would lead to total elimination of the human species. In this sense such an event would be in statistical parlance, an "absorbing barrier," beyond which the system would not recover.

Politically, socially, and technically we are not good at dealing with such big issues. Our empirical world is dominated by rare events which seem to occur in "thin tailed distributions." That is, they occur with a Normal distribution where the probabilities of rare events are extremely small, not in the fat-tailed distributions (such as the Cauchy distribution) where the tails of the distribution are much "fatter" and include much higher probabilities of ruin. Unfortunately, we have not yet been able to estimate the magnitude of the risk, let alone, the probabilities of the rare events. Taleb et al. (2014) make plausible arguments about which type of events would likely have "fat-tailed" distributions. Their claim is that systems with linear response to shocks where the shocks are limited by geography or other natural buffers (such as nuclear power plant meltdowns which could be devastating locally are typically dealt with locally) will not lead to "ruin" but systems which have a nonlinear response (such as climate change, which can cascade across boundaries and species and lead to system collapse) have a potential for total ruin. Their analysis implies the use of the "Precautionary Principle" to climate change actions leaving the choices of mitigation or adaptation as the only sensible options facing us. Marshall et al. (2015) claim a global non-linear effect of temperature on economic production.

Dealing with climate change is really about managing risk and the meaning of sustainability. From a broad geological-ecosystem point of view, sustainability implies some sort of long-term equilibrium. However, given the stochastic nature of natural events and the purpose-driven deterministic nature of human decisions, it is highly unlikely that such equilibria have ever existed, or could exist for long periods of time. As rational practical planners, we are faced with making short-term decisions (50 years or less) that will allow the maintenance of a sufficient level of goods and services in the face of increasing demand due to unplanned (rural to urban migration) or planned events (investment portfolios). Sustainability must, moreover, be achieved in the context of the massive global transformations that started in the twentieth century, and influence the magnitude and direction of resource use.

Regionally, water supplies and food production are becoming increasingly uncertain due to climate changes and also to changes in demographic patterns and economic growth. In the past, the assessment of climate variations and fluctuations of floods, and droughts was largely predictable from steady state statistical models. However, due to the interaction of the five global transitions noted above, the current approaches to risk assessment have become increasingly unreliable and we have to look elsewhere to find ways to deal with these uncertainties (see Sood and Smakhtin 2015 for problems in modeling global hydrology).

The failure of models to predict future water demands and supplies leads to a return to earlier approaches to decision making under conditions of uncertainty, and to look for simpler, more intuitively convincing models. As Taleb et al. (2014) noted, this is when concepts such as "safety" and "surprise," "robustness" and "resilience," and "unknowability, "uncertainty, and unpredictability" need to be invoked along with concepts of "risk management, fragility, and the precautionary principal". These concepts, essentially based on the "negligible possibility of ruin;"demand an approach to the future that will maximize the expected gain while minimizing the possibility of ruin to a point that it can be neglected. We do not want any nasty "surprises." We need a conservative and risk-adverse strategy to identify the boundaries of a "safe space" for humanity. Taleb et al. (2014) give an excellent exposition of the use of these concepts in the case of genetically modification organisms (GMO) in the human food chain, and to the decision to pursue nuclear power. Their analysis indicates that genetically manipulated organisms have a much larger chance of leading to ruin than nuclear power accidents, which have many more self limiting controls, and are hence much less threatening to global disaster then GMOs.

Dealing with Climate Change in the MENA Region

Apart from shutting down oil production it is unlikely that any decisions taken by the MENA region countries would have any major impact upon the global climate carbon balance. Hence for MENA the choices have to be on adaptation to climate change. Of particular concern is how climate and migration ricochet throughout the economy and affect water and energy uses, and food production in particular.

> Climate change and migration are major concerns in the MENA region, yet the empirical evidence on the impact of climate change and severe weather events on migration remains limited. Information is broadly lacking on how households in vulnerable areas perceive change in the climate, how they are affected by extreme climate events, whether they benefit from community and government programs to help them cope with and adapt to a changing climate, and how these conditions influence the decision of household members to migrate, either temperately or permanently. (Wodon et al. 2014)

Doukkali and Lejars (2015) use a social accounting model (SAM) of the Moroccan economy to measure the direct and indirect effects of energy and agricultural subsidies on government water conservation programs. This is the perfect "circular economy" where, responses to a perturbation of any of major sectors leads to "knock-on" effects which eventually circle back to either dampen or magnify the original perturbation. So, a prolonged drought, like the recent one in Syria, may lead to large scale migration of rural populations to the cities. This in turn exacerbates urban food shortages with subsequent unrest, causing a large external migration, which further drains the rural areas of agricultural labor; and the result of this is a reduction of food production and so on. Jaafar et al. (2015) suggest that such a spiral occurred in Syria when the irrigated agricultural production dropped between 15 and 30% in the Syrian parts of the Orontes Basin from 2000 to 2013. A contrary view is expressed by Lagi et al. (2015) laying the blame for food shortages on speculators driving up the food prices. One exacerbating factor was the large demands placed on US corn by the demand for ethanol fuels. Whichever the direction of causality, one sees a circular argument leading to a downward spiral of food availability and the rise of violence.

> As the New England Complex Systems Institute (NECSI) has demonstrated, short-sighted market policies in the US sparked food riots, which triggered the Arab Spring, which destabilized Syria. The millions of internally displaced Syrians and migrants fleeing civil war and ISIS are just the latest event in a worldwide chain reaction. If nothing is done to restore stability, this will not be the last disaster. (Lagi et al. 2015)

Wiebelt et al. (2014) looked at who bears the cost of climate change in Tunisia and Yemen (2010), and finds a weak negative overall impact on the Tunisian economy. At the same time, if the weather has a positive impact on agriculture (and only then) the agricultural sector may benefit from rising export prices. For

Yemen (2010), they predict 80,000–270,000 people could go hungry with rural households being the most affected by climate change. Taking a longer term view Maystadt et al. (2012) examined major intrastate conflicts (defined as conflicts with more than 1000 battle-field deaths) between, and among the Arab states from 1960–2008; They concluded that unlike other nation groupings there was an "Arab Exceptionalism" based upon the emergence of food security as the main driver of conflicts in the region. They found that Tunisia and Libya were doing well on the macro-food security measures, but need to focus on the micro-level measures focusing on child stunting, whereas Egypt and Yemen were seriously impacted by both levels of food insecurity. This may appear contradictory if, for example, Tunisia has plenty of food, enough even to export, why child stunting? This is due to the extreme inequality in access to wealth and income, making the children of the poor exposed to food insecurity.

A Look at Agricultural Land

We are bombarded with reports, papers, and articles often claiming that we will run out of agricultural land before 2050 because of climate change that leads to declining crop yields, etc. Will we run out of agricultural land need to produce the food demand of 2050? To address this question, we have to decide on what the demands on the agricultural sector will be; but this cannot be addressed without a detailed look at the components and drivers of food demand. We are immediately in a set of circular arguments which depend upon likely economic, agrarian, dietary, and technologic changes over the coming decades. In order to unravel the countervailing trends, we have to start somewhere so let us start with agricultural land, which is the essential prerequisite for developing and sustaining agriculture. The global land resources are the obvious starting point. The FAO data show that 38% of the global land surface is used for agriculture. Of this, 12% is used for crop production. Eighty percent of the cultivated area is rain fed.

Based on the FAO data, a food based-ngo (GRAIN 2012) made a case for continued support for policies that emphasized the contribution of small-farms to meeting the future world food needs. Their argument is based upon the fact that of the total global 5 billion ha of agricultural land and 600 million farms, small farms account for over 90% of the farms, and only 25% of the agricultural land, but currently supply the majority of the world's food supply. This implies that small farms are inherently more productive per unit of land than large farms. This is backed up by many economic studies by academics and development agencies. They suggest that if the policies supporting small farms were implemented and if they were given access to twice the amount of land that they currently use, the small farms could potentially double the total food supply.

A dissenting view of this conjecture by Hazell (2013) questions "is small farm led development still a relevant strategy for Africa and Asia?" In his review of the data, Hazell (2013), pointed out that during the Green Revolution era, small farm growth was seen as a "winning proposition for growth, poverty alleviation, and food security outcomes." However, under more recent economic and rural-urban migration conditions in Africa and Asia, farm sizes have continued to decrease leaving 450 million farmers with 2 ha or less. Average farm sizes have shrunk in Asia and Africa by one half from 1971 to 2006. Even in China the farm size decreased to less than 0.55 ha per household until 1960 when it has slowly risen. In Africa Hazell suggests that farms of 5 ha or less dominate the farm size distributions.

Why is this important? This is important because the small farms of today are less than one half of the size of the small farms of the 1960s and 1970s which were so successful in achieving the "green revolution," and their number is increasing while their size is decreasing. Typically 80% or more of the farm holdings in MENA are typically less than 2 ha. This shift has made it more unlikely that they will be able to take advantage of the recent higher food prices given their reduced bargaining power vis-à-vis the corporations and individuals who control the food chains to the markets and beyond.

Employment and Food

A major problem facing developing countries today is the need to create rural employment. This can be done in part by modernizing agriculture and expanding high-value export-oriented horticultural crops and by intensifying multi-season cropping in rural areas in response to new non-grain food demands. For example, the number of days of labor needed for grain production in traditional agriculture in Asia is about 20 days per acre; these are in stark contrast with labor inputs of 4 or 5 h per acre in modern mechanized grain farming in the US. However, in changing from grain to modern horticulture crops, more than 200 days of labor per acre could be generated in many developing countries. Thus, horticulture needs to be increased in the modern food production and value chains to provide rural employment and promote incentives for introducing modern methods in rural areas. In this way, contrary to popular beliefs, intensifying agriculture by increasing mechanization and changing the mix of crops grown can increase rural farm and non-farm employment.

The Water-Energy-Food Nexus: Seven Stylized Facts About the Eternal Triangle

Part of the problem of assessing policy options is that the politicians, decision makers, NGOs, and even academics have already closed their minds to many options by relying on "stylized facts" about the potentials. For the nexus there are many such facts which cloud the decision space. Some of these facts given below are correct for the Arab region, but less so for other parts of the world that are currently not yet water-stressed, and should help the discussions on policy; some are patently not correct and can lead policy recommendations astray.

1. Water is the most limiting constraint on agricultural development in the Arab region.

   Food stuff is a joint product of land, water, energy, labor, fertilizer, etc. which all combine to produce food output. Complementarity and substitutability, of the various input factors allow for each or any to be limiting factors as input availabilities change. We could also claim for example that energy is the limiting factor if electricity for agricultural use is restricted. In a similar vein we could claim fertilizer access is the limit when its supply is restricted. It is true, however, that absolute scarcity of water could mean no food production at all, but even in deserts we can substitute energy or lack of water through desalination or long-distance transport to obtain food without water.

   Too much emphasis on single factors as the cause for famine are not meaningful if the entire demand for goods and services in an economy are not considered as generators of price and demand shifts. Single factor emphasis can lead to erroneous policies and could severely restrict food output.

   Nevertheless, Tolba and Saab (2009) *predict that by 2025 only two, Iraq and Sudan, of the Arab countries will be above Falkenmark's water secure level of 1000 cubic meters per capita per year, and by 2050, all of the countries are expected to be water insecure. This makes an emphasis on water extremely important for the region.*

2. As population increases the water and energy resources needed for food production will become increasingly scarce leading to a shortfall in food production, and massive hunger.

   This assumes that there are fixed input relationships among the factors, and the economies are "closed." However as we noted above, there are substitution possibilities that can enable us to continue expanding under shortages of water by using desalinated water or improved drainage control as substitutes for the water, also increased production is possible on irrigated land areas by use of additional fertilizers as complements. Also virtual (embedded) water can be "imported" in food to avoid famine and buy time to restructure the agricultural economy.

   However, the distribution of access to food may be heavily skewed under such conditions disastrously affecting the poor. The urban and the food transitions will lead to major political stresses due to food and water shortages in the Arab region.

3. Climate change will affect regional hydrology, making rainfall and water availability much more variable; also possibly making the arid regions more arid and the temperate regions wetter.

Strong arguments exist for human caused climate change. There is a great deal of concurrence of the global climate warming, but much less on regional and local weather. Some of this is due to the data difficulties associated with the mechanics of the predictive models, and some with a paucity of data to calibrate them. This is extremely bad news for the Arab region and stresses the urgent need for restructuring the nature of agriculture in the region. The potential for these effects should be closely monitored and remedial adaptation action taken.

4. There is a fixed amount of water and land available for growing food, and hence, we cannot double food production without doubling the amounts of water and land made available for agriculture.

Here we need to avoid fixed factor solutions with no substitutability among factors. Currently all countries in the region suffer from water shortages for agricultural development; however, most countries do have sufficient water available to support domestic and industrial needs, provided that they can establish new rules for water property rights. The yield-gap analysis outlined below (YGA) also helps identify the substitution possibilities for amplifying agricultural production with scarce amounts of land and water.

5. Increasing irrigation demand for food production, coupled with increasing water demanded for rapidly rising urban and industrial water sectors, even without climate change, will create additional water stresses and severe social and economic disruption.

*This is certainly true for the Arab Region and requires significant creative planning and reapportionment of water rights to achieve the socially optimal use of water for a nation as a whole. As the Lagi, et al. (*2015*) study points out the massive dislocation that has occurred in Syria happened after the worst of the drought had passed. The need arises to identify ways of creating rural employment to stave off massive rural to urban migration flows.*

6. Due to increased rural populations, landholdings in small farm agriculture will continue to decrease in size and profitability leading to increasing immiseration of the rural poor.

This will occur unless special attention is given to improving the terms of trade between rural and urban populations. There is great concern with the levels of child stunting in Egypt and Libya even though the total amounts of food nationally are adequate, and actual farm sizes in the region are larger than in many other regions of the globe. It appears that the inability of small farmers to make a living off their crop activities is partially due to global predatory pricing in the developed parts of the world.

7. In order to protect national populations Food self-sufficiency is an appropriate policy.

This, generally outdated policy option, still appears in many of the critiques of the world food system. Politically it is important in any situation to ensure sufficient buffer stocks to tide a nation over to ride out any annual fluctuations in food supply and price. However, for land and water scarce countries relying on other natural resource (oil, minerals, tourism) may be much more reliable that large scale investments in agriculture.

Millennium Development Goals and the NEXUS

The global concern for the Nexus, and an understanding of the five global transitions, is best exemplified in the UN's 17 newly defined Millennium Development Goals (MDG). In 2015 the General assembly of the UN promulgated 17 MDGs to replace and amplify the seven MDGs chosen in the year 2000. These ambitious goals are to be achieved by 2030. The goals are very strongly defined in magnitude and timing. Six of the goals directly address the climate-water-food-energy nexus, water security, and many of the remaining goals are themselves closely related to the Nexus. The directly relevant new MGDs are:

Goal 2. End hunger, achieve food security and improved nutrition, and promote sustainable agriculture

Goal 6. Ensure availability and sustainable management of water and sanitation for all

Goal 7. Ensure access to affordable, reliable, sustainable and modern energy for all

Goal 11. Make cities and human settlements inclusive, safe, resilient and sustainable

Goal 13. Take urgent action to combat climate change and its impacts

Goal 15. Protect, restore and promote sustainable use of terrestrial ecosystems, sustainably manage forests, combat desertification and halt and reverse land degradation, and halt biodiversity loss.

Goals 2, 6, and 7 on this list directly focusing on food , water, and energy have the potential for being achieved by 2030 in the Arab Region, but the remaining ones may be much more difficult to achieve over the 15-year time frame.

Technical Fixes to Bolster Water Security

Rely on International Trade in Water and Land intensive Imports

Virtual water by far the simplest, and most used technical fix is for water-scarce countries to import the bulk of their foodstuff allowing the scarce water resource to be used for other pressing, higher valued urban and industrial supplies. Of course, doing this is not a simple political solution, in many cases the water rights still belong to the agricultural landlords, who may be unwilling to relinquish their rights. Antonelli and Tamea (2015) explore the interaction between food security and virtual water in the Middle East and North Africa and conclude that countries of the region are already overwhelmingly dependent on imported food and as a result has enabled policy makers "to minimize the scope of the water challenge and to avoid politically hazardous reforms in the water sector." Chabed et al. (2015) report on its role in food balances in Tunisia where they found that even though the population has doubled over the past four decades the water equivalent of food demand has more than tripled. This increased per capita water demand (largely due to improved diets) has been largely offset by net imports of virtual water equivalent to 35% of the total water requirement of agriculture. Talozi and Al Sakaji (2015) review the role of virtual water in Jordan and report that 55% of Jordan's available water was sent abroad as fruit exports and a much smaller amount was returned as virtual water imports.

Virtual Land like virtual water this concept relies on putting water-scarce lands to their best use; to produce cash, or industrial, crops for export, and import grain crops. For example, Egypt produces high quality cotton and imports wheat, thereby trading their cotton land for many more hectares of grain area. This allows the yield-gap potentials to be improved on the best quality and well-watered lands.

Develop Traditional Water Sources

Develop New Water Sources this is the traditional way in which food policies were underwritten in the Arab Middle East, however, over time all of the easy new source developments have been undertaken. Typically all the surface waters, excepting a couple of trans-boundary rivers, have already been fully developed as have most potential groundwater resources. A few new surface and groundwater storage sites may be developed, but they may have to rely on imports from trans-boundary rivers and aquifers, and from trans-regional sources.

Several trans-regional and trans-boundary projects have been suggested in some cases extensively studied, and in others just unexplored ideas. For example, the Mediterranean-Dead Sea (MED-DEAD), the Red Sea-Dead Sea (RED-DEAD), and the Tigris-Jordan connection through Iraq (Daines et al. 2015), represent some of the more ambitious proposals. In 2015 Jordan and Israel signed an agreement to go ahead with a $900 million World Bank-sponsored project to build an 80 million cubic meters per year desalination plant in the Gulf of Aqaba with a pipeline for rejected brine to the Dead Sea (World Bank 2012). The RED-DEAD project has largely been assumed to be more practical but as recently as 2010, Gonce (2010) CEO of the Dead Sea Vision LLC was actively promoting the $5.9 billion MED-DEAD project. The Daines (2015) project remains in the preliminary sketch phase.

Develop Non-traditional Sources

Desalination over 24 million cubic m per day is now in operation in the Arab Middle East (Water World 2013). This is over 50% of the total global capacity for desalination, and is rapidly growing. The water produced in these plants is still too expensive for most crops, but may in the future be economical for fruit and vegetable crops, particularly when they can be marketed on international markets at the season when other regions cannot grow them. New reverse osmosis, RO, plants have the advantage that they can use brackish water, recycled urban waste water, and ocean waters as their source. The advantage of relying on desalination is that, provided energy for electricity is available, then the source is 100% reliable, or essentially risk free.

The Middle East Desalination Research Center (MEDRC 2015) based in Muscat has invested heavily in regional training by conducting short courses and sponsoring students and professors involved in research projects. The associate director for development has stressed that over 50,000 engineers and technicians need to be trained for the region's bourgeoning desalination industries.

Improving Efficiency of Use

Precision Irrigation the equivalent of new water supplies are quantities of water saved by improving the efficiency of existing uses. For domestic and industrial uses there are many technology add-ons, or process changes, but for agriculture two water-saving technologies now in wide use around the world; drip, and center-pivot. For small additional costs these technologies, where appropriate, can save as much as 50% of water applications at the field level. Center pivots are popular in North Africa in Morocco, Tunisia, and Libya. They are also widely dispersed in Saudi Arabia. It should be noted, however, that as a water conservation technique these may fail because farmers often immediately bring more land under cultivation with the saved water. This is good for food production, but not as a water conservation measure.

Agronomic Research

Much more research needs to be devoted to salt tolerant and drought resistant crops. This research not only encompasses the modern genetic manipulation of crops, but also evidence-based improvements in yield of existing crops by applying Yield-Gap (YGA) methods of planning for crop and region choices (van Ittersum et al. 2013). Applying the YGA to recent data from Jordan, Morocco, and Tunisia we estimated the yield potential for moving from dryland farming to better water control and up to irrigation for wheat, barley and potatoes as follows:

Actual Yield (Ya), Water and Nutrient Limited (Yw-Ya), Water Limited (Yw), and Potential Yield (Yp) t/ha

Jordan

	Ya	Yw−Ya	Yw	Yp
• Rainfed wheat	0.95	2.48	3.43	6.91+
• Rainfed Barley	0.72	2.14	2.86	6.62

Morocco

	Ya	Yw−Ya	Yw	Yp
• Rainfed wheat	1.17	1.46	2.64	6.07
• Rainfed barley	0.99	3.00	3.99	8.56
• Irrigated potato	25.68			59.38

Tunisia

	Ya	Yw−Ya	Yw	Yp
• Rainfed wheat	0.64	2.34	2.99	6.83
• Rainfed barley	1.07	3.62	4.69	8.77
• Irrigated wheat	2.39			6.31
• Irrigated potato	15.35			23.63

These calculations suggest that there are significant potential benefits of doubling or tripling production of the important food crops if the farmers were to move to improved water and fertilizer management and up to a six to tenfold increases as they move to full irrigation and nutrient management. For example, rainfed wheat in Jordan could experience an increase from 0.95 to 2.48 t/ha by weeding, husbandry, and applying more fertilizer while still using the current water supply. A similar step in Morocco would give a smaller increase from 1.17 to 1.46 t/ha, probably reflecting the higher rainfall in Morocco. These yields do not rely upon new and innovative agricultural research, but on actual yields obtained from actual farmers in actual settings in the region.

Other research areas such as drought resistant and salt tolerant crops, already being researched at the International Centre for Biosaline Agriculture (ICBA) in Kuwait, could make major contributions toward augmenting scarce water supplies in small-scale agriculture.

Food and Value Chains

Post Harvest Losses by modernizing the food and value chains from "farm to fork" substantial quantities of water (up to 40–50% in many cases) can be saved from not producing the crops that are wasted. Rural infrastructure, roads, markets, and cold storages would speed up the process and allow the perishable goods to be better marketed by the farmers with a significant increase in retained rural value added (Rogers and Daines 2014). This has to be part of the stabilizing of the rural to urban migration. Reduction of post harvest losses is also one of the most reliable and cost effective way of conserving water—this is domestic virtual water trade.

Regulation

Pricing one of the major mysteries of water planning and management is the reluctance of governments to use pricing of water as the best tool for allocating water to its highest and most valuable uses (Rogers et al. 2002). Around the world this reluctance may be based upon realistic assumptions of the political difficulty of reclaiming water rights which have been appropriated by individuals and groups over hundreds of years. Property rights for water are deeply ingrained in many societies and are often bound up in cultural and religious beliefs. Zetland (2012) however, marshals' impressive global data to show that pricing may be finally catching on in many water utilities.

Trading closely related to pricing allowing trading of water, or water rights, among individuals and groups presupposes that the issue of water rights have been settled and that the market can identify the best price for water (or water rights). Trading of water rights goes back millennia in the Arab region, for example the aflaj in Oman (Sutton 1984) supplying and trading water for up to several thousand users have antique social and legal mechanisms for peaceably sharing water stretching back at least 2000 years. Trading has also been carried out in the US South West starting in the ninenteenth century (Howe 2000) and more recently notably in Chile and Australia (Grafton 2010). There are also large water trading schemes dating from the fifteenth century in Valencia, Spain, (Maass 1978) apparently modeled after the Moorish water sharing prctices in southern Spain.

Water Resource Command and Control

In countries with a history of strict accounting of water use by sector and strong political control, with low levels of rural corruption, it is possible to regulate water use and reallocate property rights to changing economic and physical changes. Few countries in arid regions where agricultural users were the first to appropriate the resource, have managed to regain control over property rights, have been able to exert the needed strong technical and political controls at the local level. As an example the Western states of the US have strong systems of enforceable senior and junior water rights which are rigorously controlled and adjudicated by the state courts (Howe 2000).

Legal Protection for Third Parties

Of course, the use of water by one user or groups of users can have large noticeable external effects on other users or potential users. Absent reasonable safeguards for third parties and the ecosystem, pricing, trading and strong command and control will do little to ensure the third party rights. In the US this has been rolled into the environmental protection federal and state laws.

When governments define their nexus problems too narrowly, e.g., by focusing on water conservation in just the water sector, there are often unintended consequences which may exacerbate other dimensions of the nexus, such as energy. For example, in Morocco focus on water conservation by drip irrigation led the farmers to increase their electricity demands in the process of expanding their irrigated areas with the water they had conserved leading to excessive loss to the government for both irrigation equipment and electricity subsidies (Jobbins, 2015). One needs to always be on the lookout for unintended consequences of water and energy subsidies.

Regulation of Food and Value Chains

Given the global crisis of food losses in the food chain from farm-to-fork, there is a large potential for government action in supplying infrastructure (roads, cold storages) and stabilizing local and regional markets. This is probably best achieved as joint activities with the private sector. With the appropriate infrastructure investments, the governments can rely upon the private sector to improve the efficiency of the overall food and value chains (Rogers and Daines 2014).

Summary

To get a perspective on how the nexus has grown in importance for the Arab region, we looked back at some projections made in 1994 based upon 1990 data (Rogers and Lydon 1994). At that time we estimated a 2025 total population of 492 million with an urban population of 341 million. According to the World Bank (2014a, b) the total 2010 population of MENA was only 372 million and growing at 1.8% per year. Their expected 2025 population which would be 425 million is lower than our 1994 estimate of 492million, but still on track to generate a total population of 756 million by 2050. The urban population is expected to soar from 209 million in 2010 to 378 million in 2050 (based on the assumption of more than 50% urban by 2050). These numbers indicate that even without climate change, the relationships among water, food, and energy will still be extremely important for the region. The numbers also indicate the urgency to move forward soon with restructuring agriculture in the major producers, Egypt, Iraq, Syria, Morocco, Algeria, Jordan, Lebanon, and Tunisia. From our analysis of the water security issue, we agree with most of the recommendations of our 1994 book where we stressed that the best sources of new water would be by modernizing agriculture. We also support the World Bank (2008) report on MENA which stressed that little could be done to expand water supplies to the region, but that conservation and proper uses of water in agriculture are the most important activities for dealing water and food security. In order to achieve this modernization the scope of the water institutions and agencies would have to be broadened to integrate all water uses together with the overall economy, often called Integrated Water Resources Management (IWRM).

Specific recommendations from this chapter emphasize dealing with the issue of climate change, which was not a concern in 1994, but which is now of paramount concern. In particular we like to recommend those policies and technical interventions which reduce the risk of future uncertainties. Permanent conservation, desalination, and importing virtual water all make the countries more robust with whatever happens with climate change. Crop and food production will always remain uncertain due to the vagaries of normal weather, but domestic and industrial water supplies can be made 99.99% certain when they are backed up with desalination. Softer regulatory policies can also improve food and water security by reducing demand, reducing food waste and, hence, water use.

Food, energy, and water are at the top of Maslow's Needs Hierarchy. Meeting those needs does not mean that conflicts, wars, and disease will be resolved, but that humanity will be more resilient to deal with those issues in a safer space than where it is currently located. We need to look beyond conventional agricultural management skill transfer recipes. Expanding capacity and knowledge are always good things to do, but the emphasis should be to bring in, where possible, modern commercial and corporate actors who can best transfer these skills. Finally, investments in transportation and marketing have the potential to greatly facilitate the agricultural transition to modern practices and production efficiencies. We are not searching for optimal strategies which may be risky, but look for strategies for a safe transition to a sustainable food supply. In other words, can we identify a "Safe Place for Humanity?" We believe that wise use of the strategies suggested in this chapter will achieve this goal by 2050.

References

Allan, T., Keulertz, M., & Eckart, W. (2015).The water–food–energy nexus: An introduction to nexus concepts and some conceptual and operational problems, pages 301–311. Special Issue: The Water-Food-Energy-Climate nexus in Global Drylands: The epitome of twenty-first century development. *International Journal of Water Resources Development, 31*(3).

Allouche, J., et al. (2014). *Nexus or Nirvana or Nexus Nullity? A dynamic approach to security and sustainability in the water-energy-food nexus* (STEPS Working Paper 63). Brighton.

Antonelli, M. & Tamea, S. (2015). Food-water security and virtual water trade in the Middle East and North Africa, pages 326–342. Special issue: The Water-Food-Energy-Climate nexus in Global Drylands: the epitome of twenty-first century development. *International Journal of Water Resources Development, 31*(3).

Arab Organization for Agricultural Development (AOAD). (2012). *Arab agricultural statistics yearbook*, No. 32. Khartoum: AOAD.

Bellarby, J., Foereid, B., Hastings, A., & Smith, P. (2008). *Cool farming climate impacts of agriculture and irigation potential.* Amsterdam: Greenpeace.

Breisinger, C., et al. (2010). *Food security and economic development in the Middle East and North Africa.* No. 985, IFPRI discussion paper.

Breisinger, M., et al. (2012). *Beyond the Arab awakening: Policies and investments for poverty reduction and food security* (IFPRI Food Policy Report 25).

Chahed, J., Besbes, M. & Hamdane, A. (2015). Virtual-water content of agricultural production and food trade balance of Tunisia, pages 407–421. Special Issue: The Water-Food-Energy-Climate nexus in Global Drylands: The epitome of twenty-first century development. *International Journal of Water Resources Development, 31*(3).

Dainees, S., et al. (2015). *Tigris Trans Jordan Canal Company Bond.* https://vimeo.com/147653182. SRD Research.

Doukkali, M. R., & Lejars, C. (2015). Energy cost of irrigation policy in Morocco: a social accounting matrix assessment, pages 422–435. Special Issue: The Water-Food-Energy-Climate nexus in Global Drylands: The epitome of twenty-first century development. *International Journal of Water Resources Development, 31*(3).

Dubois, O. (2015). *The case for "energy-smart food for people and climate"*. Food and Agriculture Organization of the United Nations, World Food Day-Oct 16.

Endo, A., Burnett, K., Orencio, P. M., Kumazawa, T.,Wada, C. A., Ishii, A., Tsurita, I., & Taniguchi, M. (2015). Methods of the water-energy-food nexus. *Water*, *7*, 5806–5830. doi:10.3390/w7105806.

Falkenmark, M., & Lundqvist, J. (1998). Towards water security: Political determination and human adaptation crucial. *Natural Resources Forum, 21*(1), 37–51.

Gleick, P. H. (1993). *Water in crisis: A guide to the world's fresh water resources.* Oxford: Oxford University Press.

Gonce, R. (2010). Dead Sea Vision LLC. rogonce@gmail.com www.deadseapower.com

Grafton, Q., Landry, C., Libecap, G., McGlennon, S., & O'Brien, B. (2010). *An integrated assessment of water markets: Australia, Chile, China, South Africa and the United States.* ANU Centre for Water Economics, Environmental and Policy Research Paper. Last accessed 29 Apr.

GRAIN. (2012). http://www.grain.org/article/entries/4929-hungry-for-land-small-farmers-feed-the-world-with-less-than-a-quarter-of-all-farmland

Greenwood, S. (2014, Summer). Water insecurity climate change and governance in the Arab World. *Journal Essay, Middle East Policy Council, XX1*(2).
Grey, D., & Sadoff, C. W. (2007). Sink or Swim? Water security for growth and development. *Water Policy, 9*, 545–571.
Gutierrez, E. (1999). *Boiling point: Issues and problems in water security and sanitation, water aid briefing paper*. London: Global Water Partnership.
Hallegatte, S., Shah, A., Lempert, R., Brown, C, & Gill, S. (2012, September). *Investment decision making under deep uncertainty – Application to climate change* (World Bank Policy Working Paper, WSP, 6193).
Hazell, P. (2013, December). *Is small farm led development still a relevant strategy for Africa and Asia?* www.fondation-farm.org/zoe/doc/peterhazell_original_english_paper.pdf
Hoff, H. (2011). *Understanding the nexus', background paper for the Bonn 2011 conference: 'The water, energy and food security nexus*. Stockholm: Stockholm Environment Institute.
Howe, C. W. (2000). *Protecting public values in a market setting: Improving water markets to increase economic efficiency and equity* (University of Denver Water Law Review. 357).
International Center for Biosaline Agriculture (ICBA). Kuwait.
International Fertilizer Industry Association (IFA). (2013). Climate change—A historical perspective. www.fertilizer.org
Jaafar, H. H., Zurayk, R., King, C., Ahmad, F., & Al-Outa, R. (2015). Impact of the Syrian conflict on irrigated agriculture in the Orontes Basin, pages 436–449. Special Issue: The Water-Food-Energy-Climate nexus in Global Drylands: the epitome of twenty-first century development. *International Journal of Water Resources Development, 31*(3).
Jobbins, G., Kalpakian, J., Chriyaa, A., Legrouri, A., & Mzouri, E. H. E. (2015). To what end? Drip irrigation and the water–energy–food nexus in Morocco, pages 393–406. Special Issue: The Water-Food-Energy-Climate nexus in Global Drylands: The epitome of twenty-first century development. *International Journal of Water Resources Development, 31*(3).
King, C., & Jaafar, H. (2015). Rapid assessment of the water–energy–food–climate nexus in six selected basins of North Africa and West Asia undergoing transitions and scarcity threats, pages 343–359. Special Issue: The Water-Food-Energy-Climate nexus in Global Drylands: The epitome of twenty-first century development. *International Journal of Water Resources Development, 31*(3).
Lagi, M., Bar-Yam, Y., Bertrand, K. Z., & Bar-Yam, Y. (2015) Accurate market price formation model with both supply-demand and trend-following for global food prices providing policy recommendations PNAS 2015; published ahead of print October 26.
Maass, A. (1978). *And the desert shall rejoice: Conflict, growth, and justice in arid environments*. Kreiger Publishing Company.
Marshall, B., Hsiang, S. M., & Miguel, E. (2015). Global non-linear effect of temperature on economic production. *Nature*. doi:10.1038/nature15725.
Maslow, A. H. (1943). A theory of human motivation. *Psychological Review, 50*(4), 370–396.
Maystadt, J. F., et al.. (2012). *Does food security matter for transition in Arab countries?* No. 1196, IFPRI discussion paper.
Middle East Desalination Research Center (MEDCR). (2015). News letter: Muscat.
Patel, R. (2009). Food sovereignty. *The Journal of Peasant Studies, 36*, 663–706.
Ray, P. A., & Brown, C. M. (2015). *Confronting climate uncertainty in water resources planning and project design : The decision tree framework*. Washington, DC: World Bank Group. http://documents.worldbank.org/curated/en/2015/08/24957941/confronting-climate-uncertainty-water-resources-planning-project-design-decision-tree-framework
Rogers, P. (2013). *Water security: Supplies, resources, and uncertainty, Chap. 1*. pp. 21–36 in Emirates Center for Strategic Studies and Research, Water and Food Security in the Arabian Gulf, ECSSR, Dubai.
Rogers, P., & Daines, S. (2014, February). *A safe space for humanity: The nexus of food, water, and climate* (Asian Development Bank Briefs No. 20).

Rogers, P., & Lydon, P. (1994). *Water in the Arab world: Perspectives and prognoses*. Cambridge: Harvard.

Rogers, P., de Silva, R., & Bhatia, R. (2002). Water is an economic good: How to use prices to promote equity, efficiency and sustainability. *Water Policy, 4*, 1–17.

Sood, A., & Smakhtin, V. (2015). Global hydrological models: A review. *Hydrological Sciences Journal, 60*(4), 549–565. doi:10.1080/02626667.2014.950580.

Sutton, S. (1984). The falaj – A traditional co-operative system of water management. *Waterlines, 2*(3), 8–11.

Swedish FAO Committee. (2014). *Water, food security and human dignity—a nutrition perspective*. Publication series 10: Jan Lundqvist and Jenny Grönwall, Stockholm International Water Institute and Anders Jägerskog, Sida/Swedish Embassy, Jordan.

Taleb, N. N., Read, R., Douady, R., Norman, J., & Bar-Yam, Y. (2014, October 17).The Precautionary Principle (with Application to the Genetic Modification of Organisms). NECSI archive:1410.5787.

Talozi, S., Al Sakaji, Y., & Altz-Stamm, A. (2015) Towards a water–energy–food nexus policy: realizing the blue and green virtual water of agriculture in Jordan, pages 461–482. Special Issue: The Water-Food-Energy-Climate nexus in Global Drylands: The epitome of twenty-first century development. *International Journal of Water Resources Development, 31*(3).

Tolba, M. K., & Saab, N. W. (2009). *Arab environment: Climate change*. Arab Forum for Environment and Development (AFED).

Unver, O., & Pluschke, L. (2015). *Can saving lives save livelihoods? The water-energy-food nexus and human security*, Chapter 12 in Water Development—Charting Water Wise Path, the 2015 World Water Week Report, SIWI.

US Department of Agriculture. (2015, December 1). *Climate change, global food security, and the US Food System*. USDA.

Van Ittersum, M. K., et al. (2013, March). Yield gap analysis with local to global relevance—A review. *Field Crops Research, 143*(1), 4–17.

Verner, D., & Breisinger, C. (2013). *Economics of climate change in the Arab world: Case studies from the Syrian Arab Republic, Tunisia, and the Republic of Yemen*. No. 13124 in World Bank Publications.

Water World. (2013, October 14). *Global desalination capacity growing substantially*.

Waterbury, J. (2013). *The Political economy of climate change in the Arab region* (Arab Human Development Report, Research Paper Series). UNDP.

Wiebelt, M, et al. (2010). *Climate change and floods in Yemen: Impacts on food security and options for adaptation*, No. 1139, IFPRI Discussion Paper.

Wiebelt, M., et al. (2014). *Who bears the cost of climate change? Evidence from Tunisia*. Kiel Institute for World Economy, Working Papers.

Wodon, Q., Liverani, A., Joseph, G., & Bougnoux, N. (2014). *Climate change and migration: Evidence from the Middle East and North Africa*. Washington, DC: World Bank. © World Bank. https://openknowledge.worldbank.org/handle/10986/18929

World Bank. (2010). *World development report 2010: Development and climate change*. Washington, DC: World Bank. https://openknowledge.worldbank.org/handle/

World Bank. (2011). *Red Sea–Dead Sea water conveyance study program: Question and answer sheet*.

World Bank. (2014a, September). *Middle East and North Africa Data Book, Office of the Chief Economist*. Washington, DC: World Bank.

World Bank. (2014b). *Impact of weather shocks on MENA households. MENA knowledge and learning quick notes series; no. 128*. Washington, DC: World Bank Group. http://documents.worldbank.org/curated/en/2014/07/20145326/impact-weather-shocks-mena-households

World Bank Study. (2012). Red Sea – Dead Sea Water Conveyance Study Program.

Zetland, D., & Gasson, C. (2012, October 12) A global survey of urban water tariffs — Are they sustainable, efficient and fair. *International Journal of Water Resources Development*, Property and Environment Research Center (PERC) Research Paper No. 12–20.

Water, Food and Trade as an Element of the Water-Energy-Food Nexus in the MENA Region

J.A. (Tony) Allan

Abstract The purpose of the analysis will be to provide an introduction to the architecture of the Water-Energy-Food Nexus. It will be argued that recognising this architecture could be helpful in three ways. First, it could enable public policy-makers to identify the risks of business as usual approaches to managing water and energy. Secondly, it highlights – for the key private sector players in the water, energy and food supply chains – the importance of stewarding the ecosystems of water, biodiversity and the atmosphere. Thirdly, the supply chains architecture enables water and energy scientists and professionals to engage with nexus science and nexus practice more effectively. The analysis will focus on the water, food and trade sub-nexus and the need to understand the operation of this sub-nexus in the Middle East.

Keywords Water management • Energy management • Supply chain

Why the Water-Energy-Food Nexus Now?

The purpose of this chapter is to propose a pragmatic approach to conceptualising and engaging with what has become known as the Water-Energy-Food Nexus. Humans have mobilised and consumed water, energy and food for over 10,000 years in separate and progressively longer and more complex supply chains. They have traded food for over 3000 years. In the past four decades they have encountered the consequences of their natural resource consumption impacts on water, the atmosphere and biodiversity. It was not until the second half of the twentieth century that supply chains mobilising water and energy and producing food were numerous enough and big enough to impose unsustainable impacts on the under-pinning natural ecosystems of water, the atmosphere and biodiversity. (Rockstrom 2014)

The MENA region has, however, provided evidence that there have been a number of moments in the world's environmental history when human

J.A.(T). Allan (✉)
Department of Geography, King's College London, The Strand, London WC2R 2LS, UK
e-mail: ta1@soas.ac.uk

S. Murad et al. (eds.), *Water, Energy & Food Sustainability in the Middle East*,
DOI 10.1007/978-3-319-48920-9_3

interventions proved to be unsustainable. The region is rich in solar energy – a resource which has not yet been effectively mobilised. The mobilisation and consumption of hydrocarbons has had massive impacts on the atmosphere. The region's water resource endowments do not meet the demands of its current economies. Throughout history the vulnerable Tigris-Euphrates flood-plain has proved to be easily overwhelmed by irrigation interventions. Its early civilizations were more than once brought down by mismanaging these rivers. (Jacobsen and Adams 1958) In contrast, in North Africa, during the same three millennia, African silt that for eons had built the Nile delta and its flood plains has proved to be resilient to very much bigger interventions. These hydraulic interventions were imposed to produce food and fibre for the successive civilizations of the Egyptian lower Nile as well as to feed Rome. The Nile silt was resilient, and the volumes of water available were sufficient, until the population of Egypt increased to 50 million in the 1970s in the very recent past.

In the 1970s, after 20 years of campaigning by activists deploying scientific research, the importance of installing environmentally, economically and socially sustainable ways of consuming natural resources world-wide became evident. (United Nations 1972; Bruntland Report 1987) By then negative ecosystem impacts of consuming natural resources were especially evident in the neo-liberal OECD economies of North America, Europe and East Asia.

The national and international discourse shifted from the accepted certainties of the era of industrial modernity – mid nineteenth century to 1980 – to the increased awareness of the uncertainties of late modernity after 1980. (United Nations 1992) The certainties had been based on the assumption that the rising demand for water to produce, for example food, could be forever be met by increasing supplies of water and energy. This approach has been called supply management. By the end of the 1970s it was increasingly argued that a sustainable approach must include the idea of demand management. The 'hydraulic mission' era of the late nineteenth century and the first seven decades of the twentieth century was progressively replaced by 'risk aware' policies and practices. Questions were raised first, about the costs of mobilising and consuming water in water blind food market systems and secondly about the costs of degrading water ecosystems. Thirdly, it was argued that that new policies and practices would have to be introduced via measures that are participatory and politically savvy and feasible especially in the MENA region. (Allan 2001)

The ways that natural resources are valued and consumed have to some extent been reflected in the adoption of environmentally precautionary policies in advanced economies. (EU 2015a -Water Framework Directive, EU 2015b – CAP Pillar 2, UNU-IHDP and UNEP 2012) The emerging economies – sometimes known as the BRICS economies, namely Brazil, Russia, India, China and South Africa – are beginning to adopt them. Brazil and some other South American economies very progressively.

There is already a rich literature on the toxic politics on how to mitigate and adapt to climate change with the views of OECD economies being contradicted by those of the BRICS and developing economies. (Hulme 2009; Giddens 2011)

Governments and private sector operators in West Asia and North African (MENA) economies have begun to pick up the precautionary messages and they are also beginning to adopt clean, green and sustainable policies and practices. Some NGOs have made very significant contributions to the MENA discourse on the sustainable management of the regions natural ecosystems. (EcoPeace 2015)

This gradual recognition of the risks associated with mis-managing natural ecosystems and the need to adopt policies and pratices that steward them effectively was significantly accelerated by the international financial crisis of 2008. The issue of *risk* had been integral to the shift from supply to demand management and from ecosystem blind to ecosystem aware approaches. In 2008 World Economic Forum (WEF) analysts and thought leaders examined and reported on the increasingly evident environmental and ecosystem risks being faced by the WEF corporate community. The WEF forum brings together corporate interests, but also involves scientists, NGOs, politicians and public policy-makers.

In the 2008 to 2015 period the World Economic Forum carried out surveys of its members on their perception of the intensity and likelihood of the occurrence of risks to their businesses. These sequential surveys revealed how risk perceptions have changed and especially the increasing risks associated with the ways that water, atmosphere and biodiversity are currently being consumed and abused. (WEF 2015; Conca 2015)

During this same period WEF has promoted the idea that the ways water and energy resources are being mobilised to meet the needs of society – for example to provide their food needs – are not economically or environmentally optimum. WEF adopted the term *the water, energy and food Nexus* to capture the interdependence, mutuality and complexity of managing – or rather mismanaging – them. The intent was to promote engagement that would identify and promote optimum outcomes that would optimise the consumption of water and energy to replace existing sub-optimal practices.

In 2011 WEF produced an edited collection of essays by scientists and practitioners on the Nexus thinking at the time. (WEF 2011) It revealed the absence of an accessible analytical framework. It also provided an authoritative review of the nexus hotspots – California, Texas, the Mediterranean, West Asia and North Africa, Punjab, the North China Plain and Austraiia – where the ways that water and energy were being mis-managed had brought about unsustainable ecosystem crises through the wasteful consumption of water and energy. It can be argued that conceptualising the water-environment-food nexus is a massive analytical challenge and that initial analysis is bound to be unsatisfactory. Modelling such as system effectively is probably impossible. Establishing and operating a sub-optimal one is predictable.

The challenge of handling the raw complexity is increased by the ways that the world's political economies manage and mismanage water and energy to produce food and fibre. Practices have evolved as supply chains with deeply embedded and hard to change rules that cannot deliver sustainable market regimes that both steward and protect natural ecosystems.

A number of diagrams from the World Economic Forum *Global Risks Insight Reports – 2007 to 2015* are very helpful in tracking how risk awareness has moved with respect water, energy, climate change adaptation and food security. (WEF 2015) This series of global risks reports has revealed both the increasing salience of first, water, energy and food insecurity risks and secondly, it has highlighted the water, atmosphere and biodiversity ecosystems at risk.

By 2015 the respondents to the annual WEF global risk survey believed that globally water crises had reached the highest position on the impact scale and very high position on the likelihood scale. The same WEF analysis indicates that those with West Asia and North African interests believed that water crises presented the highest business risk. The risks were higher than extreme political instability and terrorism despite the evidence of extreme instability since 2011.

Problems of Conceptualising the Water-Energy-Food Nexus and Taking into Account the Impacts of Climate Change and the Role of Trade

Ever since the link between the risks associated with mismanaging the water-energy-food nexus was highlighted about a decade ago it has been recognised that existing private sector markets and public policy arrangements are not simple systems. At the same time climate change has been shown to be just one of the pressures increasing the levels of risk associated with the complex global trading and market systems. Unfortunately communicating the certainties and uncertainties of climate change processes have proved to be very contentious. Agreement has been very difficult to achieve especially at the comprehensive international levels but also at sub-national policy levels. (Hulme 2009)

Delivering *water services* mainly via public utilities in short infrastructures that are very investment intensive, is one of the challenges. Others are integral to *the delivery of energy and food* via both long as well as via short private sector supply chains. The long chains are very complex and usually involve international trade. Like climate change, trade, is a very complicating process. It is susceptible to the distorting influences of powerful North American, European and East Asian players that impact negatively developing economies. Neither the rich exporters nor the weak developing economies that import commodities which are associated with substantial volumes of embedded water and energy have yet adopted sustainable – Nexus aware – production practices.

The Nexus concept should have recognised and adapted its agenda to cope with the very complex suite of political economy conditions to which supply chains are subordinate. It should also have taken in its stride the challenges associated with climate change and international trade. To date advocates of the Nexus approach have not begun to engage with how markets actually work or with the market rules and public policies that actually operationalise the business as usual systems. Such engagement is necessary to establish an effective market mode that would operationalise the delivery of financially and environmentally sustainable water and energy services in future. A major purpose of this analysis is to persuade advocates of the Nexus approach do engage with such market systems.

In practice the interconnected complex reality has been back-grounded by the adoption of the term Nexus. Many systems diagrams have been generated which identify and highlight the big picture via a network analyses showing nodes and links. (WEF 2011) The diagrams also highlight the importance of risks caused by pressures/trends, such as climate change, increased populations, (United Nations – DESA 2015) environmental degradation and what has emerged recently as refugee mobility. These pressures multiply the risks and amplify existing risks to the functionality and stability of the world's water services, and to its energy and food supply chains. The operational risks, such as food and water crises, are also identified in the centre of the diagram as economic, geopolitical, technological and environmental risks.

The World Economic Forum study (2011) provides a version of the global risk trends with an interconnections systems map. The 'map' is typical in that it presents an incomprehensible and a non-operationalisable version of the nexus. It correctly conveys – without the intent of doing so – the impression that the Nexus concept is beyond description. It does identify many of the major political pressures that frame any attempt to conceptualise a Nexus analysis. But it fails to identify any *operationalisable* connectivity. Nor does it identify any rules or conventions that could determine whether the connections might deliver sustainable outcomes for people, profit and the planet.

This absence of the identification of functional connectivity in the WEF and other diagrams contrasts with the partial connectivity via prices in market supply chains that reflect very inadequately the costs of inputs, trade and marketing. Unfortunately existing market accounting systems, despite the detail they record of some costs and careful auditing by a vast accounting industry, are flawed as means of operationalizing Nexus goals and principles by their blindness to the value of ecosystems. What could be a complete record of costs is in practice an inadequate record of connectivity.

Those operating these supply chain markets comply with what in practice are inadequate reporting and accounting rules that fail to capture what accountants term the 'materiality' of information. (IASB 2015) In this case the materiality of for example, the costs of mobilising water inputs in crop production and the externalities of mismanaging water ecosystems. In the world of accountants it is authoritatively stated that – information is material If omitting it or misstating it could influence decisions that users make on the basis of financial information about a specific reporting entity. (IASB 2015) Would that those who make accounting rules and the professional accountants who police them would step up and install rules that capture the costs of natural resource consumption in energy and food supply chains. (Allan et al. 2015) The argument being made here is that it is more likely that the costs that need to be reflected could be captured in reformed accounting systems rather than in any attempt to establish rules based on the connections that might be evident in diagrams.

The values of natural resource inputs such as water and energy are doubly difficult to track because subsidies and taxes have been put in place – often with good intent – in both the energy and food supply chains. Their existence has made it even more difficult to provide financial signals to consumers of the consequences of their consumption choices and of their wasteful behaviour.

Lessons can be learned from the attempts to incorporate the risks and costs of ecosystem stewardship into the three supply chains. They have been learned, for example, in the attempts to mitigate and adapt to targets identified by climate change science. Those operating private and public supply chains, as well as legislators, the media and consumers have been involved. The first painful lesson has been that the science of modelling climate change and the impacts of climate change are not neutral political processes. (Hulme 2009; Giddens 2011)

The processes have proved to be very contentious partly because there is not just one climate change model. There are over 20 major integrated assessment models (IAMs) and their predictions do not agree. They agree on the trend in global temperature. But they do not agree on future levels of rainfall in the world's numerous major catchments. (Gosling et al. 2010, 2011; Gosling and Arnell 2011, 2013) An important issue which it is not possible to address in detail in this analysis is the difficulty of modelling future rainfall. The MENA region will almost certainly have to cope with lower levels of rainfall. But the major source of blue water in the region is the River Nile which brings water from outside the region. Three Nile catchment models predict a reduction in future rainfall. Two others predict that there will be an increase. (Conway 2005)

Meanwhile this author has for decades pointed out that societies in the MENA region have coped with the much greater challenge of the trebling of MENA population in less than half a century. Some hot-spots in the region have recorded very steep increases in population. The population of Israel plus Palestine has increased sevenfold since 1948 and will increase tenfold. Since the 1970s a version of food security has been enjoyed by the MENA region economies by importing from global markets water intensive food commodities. (Allan 2001) Future water security can only be enjoyed by MENA economies if global trade in food can be conducted in stable global politics and in the absence of any lengthy disruption of trade.

Operational Supply Chains – A Pragmatic Approach to Identifying How Markets and Public Policy Actually Allocate Water and Energy in the MENA Region

The challenge of delivering water services, energy services and food is unarguably a very complex global process. It involves fair and unfair trade and a capacity to cope, or not, with the predictable trends and the unpredictable volatilities associated with climate change. These service-delivering activities are even more complex if the services have to be delivered in ways that effectively steward water ecosystems, the atmosphere and biodiversity. In attempting to address this very significantly expanded agenda the Nexus approach has a much bigger task than the conventional one of meeting global demands for water services, energy and food sustainably. In addition this bigger Nexus agenda and approach would have to be installed in ways that are politically feasible in the face of deeply entrenched bad practices. The three supply chains that currently populate the WEF Nexus have evolved over long periods. One of them – the food supply chain market system – has evolved over millennia. Throughout the long evolution of the food supply chain system it has been assumed that ecosystems did not need to be stewarded.

Those operating food and energy supply chains have been blind to the value of ecosystems. Consumers, who want food and energy to be cheap are willingly blind to the value of natural resources and of well stewarded ecosystems. To take them into account would increase food prices. They are also unwittingly blind to the role of international trade in accessing the natural resources on which they have progressively come to depend. *International trade* has since the 1950s been an increasingly important element in providing a version of food security in the MENA economies. (Allan 2001) All the MENA economies are net food importers. That is they are 'importers' of the environments of other economies. The MENA economies have benefitted very significantly from the willingness of food exporters to not include the costs of degrading their ecosystems.

This section highlights the problem of installing natural resource aware policies, and introducing rules and practices that would establish a Nexus inspired approach to delivering sustainable water services, sustainable energy and sustainable food to the world's consumers – including to those in the MENA economies. The newly articulated Nexus approach requires that water and energy be consumed in effective production and trading systems which also steward water, the atmosphere and biodiversity. Attempts to install the Nexus approach could be attempted via a radical, top down, implementation of new rules that modellers assume could optimise the integration and the inescapable mutuality of water and energy provision and consumption. No such attempt has been made to date. It is argued here, however, that an incremental augmentation of the existing reporting and accounting rules in water, energy and food supply chains is a much more feasible approach. It could be the way to capture the mutuality of water and energy resource consumption together with the necessary recognition of the value of ecosystems.

It has been established in the analysis thus far that water, energy and food supply chain practices have evolved based on very dangerous assumptions. For example one dangerous assumption is that water and water ecosystem impacts should not be valued and energy should be undervalued. The consequence is that energy and food supply chains are not effectively regulated. They lack effective supply chain accounting practices. In addition the global trading regime – where transactions on water intensive food and energy take place – are unfair. It should be noted that the value of water in *the water services supply chain* is in some cases fully recognised and the trend world-wide is to reflect the full economic cost of water in the prices charged to domestic and industrial consumers. In the MENA region, however, there are major exceptions to this global trend. The leaders of oil rich MENA economies cannot persuade their citizens that they are not entitled to free water and to nearly free energy.This outcome epitomises the elemental politics associated with the provision of water and energy services.

A feature of the past two decades has been attempts to come to terms with the challeneges of accounting for environmental assets. Natural capital accounting rules are being discussed at the national level. (Natural Capital Coalition 2015) Reporting and accounting rules are being reviewed in private sector food and energy markets (KPMG 2014a, b, c; Trucost 2015; TEEB 2015). There has also been a proliferation of national and international bodies researching the role of accounting for natural capital, namely The World Bank and IFC, UNEP and OECD (McPhail et al. 2012) confirms the Nexus approach could be provided with knowledge and alternative accounting systems. But so far these attempts to conceptualise and operationalise natural capital accounting have not been communicated in the language of accounting professionals.

The analysis of natural capital is made difficult by the very different characteristics of water, energy and food. The three supply chains are different in many ways. They differ in the types, volumes and proportions of water and energy mobilised to produce food. For example, there are two main types of water consumed in society's economic systems. But there are at least 12 types of energy. (Allan and Matthews 2016) The food supply chain accounts for about 92% of global water consumption. (Hoekstra 2008) But only about 30% of emissions that impact the atmosphere. Domestic and industrial water users consumers account for about 8% of water consumption but they account for at least nearly 80% of emissions.

A feature of MENA and global natural resource accounting is its blindness to one of the major sources of water. Green water, otherwise known as effective rainfall is ignored. A significant proportion of MENA food consumption depends on the green water (effective rainfall) mobilised within the region. This regional green water is ignored in almost all analyses of the region's water consumption. (Besbes et al. 2014) In addition, about 90% of the water 'imported' in food commodities is also green water. The MENA region is unique, however, in consuming such a high proportion of its water consumption as blue (surface and groundwater). This high proportion of blue water consumption is accounted for by the exceptionally high percentage of blue water consumed in Egypt's economy. About 98% of the water consumption by its 90 million population is derived from blue water compared with the world average of 33%. (Hoekstra 2008)

At this point in this review of how water is allocated to different sectors in the MENA region the reader is asked to consider the feasibility of installing Nexus systems. A nexus inspired system would replace existing supply chain based arrangements to promote the protection of water and atmosphere ecosystems and biodiversity as well as incentivising the efficient consumption of water resources and energy

It is argued here that a nexified system would be too difficult to install. Much more feasible would be the incremental installation of reporting and accounting rules in the albeit sub-optimal supply chain based systems. A radically different and untested nexus inspired system would not be digestible. Calls to deploy new, radical and comprehensive nexus approaches that are ungrounded – that is where the proposed system is not based on existing supply chains – are unsurprisingly not yet in evidence. Engagement with nexus concepts is restricted to trying to map the interacting elements of the very complex systems. There have been no attempts to identify new accounting systems or new governance structures.

Three conditions have brought about this outcome. First the delivery of water, energy and food services is – as has already been pointed out – very complex. Secondly, the systems of delivery have not attracted the modelling effort needed to understand them. Thirdly, installing the Nexus approach is very challenging politically. Attempts to couple market and consumer behaviour to sound principles of ecosystem security and economic efficiency, about which there are elemental differences of opinion and political position taking, is a proven graveyard of principled policy intent.

It is argued here that a much more feasible and practical is an approach that is very deeply informed about existing operational landscapes. For example, with respect to food it would be necessary to know who manages water resources. Farmers are key. They are at the beginning of the food supply chain are know the risks associated with food and fibre production and marketing. They manage all the green water consumed in food production and about 60%, at least, of the blue food-water. (Hoekstra 2008) Also essential, and again best know by practitioners, is knowledge of evolving and potentially best agronomic practices. They know about current innovation in the areas of equipment, seeds and rotations as well about bio-technologies and the control of pests and weeds including about the current contention over weed management. Farmer knowledge should be integral to any attempt to nexify water, energy and food supply chains and capture the costs of stewarding ecosystems. Other players in the food supply chains – the corporate traders, the seed, fertilizer, pesticides, equipment corporates and the super-markets are also uniquely informed on the existing market practices and the feasible practical politics of incremental reform.

The communication of ideas and practices that take into account nexus principles of efficiency and stewardship is also very imoportant. Nexus inspired Interventions will be successful to the extent that they are communicated in the language of the accountant and the investor rather than the systems analyst. Those who operate water, energy and food supply chains comply with contracts that are as yet blind to the value of ecosystems and their stewardship in complex multi-risk environments. Investors and accountants are the agents that could devise the incentives and regulations that are needed to couple market behaviour and deliver outcomes that steward ecosystems by capturing their value.

An under-researched nexus systems approach to allocating and consuming water and energy is unlikely to have the necessary reforming traction. The literature generated so far is not intimate with supply chain practices and it written in a language register that does not engage effectively with those operating markets, legislating and levying taxes and tariffs, and installing subsidies that could incentivise the reduction of waste and ecosystem stewardship.

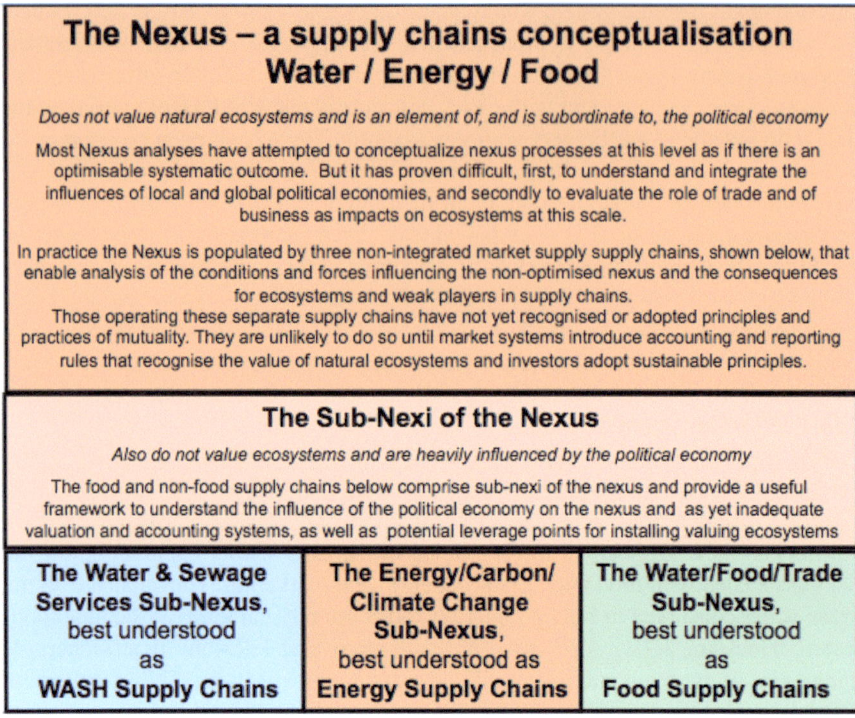

Fig. 1 A water services, energy and food supply chains approach to conceptualising the Nexus and its relationship with political economies in which Nexus and Sub-nexus markets and related public sector processes are embedded (Source: Based on Allan and Matthews 2016)

An alternative approach to conceptualising the nexus is shown in Fig. 1. It conceptualises a nexus that comprises – but does not yet integrate – three separate supply chains. They deliver water services for domestic and industrial consumers and they deliver energy and food in local and global economies.

Concluding Comments

The analysis has critically assessed the current very prominent international debate on the analytical framework and agenda for action known as the *Nexus of water, energy and food production* (Dodds 2016) and the related processes of climate change and trade. It has shown that the debate is timely – especially in the MENA region. In the MENA region the degradation of the ecosystems of water, the atmosphere and biodiversity and the contradictions of the ways that water and energy are consumed highlight the need to install new approaches.

It has been shown that the attempts to nexify water and energy consumption and management have so far not gained currency in MENA or globally. It has been strongly argued here that the adoption of Nexus inspired approaches will require the involvement of supply chain practitioners, and especially farmers. They are necessary because they are the deeply informed players who understand existing practices that supply water services, energy and food. They are also intimate with the intense politics integral to nexification. This process would require the adoption of an ambitious integrated approach that aims to couple reformed practices with principles of economic efficiency and environmental stewardship.

Any attempt to reconfigure radically the existing supply chains to comply with a Nexus inspired system would be a very costly and very high risk project in MENA as well as in the rest of the world. First, a very expensive research effort just as major as the Integrated Assessment Modelling of climate change science would be necessary. Of which there is no sign. Secondly, our inadequate market systems with their partially blind reporting and accounting systems integrate an extraordinarily vast and diverse range of resources, commodities, financial services and professional knowledge. Markets – although not yet configured to promote mutuality and the stewardship of ecosystems are regulated by public sector governance. These long evolved market systems are very firmly in place socially, economically and technically. Any attempt to nexify the three supply chains should be first, an incremental process. Secondly, it must engage those who operate existing supply chains – private and public operators, trades unions, regulators, legislators and accounants – in ways that enable them to develop approaches that are politically and administratively feasible.

References

Allan, J. A. (2001). *The Middle East water question: Hydropolitics and the global economy*. London: IB Tauris.

Allan, J. A., & Matthews, N. (2016). The water, energy and food nexus and ecosystems: The political economy of food and non-food supply chains. In F. Dodds (Ed.), *The water, food, energy and climate Nexus: Challenges and an Agenda for action* (pp. 78–90). London: Routledge.

Allan, T., Keulertz, M., & Colman, T. (2015). *The complexity and urgency of water: Time for the accountancy profession to step up*. New York: International Federation of Accountants. https://www.ifac.org/global-knowledge-gateway/viewpoints/complexity-and-urgency-water-time-accountancy-profession-step

Besbes, M., Chahed, J., & Hamdane, A. (2014). *Water security of Tunisia*. L'Harmattan. http://www.harmattan.fr

Bruntland, G. H. (1987). *Report of the World Commission on environment and development; our common future, Bruntland report*. New York: UN Documents.

Conca, K. (2015). Which risks get managed? Addressing climate risks in the context of water governance. *Water Alternatives, 8*(3), 301–316.

Conway, D. (2005). From head water tributaries to international river: Observing and adapting to climate variability and change in the Nile Basin. *Global Environmental Change, 15*, 99–114.

Dodds, F. (Ed.). (2016). *The water, food, energy and climate Nexus: Challenges and an Agenda for action*. London: Routledge.

EcoPeace. (2015). see http://foeme.org/www/?module=publications&project_id=23

EU. (2015a). *The EU water framework directive and subsequent daughter frameworks*. http://ec.europa.eu/environment/water/water-framework/index_en.html
EU. (2015b). *Second pillar of the CAP: Rural development policy*. http://www.europarl.europa.eu/atyourservice/en/displayFtu.html?ftuId=FTU_5.2.6.html
Giddens, A. (2011). *The politics of climate change*. Cambridge: Polity Press.
Gosling, S. N., & Arnell, N. W. (2011). Simulating current global river runoff with a global hydrological model: Model revisions, validation, and sensitivity analysis. *Hydrological Processes, 25*, 1129–1145.
Gosling, S. N., & Arnell, N. W. (2013). A global assessment of the impact of climate change on water scarcity. In N. W. Arnell (Ed.), *Cimate change*, Special Issue on "The QUEST-GSI Project" – Springer Link.
Gosling, S. N., Bretherton, D., Haines, K., & Arnell, N. W. (2010). Global hydrology modelling and uncertainty: Running multiple ensembles with a campus grid. *Philosophical Transactions of the Royal Society – Maths Physics and Engineering Science, 368*, 4005–4021.
Gosling, S. N., Taylor, R. G., Arnell, N. W., & Todd, M. C. (2011). A comparative analysis of projected impacts of climate change on river runoff from global and catchment-scale hydrological models. *Hydrology and Earth System Sciences, 15*, 279–294. doi:10.5194/hess-15-279-2011.
Hoekstra, A. Y. (2008). *Globalization of water: Sharing the Planet's freshwater resources*. Oxford: Blackwell.
Hulme, M. (2009). *Why we disagree about climate change: Understanding, controversy, inaction and opportunity*. Cambridge: CUP.
International Sustainability Accounting Standards Board (IASB). (2015). Definition of materiality in the accounting profession.
Jacobsen, T., & Adams, R. M. (1958). Salt and silt in ancient Mesopotamian agriculture: Progressive changes in soil salinity and sedimentation contributed to the breakup of past civilizations. *Science, 128*(3334), 1251–1258.
KPMG. (2014a). *Identifying natural capital risk and materiality*. London: KPMG.
KPMG. (2014b). *Business and investors: Providers and users of natural capital disclosure*. London: KPMG.
KPMG. (2014c). *Natural capital and the accountancy profession: Applying traditional skills to new thinking and practice*. London: KPMG.
McPhail, A., Lucossol, A. R., & Perry, C. (2012). *Achieving financial sustainability and recovering costs in Bank financed water supply and sanitation and irrigation projects*. Washington, DC: The World Bank.
Natural Capital Coalition. (2015). http://www.naturalcapitalcoalition.org/
Rockstrom, J. (2014). Planetary boundaries: A safe operating space in the Anthropocene. http://www.climate-service-center.de/imperia/md/content/csc/j/csc_jt2014_rockstroem.pdf
TEEB. (2015). http://www.teebweb.org/about/
Trucost. (2015). http://www.trucost.com/published-research/99/natural-capital-at-risk-the-top-100-externalities-of-business
United Nations. (1972). *UN Conference on the human environment*. Stockholm.
United Nations. (1992). *Our common future*. Report on the Rio Conference on Environment and Development.
United Nations – DESA. (2015). *World population prospects: 2015 revision*. New York: United Nations DESA.
UNU-IHDP, & UNEP. (2012). *Inclusive wealth report: Measuring progress to toward sustainability*. Cambridge: CUP.
WEF. (2011). *Water security: The water-food-energy-climate nexus, World Economic Forum*. Washington, DC: Island Press.
WEF. (2015). *Global risks report – Tenth edition*. Geneva: WEF.

Water and Water Supply in The MENA: Less of the Same

John Waterbury

Abstract The MENA is the region of the world likely to be the most highly and negatively impacted by climate change. The challenges of the coming century will be severe but also familiar to the policy-makers and technical experts of the region. The challenges have characterized the region throughout history. They differ now only in degree not in kind.

The challenges do not lie in technology: technological solutions are known and understood. The challenges lie in public policy which is opaque and poorly understood. The public policy challenge is compounded by the fact that four major Arab states—Libya, Yemen, Syria and Iraq—are no longer capable of coherent and comprehensive policy-making.

We ask, what are the parameters of ground water policy and management? What is the role of pricing in the allocation of water? How will states go about organizing their trade for virtual water and agricultural produce?

Policy change and innovation is often driven by crisis. Is that the case in the MENA? The major instance of crisis-driven change was evident in the 1970s and 1980s as the region's states grappled with structural adjustment. By contrast the crises looming in the water and food sectors are familiar and can be met through familiar policies and without destroying existing political coalitions. Egypt, Jordan, Morocco and Saudi Arabia have all undertaken policy initiatives to re-prioritize water use in their economies.

Finally we review the three major transboundary water courses in the region: the Nile, the Jordan and the Tigris-Euphrates. The possibility of uncoerced cooperation in these basins is remote. We explore why that is the case. The best policies now are for riparians to pursue domestic programs of enhanced water efficiency that may stand as benchmarks once, if ever, serious inter-state negotiations begin.

Keywords Water supply policies • Renewable water strategies • Water Management

J. Waterbury (✉)
American University of Beirut, Beirut, Lebanon
e-mail: waterbury@aub.edu.lb

S. Murad et al. (eds.), *Water, Energy & Food Sustainability in the Middle East*,
DOI 10.1007/978-3-319-48920-9_4

The Argument

The twenty-first century has long been touted as the Arab world's rendez-vous with its water destiny. The region, however, has been grappling with that destiny for centuries. More recently, but well before global warming had captured international attention, rapid population growth and economic development had exacerbated older problems of managing scarce water resources. In this respect the MENA shared the challenges of all the world's semi-arid regions, including Australia, the US Southwest and Mexico, the African Sahel, and South Asia. By virtually all estimates these regions will face acute water shortages in the coming decades. The shortage will be a function partially of increased demand but more importantly of dwindling supply, following the climate change adage that the wet will get wetter and the dry will get drier.[1] That is the bad news.

The good news is that the challenges of the twenty-first century for the Arab region are ones of degree, not of kind. The region has grappled with them for millennia and in several societies has built considerable expertise in coping with them. Both expert human resources and, more rarely, appropriate public policies are in place to deal with them. In other words the Arab world is in familiar territory.

The great bulk of water extractions—ca. 85%—are devoted to irrigation and agricultural production. There is little doubt that the agricultural sector offers the most cost-effective targets for improved efficiency in water use—the low-hanging fruit so to speak. Public policy should therefore focus on water efficiency in the agricultural sector.

The central argument of this paper is that the technology and, indeed, the financial resources are or will be available to meet the challenges. What is not in place, yet, are the public policies to make effective use of expertise and technology. Moreover I believe that the scope of the challenges and the technological tools available to deal with them are reasonably well understood while the public policy process remains something of a black box.[2]

There are surely those who understand the policy process by which critical decisions were taken. National experts and academics were often involved but also bound by confidentiality rules. So too experts from the international financial institutions, regional banks, and multi-national engineering consulting firms. Few of these experts have ever committed to paper or public electronic files what they know. The often-invoked "securitization" of the water and WEF dossiers makes open and honest policy evaluation and analysis very dangerous.[3]

The point is that all the advice proffered to the states of the Arab world with respect to water and the environment will not go very far if the advice is not premised on how policies may actually be made. In sections "The public policy process" and "Policy Foci" below I explore this point in greater detail.

---

[1]This paper draws on Waterbury (2013, 2014).

[2]AFED has published a series of reports that outline the technical challenges and solutions in detail. See especially AFED 2009, 2014.

[3]A model of an expert policy brief with detailed policy recommendations, including both positive and negative impacts, can be found in Bricheri-Colombi (2011).

Structural Changes in the Arab World

There are certain processes under way in the Arab region that will significantly affect policy-making.

(a) The most striking structural change is that since 2011 four Arab states have become dysfunctional if not failed: Yemen, Libya, Iraq and Syria. All of them are major regional players in the WEF—water-energy-food nexus. None has a government that fully controls its territory nor fully monopolizes the use of force. They have ceased to be effective policy-making units. Iraq is the least dysfunctional of the four.

To the four we should probably add Lebanon. Lebanon has the trappings of a normal state, but for nearly 2 years it has been unable to elect a President and its legislature has been paralyzed. Even in the face of a major crisis—the garbage crisis of 2015/2016—it has been unable to take policy action. And the state does not monopolize the use of force in the country.

In Syria this has had the added, negative effect that one of the Arab world's leading agricultural research units, ICARDA, located in Aleppo, has seen its important programs and research seriously disrupted.[4]

The failed states affect the prospects for regional cooperation. It is not simply that they are not responsible regional actors, but also that they have become the cockpits in which bitter regional and sectarian rivalries are being played out. As a result the prospects for any effective regional efforts to address problems of agricultural trade, water, and climate change are extremely dim.

(b) The Arab region is not a major contributor of GHG emissions and thus has little reason to engage in mitigation efforts (i.e. to reduce its GHG emissions). By the same token the region cannot affect the provision of the public "bad" that is global warming. That "bad" is provided by the big emitters—the US, China, India and Western Europe. It directly affects the Arab region's water resources.

The Arab region will by and large focus on adaptation to climate change with its attendant decrease in water supply.[5] In this respect it is important to note that mitigation is fundamentally a collective action problem: unless several major emitters of GHG act together to reduce their emissions, the public bad will still be provided. Adaptation, however, can be carried out by individual countries or even by sub-national jurisdictions such as cities or governorates.

---

[4]ICARDA is the International Center for Agricultural Research in the Dry Areas and a member of the world-wide chain of research centers known as CGIAR: the Consortium of International Agricultural Research Centers.

[5]Arab states may engage in mitigation for reasons indirectly related to global warming itself. Countries of the GCC are moving into renewable energy in order to conserve petroleum for export. The substitution effect may be the most pronounced in energy-intensive desalination plants. Arab oil importers were incentivized to move into renewables when petroleum prices rose sharply at the beginning of the new millennium.

(c) Will the locus of decision-making change? The MENA, with the exception (perhaps) of Lebanon, has always been state-centric. Policy decisions are made centrally and subverted locally. Will environmental and water-related decisions increasingly be made by other units—cities? regions? Cities will hold 70% of the world's population probably within 30 years. They account for 70% of GHG emissions (Bloomberg 2015). They use a great deal of water and provide large volumes of waste water that can be used in some types of agriculture. California or the trans-province basin of the Murray-Darling in Australia are all effective decision-making entities. Governments have proven to be cumbersome, slow, and easily paralyzed because their constituents are so diverse and embody rival interests. In failed states sub units must take on a policy-making role, but, in general, the MENA does not have robust systems of delegation to municipalities or any other sub-state units.

That may be changing. The locus of decision-making will vary with the nature of the problem. Bricheri-Colombi (2011) sees "canal command areas" in Egypt as deciding on the allocation of fixed shares of irrigation water. Similarly for several years Morocco has been experimenting with Hydraulic Basin Agencies (ABHs) to manage groundwater resources and to negotiate and oversee groundwater exploitation contracts (*contrats de nappe)* (see Loudyi and Oubalkace 2015; Alaoui 2013).

Increasing productivity in marginal, rainfed areas involving the poorest populations in the region will almost surely fall to public authorities. But could those authorities be "provincial"—say Darfur in western Sudan where control of local resources has always been a source of conflict?

(d) The private sector is becoming an increasingly important actor and interest group in the agricultural and agro-business sectors. This will include regional investors from the Gulf and Saudi Arabia. This trend will intensify in the coming years. The states of the region will have to become increasingly sensitive to the issues of policy incentives for private actors.

(e) The Arab world's agricultural sectors will become much more capital intensive, focused on high value crops, much of them for export. This will exacerbate the problem of "dualism" in the agricultural sector where one finds a capital-intensive, high tech, profit-driven sector alongside a smallholder sector with much lower productivity and incomes encompassing much of the country's impoverished population.

The marginalized rural sector tends to be hard on the environment and inefficient in its use of surface and ground water. Giant countries like Brazil and India display now this kind of dual rural economy. Erasing that duality will require active public policy interventions.

(f) The region is already the most dependent in the world on agricultural imports. This is how it compensates for its relative lack of water resources. That dependency will be aggravated in the coming decades by a combination of population growth and global warming. It presents a formidable challenge to individual states but also an opportunity for some regional solutions.

The Challenges

Per Capita Resources

Of the world's 33 most water stressed countries 14 are in the MENA region (see Fig. 1 and Vidal 2015), yet there is still wide variance in its per capita distribution. AFED (2014) forecasts the following averages for 2050:

Iraq	1077 m^3
Lebanon	962
Morocco	739
Sudan	709
Syria	508
Egypt	464
Jordan	94
Libya	81
KSA	53

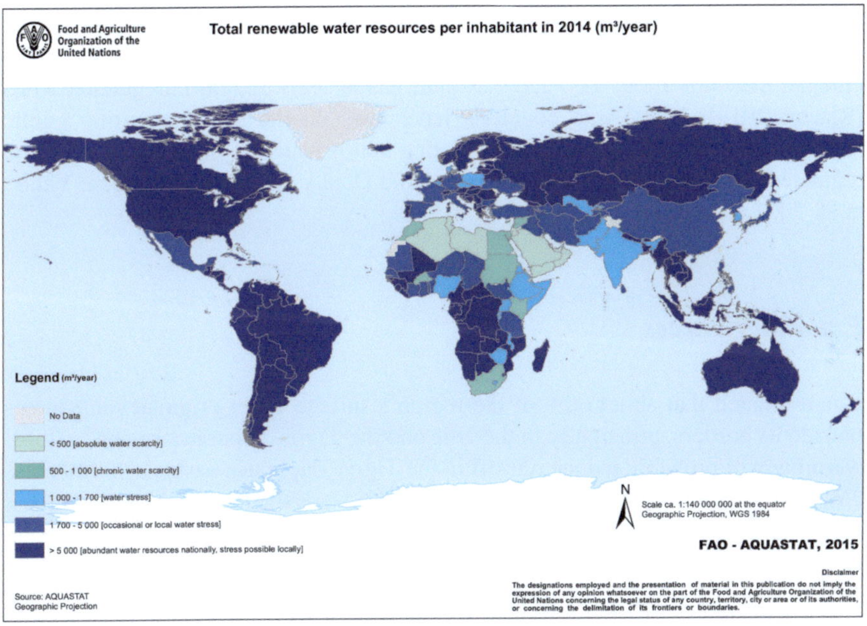

Fig. 1 Total renewable water resources per inhabitant in 2014 (m^3/year) original from FAO Aquastats (Reproduced by permission from: Shanta Devarajan, 2014 "Corrosive Subsidies" Middle East and North Africa Economic Monitor, (October), World Bank, Washington, DC. doi:10.1596/978-1-4648-0442-7 License: Creative Commons Attribution CC BY 3.0 IGO)

Table 1 Change in average temperature and agricultural output

	Change in average temperature			Change in output 2080	
	Present	Future	Change		
Country	1961–1990	2070–2099		Without carbon fertilization	With carbon fertilization
Algeria	22.67	27.81	5.14	−36.0	−26.4
Iran	17.26	22.63	5.37	−28.9	−18.2
Iraq	20.86	26.16	5.30	−41.1	−32.2
Saudi Arabia	24.57	29.3	4.73	−21.9	−10.2
Syria	17.48	22.19	4.71	−27.0	−16.0
Yemen	23.77	27.72	3.95	−28.2	−17.0
Morocco	17.43	21.91	4.48	−39.0	−29.9

Source: Cline (2007)

Water Mining

Several countries are "mining" renewable water resources at unsustainable rates. Ground water levels are falling precipitously as withdrawals far exceed annual re-charge from rain fall and snow melt (which are themselves falling).

The over-pumping of aquifers and the proliferation of private, unlicensed and unmonitored wells is increasingly recognized as a major threat to sustainable agriculture in the region. Well over half of Syria's 200,000 private wells are "illegal" (Droubi 2009: 22), and Lebanon has at least 50,000 unregulated wells (Riachi, 2014). Most Arab countries have laws on the books asserting public 'ownership' of aquifer water and stipulating that private use must be licensed and regulated (see Bruch et al. 2007 in general and Doukkali 2005 on Morocco; Venot, et al. 2007 on Jordan).

External Sources

It is estimated that about 60% of the region's surface water originates in sources outside its borders, principally in the Nile and the Tigris-Euphrates basins.[6] Global warming will probably reduce run-off in the Tigris-Euphrates basins by as much as 25%. The impact of global warming on the Nile watershed is less certain. Jeuland and Whittington (2014: Table 1) present the range of estimates on precipitation and run off. Even though there is a possibility of increased precipitation in the Nile

[6]The border of Mauritania, a member of the Arab League of States, and Senegal is formed by the Senegal River which arises in the Guinea highlands. In calculating Mauritania's per cap. Annual water resources the flow of the Senegal is used in the numerator giving that otherwise arid country the highest per cap. Average in the region—3147 m^3 in 2011 as opposed to Iraq's 2666 m^3.

watershed, the proliferation of new storage sites coupled with rising temperatures will increase rates of surface evaporation at various reservoirs.

Land Use

Agricultural land is distributed among irrigated surfaces (14–15 million ha.s),[7] 35–40 million ha.s of rain-fed cultivated land (including 5 million ha.s of permanent crops), and about 35 million ha.s of rangeland (Sadik 2014).

These surfaces will vary according to the levels of rainfall which in turn vary widely. On average about 85% of all water withdrawals go to irrigated agriculture. Average irrigation efficiency in the Arab world is 51% compared to a world average of 57%. Efficiency levels approach 70% in Egypt, Algeria and Tunisia. If all other Arab countries achieved that ratio ca 50 bcm annually would be "saved". By contrast the yield gaps in rain-fed areas are very substantial. Yields are around 60% of world averages. If the main regional cereal producers could raise their averages to the world average then regional production would increase from 21 million to 68 million tons (Sadik 2014:33).

Import Dependency

The Arab region produces about 75% in value of what it consumes of agricultural products but in terms of cereals that ratio drops to about 25%. Its cereal import dependency is by far the highest of any region in the world. This works out to ca 70 million tons/annum of cereal imports worth around $50 billion (Khouri and Byringyro 2014: 110). Arab countries are net exporters only of fruit and vegetables, although Syria briefly became a cereal exporter before 2011. Egypt is the world's largest importer of wheat. Wheat production in Arab countries has grown at an annual rate of 3% between 1961 and 2012, mainly driven by the yield growth achieved before 1990 The wheat output growth rate drastically slowed since then due to the sharp decline in wheat yields since 1991 (Shideed et al. 2014:84) (Fig. 2).

The region's cereal import dependency is unnerving. International commodity markets are volatile as was shown especially in 1973 and then again in 2008/2009. There is a real question when markets next tighten whether or not there will be adequate supply at any price especially when exporting countries slap bans on

---

[7]Brazil's National System of Protected Areas was established in 2000 after more than a decade of debate in the Congress, building on public consultation across society and academia and representing a major contribution toward Brazil's international environmental commitments, including UN conventions on biological diversity and climate change. The authors go on to say that the NSPA is being dismantled without public debate or review and despite substantial protest (Gibbs et al. 2015).

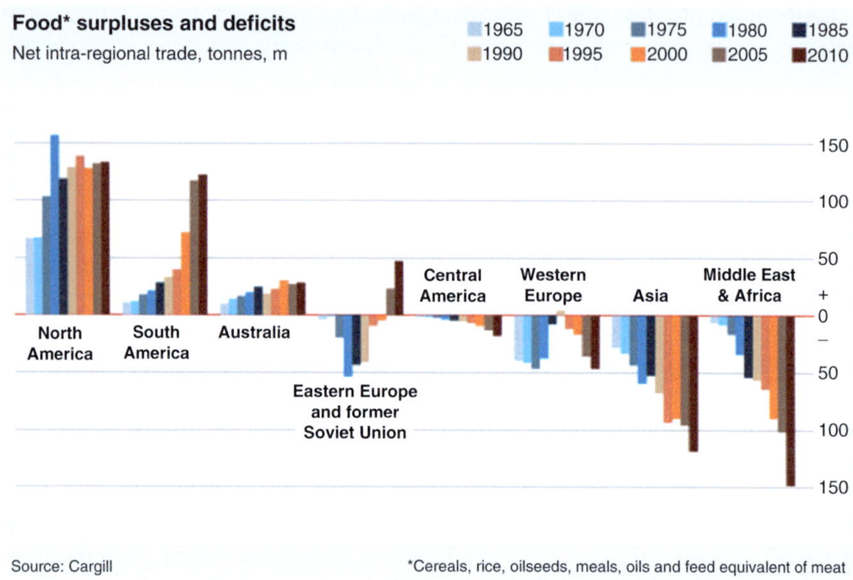

Fig. 2 Food surpluses and deficits (Reproduced by permission from: "Food Surpluses and Deficits" from "How to Feed a Planet", The Economist, May 28, 2012)

specific exports in order to hold down domestic prices and protect domestic consumers.

Drought

Along with other extreme, weather-related events, drought is becoming more common. Morocco offers a striking example. According to Alaoui (2013) drought has become a structural phenomenon in Morocco. Of the last several sequences of drought, the most severe were recorded for the periods 1944–1945, and then not until 1981–1985, 1991–1995, 1998–2001 and 2006–2007.

Dualism

As we shall see in greater detail, the logic of the region's food trade will favor capital-intensive, high tech agriculture. By contrast the fact that 70% of the region's poor are concentrated in rural areas dominated by rain-fed agriculture will push policy-makers toward investment in low-productivity agriculture especially through various subsidies on inputs, including energy.

Non agricultural Water Use

While the low-hanging fruit in water savings lies in the agricultural sector, municipal and industrial water use are growing as a proportion of total water use. Indeed in Jordan municipal and industrial use are approaching that of agriculture, and Palestine and Lebanon are trailing close behind.

The Public Policy Process

In an ideal world (at least from the point of view of democratic theory), policies would be made through a process initiated by politicians seeking election. They would suggest to voters how these priorities should be defined and set. Once a critical mass of voters is convinced of a politician's argument, he or she will be elected and if there are enough like-minded elected representatives of society, and if the executive reflects this electoral consensus, then policies will be legislated and implemented. Interest groups all along the way will try to influence, and in some instances block, the policies under discussion. Again ideally, the debate will be shaped by hard evidence and science regarding the specific issues. When interests and science clash, as they often do with respect to the environment, poor policies or no policies may emerge.[7]

In the Arab world, where authoritarian regimes predominate, a crucial actor in this process, legislatures that actually debate and make laws, is absent. In addition, the scientific evidence that should inform public debate may be considered a state secret, and the public may learn of policy debates only after major decisions have been taken by the executive.

The weakness of legislatures means that interest groups do not engage in lobbying of a kind common in western democracies but rather intervene directly with the executive to secure desired policies or ignore or bribe their way around laws and regulations after they are issued. In other words, stakeholders may influence policies *after* they are made, rather than before, by subverting or ignoring them.

In contrast to legislatures, the donor/financial community and non-regional state actors play a major role in the policy process. Their leverage comes through the support they can offer in times of stress or crisis. That support may take the form of investment, financial lines of credit or military/security backing, or a combination of all three. The support is contingent on policies (sometimes maintaining the status quo, sometimes representing sharp departures) that these third party actors deem desirable. These actors focus on policies *before* they are made, seeking to shape their form and substance. They are generally alert to the distortions that may be introduced in the implementation phase.

As noted above, there are many academics, members of NGOs or IFIs, technical advisors, and public officials who do understand at least parts of the policy-making process but feel unable to comment on it in any detail. For the rest of us we risk two major analytic traps. First we observe the outcomes of policies and then read backwards to assume that the policy's beneficiaries were responsible for its adoption. Our understanding of cause and effect may be off target. Similarly it is common to invoke crisis as the key variable in policy adoption. Whatever context produces the policy is thus by definition one of crisis.

There is general consensus that the political systems of the Arab world lack strong institutions of accountability. It is easy to assume that that affords authoritarian leaders the ability to make unpopular decisions and to take bold action. But weak accountability is a two-edged sword. Along with bold action leaders may remain relatively passive even in the face of discontent or immanent crisis. I argue that weak accountability in the Arab world is more likely to lead to inertia than to bold action.

There was, however, a long era, not yet over, in which Arab authoritarians, at least in the oil-poor, labor rich countries of the region, broke sharply and repeatedly with business as usual. It involved the "fiscal crisis" of the Arab state first manifest in the 1970s, and recurring periodically in the decades since then. Leaders dismantled long-standing social pacts, narrowed their coalitional support, and surrendered some of the levers of state control over their economies. There was plenty of popular protest during these decades, but the leaders adopted policies of change more in response to the pressure of the donor community than in response to their own people.

So we need to ask ourselves whether or not the water and environmental "crisis" facing the region will have similar effects? Is there sufficient consensus and sense of urgency in the donor/IFI/international banking communities to push leaders toward bold action?[8] Could it be that, unlike the financial crises of the last decades, the water and environmental crises can be dealt with by doing what many states have been doing for a long time, only doing it better? My tentative answer to this question is 'yes'. Unlike the response to the financial crises, the water/environment crisis does not require bold and painful departures from business as usual. It will unfold at a slower pace than the financial crises. It will however involve policies that affect the welfare of significant interests and constituents through scarcity pricing, elimination of energy subsidies, and tolerance of poverty in the low-productivity sector of dual agricultural economies.

[8]We need to keep in mind that some of the most powerful sources of financial leverage in the region today come from within the region, principally the GCC led by Saudi Arabia. Those countries are at best ambivalent about policies to move away from petroleum but have no reason to oppose greater efficiency in the use of water. Whatever their preferences, the current sharp drop in world petroleum prices may reduce their leverage.

Bold Departures

Looking across the coming decades it is easy to discern a growing water and environmental crisis or crises in the Arab region. But that kind of crisis is much different from a Moody's downgrade of sovereign debt or a sudden surge of capital flight. A crisis playing out over decades does not usually inspire politicians to take action.

Nonetheless a number of Arab countries have taken fairly bold action in addressing water challenges. The common theme has been to re-prioritize sectoral water use, putting agriculture third in line after municipal and industrial use.

Saudi Arabia (KSA) in the 1980s had subsidized wheat production through private farms relying on pivot irrigation and non-recharging fossil water deposits (inter alia see Woertz, 2013) The KSA's abandonment of water-intensive, subsidized wheat production came about in phases. It had been clear for some time that the schemes were neither financially nor environmentally sustainable. The leaders of the KSA were sufficiently impressed by the trend lines in water use to be induced to confront powerful and favored private actors who had benefitted from earlier policies. In 2008 it was resolved to phase out all water-intensive crops by 2016. (Sadik 2014:28). We can only guess at the decision-making process that went on in the KSA.

In 2008, after years of drought, the Kingdom of Jordan issued a new national water strategy to the year 2022. The most startling aspect of the strategy was the relegation of agriculture to third place in the priority of water use, after human and non-agricultural needs This strategy did not require parliamentary debate and approval. It was an executive act, and as such a relatively rare instance of the use of authoritarian leverage for a bold initiative. It is well known that districting for parliament in Jordan favors rural constituencies, and that the commercial growers of the Jordan Valley constitute a powerful interest as do the tribal cultivators of the high plateau.

There is therefore speculation that this top-down strategy may not survive the *ex post* machinations of farm lobbies to dilute or circumvent it. Efforts to meter agricultural water use and to introduce significant charges for use have met with a broad range of largely successful efforts to disable the meters or to pump above quotas (see Venot et al. 2007).

Similarly in both timing and content Morocco has moved to protect non agricultural demand for water.[9] Beginning with a basic water law in 1995 (Law 10) river basin agencies were created to manage water resources within their watersheds. A basic guideline was and is that 2 years' supply of municipal water in the basin must be stored before diversions to agriculture can take place.

In April 2008 Morocco adopted its Plan Vert (Green Plan) that consolidated and extended policies begun in 1995 (see Morocco, 2008). On July 30, 2009, King

[9]I believe it is a matter of coincidence and not of structure that the KSA, Jordan and Morocco are all monarchies and ready to take bold action in agricultural policy.

Mohammed VI, in his speech from the throne, called for a national charter for the protection of the environment and natural resources. His call was prompted by a report of the High Commissioner for Water, Forests and Anti-Desertification that warned that global warming was causing annual losses to the Moroccan economy equivalent to 2% of GDP (al-Hayat, Aug. 20, 2009; see also Doukkali: 2005).

The World Bank claimed some credit for this process in Morocco and elsewhere. In Making the Most of Scarcity (2007: 118) the World Bank authors summarized the process across the MENA region:

> Since 2002, local and international experts, in collaboration with the Mediterranean Environmental Technical Assistance Program and the World Bank, have calculated the costs of environmental degradation in several MENA countries, combined them, and expressed them as a share of each country's GDP. They presented these results to the Ministries of Finance and Economy as well as relevant line ministries... These simple but powerful messages have been one factor for catalyzing important changes. After seeing these figures, the government of Algeria increased its budget for environmental protection by US$450 million and revised its environmental investment priorities.

In 2005, Egypt adopted its National Water Resources Plan (NWRP). The Plan recognized the growing demand for water in municipalities and in industry. While it was claimed that all sources of demand could be met, "in many parts of the plan, it is explicitly or implicitly stated that the water can only be made available by reducing supplies to existing farmers, to the point where their incomes are reduced by 20%" (Bricheri-Colombi: 2011).

Policy Foci

The policy domains described in this section are not mutually exclusive. Policies formulated with respect to one domain may have both positive and negative impacts in others.

Sustainable Intensification

The Arab region manifests significant yield gaps between its averages and best world practice. With the exception of Egypt, the efficiency with which irrigation water is used also lags behind world averages. Intensification is the process by which these gaps can be overcome.

Intensification is not necessarily sustainable. Sustainable intensification can be defined as making maximum use of a unit of production without permanently or substantially impairing its productive capacity for future generations (Robertson, 2015). To the extent the Arab region focuses on sustainable intensification it will move away from large scale water storage projects and land reclamation and focus more on the institutions of demand-management.

At present, however, we are in a policy era where old, supply-side projects co-exist with new demand-management policies. Hundreds of new dams are being built in the major international river systems that supply the Arab region as well as within specific countries like Morocco, Iraq (if some stability is restored), the Sudan, etc. Similarly, horizontal expansion through land reclamation will proceed even as efforts to increase productivity on existing land grow in importance.

The issue of sustainability arises principally with respect to in-puts. Past experience has shown that intensification often leads to higher in-puts of agro chemicals and therefore of chemical run-off while at the same time reducing the micro-nutrient content (iron, zinc, protein) of cereals (De Fries et al. 2015). For the environment as a whole that is not sustainable.

Intensification will also involve the introduction of genetically-modified seeds and other cultivars. These raise big sustainability issues. If single varieties are widely adopted that will increase vulnerability to variety-specific pests and diseases. If single varieties are coupled with pest- and disease-resistant genes, one risks the emergence of 'super pests'.

Still, it seems inevitable that GMOs that are drought-resistant and tolerant of high heat will be vital in bridging food gaps in the semi-arid tropics, including the Arab world.[10] In this respect, as Shideed et al. (2014: 82) points out, national capacity for agricultural R&D is the most important factor explaining the sustainable long-term agricultural productivity growth in some countries and not in others. Countries with national research systems capable of continuously producing new technologies adaptable to local farming systems generally achieve higher growth rates in agricultural total factor productivity.

In considering water efficiency we must keep in mind that "waste" is not always bad. In irrigated agriculture over-irrigation produces large amounts of drainage water that are then used for other purposes in agriculture and elsewhere. Greater efficiency may harm the often-anonymous end users. By contrast untreated municipal waste water is often simply dumped unused because without treatment it is unusable.

That said, one set of intensification targets is aimed at increasing yields per unit of water. The most direct route to this goal is to "close" irrigation water delivery systems by replacing unlined, open canals with pipes or lined canals and introducing drip or sprinkler irrigation to deliver the water to the fields and plants. A further step is to implant sensors in the soil to monitor moisture levels and optimize water delivery to meet plant needs.[11]

---

[10]For example, corn and sugar cane have a carbon dioxide concentrating mechanism called C4 photosynthesis. This could be bred into crops like wheat and rice to increase their water efficiency (Michalak and Field 2015:13).

[11]In the 2008 Maroc Plan Vert (see Morocco, 2008) it is planned to convert 500,000 ha.s to sprinkler irrigation, with another 200,000 to be added in a later phase.

Rainfed Agriculture

Most of the Arab region's arable land and most of its coarse grain production is concentrated in its rain-fed zones. Inasmuch as the region's average rain fall is low, the rainfed zones are extremely vulnerable to seasonal and annual fluctuations in rain fall. Six years of drought in Syria prior to 2011 uprooted an estimated three million people. Zones of unpredictable rainfall are home to the highest rates of national poverty.

State authorities must develop policies that cater to both irrigated and rainfed agriculture. The latter are the most difficult because they involve poor, often scattered populations. To serve them public authorities must engage in the unglamorous and uncertain tasks of institution-building as through water user associations, the extension of micro credit and the provision of crop and livestock insurance. For this process to work the ability of the better-off to buy officials must be curtailed and some level of meaningful accountability established. This is a tall order as many cases studies show (on Jordan, Venot et al. 2007; on Morocco, Loudyi and Oubalkace 2015; on Egypt, Barnes 2015).

There are great opportunities to raise average crop yields in these zones especially through supplemental irrigation. Before the collapse of the Syrian state beginning in 2011, supplemental irrigation led to the doubling of wheat yields (Shideed et al. 2014). Rain harvesting, local waste water re-use, no-till farming, local solar power units for water pumping, improved livestock breeding and range management all could enhance productivity substantially. It is possible but not yet clear if bio-fuel production is economically feasible in rainfed zones. That is an exciting possibility but also one that might re-enforce the dualism of the rainfed sector.

Dualism

National statistics tell the story: in Morocco agriculture accounts for 14% of GDP and 42% of employment; in Egypt it is 14% and 29%; in Yemen it is 15% and 55%. This means a large segment of the work force is locked in low-productivity work.

This is a chronic problem in most agricultural systems and spans both the rainfed and the irrigated sectors. Crudely put it pits big, capital intensive, large scale agriculture against capital-poor, small-scale, low-productivity agriculture. There are powerful market and strategic forces that favor "big ag" and political realities that generate some attention to the poor.

For example dualism is directly related to the best options for the region to trade for its food (see 5f) "The logic of integrating agricultural production into regional and global supply chains *is to promote sophisticated, agile, large scale agriculture.*

This is the kind of agriculture that can assure the quality of its product and also adjust to changing market conditions" (Khouri and Byringyro 2014: emphasis added). But domestic political logic and the concern of the development community for equity will push policy makers to focus on small-scale farmers or communities of farmers. Political elites try to serve both constituencies but at the expense of coherent or affordable policies.

The rural poor do not carry much political weight except perhaps when they move to cities. In some regimes, such as Syria's, which once relied on a peasant base, powerful political allies are now found in urban economic elites with privileged access to the state. The rural poor will better their lot mainly by leaving the countryside.

We may therefore expect the growing weight of "big ag" in the policy-making process. It may be predominantly national, as in the agro-interests of the Jordan Valley, or international/regional as in big commercial schemes such as the Toshka project in Egypt.

Regulating Aquifers

This may be the single most important policy issue facing decision makers in the Arab region. Groundwater is a crucial link between rainfall and irrigation, and it captures seepage from irrigation canals and drainage systems. Everywhere aquifers are being grossly over-exploited. If current extraction rates are maintained, "running dry" can be forecast. It is 2050 for Jordan (Glatzel 2013). In Morocco, Le Plan Vert states (Morocco 2008:21); "5 bassins hydriques sur 8 sont potentiellement déficitaires d'ici 2020: les pompages massifs d'eau à usage d'irrigation privée menacent l'existence de certaines nappes majeures". Everywhere unlicensed and unregulated wells proliferate (see also Doukkali 2005).

In most Arab countries the state owns sub soil assets and resources. In theory this should make regulating groundwater extraction feasible, but I know of no significant success stories. Jordan tried to meter wells and tax amounts above a licensed rate of extraction, but farmers tampered with the meters and "influenced" inspectors (Venot et al. 2007).

Technology may provide a partial solution. Brazil was able to enforce limitations on soy cultivation and hence deforestation by satellite surveys of cropped areas with detail to the farm level. Farmers who exceeded their permitted acreage were subject to stiff fines (Gibbs et.al. 2015). Aerial surveys coupled with matching pumping permits to marketed produce may afford accurate monitoring but nothing can replace an honest enforcer.

In groundwater extraction, we have an instance where crisis seems the appropriate term. We shall see if Arab policy-makers respond to it (the KSA did in 2008) or if they allow a kind of triage to proceed in the countryside.

Markets

Despite cultural and derivative political reservations about pricing water, policy-makers have experimented with costing schemes to reflect its scarcity and to shape demand for it. Charges may be levied to off-set the costs of delivery or levied on the amounts of drainage water that reflect over-irrigation. Municipal and industrial water is often metered. It may be that because water is vital to life and has no substitute, all humans should receive *gratis* a certain minimum after which allocation will be determined by supply and demand. Terry Anderson (2015) goes so far as to claim that "sustainability requires profitability for survival" (see also Babiker and Feheid, 2011).

In recent crises markets have been critical in allocating water supply in the face of unsatisfied demand. During the severe drought of 2000–2008 tradable water rights in Australia's Murray-Darling basin led to transfers to the highest bidders (rice farmers traded rights to grape farmers and to cities (Briscoe 2015). Similar trading is going on now in California. One could imagine over-exploitation of aquifers being dealt with through trade in water rights. It is easy to see how norms of equity might be jeopardized in such trade but it may be the most effective way to deal with scarcity.[12]

Such markets will not work unless the infrastructure to store and transport water is in place. Who will provide the infrastructure—the state or the private sector? Who will pay for it—the consumer or the tax payer? Should provision of water be considered a public utility or a factor of production? It is doubtful that this question can be answered satisfactorily. It is surely both.

Trading for Food and Virtual Water

Food security in the Arab region cannot remotely be contingent on self-sufficiency. The Sudan and a peaceful Syria or Iraq might plausibly talk of self-sufficiency but

[12]Loudyi and Oubalkace (2015) describe the allocation system in Morocco as it is. "The allocation of water resources between different competing, and most often conflicting, uses is done at the global level through the Master Plan of Integrated Water Resources Management (PDAIRE) and the National Water Plan (PNE).

When water rights are not applicable (emphasis added), in practice and on a larger scale, the distribution between same-category users follows the rule of "first come, first served", particularly for groundwater. Despite the regime of water allocation and the obligation to respect the requirements of the PDAIRE, and in the absence of strict control, this practice leads to reducing the responsibility/accountability of water users who prefer to serve their immediate individual interests over the collective and long-term interests of their community. This also causes the over-exploitation of groundwater and widens the gap between traditional uses of poor peasants and modern uses of rich farmers".

for the rest trade strategy will be a crucial part of managing water and land resources.

The crux of the matter is to move virtual water embedded in agricultural products including livestock from wetter to drier regions. In that respect sub-Saharan Africa (SSA) is particularly promising. Not only can it boost production enormously by raising average yields, but it still has room for significant horizontal expansion (see Gilmont, 2016). However, countering this optimism is the fact that most of the projected world population growth in the coming century will occur in SSA—perhaps a billion additional people. That new demand will surely reduce the amount of exportable surplus.

Part of the Arab region's trade strategy must focus on the region itself. "To be effective, intra-regional cooperation in food security requires an approach based on the harmonization of national agricultural strategies and policies, implementation of agricultural practices, regulations, measures and incentives conducive to the efficient use of resources with special attention to the improvement of the management of shared regional water resources." (Sadik 2014:41).

The likelihood of regional cooperation is today (2016) more remote than at any time since about 1961 when the United Arab Republic broke apart. Failed states, sectarian bitterness and conflict, and traditional geo-political rivalries present formidable obstacles to the pursuit of regional, collective self-interest.

The Arab region is more comfortable in dealing with trade partners outside its boundaries than within. The one exception has been the Sudan which, since the early 1970s, has been periodically touted as the granary of the Arab world. Some investment has flowed there, but in general Sudanese potential has remained just that. Civil strife, financial crisis, poor infrastructure and bureaucratic red tape have rendered the Sudan much less attractive than originally thought.

Small city states, like Abu Dhabi and the GCC in general, may be relatively relaxed because they have substantial foreign exchange reserves, and their populations are small in terms of world supply. In Abu Dhabi "Food security in the future will be achieved through effective and fair international agreements and trade with food exporting countries, combined with the capacity to increase production domestically when food supply from food exporting countries is constrained" (Razan Khalifa al Mubarak in Sadik 2014:25).

Other oil-rich countries, especially the KSA, are exploring investment and land purchases or leases in countries with substantial exporting potential, including the Sudan, Ethiopia, Madagascar, and Thailand. It is estimated that in Africa alone some 7 million ha.s are under some sort of contract with Arab investors. However for the same reasons that have impeded investment in the Sudan not much has actually happened (Woertz 2013). The sharp decline in world petroleum prices in 2015/2016 makes any up-tick in such investments unlikely. In the summer of 2015 the KSA reached agreement with Russia to invest $10 billion in its agricultural sector. Will it be able to honor that pledge?

Egypt, with 90 million inhabitants, is another matter altogether. In future periods of tight world markets, as in 2008, it is not obvious where it will find international supply to meet domestic demand nor where it will find the foreign exchange to pay for the imports. In the more distant future, it may have a natural gas bonanza which will help cushion it against the likelihood that agricultural and energy prices will tend to move together.

Desalination

Municipal water supply is already produced by desalination in GCC countries and it will spread to severely water-stressed countries like Jordan and Palestine. Siddiqi et al. (2013) estimates that in Jordan desalinated water supply will increase to 520 million m^3 and will constitute one third of the total water supply for the country.

Coastal cities that lie close to the water source and at altitudes that minimize pumping will surely experiment with desalination in the future. More problematic are in-land cities at fairly high altitudes where pumping costs may be prohibitive: Sana'a, Amman, Jerusalem and Damascus come to mind. In all situations, the key policy question is how to recover the costs of desalination and pumping. Subsidizing supply will encourage over consumption. Charging for the real costs of delivery will hurt the urban poor.

Waste Water Treatment

There is great scope for recovering and treating agricultural drainage water and municipal waste water. Treated drainage and domestic waste water cannot be used in all forms of agriculture, let alone for domestic consumption, but it can be used to recharge aquifers, to irrigate tree crops and some coarse grain crops. Increased irrigation efficiency will reduce the amount of drainage water, but that is to be welcomed as the efficiency savings will be of higher quality water.

The Arab region produces annually about 15 bcms of municipal waste water less than half of which is treated. The volumes will grow with urban growth, so this should become a significant source of usable water.

International Hydropolitics

There are three major international drainage basins in the Arab world and a few minor ones, like the Assi/Orontes (Lebanon, Syria, Turkey). The three major basins comprise several tributaries which also cross national boundaries. Only one basin, the Jordan, is contained in the Arab region. The other two—the Nile and the Tigris-Euphrates—find their sources and receive much of their water from outside the

Arab world. So the Arab world is "hydropolitically" vulnerable to non-Arab neighbors and also to declining rainfall in the major watersheds.[13]

The three basins vary enormously in their basic characteristics. In the Tigris-Euphrates basin the mid-stream state, Syria, and the downstream state, Iraq, have been accomplices to their own growing water penury and to Turkey's uncontested dominance over the system. In the Nile Basin, Egypt, the downstream state has long been (but no longer is) the dominant power that has imposed a regime on the river since 1959. History and the 1959 Treaty between Egypt and the Sudan, makes Sudan, the midstream state, an ally of Egypt while agricultural potential and a need for electric power make it an ally of Ethiopia. The Jordan basin lies within the Arab region, but the regime-maker, Israel is a non-Arab player. The striking feature of the Jordan basin, including the Yarmouk river, is how little water is at stake. Even if a 'super fair' formula for allocating the Jordan's waters were agreed to, the inhabitants of the basin would remain among the most water-poor in the world. For that reason desalination for human water consumption is a vital option in the Jordan basin, unlike in the Nile and Tigris Euphrates basins.

It is important to note that there are major aquifers that straddle boundaries in the region. The Nubian sandstone aquifer is shared by Egypt, Sudan, Chad and Libya, and there is a large aquifer shared by Turkey and Syria, partially recharged by Turkish drainage water. There are no agreements among the aquifer riparians governing their exploitation.[14]

Similarly, there are no multi-lateral agreements governing use of shared waters among riparians in any of the three basins. There are a number of bi-lateral agreements and understandings such as the 1959 agreement between Egypt and the Sudan for the "full utilization" of the Nile, agreements between Turkey and Syria and Turkey and Iraq, and bi-lateral agreements between Israel and Jordan and Israel and the Palestinian Authority. Bi-lateralism is no accident. Dominant powers in international basins avoid multi-lateralism in which the weaker stakeholders might coalesce to challenge the powerful.

The result is sub-optimal economic and social outcomes in the use of the water. It is not difficult to work out cost/benefit analyses that show the economic pay-offs of cooperation (Rogers 1997 and Jeuland et al. 2016).

Despite the demonstrated benefits of cooperation, political leaders often do not opt for them. I have explored the reasons why in detail elsewhere (Waterbury 2011, 2014). Suffice it to note that benefits are delayed in time while costs may be incurred up front, other bi-lateral issues (terrorism, refugees) may appear more pressing, and

---

[13]There appears to be more certainty about diminished rainfall in the Tigris-Euphrates than in the Nile basin where estimates vary. Inter alia see Jeuland and Whittington 2014.

[14]in 2010 the United Nations International Law Commission drafted articles for a convention governing trans-boundary aquifers and in December 2015 the UN General Assembly approved a draft. http://www.unesco.org/water/news/transboundary_aquifers.shtml

In 2013, under the auspices of the IAEA and the UNDP, representatives of Egypt, Sudan, Chad and Libya adopted an action program for the Nubian sandstone aquifer (IAEA 2013). There has been no follow up on this initiative, and Libya is no longer an effective participant.

sanctions for non compliance are weak or lack credibility. The "regime" by which waters are allocated and used (or abused) may be imposed by the most powerful riparian (Turkey, Egypt and Israel in their respective basins). Coerced cooperation is not true cooperation which must be voluntary and based on self-interest.

Cooperation need not be about allocation. It can focus on specific benefits such as water quality, flood control, silt capture, mitigation of scouring, etc. Jon Martin Trondalen (2008), for example, has proposed a desalination plant on the Euphrates located in Syria but designed to help Iraq. Throughout history Iraq has suffered from poor drainage and high soil salinity in the lower Tigris-Euphrates. Arguably addressing that issue could reap rewards in agricultural production that would be as important as assured water supply from Turkey.

The Nile Basin

The current water regime in the Nile Basin was established in 1959 in a treaty between Egypt and the Sudan that divided the total flow of the river (as measured at Aswan) between them. There was no water allocation for any of the remaining eight riparians including Ethiopia which supplies about 85% of the river's total discharge. The allocation between Egypt and the Sudan was based on an estimated annual average discharge of 84 bcms. That estimate has proved too low in subsequent decades. In addition the Sudan was unable to use its full allocation. The result was a windfall for Egypt, increasing its legal share of 2/3 of the Nile's discharge by as much as 10%. The windfall has become part of Egypt's basic usage. If we see a significant decline in Nile discharge due to global warming (see footnote 13) and to the construction of a series of dams in Sudan and Ethiopia, Egypt's agriculture will be seriously impacted.

Although Ethiopia has historically been a weak actor in the Nile basin, since 2011 it has challenged Egypt's hegemony, perhaps taking advantage of Egypt's political turmoil. In that year it announced, without any prior agreement, that it would go ahead with construction of the Grand Ethiopian Renaissance Dam (GERD) near its border with Sudan for the purpose of generating hydro-electricity. It implied that it would not divert stored water for irrigation. Subsequently and again without prior agreement, Ethiopia vastly increased the reservoir's storage capacity over the initial design. In 2015 with the dam half completed, Egypt largely acquiesced in its construction and entered into negotiations on the dam's filling and operating rules.

It is possible, as the Ethiopians claim, that the reduction in storage losses that the new dam will achieve will off set any net drop in the flow of the Nile that the GERD might cause. As is the case in the Tigris-Euphrates, the immediate issue may be the rate at which the reservoir is filled. Ethiopia is experiencing severe drought just when filling is due to begin.

As noted above, Egypt is already at a point where it may have to re-allocate water between agricultural and non agricultural uses such that farm income is significantly lowered. Egypt will be lucky if the GERD does not further complicate that problem. In the longer term, the more worrisome stakeholder is the Sudan. It has already embraced the GERD from which it may buy power for Khartoum-Omdurman, and it will benefit from the GERD's ability to trap silt which would otherwise eat into the storage capacity of Sudanese reservoirs. More importantly, the Sudan has vast potential for horizontal agricultural expansion in the land lying between the White and Blue Niles. Now that the Southern Sudan has become independent, taking with it much of the Sudan's oil deposits, the new Sudanese state may have little choice but to pursue horizontal agricultural expansion aggressively.

The Tigris-Euphrates Basin

Turkey remains the uncontested power in the Tigris-Euphrates basin based on its military and economic supremacy. It would have been logical to expect Syria and Iraq to cooperate in countering Turkish plans to exploit its waters but the peculiar rivalry of the two Ba'athi regimes that controlled both countries for decades precluded any such cooperation. Indeed, the only example of international armed conflict breaking out over water came in 1975 when Syria began to fill the reservoir at the Tabqa dam on the Euphrates, thereby suddenly reducing Iraq's water supply. Conflict was avoided as it was between Egypt and Ethiopia in 2012/2013.[15]

In their current state of turmoil, Syria and Iraq can exert virtually no counter pressure to Turkey's GAP (South East Anatolian) project. The GAP includes 15 major dams on the Tigris and Euphrates rivers, built or projected, and an outer target of irrigating 1.5 mn ha.s. Turkey currently has enough storage capacity in the headwaters to impound the Euphrates for 2–3 years (Shamout 2015). In the summer of 2014 Turkey literally turned off the flow of the Euphrates across its border with Syria (Fig. 3).

The flow of the Euphrates is already down 40% below the averages of the early 1970s. Groundwater depletion in the Tigris-Euphrates basin is second only to that of the Indus basin (Voss et al. 2013). The failed or quasi failed states in Iraq and Syria mean that Turkey will have its way in the Tigris-Euphrates (and so too Iran to a lesser extent). It will be able to pursue GAP unimpeded. A basic regime in which Syria and Iraq see their "normal" river flow go down by ca 40% will be the new status quo. Syria and Iraq may become so desperate for investment and markets in a rebuilding phase that confronting Turkey over the Tigris-Euphrates may not seem worth the candle.

Turkey will try to assure its neighbors that certain minimum flows will be guaranteed (500 cubic meters per second) at the border with Syria) and that both countries can buy hydropower and agricultural produce from the GAP both of

[15]In 1964 Israel intervened militarily very early to thwart Syrian efforts to divert the headwaters of the Jordan thereby impeding Israel's National Water Carrier scheme. To my knowledge this is the only instance of inter-state armed conflict over water in the region, although some might claim that Israel's occupation of southern Lebanon in 1980 was aimed at gaining control of the Litani River.

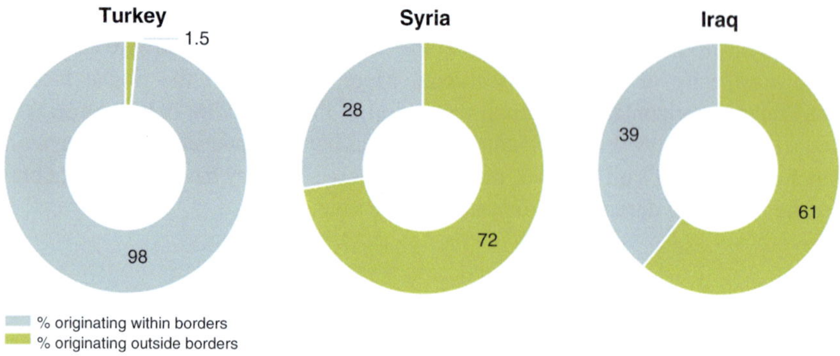

Source: FAO (2015), AQUASTAT Database, estimates for 2012.[10]

Fig. 3 The dependence of Turkey, Syria and Iraq on transboundary water resources (Reproduced by permission from: Shamout, M. Nouar with Glada Lahn (2015) The Euphrates in Crisis: Channels of Cooperation for a Threatened River, Chatham House, Research Paper: Energy, Environment and Resources, April, p.11; 020 7957 5700)

which, the Turks claim, can be produced more efficiently in Turkey than downstream. The cumsec deal was reached in 1987 at a time when Turkey was worried about Syria's ability to host the Kurdish Workers Party (PKK). In 1990 Syria agreed to release to Iraq 58% of whatever water it received from Turkey.

Like Egypt in the Nile Basin, Turkey ponders the trade-offs of having economically weak neighbors that cannot muster much of a challenge to the dominant power as opposed to dynamic neighbors who offer markets and investment opportunities to that power. History would suggest a preference for weakness, reflecting the parallel preference for non-cooperation.[16]

The Jordan Basin

The regime in The Jordan basin has been set by Israel since 1964. At that time it diverted water from Lake Tiberias (or Kinneret) to its National Water Carrier, allowing it to expand irrigated agriculture especially in the Negev. Over time the flow of the river in its lower reaches, before reaching the Dead Sea, dropped from ca. 1.3 bcm (compared to the Nile's 84 bcm.s) to about 100 mcms today (half a day's flow of the Nile). In addition, through its occupation of the West Bank and Jerusalem Israel is able to control the licensing of new wells thereby protecting the movement of groundwater from the plateaus to the coastal plain as well as to Israeli settlements in the occupied territories. The peace agreement with Jordan in 1994

[16]The three riparians have even failed to establish a joint technical committee to exchange basic data and carry out analysis of projects and alternative uses. Since 1920 the establishment of a technical water- management committee has been agreed three times in principle but has never materialized in an effective form (Shamout 2015).

provided Jordan with some additional summer water in exchange for a like amount of 'surplus' water in the Yarmouk in the winter.

There are five riparians in the basin: Lebanon, Syria, Jordan, Palestine and Israel. There is an agreement between Syria and Jordan on the Yarmouk, but Syria has consistently violated it. There are no agreements involving Lebanon.

The Red Sea-Dead Sea Conveyance Project held out some promise of bringing the main parties—Israel, Palestine, and Jordan—together through a tripartite agreement. This project was originally designed to pump Red Sea water to a desalination plant near the Dead Sea which is 400 m below sea level. Dropping the pumped water from the heights above the Dead Sea would have generated the power to desalinate the water. The water would have gone to consumers in Israel, Jordan and Palestine. The brine from desalination would have been pumped into the Dead Sea to halt its rapid desiccation.

After feasibility studies the project was scaled back due to the high cost of the original design. In 2013 in a memorandum of understanding it was decided to go ahead with a desalination plant at Aqaba. Its brine would be pumped to the Dead Sea and some of its treated water would be sold to southern Israel. Israel in turn would sell water from its own desalination plants to Palestine and Jordan (The Source 2015; Burgan 2015; Zack 2015). It is instructive to note that Israel now produces about 750 mcms a year of desalinated water which is well over half the historic flow of the Jordan and far more than its current flow.

The regime imposed by Israel has grave and looming consequences for Palestine, especially Gaza. The over-exploitation of the aquifers shared by Israel and the West Bank could lead within a matter of years to the depletion of most Palestinian wells. Agriculture would then depend almost entirely on rain. For Gaza the future is even more bleak. Its aquifer has been seriously depleted such that sea water intrusion has compromised what little remains. Desalination is the only hope for Gaza's water future.

Conclusion

Un-coerced, negotiated cooperation in the Arab region's transboundary water courses is not likely in the foreseeable future. In the meantime, expert communities must engage in policy-analysis. For example it has been suggested that Syria and Iraq (either through public entities or through private investors) become stakeholders in Turkey's GAP. Iraqi and Syrian investors would not only share in profits but also assure a supply of agricultural produce and hydropower to the downstream states. Similarly, Egypt could consider investing in the Sudan's horizontal agricultural expansion on the same basis. In other words these water courses offer an opportunity to find effective solutions for the Arab region's food security challenges.

In addition, the absence of cooperation should motivate policy-makers to establish best-practice benchmarks in their domestic agricultural sectors; that is, pursuing sustainable intensification and greater irrigation efficiency so as to establish benchmarks that could be used in negotiations with other riparians over allocation.

Transboundary water markets are a long way away. The infrastructure of storage and delivery is not in place, but, more important, for the reasons cited above, the will to build transboundary markets is lacking. However, no matter how improbable, we may be witnessing the first steps toward a regional market among Israel, Palestine and Jordan. The problem is that it is largely imposed.

General Conclusion

There are grounds for pessimism, if not despair, in contemplating the Arab region's water future. Its population is still growing, albeit at a declining rate, while it is becoming drier. To the extent regional cooperation is important, it is today characterized by bitter conflict. Four important Arab states are politically crippled and physically damaged.

But there is some basis for optimism. Rebuilding Syria, Iraq, Yemen and Libya may offer an opportunity to examine in detail their strategies for water supply and agriculture. It is good in that respect that Turkey and Iran are both likely to be active players in reconstruction in Iraq and Syria. The reconstruction process may offer opportunities for broader regional understandings and platforms for formal cooperation.

Domestic policies dealing with water resource challenges will entail improving on existing policies rather then moving into new, uncharted policy territories. The region does face an unfolding water resource crisis, but, unlike the financial crises of the last decades of the twentieth century, it will not require shredding political alliances to deal with it.

The fact that the policy frameworks are already in place means also that so too are the expert communities, often with established links to international know-how, to implement the policies. Those expert communities are more experienced in supply-side policies and projects than in policies of demand management. The latter are much more about social engineering and structuring incentive systems than they are about physical infrastructure and civil engineering. Nonetheless demand management has been under way for many years. It is more a question of how to do it rather than whether to do it.

There are substantial equity issues at stake but they too are not new. Most poverty in the Arab region is rural. It reflects the low productivity segment of the rural economy. This segment represents a great opportunity for sustainable intensification aimed at reducing the yield gaps between it and the better performing segment. There will be tension, probably unresolved, between policies focused on helping the rural poor raise their productivity and encouraging the commercial private sector to move in with state encouragement.

If trade strategy prevails, we will see 'big ag' calling the tune. It can develop the specialty agriculture targeted at foreign markets with high standards that will help the Arab region trade for the food it cannot or will not produce. It will depend on farmer-investors backed up with a panoply of services and infrastructure in storage, transport, quality control, insurance, and credit. "Big ag" should certainly pay for the water it receives either through the metering of wells or direct charges for state-provided irrigation water.

Organizing collective action among the rural poor in the Arab region dates back to the land reform era of the 1950s and 1960s. Organizing water users associations or devising "contrats de nappe" is merely an extension of these earlier efforts. The problem then as now is that central, autocratic powers are reluctant to encourage real autonomy in such units so that they remain subservient to state authority. Without some real autonomy in decision-making it is not clear what local units can accomplish.

Governments in the region have an obligation to assure that every citizen has access to adequate water for life functions and hygiene. Above that minimum market signals should play a major role in deciding who uses the remainder and how. Taxes, like "sin" taxes on alcohol and cigarettes, can be levied on uses that society regards as frivolous. It would be nice if society had institutional mechanisms through which to express its preferences.

References

"A New Path to Peace". *The Source*, IWA 2015. http://www.thesourcemagazine.org/a-new-path-to peace/

Abou-Hadid, A. (2014). Impact of climate change on food security. In A. Sadik, M. El-Solh, & N. Saab (Eds.) *Arab Environment: Food security. Annual report of the Arab forum for environment and development, 2014* (pp.130–154). Beirut: Technical Publications.

AFED. (2014). A. Sadik, M. El-Solh, & N. Saab (Eds.), *Arab environment: Food security. Annual report of the Arab forum for environment and development, 2014*. Beirut: Technical Publications.

Alaoui, M. (2013). Water sector in Morocco: Situation and perspectives. *Journal of Water Resources and Ocean Science, 2*(5), 108–114. doi:10.11648/j.wros.20130205.18.

Anderson, T. (2015). Dynamic markets for dynamic environments: The case for water marketing. *Daedalus, 144*(3), summer, 83–93.

Arab Forum for the Environment and Development (AFED). (2009). In M. Tolba & N. Saab (Eds.), *Arab environment: Climate change: Impact of climate change on Arab countries*. Beirut: AFED.

Arab Human Development Report Papers. (2010, January) *Mapping climate change issues, initiatives and actors in the Arab region*. Paper 10,02, New York: UNDP Regional Bureau for Arab States.

Babiker, M., & Feheid, M. (2011, May). *Climate change policy in the MENA region: Prospects, challenges and the implication of market instruments*. ERF Working Paper 588.

Badraoui, M. (2014) .The Green Morocco Plan: an innovative strategy of agricultural development. In A. Sadik, M. El-Solh, & N. Saab (Eds.) *Arab environment: Food security. Annual report of the Arab forum for environment and development, 2014* (pp. 34–35). Beirut: Technical Publications.

Barnes, J. (2015). *Cultivating the Nile: The everyday politics of water in Egypt*. Durham: Duke University Press.

Bloomberg, M. (2015). City century: Why municipalities are the key to fighting climate change. *Foreign Affairs*, *95*, Sept–Oct, 116–124

Bricheri-Colombi, S. (2011, April). *Matching water supply and demand*. Government of Egypt, Ministry of Water and Irrigation, coordination project NWRP-CP, technical report 57.

Bricheri-Colombi, S. (2004) "Hydrocentricity: A limited approach to achieving food and water security", Water International, 29, n.3, pp. 318–328

Briscoe, J. (2015). Water security in a changing world. *Daedalus*, *144*(3) summer, 27–34.

Bruch, C., et al. (2007, December). Legal frameworks governing water in the Middle East and North Africa. *International Journal of Water Resources Development*, *23*(4), 595–624.

Burgan, B. (2015, December 27). Is the red-dead correct? *Jordan Times*.

Cline, W. (2007). *Global warming and agriculture: Impact estimates by country*. Washington, DC: Peterson Institute.

De Fries, R., et al. (2015, July 17). Metrics for land-scarce agriculture. *Science*, *349*, 238–240.

Díaz-Bonilla, E. (2015, August). *Macroeconomics, agriculture and food security: A guide to policy analysis in developing countries*. IFPRI issue brief 87. Washington, DC: International Food Policy Research Institute.

Doukkali, M. (2005). Water institutional reforms in Morocco. *Water Policy, 7*, 71–88.

Droubi, A. (2009, August 4). *Syria: Country study*. In: Climate change, water, and the policy-making process in the Levant and North Africa conference. Issam Fares Institute for Public Policy and International Affairs, The American University of Beirut.

Fishman, C. (2015, August 14). How California is winning the drought. *New York Times*.

Gebbers, R., et al. (2010) Precision agriculture and food security, *Science*, *327*, 828–830

Gibbs, H. K., et al. (2015). Brazil's Soy Moratorium. *Science, 347*(6220), 377–378.

Gilmont, M. (2016). Water resource decoupling in the MENA through food trade as a mechanism for circumventing national water scarcity, *Food Security*

Glatzel, K. (2013, May). *Private sector responses to water scarcity in the Hashemite Kingdom of Jordan*. A thesis submitted to Imperial College London for the degree of Doctor of Philosophy.

Hreiche, A. (2009, August 4). *Climate change in the Levant and North Africa Region: An assessment of implications for water resources, regional state awareness, and preparedness*. Climate Change, Water, and the Policy-making Process in the Levant and North Africa, ppt. Issam Fares Institute for Public Policy and International Affairs, the American University of Beirut.

IAEA & The UNDP (2013) *Regional strategic action programme for the Nubian aquifer system*. Vienna

Jeuland, M., & Whittington, D. (2014). Water resources planning under climate change: Assessing the robustness of real options for the Blue Nile. *Water Resources Research, 50*. doi:10.1002/2013WR013705.

Jeuland, M., Wu, X., & Whittington, D. (2016, forthcoming) Infrastructure development and the economics of cooperation in the Eastern Nile. *Water International*.

Jobbins, G., Kalpakian, J., Chriyaa, A., Legrouri, A., & Mzouri, El Houssine El . (2015). To what end? Drip irrigation and the water–energy–food nexus in Morocco. *International Journal of Water Resources Development, 31*, 1–14. doi:10.1080/07900627.2015.1020146. http://dx.doi.org/10.1080/07900627.2015.1020146

Khouri, N., & Byringyro, F. (2014). Developing food chains. In A. Sadik, M. El-Solh, & N. Saab (Eds.) *Arab environment: Food security. Annual report of the Arab forum for environment and development, 2014* (pp. 102–129). Beirut: Technical Publications.

Loudyi, D., & Oubalkace, M. (2015). Water governance in Morocco. *Hydrolink, 4*, 122–125.

Michalak, A., & Field, C. (2015). Water, climate, energy, food. *Daedalus*, *144*(3), summer, 7–17.

Morocco. (2008, April). *Ministry of Agriculture, Plan Maroc Vert: premières perspectives sur la stratégie agricole*, Rabat. http://www.agrimaroc.net/Plan_Maroc_Vert.pdf

Riachi, R. (2014, October). *Beyond rehashed policies: Lebanon must tackle its water crisis head-on*. Lebanese Center for Policy Studies.

Robertson, G. P. (2015). A sustainable agriculture? *Daedalus, 144*(4), Fall, 76–89.

Rogers, P. (1997). International river basins: Pervasive unidirectional externalities. In P. Dasgupta, K.-G. Maler, & A. Vercelli (Eds.), *The economics of transnational commons* (pp. 35–76). Oxford: Clarendon Press.

Sadik, Abdul-Karim. (2014). The state of food security and agricultural resources. In A. Sadik, M. El-Solh, & N. Saab (Eds.) *Arab environment: Food security. Annual report of the Arab forum for environment and development, 2014* (pp. 12–43). Beirut: Technical Publications.

Salman, M. A. S. (2016, January 7). Sudan: Renaissance Dam—Limelight on the Fourth Ministerial Meeting and Khartoum Document. *AllAfrica*. http://allafrica.com/stories/201601080844.html

Shamout, M. N., with Lahn, G. (2015, April). *The Euphrates in crisis channels of cooperation for a threatened river*. Chatham House, Research Paper: Energy, Environment and Resources.

Shideed, K., Theib, A., & Van Ginkel, O. M. (2014). Potential of rainfed agriculture and smallholder farmers in food self-sufficiency. In A. Sadik, M. El-Solh, N. Saab & AFED (Eds.), *Arab environment: Food security. Annual report of the Arab forum for environment and development, 2014* (pp. 75–101). Beirut: Technical Publications.

Siddiqi, A., et al. (2013). Bridging decision networks for integrated water and energy planning. *Energy Strategy Reviews*. http://dx.doi.org/10.1016/j.esr.2013.02.003

Solh, M. E.(2014). The role of science and technology in enhancing food security. In A. Sadik, M. El-Solh, & N. Saab (Eds.) *Annual report of the Arab forum for environment and development, 2014* (Arab Environment: Food Security) (pp.44–73). Beirut: Technical Publications.

Stone, R. (2015, September 4). Saving Iran's Great Salt Lake. *Science, 349*(6252), 1045–1047.

Tilman, D., & Clark, M. (2015). Food, agriculture and the environment: Can we feed the world and save earth? *Daedalus, 144*(4), 8–23.

Trondalen, J. M. (2008). *Water and peace for the people: Possible solutions for water disputes in the Middle East*. Paris: UNESCO.

United Nations. (1997, May 21) *Convention on the law of the non-navigational uses of international watercourses*, New York. https://treaties.un.org/Pages/ViewDetails.aspx?src=IND&mtdsg_no=XXVII-12&chapter=27&lang=en

Varis, O., & Abu-Zeid, K. (2009). Socio-economic and environmental aspects of water management in the 21st century: Trends, challenges and prospects for the MENA region. *International Journal of Water Resources Development, 25*(3), 507–522.

Venot, J.-P., Molle, F., & Hassan, Y. (2007). *Irrigated agriculture, water pricing and water savings in the lower Jordan River Basin (in Jordan)* (Comprehensive assessment of water management in agriculture research report 18). Colombo: International Water Management Institute.

Vidal, J. (2015, August 27). Middle East faces water shortages for the next 25 years, study says. *The Guardian*.

Voss, K. A., Famiglietti, J. S., Lo, M., de Linage, C., Rodell, M., & Swenson, S. C. (2013). Groundwater depletion in the Middle East from GRACE with implications for transboundary water management in the Tigris-Euphrates-Western Iran region. *Water Resources Research, 49*. doi:10.1002/wrcr.20078.

Waterbury, J. (2002). *The Nile Basin: National determinants of collective action*. New Haven: Yale University Press.

Waterbury, J. (2011). You can't get there from here: Theoretic puzzles of collective action. In T. Tvedt, B. Chapman, & R. Hagen (Eds.), *A history of water: V.3 water, geopolitics and the new world order* (pp. 28–48). London: I.B. Tauris.

Waterbury, J. (2013). *The political economy of climate change in the Arab Region*. United Nations Development Programme, Regional Bureau for Arab States, Arab Human Development Report Research Paper Series.

Waterbury, J. (2014). Puzzles of commitment, compliance and defection in collective action. Workshop on "Water politics and regional stability," 21–23 May, 2014, Lee Kwan Yew School of Public Policy.
Wheeler, T., & von Braun, J. (2013). Climate change impacts on global food security. *Science, 341*, 508–513.
Whittington, D., Waterbury, J., & Jeuland, M. (2014). The Grand Renaissance Dam and prospects for cooperation on the Eastern Nile. *Water Policy, 16*(4), 595–608.
Woertz, E. (2013). *Oil for food. The global food crisis and the Middle East*. Oxford/New York: Oxford University Press.
World Bank. (2007). *Making the most of scarcity: Accountability for better water management in the Middle East and North Africa, MENA development report*. Washington, DC: World Bank.
World Bank. (2013). *Adaptation to climate change in the Middle East and North Africa Region*. Washington, DC. http://web.worldbank.org/WBSITE/EXTERNAL/COUNTRIES/MENAEXT/0,,contentMDK%3A21596766~pagePK%3A146736~piPK%3A146830~theSitePK%3A256299,00.html
Zack, L. (2015). Water Blues. In: *Middle East report*, Fall, pp. 32–35.
Zereini, F., & Hötzl, H. (Eds.). (2008). *Climatic changes and water resources in the Middle East and North Africa*. Berlin: Springer-Verlag.
Zubari, W. (2014). Virtual water trade as a policy instrument in contributing to the achievement of food security in the GCC countries. In A. Sadik, M. El-Solh, & N. Saab (Eds.), *Arab environment: Food security*. Annual report of the Arab Forum for environment and development, 2014. Beirut: Technical Publications, Annex 3.

Water Conservation in the Arab Region

Abdin M.A. Salih and Gamal M. Abdo

Abstract The Arab region covers an area of over 13 million square kilometers, with almost 90% of it is either arid or extremely arid with very little precipitation, extremely high evaporation and almost no vegetation cover. The region is classified in many international reports as the poorest region in the world in the context of renewable water resources and critical water scarcity, which hinders the socio-economic development of many countries in this region. The rapidly increasing population has reduced the per capita share of renewable water to less than the poverty line of 1000 m^3/(capita·a) and, in many Arab countries, to less than the extreme poverty line of 500 m^3/(capita·a). This has led to over-exploitation of non-renewable groundwater and desalination of salty water in many countries with considerable costs and contamination of many renewable sources. Atmospheric processes responsible for aridity in the Arab region are projected to intensify due to climate change, resulting in an alarming decrease in precipitation and increase in evaporation rates. It is recognized that water security is a key element to achieve food security, socio-economic development and ultimately political stability in the region. Hence, various efforts have been exerted to identify key problems and suggested solutions. The Arab Water Ministers Council of the Arab League, as well as Reports of the Arab Forum for the Environment and Development (AFED) and the recommendations of Regional Meetings of the Arab National Committees of the International Hydrological Programme of UNESCO (IHP), among others, have all made similar recommendations on the need to address the issues of water scarcity in the Arab region. However, water conservation has been endorsed as an important area for coping with water scarcity in the region. There are many definitions of water conservation in the scientific literature, and many areas of action including huge water savings from irrigation, industrial use, and domestic use as well as methods and approaches for augmenting water supply through non-conventional practices such as water harvesting and waste water reuse. In this paper, a review is provided for definitions, methods and impacts of water conservation and its role in alleviating water scarcity in the Arab region.

A.M.A. Salih • G.M. Abdo (✉)
Water Research Centre, Faculty of Engineering, University of Khartoum, Khartoum, Sudan
e-mail: abdeensalih@gmail.com; gabdo2000@yahoo.com

S. Murad et al. (eds.), *Water, Energy & Food Sustainability in the Middle East*,
DOI 10.1007/978-3-319-48920-9_5

Keywords Water conservation • Fresh water resources • Water loss reduction

Introduction

Water scarcity is among the main problems to be faced by countries of the world in the twenty first century. Water use has been globally growing at more than twice the rate of population increase in the last century and an increasing number of regions are chronically short of water (FAO 2007). The Arab region is among those in which water scarcity is becoming a serious constraint impeding the socio-economic development of many countries in the region. The region belongs to one of the most arid areas of the world, characterized by scare, unevenly distributed and highly variable water resources. Although there are considerable differences in estimated total quantities of renewable water resources in this region, the figure of 355 km^3/year has been circulated in various reliable sources. With regards to water availability and the per capita share of renewable water resources, the region is classified in many international reports as the poorest region in the world, as presented in Fig. 1 (AFED 2010). The problem has been continuously magnified by vastly expanding populations and growing water demand to meet the great socio-economic development of the region in the last few decades. Urbanization, industrialization and the expansion of irrigated agricultural lands to achieve food security have all contributed to a dramatic and unsustainable increase in water consumption. Frequent droughts, in conjunction with an overuse of groundwater from major aquifers, have greatly reduced the availability of both renewable and non-renewable water resources. Saline intrusion into fresh water aquifers and pollution from urban activities are also common phenomena in the Arab region that lead to a deterioration of water quality and reduced fresh water availability. Due to the aforementioned factors, the per capita share of renewable water has been reduced to less than the poverty line of 1000 m^3/capita/year and in some Arab countries, to less than the extreme poverty line of 500 m^3/capita/year. According to IFAD (2009), the per capita share in the year 2025 will be reduced to only 15% of what it was in 1960, when it stood at 3300 m^3/capita/year. An important element that makes the water situation in the Arab region even worse is the trans-boundary nature of its water resources. According to available information, out of total renewable water resources of 355 m^3/year, more than 60% originate from sources outside the boarders of the Arab region through international rivers such as the Nile, Euphrates-Tigris, and Senegal Rivers (AFED 2010). Yet there are no formal agreements or institutional mechanisms for joint management of shared water resources in the Arab region. A major challenge facing the Arab region is therefore to shift the existing image of water competition and possible friction, to a more constructive image of water for national and regional development. Vulnerability of water resources to climate change and adaptation to scarcer water availability is another important challenge facing the Arab region. Recent reports (AFED 2010) warned that climate change will be responsible for a 25% decrease in precipitation and 25% increase in evaporation in the Arab region by the end of the century,

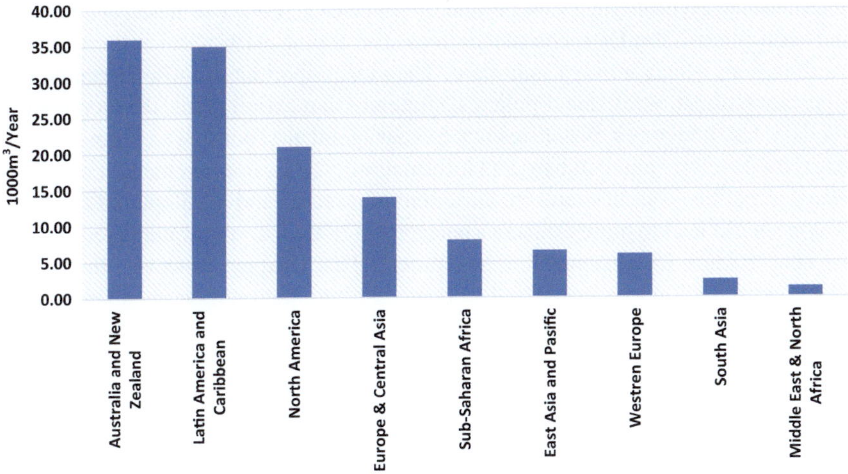

Fig. 1 Actual renewable freshwater resources per capita by region (Source: AFED 2010)

leading to severely negative impacts on water resources. In addition to the aforementioned challenges, most of the countries in the Arab region suffer from lower levels of education and capacities to deal with water resources scarcity in a sustainable manner compared with countries from other regions of the world (UNESCO 2010).

Water utilization in the Arab region is distributed among three main sectors; agriculture, domestic and industry. According to the data provided in AFED (2010), the agricultural sector is the highest consumer of water in the Arab region utilizing on average 83% of the natural renewable available water resources. The remaining percentage is distributed amongst the domestic and industrial sectors as 10% and 7%, respectively. Data on water availability and water utilization also show that some countries such as Saudi Arabia, Bahrain, Kuwait, Qatar and UAE are using much more water than their natural renewable water resources. These countries have historically been faced with extreme water shortage and they have eventually resorted to more expensive sources, namely desalination and non-renewable groundwater, to provide reliable water supplies to meet their demands. Regarding future water demand in the Arab region, Salih (2011) reported that if the current water use patterns in the region remain unchanged, the water demand for the year 2030 is expected to increase to about 670 BCM compared to the total renewable water resources of 355 BCM.

Many relevant organizations in the region consider water security as a key element for food security and ultimately political security, and hence various efforts have been exerted to identify the key water problems and suggestion of possible solutions. One could mention here the water security documents of the Arab Council of Minister of Water Resources of the Arab League; the Reports of the Arab Forum for the Environment and Developments (AFED); and the recommendations of the 13th regional meeting of the Arab National Committees of the

International Hydrological Programme of UNESCO (IHP). All these reports came out with almost similar recommendations, for coping with water scarcity in the Arab region, which are very important. The most important of those recommendations are realization of sound policies, strategies and action plans, development of models for integrated water resources management, sound water governance, capacity development, database and information systems improvement, technology transfer, public awareness and enhancement of research and development facilities. These should lead to more efficient management of their scarce water resources and its conservation for sustainable development. This Chapter, however, focuses on water conservation as a viable and strategically important option to enhance water availability and sustainability for socioeconomic development in the Arab region.

Definition of Water Conservation

According to the Collin English Dictionary, the word conservation means:

1. The act or an instance of conserving or keeping from change, loss, injury, etc.. . .
2. A protection, preservation and careful management of natural resources and the environment.

A more or less similar meaning has been given by the American Heritage Dictionary of the English language as:

1. Preservation, protection or restoration from loss, damage or neglect of wildlife and natural resources such as water, forest or soil.
2. The maintenance of a physical quantity such as energy or mass during a physical or chemical change.

As for the term water conservation, there are also many definitions in the scientific literature all of which aim at improving water management practices to enhance the beneficial use of water while preserving its quality (Vickers 2002). The Alberta Water Council (1977) within its project "Water Conservation, Efficiency and Productivity Principles, Definitions, Performance Measures and Environmental Indicators" has reviewed definitions of water conservation from different credible sources, modified them as appropriate and adopted the following definition for water conservation:

1. Any beneficial reduction in water use, loss, or waste.
2. Water management practices that improve the use of water resources to benefit people or the environment.

According to a UN report (1997), water conservation refers to the preservation, control and development of water resources, both surface and groundwater, and the prevention of pollution. More generally the Environmental Engineering Dictionary defined water conservation as the physical control, protection, management, and use of water resources in such a way as to maintain crop, grazing, and forest lands, vegetative cover, wildlife, and wildlife habitat for maximum sustained benefits to people, agriculture, industry, commerce, and other segments of the national economy. In the context of this chapter, water conservation can in general be categorized into two main aspects:

1. Water saving through water loss reduction (WLR) in the use of conventional water resources for the different purposes;
2. Increasing the use of the non-conventional water resources

Need for Water Conservation in the Arab Region

Water conservation has been recommended by many researchers as a sound and cost effective option to cope with water scarcity and achieve food security in the Arab region (AFED 2010; Abdo et al. 2010). The topic is receiving considerable attention worldwide and there had been many international workshops and conferences discussing various aspects of water conservation including technical solutions, and economic, institutional and capacity development issues (UNU 2008, 2009). Consequently, there exists a wide range of experiences and examples of success stories on water conservation from different parts of the world. An excellent review of the experiences of many countries from different parts of the world on water conservation and water loss reduction in the municipal, agricultural and industrial sectors is given by Salih (2011).

The Arab region is known for its inefficient utilization of water and significant water wastage in individual sectors. In conventional irrigation systems, which are common in the Arab region, the overall efficiency could be as low as 20% compared with 60–70% in advanced systems such as drip and sprinkler systems (UCO 2003). Also, a substantial volume of water is lost daily in water supply distribution networks. Previous studies estimate the percentage of losses of produced water in some countries of the Arab region between 30 and 70% (AFED 2010). Realizing the water scarcity situation in most of the Arab countries and the challenge to meet the increasing water demand, water conservation has been recommended by many researchers as a sound and viable option to harness substantial volumes of water to meet the demand. It is therefore suggested in this chapter that Arab countries should primarily direct attention towards water conservation and accordingly and make more effort to develop polices, strategies and action plans for implementing appropriate programs. Recently, there has been a growing worldwide concern about water conservation, and the concept has now become an essential practice even in areas with abundant water resources. Consequently, water conservation programs are now being implemented in many parts of the world as an essential step towards development in all sectors, such as the agricultural, industrial and municipal sectors. In the Arab region, water use efficiency is known to be very low especially in irrigation. Therefore a shift in water use practices is urgently needed in order to secure more water to meet the increasing needs for development. The shift in water uses has already been reflected in developing water saving programs which started to be implemented in many countries of the Arab region. Some of these programs include irrigation management projects aiming at increasing irrigation efficiency. Water loss reduction in distribution networks is also an important practice that could conserve large volumes of water. The use of non-conventional water resources is also receiving considerable attention worldwide and efforts are underway in several Arab countries to develop these resources. Recycling of wastewater and agricultural drainage provide additional water source that could be allocated for irrigation in exchange for fresh water diverted from agriculture to meet the growing demand for urban and rural water supply. Water harvesting is also one of the most effective practices for non-conventional water use in the Arab region. Large potential and extensive experience in various aspects of water harvesting already exist in the region. The above mentioned water conservation methodologies, together with some others, will be elaborated in the subsequent sections of this Chapter. It is to be noted that water conservation methodologies are in many cases interlinked and can have negative impacts on each other. An example is that of groundwater aquifers depending on their recharge from irrigation canals that may be lined under a new water conservation program. Therefore adoption of an integrated approach in addressing the technical, environmental and socio-economic impacts of water conservation projects is highly needed.

Water Conservation Through Water Loss Reduction (WLR)

Water Conservation in the Agricultural Sector

As has been mentioned previously, the agricultural sector is the major consumer of water in the Arab region utilizing on average more than 80% of the available renewable water resources. Over the past three decades, there has been a significant expansion in irrigated agriculture in the Arab region in an effort to achieve food security. However, with the current irrigation management practices, water losses in most of the countries are very substantial leading to low irrigation efficiency. Available estimates for some Arab countries indicate that the average irrigation efficiency ranges between 30 and 45%.(AFED 2010). This is due mainly to the traditional belief seeing water as a basic necessity and human right and the failure to recognize the economic value of water and the real cost of providing its services. Recently there has been a growing concern on the inefficient use of water resources in the agricultural sector among many specialized regional and international organizations such as the Food and Agricultural Organization (FAO), International Commission for Irrigation and Drainage (ICID), United Nation Educational, Scientific and Cultural Organization (UNESCO) and Arab Organization for Agricultural Development (AODA), in addition to many other scientific institutions. The general consensus is that such a wasteful use of scarce water resources in agriculture in the Arab region should not continue and more efforts should be directed towards water conservation practices. Many studies indicate that agricultural production in the Arab region could be doubled with the same amount of water that is currently being used, if better irrigation management practices are adopted (Abdo et al. 2010). Water conservation in the irrigation sector could be achieved through the use of better on-farm water management systems that reduce irrigation water distribution losses, changing cropping patterns, improving irrigation scheduling, and adopting modern irrigation technologies such as sprinkler or drip irrigation. AFED (2010) reported that the use of sprinkler and drip irrigation systems in some Arab countries has reduced water losses considerably. For example in Morocco and Jordan, the efficiency level has increased to about 70% due to the adoption of these techniques. Furthermore, the use of drip irrigation in many parts of the Arab region has increased agricultural productivity. In addition to their potential for increasing water use efficiency, drip and sprinkler irrigation technologies can also provide opportunities to cultivate low quality lands, sandy and rocky soils, and enable some countries to change cropping patterns by shifting from high water consuming, low value crops to low water consuming, high value crops. However, due to the high purchase prices and high operation and maintenance costs of these technologies, their agronomic and economic feasibility should be justified. A number of successful applications of water conservation programs in the irrigation sector from the Arab region are discussed in an important document on water conservation published by the Azzamil Group Chair for Conservation of Electricity and Water, King Saud University, Saudi Arabia (KSU 2011). For example in Syria, the introduction of modern irrigation systems resulted in saving 58% of the water used and in increase of productivity of about 35%. In Egypt, improved irrigation management practices have increased the productivity by 30–40% with the same amount of water, reduced the irrigation period by 50–60% and pumping period and cost, and improved water availability along the whole length of irrigation canals.

The following are some recommendations from the Arab Organization for Agricultural Development strategy for Water Policy Reform in the Arab Region with regards to water conservation in the agricultural sector (AOAD 1997):

1. The Arab Region must focus on the improvement of water demand management, particularly in the agricultural sector. Improvement of agricultural water demand management includes modernization of irrigation networks; improved irrigation scheduling; modification in cropping patterns through the use of crops with lower water requirements; and application of extensive supplementary irrigation in rain fed farming to alleviate the need for continuous expansion of permanently irrigated schemes.
2. Improved irrigation scheduling both at the system and the farm levels needs to be given high priority to ensure that, within the constraints of system design and management capabilities, optimum crop water requirements are met with minimum water losses while avoiding soil salinization and water logging. Such programs have the potential for significant water savings at relatively reasonable costs.

3. Application of irrigation scheduling in response to soil moisture measurements in association with the adoption of modern irrigation systems has resulted in reducing irrigation rates in many parts of the world while at the same time increasing productivity.
4. Modern irrigation techniques need to be carefully selected and adapted to the local physical, agronomic and socio-economic environment, as well as to the technical and managerial skills of local farmers. Upgrading existing irrigation schemes should in most cases be preceded by pilot trials for alternative design concepts. Costly improved technologies can only be justified if their agronomic and economic potential is fully exploited.
5. It is apparent that in most of the Arab region there is still a lack of economic and fiscal incentives for irrigation improvement. Hence high priority should be given to improved management of irrigation water by encouraging farmers to invest in water-saving technologies and to cultivate crops with low water demand.
6. Such measures need to be supported by appropriate irrigation and agricultural research programs for identifying better irrigation practices and more appropriate crops. Moreover, nationwide irrigation extension programs should be promoted in parallel with the introduction of irrigation-saving techniques and the upgrading of O&M of the existing irrigation systems.
7. A Regional Water Security Master Plan which includes water reform policies that aim at achieving the required Arab water and food securities can be formulated to achieve the following (AOAD 1997):

   - Improve the quality of research and study regarding the use of modern agricultural technologies appropriate to local conditions, maintain modern irrigation systems, identify crop water requirements and cropping patterns for the purpose of benefiting from available water resources, and attaining Arab food and water securities.
   - Intensify water extension services and increase their efficiency especially in irrigated agriculture, which is considered a key element in efficient water use in Arab agriculture, and make use of all necessary means to build and improve national capacity regarding this issue.
   - Search for methods and tools that encourage cooperation among all government agencies and the private sector for the purpose of preparing and implementing development programs and plans in the agricultural and water sectors; and to supply the necessary support for Water Users' Associations (WUAs) which will help in realizing efficient water use at the field level.
   - Provide tools for the purpose of finding and strengthening links among institutions, extension services and research infrastructures on the one hand, and farmers' organizations on the other hand, and all that might contribute to attain public and common goals related to policies of best use of water and land resources.

- Strengthen Arab cooperation for the purpose of protecting water quality; stressing environmental issues in agricultural and water policies; and taking all measures in order to protect and conserve land and water resources from environmental degradation within a framework of balanced agricultural policies that help attain goals for a sustainable agriculture and rural development that meets present and future generation's needs.

Water Conservation in Municipal Water Supply

There is an increasing world-wide attention to water conservation and loss reduction in municipal water supply systems in order to save water and use it to expand supplies for the increasing population. It is estimated that approximately 45 million m^3 of drinking water are lost in the world's water systems every day which if saved, could serve nearly 200 million people (UNU 2008). Reducing water losses in urban water supply networks could therefore make a substantial contribution in reducing the number of people without sustainable access to clean water. Also, substantial volumes of water could be lost inside the house through washing machines, dish washers, showers, taps, sinks and toilets, or outside in car wash and garden watering. Garden watering is estimated to consume about 50% of the total household consumption (Abdo et al. 2010), of which significant quantities are lost due to inappropriate watering practices. In addition to reduction of revenue, water losses from distribution systems could also contribute to rising groundwater levels in urban centers leading to many geotechnical and environmental problems. A good example in this is the case of Riyadh city in Saudi Arabia (Hamadto 2009).

There are no reliable published estimates on the percentages of losses in water supply networks in countries of the Arab region, but many researchers indicate that they are high. Salih (2011) presented some estimates for selected cities in Egypt and Jordan, where the losses range between 30 and 60% of produced water. However, these losses have significantly been reduced after implementation of conservation programs.

Water conservation methods in domestic water supply include:

1. Water loss reduction in the water supply system
2. Use of household water saving devices
3. Metering

Figure 2 shows the water balance table proposed by the International Water Association (IWA) and the American Water Works Association (AWWA) to assist water authorities in tracking distribution system losses (EPA 2010). In the AWWA/IWA methodology, all water that enters and leaves the distribution system can be classified as belonging to one of the categories in the water balance table. The losses to be reduced are due to leakage from transmission and distribution mains, leakage and overflows from storage tanks, and leakage from service connections up to and including the meter. Preventing or repairing these losses usually requires an

<table>
<tr><td rowspan="5">System input Volume</td><td rowspan="2">Authorized Consumption</td><td>Billed Authorized Consumption</td><td>Billed metered consumption
Billed Un-metered Consumption</td><td>Revenue water</td></tr>
<tr><td>Un-billed Authorized Consumption</td><td>Un-billed Metered Consumption
Un-billed Un-metered Consumption</td><td rowspan="4">Non-Revenue Water</td></tr>
<tr><td rowspan="3">Water losses</td><td>Apparent losses</td><td>Unauthorized Consumption
Meter Inaccuracies
Systematic Data Handling Errors</td></tr>
<tr><td rowspan="2">Real losses (Physical Losses)</td><td>Leakage in Transmission & Distribution Mains</td></tr>
<tr><td>Storage Leaks and Overflow from storage Tanks, Service Connection Leaks Up to meter</td></tr>
</table>

Fig. 2 IWA/AWWA water balance table (EPA 2010)

investment. However there is a point at which it costs more to control leakage than is economically justifiable. Therefore a balance must be maintained between water loss reduction and costs associated with water loss reducing measures (EPA 2010).

House-hold water saving could be achieved through the use of waterless utilities such as low flow showers, ultra-low flush toilets, waterless washing machines and dish washers. According to CUWCC (2000), water saving from using the above mentioned devices is estimated at an average of 15% of the total domestic water use.

Metering water consumption is an aspect of the water supply operations and it is an effective means to encourage water conservation by making customers aware of their usage. Meters also make it possible to charge customers based upon the quantities of water that they consume and therefore make billing fair for all customers.

Privatization of water supply sectors has recently emerged and has been adopted by many countries worldwide as an important direction towards effective and sustainable supply systems. Private water supply companies tend to impose water pricing policies that maximize their profit. However, when implementing privatization, water charges should be affordable and should take into consideration the poor and the needy. This is an important issue that should be seriously looked at especially in water scarce countries such as the Arab countries.

Figure 2 indicates that the nonrevenue water generally constitutes a major portion of the total system input. In the Arab region, the proportion of nonrevenue water varies from one country to another as shown in Fig. 3 (UNDP 2013). Hence, one of the municipal sector's major challenges is to reduce nonrevenue water in the distribution network. The World Bank defines nonrevenue water "as the difference

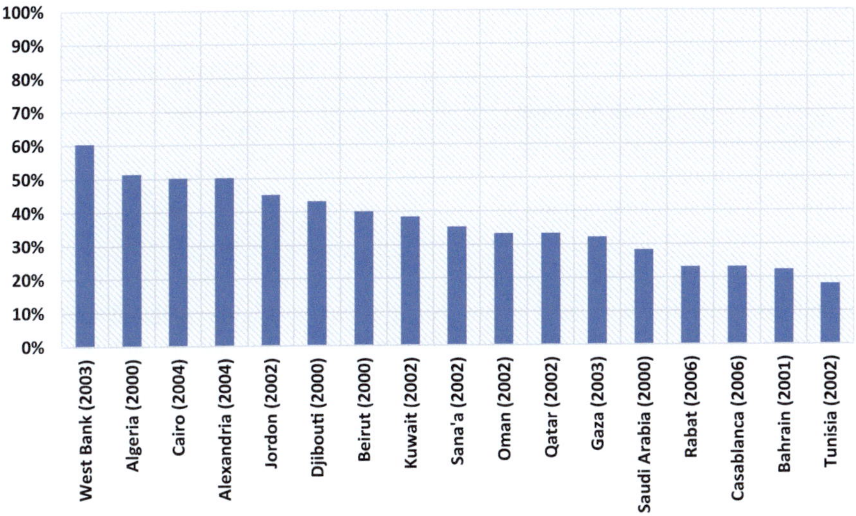

Fig. 3 Nonrevenue water in water supply utilities in selected Arab countries and cities (UNDP 2013)

between the amount of water delivered by the water utility and the amount actually billed". Nonrevenue water includes distribution network losses by leakage, illegal water use and inaccurate metering. The nonrevenue water can reach more than 60% in poorly maintained distribution networks in some Arab cities. The volume of nonrevenue water in Arab countries, ranging from 15% to 60%, greatly exceeds that in developed countries, where it ranges from less than 10% for new systems to 25% for older systems. Governance schemes need to take nonrevenue water into account particularly in water-scarce Arab countries. This lost water carries a high opportunity cost, equivalent in the Gulf countries to the cost of desalination and pumping (UNDP 2013). Reducing nonrevenue water by improving water distribution systems should thus be a major step for water conservation.

Water Conservation in Industry

Industrial activities use substantial amounts of water amounting to about 22% of the world's consumption of fresh water (Abdo et al. 2010). The percentage use is considerably higher in industrial countries and can even exceed agricultural use in some countries. In industrial facilities, water is used in a wide range of activities. Common uses include incorporation in the final product, washing or rinsing of raw materials, intermediates or final products, preparation of solvents or slurries, cleaning of equipment and space, removing or providing heat and meeting hygienic and domestic needs. Since most of these uses are non-consumptive, large volumes

Table 1 Water-saving potential in industry (AFED 2011)

Efficiency measurers	Potential Saving (%)
Closed loop reuse	90
Closed loop recycling with treatment	60
Automatic shut-off valves	15
Counter-current rinsing	40
High-pressure, low-volume upgrades	20
Reuse of wash water	50

could be generated as waste water which could be highly polluted and may endanger the environment if not properly treated. Significant amounts of water could be saved from treated industrial effluents.

As has been mentioned previously, the industry's share of overall water use in the Arab world is relatively small. However, industrial demand for water is also rising following similar trends in the agricultural and municipal sectors. Therefore, efforts should be directed towards enhancing water efficiency in industrial activities. Experience from around the world shows that adopting a systematic approach to water efficiency could reduce water consumption by 20–50%, and up to 90% when more advanced measures are implemented (AFED, 2011). Table 1 lists a number of industrial efficiency measures and their associated water-saving potential. As can be seen, the fraction of water that can potentially be saved by adopting closed loop re-circulating systems can be as high as 90%.

Augmenting Supply through Non-conventional Water Resources

Wastewater Reuse

Wastewater reuse is believed to be one potential intervention strategy for developing non-conventional water resources. The scarcity of water and the need for protecting the environment and natural resources have motivated Arab countries to introduce wastewater treatment and reuse as an additional water resource in their national water resource management plans. Since agriculture is the main water consumer, the extended reuse of treated wastewater for irrigation could contribute considerably to the reduction of water stress in the region. However there are economic, social, institutional, health, and environmental constraints that need to be considered to ensure sustainable and safe re-use and recycle of wastewater. Based on estimates by the Food and Agriculture Organization of the United Nations (FAO, 2009) the volume of wastewater generated by the domestic and industrial sectors in the different Arab countries is estimated as about 13 BCM/year. Figure 4 shows the volume of wastewater produced and treated in some Arab countries for the year 2009–2010 as an example (UNDP, 2013). It has been reported that the ratio

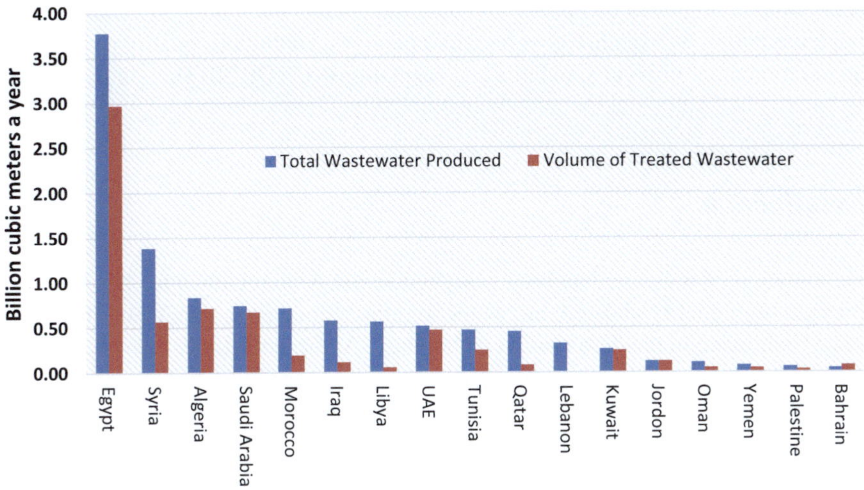

Fig. 4 Wastewater produced and treated in some Arab countries, 2009–2010 (UNDP 2013)

by volume of treated wastewater to that generated in the Arab region is higher than Asia, Latin America/Caribbean and Africa (AFED, 2010). Within the Arab world, Egypt and Tunisia are the leaders in the area of wastewater reclamation and reuse. Reuse of agricultural drainage water could also save huge volumes of water in the Arab region. This is practiced on a very large scale in Egypt, where 5 BCM m^3 of agricultural drainage water (equivalent to 10% of the total water resources) are reused annually after mixing with freshwater. Reuse of drainage water is practiced on a more limited scale in Iraq, Saudi Arabia, and Syria (WWF, 2006). However, the potential of wastewater reuse in the Arab region is large.

Gray water constitutes an important source of nonconventional water in the Arab Region. The term "grey water" refers to domestic waste water generated from less polluted sources, such as kitchen sinks, washing machines, dish washers, hand-washing basins and showers. Grey water, which consitutes around 50–70% of domestic wastewater, is recognized as a potential water saver and demand management tool (UNDP 2013). Grey water can be generated on one's premises treated by grey water kits which consist of four connected barrels. The grey water flows from the house gravitationally into the barrels, where treatment occurs in stages; the fourth barrel receives treated water, clean and ready for reuse. Domestic grey water users include not only private homes but also mosques, kindergartens and gardens. More recently, the Lebanese Ministry of Energy and Water incorporated grey water into its Ten-Year Water Plan for Lebanon. To explore grey water potential in the region, the Canadian International Development Research Centre has supported grey water treatment in Jordan, Lebanon, State of Palestine and Yemen, equipping more than 2000 houses with grey water treatment systems over the period 1998 to 2008. These projects revealed not only the high regional potential for using such forms of treated wastewater but also the absence of any health risks. Annual economic saving was estimated at more than $300 per family, but successful implementation will require government incentives, continuous quality monitoring and enforcement of local standards and regulations (UNDP 2013).

Desalination

Desalination of seawater has been practiced in the Arab region since the 1950s to provide high quality and reliable water supplies. Since then, significant advances have been made in the desalination technology. At the same time, there has been a significant expansion in the desalination capacity of the Gulf Cooperation Council (GCC). Figure 5 shows the volume of desalinated water in selected Arab countries in the year 2010 and projected to 2016 (UNDP 2013) indicating that the GCC countries are the leaders in this field. According to the GCC water statistics, the current total use of desalinated water is estimated at about 4 BCM/year and the desalination capacity is over 11 BCM/year (GCC 2008). Therefore, there is a big role for water desalination to play in coping with water scarcity in the Arab region. However, there are major issues of concern that need to be investigated such as the economic and environmental aspects of water desalination and the participation of the private sector in managing and operating desalination facilities.

Water Harvesting

Many researchers believe that management and effective utilization of rainwater through water harvesting (WH) projects is a key solution to the water scarcity problem in the Arab region. WH technology is especially relevant to the region where the problems of environmental degradation, drought and population pressures are most evident. WH is now receiving a growing awareness and considerable worldwide attention as a most suitable, practical and economical means to bridge the gap between water needs and available supply in many parts of arid and semi-arid areas of the world (Abdo and El Daw 2006). Due to the significant importance of the subject, many regional and international organizations such as: FAO, WFP, AOAD and UNESCO take a leading role in fostering the use of WH techniques in the Arab countries and other parts of the world. They also take the responsibility for the successful implementation of WH systems and for supporting in-depth pilot studies aiming to improve the efficiency of traditional techniques. In addition to water supply augmentation, WH can also have great potential in protecting the natural environment from degradation. Nasr (1999) discussed the problem of desertification in the Middle East and North Africa (MENA) and demonstrated the potential of using WH to combat desertification. The potential of WH in the management of groundwater recharge through artificial recharge methodologies is

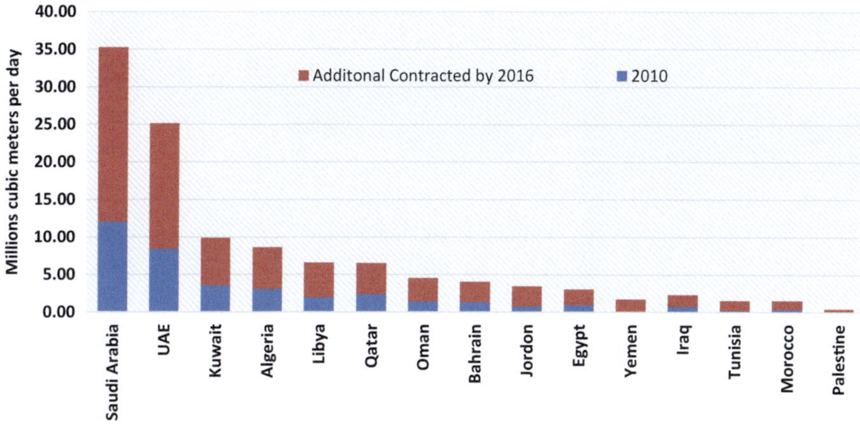

Fig. 5 Total contracted capacity of desalination plants in some Arab country (UNDP 2013)

also great in the Arab region where there are numerous alluvial aquifers underlying Wadi systems.

The fact that rainfall is very meager in the Arab region and that one millimeter of harvested rainfall is equivalent to one liter of water per square meter, suggests the importance of WH apart from the quantity of rainwater collected. It should be noticed that about 83% of the Arab world area, apart from Sudan and Somalia, receives less than 100 mm/year rainfall and about 13% receives 100–400 mm/year rainfall while not more than 4% exceeds 400 mm/year rainfall (El Kady 2003). Despite the long experience and the extensive implementation of WH in the Arab states, the amount of rain effectively utilized is very small, though the total volume amounts to 1500 BCM. Therefore, the available potentialities of the technique are still great and promising. In Jordan, for instance, Abu-Zreig (2003) states that out of a total rainfall of 8.5 billion m^3/year only 5% of this amount is being beneficially used. According to Al Ghariani (2003), the total precipitation in Northern Libya exceeds 30 billion m^3 out of which only 3% is effectively utilized. Figure 6 shows the rainfall potential in some of the Arab countries and Fig. 7 shows the percentage utilization through WH projects, which does not exceed 13% at the best as in the case of Moroco. This again demonstrates the great potentials and opportunities for the future implementation of WH projects. Raising the percentage of utilization will certainly be the key factor in bridging the ever increasing gap between water supply and demand in the Arab world. However there are some problems and constraints that hinder the use of the technique. These are mainly classified as technical, socio-economic, environmental, financial and institutional constraints. Abdo and Eldaw (2006) discussed these problems and suggested some actions to solve them.

Salih and Abdalla (2014) gave a historical background on ancient innovations in irrigation and water resources management in the Arab Region and showed how the ancient inhabitants had successfully reversed the scarcity challenges and constraints into opportunities utilizing the available knowledge at the time to devise

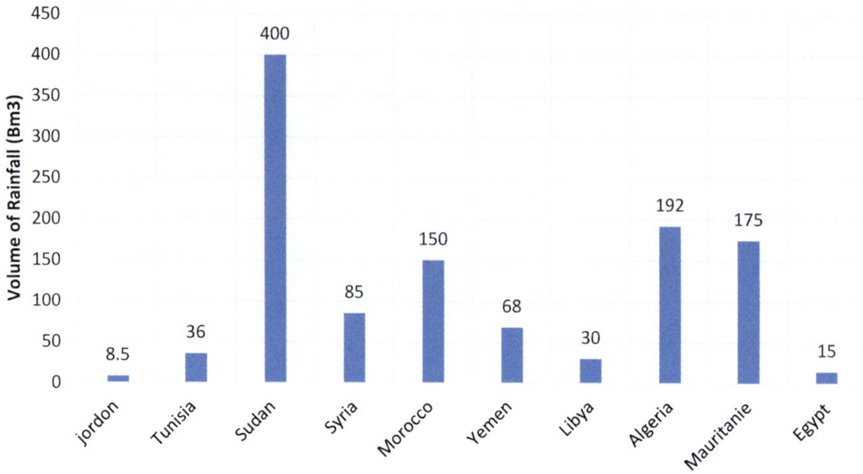

Fig. 6 Amount of rainfall water in some Arab countries (WRC 2016) (Adapted from AOAD 2002)

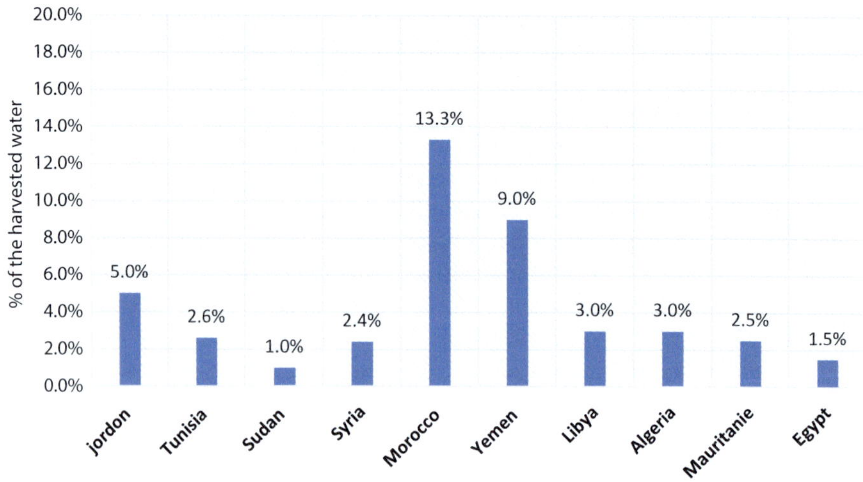

Fig. 7 Percentage of harvested water in some Arab countries (WRC 2016) (Adapted from AOAD 2002)

innovative coping mechanisms reflecting great ingenuity. Few examples of these innovations have been summarized in this paper including some forms of water harvesting that had been in use in the region (Aflaj, basins, dams, retarding structures, dew collection, etc.). These technologies have developed considerably in the last few decades in most parts of the world. Figure 8 gives a brief summary of the wide potential of these techniques which utilized moisture, rainfall, floods and groundwater resources. However, groundwater harvesting through Aflaj systems,

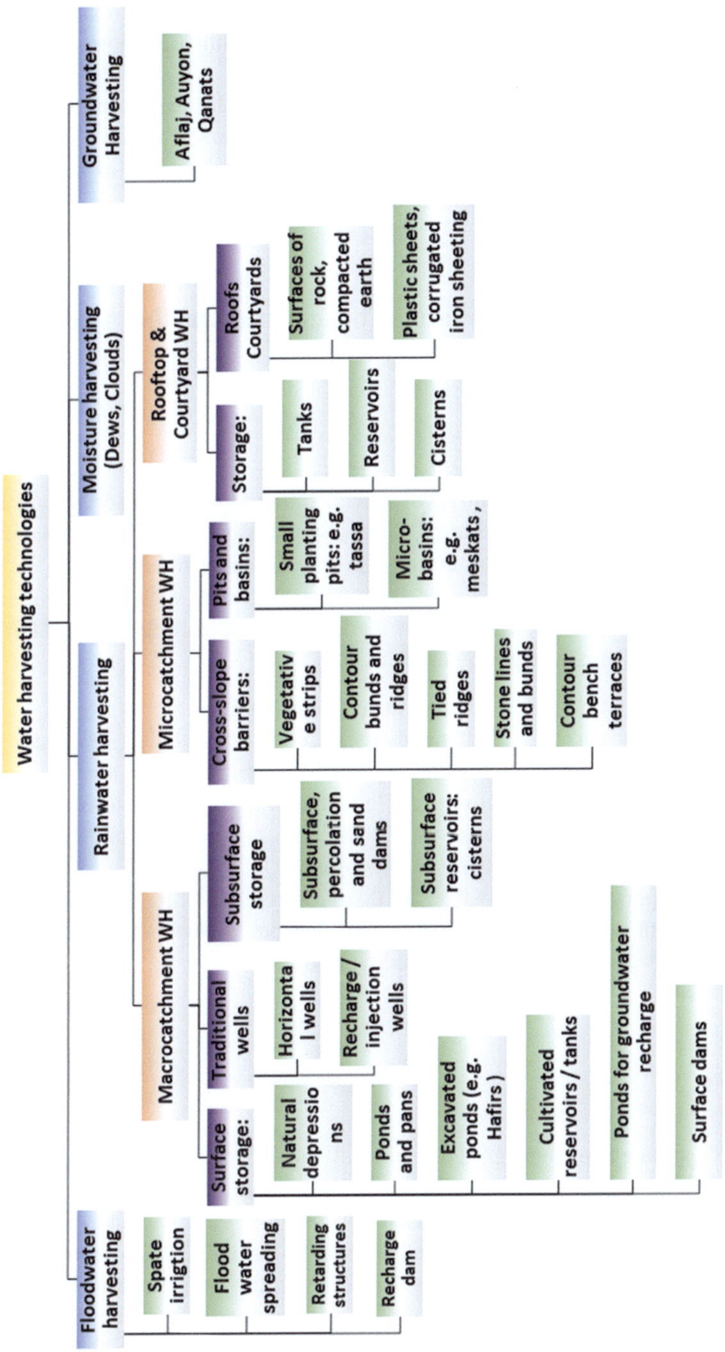

Fig. 8 Water harvesting technologies (WRC 2016)

which was utilized historically in about 35 countries in the world, has currently suffered greatly (WRC 2016) due to utilization of pumps for abstracting groundwater. Nonetheless countries like Iran still receives annually close to 9 BCM through their ancient system.

WRC (2016) has recently drafted a policy brief on water harvesting in the Arab region, in response to a request from the Arab Water Council which demonstrates the great potential of water harvesting in the Arab region as well as some policy statements for the sustainable management of water harvesting.

Successful Examples for Water Conservation in the Arab Region (UNDP 2013)

Tunisia

Tunisia was one of the first countries to adopt a national water conservation policy and strategy for urban and agricultural water use and water resources management. Water demand for irrigation has stabilized despite droughts and increasing agricultural development. The policy has also ensured the water supply for tourism, a source of foreign currency, and cities, a source of social stability. The underlying principles of the Tunisian strategy are:

- A transition from isolated technical measures to an integrated strategy.
- A participatory approach that makes users more responsible (960 water user associations were created, covering 60% of the public irrigated area).
- A gradual introduction of reforms and adaptation to local situations.
- Financial incentives to promote water conservation equipment and technologies (40–60% subsidy for purchasing such equipment).
- Income support for farmers, allowing them to invest and hire more labour.
- A transparent and flexible pricing system that complements national goals of food security and allows a gradual recovery of costs.

Morocco

Morocco's strategy for managing agricultural water demand focuses on the comparative cost of saving 1 m^3 with the cost of developing an equivalent amount of new water resources. Adopting drip irrigation costs less than developing new water sources. The water saved is optimized by improving market gardening and tree growing yields. The productivity gains have been profitable, generating extra added value. Benefit-cost analysis reveals a return of more than 30% of the capital costs. The benefits of the new strategy are not only economic but social (increase in farmers' income) and environmental (reduction of water abstraction).

Water Conservation Awareness

Awareness and understanding the need to use water efficiently at all stages is an important element in water conservation and water resources management at large. There is already a growing recognition of the importance of social norms and attitude towards water conservation. Many recent policy documents emphasize the importance of awareness raising to influence those norms and values towards a more sustainable use of water resources. There is evidence now that conservation awareness campaigns can effectively reduce water demands. ESCWA (2001) provided a useful guide to promoting awareness in water conservation in which the framework for water conservation awareness is explained and includes preparation of a water conservation strategy, methodologies for implementation as well as monitoring and evaluation of effectiveness. Also Wouter and Steenbergen (2002) is a useful document regarding innovative and attractive ideas for water awareness campaigns including water conservation awareness. The document brings together many of these ideas from a variety of organizations working in a diverse range of countries and it can be useful as a resource for a variety of people including water professionals, policy makers, NGO's and people working on water publicity campaigns and educational institutions.

Conclusion

The Arab world is facing one of the severest water scarcities in the world. Water availability is decreasing and water shortages are rapidly growing, threatening food and national security of almost all states. The water sector is characterized by inefficiencies in allocation among alternative competing uses with significant wastage of water in individual sectors. This calls for the urgent need to reform current practices of water management and shift to one of improving water management and rationalizing water consumption to meet the challenges over the coming decades. World experience indicates that water conservation can be a very successful approach to enhance the efficient use and augment water supplies for all sectors. It has been shown in this chapter that the potential for water conservation in the Arab region is great especially with regards to water saving in irrigation and water harvesting. Therefore several actions should be incorporated in integrated water conservation programs in all Arab countries. Strong political support and adequate capacity (human, institutional and enabling environment) are essential to facilitate implementation of these programs. It is also necessary to increase public awareness of both politicians and society on water scarcity and its implications, and emphasize the need to return to traditional and cultural values of water conservation.

References

Abdo, G., & Eldaw, A. (2006). *Water harvesting experience in the Arab world*. Proceedings of regional workshop eon management of aquifer recharge and water harvesting in arid and semi-arid regions of Asia, December, 2004, Yazd, Iran.

Abdo, G., Salih, A., & Al-Sheikh, A. (2010). *Water conservation as a tool to achieve water security in the Arab region*. Water security conference, Arab engineers union, Khartoum, Sudan, 23–24 October, 2010.

Abu-Zreig, M. (2003). *Water harvesting methods in Wadi System*. Proceedings of the Second International Conference on Wadi Hydrology, 1–4 July, Amman, Jordon.

AFED (Arab Forum for Environment and Development). (2010). *Arab environment: Water report*. Beirut.

AFED (Arab Forum for Environment and Development). (2011). *Water efficiency manual*. Beirut.

Al Ghariani, S. (2003). *Indigenous system of rainwater harvesting and runoff agriculture in Libya*. Proceedings of the second international conference on Wadi hydrology, 1–4 July, Amman.

AOAD. (1979). *Cairo declaration of Arab cooperation principles regarding use, Development and protection of Arab water resources*. First Arab Ministerial conference for agriculture and waters, Arab organization for agricultural development, Cairo in March–April 1997

AOAD (Arab Organization for Agricultural Development). (2002). *Enhancing WH technology in the Arab Region*. Khartoum, Sudan (in Arabic).

AWC. (1977). *Water conservation, efficiency and productivity definitions*. Alberta: Alberta Water Council.

CUWCC (California Urban Water Conservation Council). (2000). *Best management practices costs and saving study*. Sacramento.

El Kady MM. (2003). *Water scarcity from problems to opportunities in the Middle East*. NWRC Publication.

EPA (Environmental Protection Agency). (2010). *Control and mitigation of drinking water losses in distribution systems*. Washington, DC: Office of Water.

ESCWA. (2001). *Water conservation: A guide to promoting Public Awareness, Economic and social commission for Asia and the Pacific* (Water Resources Series No 81). New York: UN.

FAO. (2007). *Coping with water scarcity*. Challenge of the twenty-first century, World Water Day 2007, UN-Water.

FAO. (2009). AQUASTAT, 2009. http://www.fao.org/nr/water/wwap/aquastat/main/index.stm

GCC (Gulf Cooperation Council). (2008). GCC Water statistics report. Riyadh.

Hamadto, A. (2009). *Rising groundwater levels in Riyadh*. Regional Workshop on Groundwater Management in Arid and Semi-Arid Regions, 27–28 Sept, Khartoum.

IFAD (International Fund for Agricultural Development). (2009). *Fighting water scarcity in the Arab countries*. Rome.

KSU (King Saud University). (2011). *Conservation of water use*. Chair of Azzamil Group for conservation of electricity and water. Saudi Arabia.

Nasr, M. (1999). *Assessing desertification and water harvesting in the Middle East and North Africa*. No. 10 Policy Implications, ZEF, Bonn.

Salih, A. (2011). *World experience in water conservation*. In: KSU publication on conservation of water use, Chair of azzamil group for conservation of electricity and water, King Saud University, Saudi Arabia (in Arabic).

Salih, M. A., & Abdalla S. H. (2014). *Ancient innovations in irrigation and water resources management in North Africa and West Asia*. IWA regional symposium on water, wastewater and environment: Traditions and culture. 22–24 March 2014, Patras.

UCO (UNESCO Cairo Office). (2003). *Integrated water resources management in the Arab region: Action Plan for 2005–2030*. Cairo.

UN. (1997). *Water Conservation Definition*. Glossary of Environment Statistics, Studies in Methods, Series F, No. 67, United Nations, New York, 1997.

UNDP. (2013). *Water governance in the Arab region managing scarcity and securing the future*. United Nations Development Programme, Regional Bureau for Arab States

UNESCO (United Nations Educational, Scientific and Cultural Organization). (2010). *UNESCO science report 2010*. Paris: UNESCO Publishing.

UNU (United Nations University). (2008). Proceedings of the international workshop on drinking water loss reduction: Developing capacity for applying solutions. Bonn: UNW-DPC.

UNU (United Nations University). (2009). Proceedings of the 2nd regional workshop on water loss reduction in water & sanitation utilities South East Europe Countries. Sofia.

Vickers, A. (2002). *Handbook of water use and conservation*. Amherst: Water Plow Press.

Wouter, S., & Steenburgen, F. (2002). *Ideas for water awareness campaign*. Stockholm: The Global Water Partnership.

WRC. (2016). *Policy brief on water harvesting in the Arab region*. Water Research Centre, Faculty of Engineering, University of Khartoum.

WWF. (2006). Fourth World Water Forum, Mexico City.

State of the Art and Future Applications of Desalination Technologies in the Middle East

Corrado Sommariva

Abstract Traditionally desalination has been associated to the Middle East and North Africa economies. However the availability and security of water supplies is today a growing concern and policy priority, both in the traditionally supply-constrained market of the Middle East and, increasingly, in other regions of the world. Water shortage is not only a phenomenon limited to the Middle East, and several large scale desalination projects have been awarded in other areas of the world.

Desalination volumes have nearly doubled since 2000 and it is expected to triple by 2020. Desalinated water supply has grown from 9.8 billion m^3/year in 2000 to 18.1 billion m^3/year in 2008, reflecting an 8% compound annual growth rate (CAGR). As water stress increases and desalination use expands outside of early-adopting areas like the Middle East, it is forecast that desalinated water volumes will reach 54 billion m^3/year in 2020.

Desalination is now used in more than 120 countries around the world. Several large scale projects demonstrated during the last 30 years that it is now technically and economically feasible to generate large volumes of water of suitable purity through the process of desalination of seawater, brackish water, and water reuse. In the past the cost for seawater desalination was below US$ 0.50/m^3 in many projects, however due to material cost increase the cost of desalination has subsequently increased to US$ 1–1.5/m^3. The present chapter aims at illustrating various state of the art desalination technologies adopted for main industrial projects as well as new emerging technologies aiming at a more sustainable generation of water.

The present chapter also makes a comparison among different technologies based on energy consumption and association with power generation.

Keywords Desalination technologies • Reverse osmosis • Brackish water

---

C. Sommariva (✉)
ILF Consulting Engineers, Al Reem Island, Sky Tower, Unit 36-01, 36th Floor, 73250, Abu Dhabi, United Arab Emirates
e-mail: corrado.sommariva@ilf.com

S. Murad et al. (eds.), *Water, Energy & Food Sustainability in the Middle East*,
DOI 10.1007/978-3-319-48920-9_6

Introduction

Desalination is defined as a water treatment process that removes salts from water. Desalination processes can be used in various applications including:

- Municipal desalting of brackish or seawater for drinking water production.
- Industrial and commercial applications for production of high-purity boiler feed water, process water, bottled water, and for zero discharge applications; and production of water for industries including the pharmaceutical, electronics, bio/medical, mining, power, petroleum, beverage, tourism, and pulp/paper industries.
- Treatment of wastewater for reuse applications.

Historically, large scale desalination has mainly been built in the Gulf region where there is no alternative for public water supply. Nearly half the world's desalinated water is produced in the Gulf.

Desalination has provided a reliable source of fresh water to the growing population and economies in the Arabian Gulf region for nearly half a century and it is the main method of supplying the region with potable water.

In the Middle East, desalination of seawater is the only new and economically sustainable source of fresh water. No real development in society or industry in the area would have been possible without the parallel development and implementation of desalination.

Today, desalination is a critical component of sustaining life and economy in the Middle East and Gulf region. Some economies in the Gulf rely on desalination to produce 90% or more of their drinking water, and the overall capacity installed in this region amounts to about 40% of the world's desalinated water capacity.

Today, desalination has become a solution to water problems well beyond the traditional arid areas of the Middle East and North Africa (MENA) and has now become an accepted water treatment process around the world as a price-competitive option for more communities.

This trend continues as the cost of desalination is decreasing with respect to the level of new supplies using conventional means. With the emergence of desalination as part of mainstream water resource management in many parts of the world, the industry has paid increasing attention to reducing energy consumption and increasing environmental responsibility in its practice.

Desalination Installed Capacity and Market

According to the 25th IDA/GWI Worldwide Desalting Plant Inventory, the total global contracted capacity reached 80.47 million cubic meters per day (m^3/d) as of August 2012, with 632 new plants added during the period from mid-2011 to August 2012. Global online desalination capacity was computed accordingly as of 74.8 million m^3/d, compared to 47.6 million m^3/day at the end of 2008.

There are more than 16,000 desalination plants around the world, and the period between 2010 and 2015 has seen a 57% increase in the capacity of desalination plants online.

The growth of the market for desalination reflects the fact that coastal communities are increasingly turning to the sea to meet their drinking water needs, while inland there is a tendency for groundwater to become increasingly brackish over time. Around 60% of desalination capacity treats seawater; the remainder treats brackish and less saline feedwater.

The largest thermal desalination plant in the world the 1,025,000 m^3/d Ras Al Khair project in Saudi Arabia, which uses both membrane and thermal technology.

The combination of lower cost desalination technologies and increased water scarcity has pushed for the establishment of big desalination plants outside the Gulf. The largest membrane desalination plant in the world – the 444,000 m^3/d Victoria Desalination Plant in Melbourne Australia – came online in 2012, but it will be soon surpassed by the 500,000 m^3/d Magtaa plant in Algeria, and the 510,000 m^3/d Soreq plant in Israel.

Desalination plants are generally long life assets. These plants are strategic assets requiring large investments. With an often abrupt industrial and demographic growth pattern in the GCC countries, desalination planning is very difficult (Sommariva et al. 2001a; Al Zahrani et al. 2004).

The planning of a desalination project is an extremely delicate process. Generally, the period between the inception of a project – the phase that includes a feasibility study, technical specifications, tendering process, etc. – and the first water production requires a minimum of three to a maximum of eight years. In this scenario, increasing capacity is a process that needs to be explored well in advance, and proper master planning is essential.

The investment in a desalination plant is generally amortised in the timeframe of 20–30 years, and therefore many technical parameters that may have an influence on today's decisions can drastically change during the asset's lifetime; these include the cost of energy, cost of chemicals, availability of steam and power, and level of O&M expenditures.

Continuing efforts to reduce the energy footprint is one of the most important aspects of the desalination technology. In particular, thermal desalination technologies have been traditionally less efficient from the energy footprint point of view than membrane processes, such as seawater reverse osmosis or "SWRO". The two major thermal desalination processes are Multi-effect Distillation (MED) and Multi-stage Flash (MSF).

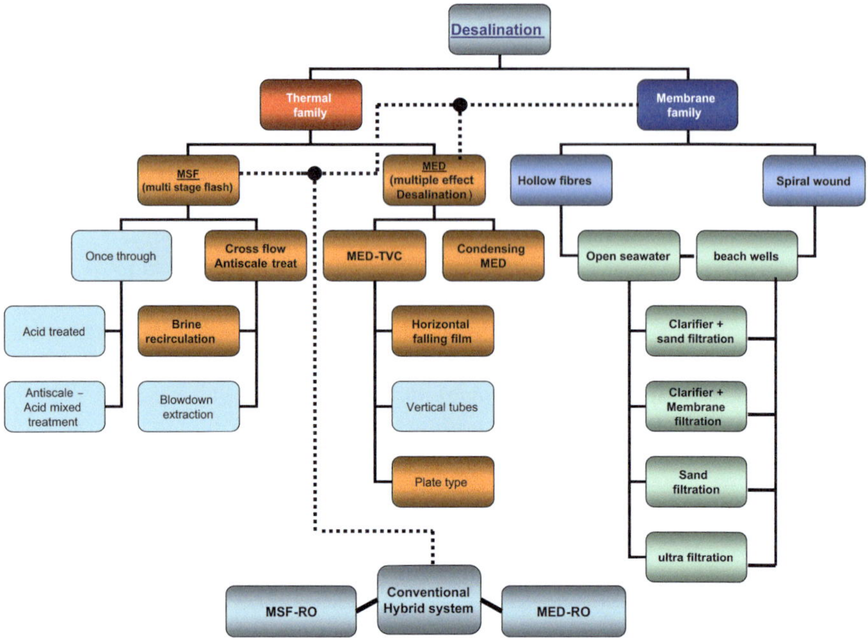

Fig. 1 Schematic illustration of desalination technology categories

Desalination State of the Art Technologies

The Family of Desalination Processes

The diagram shown in Fig. 1 illustrates schematically the family of desalination technologies that are adopted for large-scale production of water.

Commercially proven technologies include:

- Evaporative (distillation) processes
- Membrane (osmotic) processes

A combination of evaporative and membrane processes is the so-called

- Hybrid process (Sommariva and Awerbuch 2005)

Both evaporative and membrane technologies require a driving force (or driving potential) necessary for the separation process, hence they require the input of energy under various forms.

For evaporative processes the driving potential to achieve the separation of pure water from brine is the temperature difference between the hottest stage and the cooler stage, while for membrane processes pressure is the driving force.

The items in Fig. 1 marked with bold character and with colour effect are the so-called modern state of the art technologies. These technologies are available in the industry today and are operated commercially.

The remaining items indicate technologies and processes that have generally become obsolete and although still surviving in old installations not yet retired form operation, they are not specified for new projects tenders.

Thermal Family

The thermal desalination family is composed of evaporative processes.

These processes use thermal energy to produce distilled pure water from sea or brackish water. Evaporative processes rely on a phase change from liquid (in this case brine) to vapor. In this process only the water molecules pass to the vapour phase leaving the other constituents behind in the liquid. The two dominating systems that have evolved are Multi Stage Flash (MSF) and Multiple Effect Distillation (MED).

Multi Stage Flash Technology (MSF)

The MSF desalination plant is a process of compact modular construction and well-proven operational feedback in large scale industrial operation since the 1950s.

MSF technology is now considered a mature technology and its thermodynamic design continues to benefit from the operational feedback of installations that have been in operation for a long time and its performance was beyond the expectations that were projected at design stage.

The first MSF design was based on a long tube configuration with an acid dosing scale control method.

Figure 2 shows schematically the main difference between a long tube and a cross flow MSF plant. In the long tube configuration, flashing brine indicated as green arrows in the drawing flows parallel to the tube bundle which crosses each stage partition wall but in the opposite direction to the recirculating brine.

In the cross flow arrangement the tube bundle is generally located in the middle of the flash chamber and each stage tube bundle is connected by water boxes external to the vessel.

Whilst long tube arrangements present the advantage of having a large number of stages with relatively low additional costs, the expansion in size of this pattern is limited by the tube length and by the stage width.

However several long tubes MSF are still surviving in some plants in Europe and the Middle East. The scale control has been changed to mixed acid and antiscale dosing.

On the other hand, particularly with the increase of plant capacity and unit size, the cross flow design proved not to be competitive with respect to the cross flow

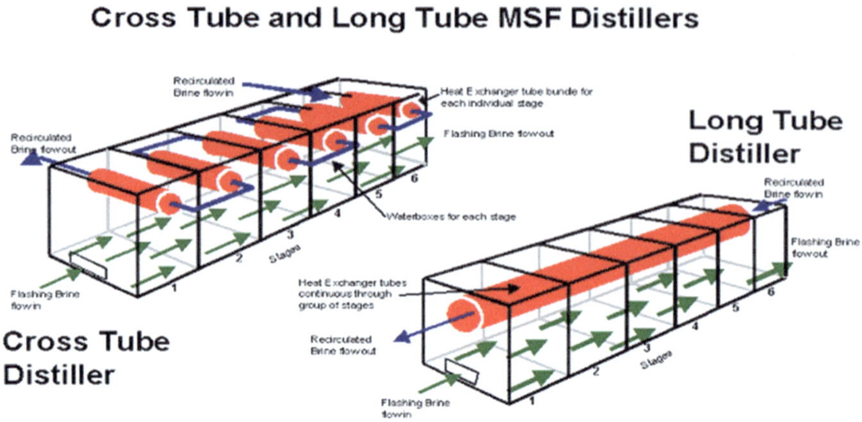

Fig. 2 MSF long tube and cross flow configuration

design and this configuration was abandoned since 1990 despite the lower specific power and heat consumption that the long tube design could offer compared to MSF (25th IDA/GWI Worldwide Desalting Plant Inventory; Sommariva 2010).

With the cross flow design the unit size has steadily increased over the years up to today's maximum of around 20 MIGD (3788 t/h) though larger units shall be considered for new projects as long as the tube length will not compromise the cross tube configuration.

The clear advantage offered by this configuration is the long life of the assets that has shown to reach 30 years and above using carbon steel material but will definitely exceed 40 years with modern material selections. Service and chemical costs are also relatively low.

Cross flow MSF distillers can be designed for a range of performance ratio (between water production and steam consumption), with a practical limit of about 11:1 (Wade et al. 1999).

Capital cost increases with performance ratio, due to the larger heat transfer surface area needed, and greater number of stages. The optimum value is usually in the range 7–9, depending on energy cost.

The operating temperature of the MSF technology is quite high and generally reaches a 110–112 °C top brine temperature. Steam is supplied from a steam turbine plant (at around 2.5 to 3.0 bar) and heat recovery boilers or a dedicated boiler plant (at around 15 to 20 bar) (Al Zahrani et al. 2004).

From the energy consumption and capital cost view point, MSF is the least efficient among the desalination processes. However in its present form of the multi stage/brine recirculation/cross tube arrangement, the technology has proven itself in long term practice to have solved the problems of reliability, scaling, chemical consumption, and unit size progression which severely limited earlier MED designs.

Until the early 2000s, MSF desalination was the main desalination technology in the GCC for plant unit size of 15 MIGD and greater and was generally combined with power generation plants. However, from the year 2010 onwards this technology became obsolete owing to high energy and CAPEX requirements.

Multiple Effect Technology (MED)

MED technology has been one of the first technologies adopted for seawater desalination.

This technology was initially very successful because of its ability to generate water with high performance ratio and the low operating temperature allowed moderate scale formation.

The first generation of MSF plant encountered severe scaling problems related to the high operating temperature and acid base scale control has always posed problems of handling and safety.

The scaling problem in MSF plants was gradually overcome by the development of sponge ball cleaning systems and specific anti-scale chemical products. Consequently, large capacity MSF plants were replaced by smaller installations using MED technology, capable of being installed in remote areas. Nowadays MED technology is the principal distillation alternative to MSF.

The main difference between MED and MSF is in the method of evaporation and heat transfer. In MED plants, evaporation occurs from a seawater film in contact with the heat transfer surface, whereas in MSF plants only convective heating of seawater occurs within the tubes and evaporation takes place from a flow of brine 'flashing' in each stage to produce vapour.

MED desalination plants are generally built in units of about 500 to 46,000 m^3/d (0.1 to 10 MIGD). A dramatic increase in the unit size has been observed in the last 5 years and this has allowed MED technology to gradually take over the market shares belonging to MSF technology.

The configuration in the MSF process is unavoidably rigid due to the fact that each stage shares a partition wall and main structural elements with adjacent stages. Such a constraint would not apply in an MED process which offers the possibility of modifying basic flow configuration in many more patterns as opposed to MSF technology.

The performance ratio between water production and steam consumption of straight MED plants is approximately equal to the number of effects minus 1 to 2. For a modern project, a 10:1 performance ratio plant is typical. Accordingly, the number of effects expected would be around 12. This is much lower than in equivalent MSF plant. The smaller number of effects in MED plants results in capital cost savings.

The thermal compression of the vapour (TVC) from the low temperature stages to the first effect of the MED process offers the possibility of increasing the performance ratio of the unit by recovering the latent heat of the steam that is

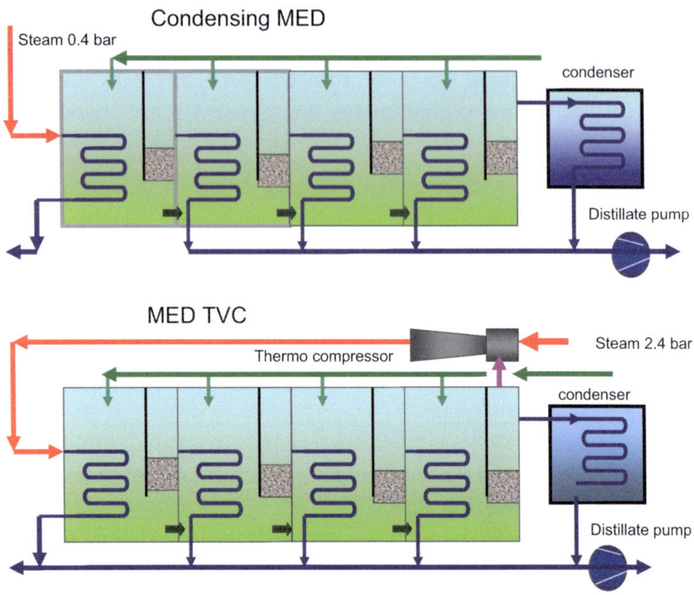

Fig. 3 Illustration of condensing MED unit and MED-TVC schematic process differences

thermo compressed to the first stage. On the other hand, this solution increases the steam extraction pressure and as the desalination unit is matched with a steam turbine the inherent power losses also increase.

The conceptual difference between a condensing MED process and a MED-TVC is illustrated schematically in the Fig. 3.

Internal power consumption of MED plants is lower than MSF, as there is no requirement to recirculate large quantities of brine. The combination of a higher performance ratio and lower power consumption results in lower overall energy costs.

Membrane Family

Membrane processes may be applied to a variety of raw water from brackish water to Gulf seawater and recently membrane processes have been successfully applied to the treatment of waste water.

The membrane acts a barrier between two phases that permits preferential and selective crossing of one or more kinds of fluid mixtures from one phase to the other (US Patent 3,133,132)

The driving forces for membrane separation may be different such as:

- difference in pressure,
- difference in concentration,
- difference in chemical potential.

Typically industrial reverse osmosis (RO) – ultrafiltration (UF) processes are pressure driven. In RO processes electric energy is used to pump seawater (or brackish water) through a series of semi permeable membranes to obtain a low salinity permeate as a product.

Thermal desalination membrane processes (with exception of membrane distillation currently applied only on small scale projects) do not rely on phase change but rather on the size and transport mobility of water molecules through a permeable membrane.

SWRO Technology

Reverse osmosis is used for the separation of fresh water from seawater or brackish water.

In 1963 Loeb and Sourirajan at the University of California, in Los Angeles (US Patent 3,133,132), developed the first synthetic RO membrane. In RO, permeate passes from the feed to the product side of the membrane when a pressure exceeding the osmotic pressure is applied. This 'reverses' the natural osmotic effect and concentrates salt ions into a waste concentrate stream. However, high pressure energy-intensive pumps (up to 60–70 bar) are required to drive the process.

By the 1970s larger scale commercial RO and ED/EDR systems began to be used more extensively. Initially, in brackish applications, RO had to compete against the now established electrodialysis (ED) technologies. Furthermore early RO was complicated and not always reliable.

The growth of RO was due to market standardization of the spiral wound membrane module, and the introduction of thin film composite (TFC) membranes to replace earlier cellulose acetate materials.

In the late 1970s and 1980s the development by Dow of the FilmTec TFC polyamide membrane brand resulted in process improvements including lower operating pressures, higher fluxes and higher salt rejection, which helped to reduce energy consumption and pumping pressures.

In the 1970s, the introduction of the isobaric energy recovery technology significantly reduced the operating costs of seawater RO.

By the 1980s, desalination technology had become a fully commercial enterprise and by the 1990s, the use of RO desalination technologies for municipal water supplies had become commonplace.

In this moment RO is used for:

- Desalination
- Industrial waste water treatment
- Food processing
- Production of ultra-pure water for electronic and food farms

The adoption of membranes for seawater desalination by reverse osmosis (SWRO) in the way the industry operates nowadays was put in practical use in the late 1970s. However the initial RO membrane modules were very expensive and they had very small capacity.

The membrane industry since then has continuously improved both in performance and cost, and RO became adopted in large sized plants.

SWRO membranes fall into two main categories, hollow fine fibre (HFF), and spiral wound (SW).

The use of HFF modules made from cellulose triacetate or aromatic polyamides is now limited exclusively to seawater desalination.

These modules incorporate the membrane around a central tube, and the feed solution is quite slow. As much as 40–50% of the feed may be removed as permeate in a single pass through the module.

Since then RO technology has made great progress in recent years, increasing in reliability and service factors and has become the technology of choice wherever there is a need for a stand alone desalination plant. The recent success of SWRO is derived by the lower sensible energy footprint compared to thermal desalination.

The application of this technology in the Middle East has been increasing lately and SWRO is recently taking over thermal desalination also in cogeneration plants.

In large scale SWRO plants, defined as those producing over 250,000 m^3/d of fresh water, it is clear that technology will continue to see a further reduction in energy costs per unit of desalinated water, due to more efficient energy recovery devices, the introduction of new membrane types that are more resistant to fouling, larger membrane housings, and faster flows in RO plants (Sommariva 2004).

Energy Requirements

Desalination plants are energy intensive and the significant increase in fuel-energy and material costs that was experienced in the years 2006 and 2007 had a dramatic impact on the capital and operational costs of desalination and power plants.

All seawater desalting processes – multi-stage flash (MSF), multi-effect distillation (MED), and SWRO – consume significant amounts of energy.

The energy input for membrane processes is provided by electric power that is required for the major process pumps and equipment. For "thermal desalination" the energy input is both provided by the power required for the process pumps and by the heat that is the driving force of the distillation process (Sommariva 2008; Sommariva et al. 2001b).

The heat requirements of thermally-driven processes are usually met for large installations through the development of co-generation plants, where combined power and water production is achieved using the steam extracted from the turbine at a suitable pressure to produce distilled water through an evaporation process.

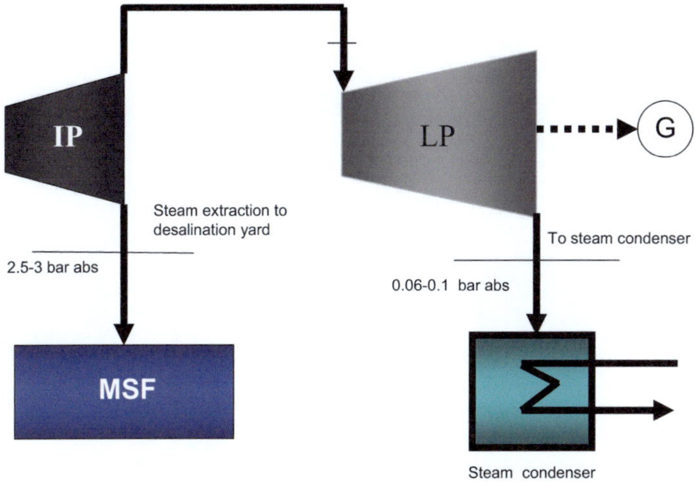

Fig. 4 Cogeneration of power and water: schematic diagram of condensing steam turbine with steam extraction feeding an MSF desalination system

The scheme that is applied is schematically indicated in Fig. 4 for a traditional condensing steam turbine configuration and in Fig. 5 for a combined cycle with back pressure steam turbine.

This process is largely energy intensive as the steam required to drive the desalination plant is extracted at a pressure of about 2–2.5 bars and could be utilised to produce substantial additional power in the steam turbine (Sommariva 2010).

The overall energy requirements according to the technology can be summarised in the Table 1. These values are based on the current state of art and include the energy requirement necessary not only for the desalination plant but also for its auxiliaries such as seawater intake, remineralization system, and potable water storage seawater chlorination.

Obviously, with reference to the second law of thermodynamic, it is impossible to compare precisely the heat and power on the same basis. Therefore the widely accepted method to align electric power and thermal energy input to the desalination plant is the reference cycle method.

Therefore the overall energy requirement has been compared with the reference cycle method. With this method, the energy associated to the steam extracted by the desalination plant is considered in terms of equivalent loss of electric power that would otherwise be rendered by the steam extracted in the power generation yard.

Table 1 summarizes overall electric energy consumption for the state of the art desalination technologies based on the current industrial configuration applicable.

As it can be seen from Table 1 after aligning the thermal and electric input to the desalination process, the difference in the energy input per unit of product water between thermal and membrane technologies is quite substantial.

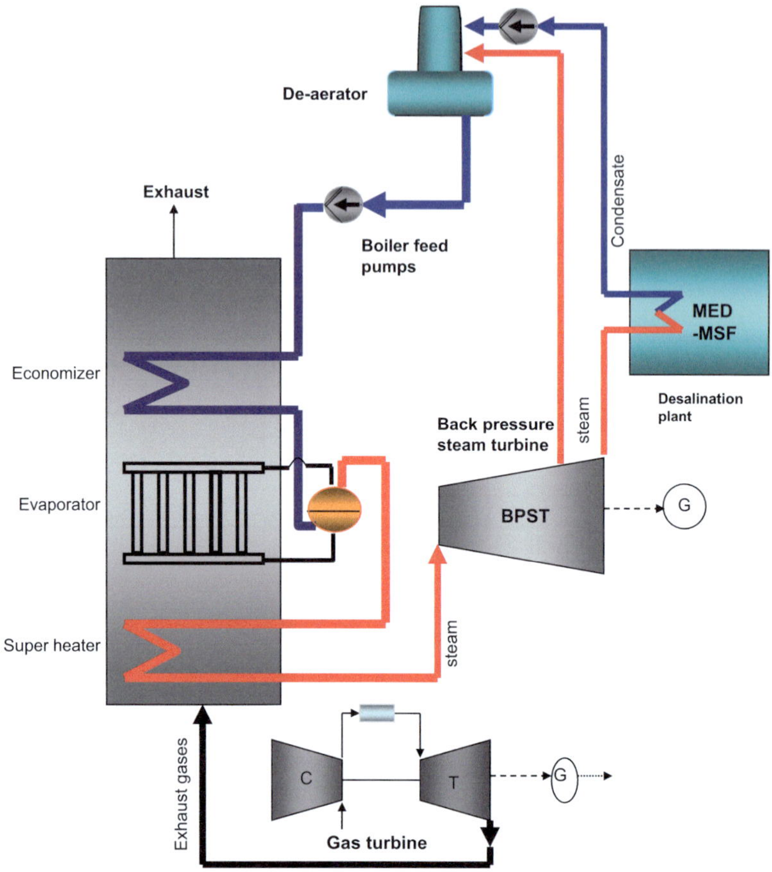

Fig. 5 Power generation combined cycle with bottoming MSF-MED schematic diagram

For general desalination projects the energy consumption of RO is considered to be lower than that for thermal processes such as MSF and MED. An exception to this is given by the low temperature condensing MED technology. This technology does not require steam for thermo-compressors and uses a 400 to 350 mbar(a) steam turbine exhaust to match the inlet temperature required for the MED units. With this configuration the thermodynamic losses are kept to a practical minimum. In this scenario MED provides a very similar equivalent power consumption to RO and MED power consumption improves. This concept was applied to relatively small plants and the main challenge is the large volume that is required for the high vapour specific volumes at low operating temperature.

However this project has particular circumstances whereby the energy consumption of MED is evaluated to be competitive with RO. These circumstances are:

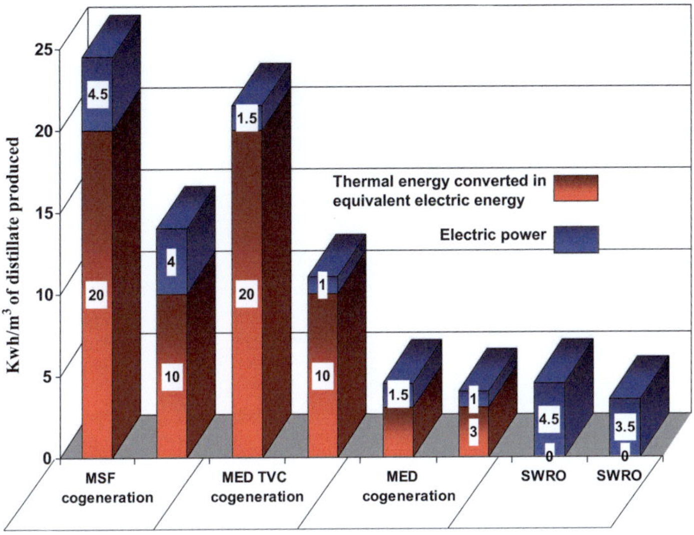

Fig. 6 Desalination technologies thermal and electric power specific energy consumption

This difference becomes even more significant as the efficiency of the power cycle decreases and the heat rate (i.e. how much fuel is burnt to produce a kW of power) decreases.

In addition to the values indicated above the difference in the power requirements can become even larger if the steam for the thermal desalination plant is generated through auxiliary boilers or with large supplementary firing that decrease the plant heat rate.

Nowadays this configuration is normally avoided at the planning and design stage. However there are still several "stand alone thermal desalination plants" – some of large capacity – still in operation.

Furthermore power plants are generally sized to meet the peak power requirement that occurs during the summer period. Steam requirements to desalination are designed to match this situation.

Unlike water demand, power demand drops dramatically in the winter season. This creates a mismatch between the steam available from the power cycle and steam actually necessary to produce the required water capacity. The additional steam needs therefore to be produced bypassing the power plant as schematically shown in Fig. 7.

The power loss for steam extraction in these operational scenarios is very high and can reach up to 40 kWh/m^3 of product water.

This very high energy input is the reason why recently, even in cogeneration projects in the Middle East, SWRO has been preferred to thermal technology.

In particular it has often been proven to be more convenient taking full advantage of the steam available from the power cycle and condensing it in the power

Table 1 Desalination technologies energy consumption thermal and electric power cogeneration

	Specific electric power	Specific heat consumption	Steam Extraction pressure	Thermal energy	Equivalent power loss	Total Energy requirements
	Kwh/m^3	kJ/kg	Bar abs	Thermal kwh/m^3	Electric kwh/m^3	kwh/m^3
SWRO(Mediterranean Sea)	3.5	0	N.A.	0	0	3.5
SWRO (Gulf)	4.5	0	N.A.	0	0	4.5
MSF	4–5	287	2.5–2.2	78	10–20	14–25
MED-TVC	1.0–1.5	287	2.5–2.2	78	10–20	11–21.5
MED	1.0–1.5	250	0.35–0.5	69	3	4–4.5

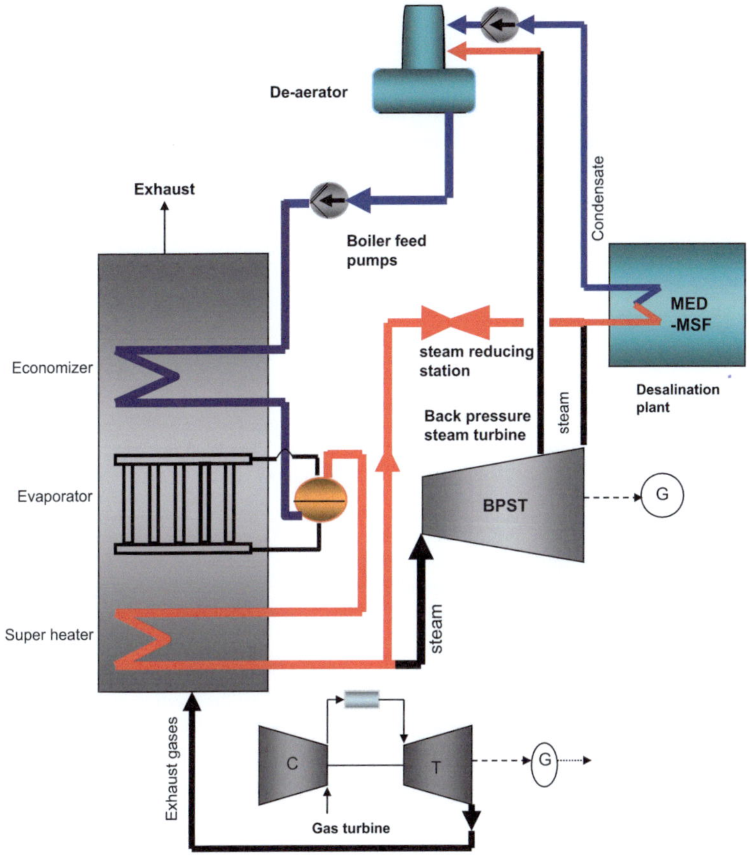

Fig. 7 Power generation combined cycle with bottoming MSF-MED schematic diagram with steam turbine isolation in winter conditions

plant condenser rather than extracting the steam to drive a thermal desalination process.

The additional power that is rendered by the steam turbine in this manner would be more than sufficient to drive a SWRO process and the electric output generated by the plant will be higher.

The viability of this solution that has been adopted in several cogeneration sites in the Middle East is generally subject to the seawater quality at the site and the extent of pre-treatment that are required for the SWRO process.

Although overall energy consumption of thermal desalination plants is greater than that of SWRO, a fair comparison would take into account that thermal desalination plants are generally installed in a co-generation scheme. That is because part of the energetic cost required to pump seawater to the thermal desalination plant would be saved by pumping seawater to the steam condenser which in turn condenses the steam coming from the back pressure of the condensing steam turbine. In this system, the energy required to produce an equivalent amount of potable water as an SWRO could be estimated in the range of 0.3–0.5 kwh/m^3 of product water. In a fair comparison of these technologies this energy amount could be credited to the electrical power demand of thermal technologies.

The energy consumption for membrane technologies is strictly related to the nature of the membranes that are employed for the desalination or purification system. New membranes are continuously developed with lower trans-membrane

Table 2 Innovative desalination technology development outlook

Thermal			
	Energy requirement		Energy optimisation Development outlook
Process	Thermal [kJ/kg]	Electric energy [kWh/m^3]	Notes
Low energy application to MED technology	200 Required at 70 °C in form of hot water or steam therefore at low exergy value	1.0–1.5	Relatively limited. However the thermal energy footprint could be reduced to 150 kj/kg.
LTD desalination[b]	250 kj/kg Required at 70 °C down to 50 °C in form of hot water or steam	0.8–3.0[a]	Potentially very high. However the thermal energy footprint could be reduced to 100 kj/kg.
Membrane distillation	300–400 kj/kg Required at 70 °C down to 50 °C in form of hot water or steam	1–2.0[a]	Potentially very high. However the thermal energy footprint could be reduced to 100 kj/kg with multistage installation and proper development of MD membranes
Forward Osmosis With associated thermal energy for draw solution separation	80–100 kj/kg Required at 90 °C in form of hot water or steam	2–3	Specific power consumption development outlook could decrease to 1–1.5 through the development of a dedicated FO membrane
Pressure retarded Osmosis	**80–100 kj/kg Required at 90 °C in form of hot water or steam**	**1–2**	**Tested in pilot application with waste water could potentially offer some substantial energy savings**
Forward Osmosis	0	4–4.9	The values of 4–4.9 have been actually tested and verified in the semi industrial plant operation the development outlook suggests that a further reduction to 3 kwh/m3 is feasible in the short term

[a]The specific electric consumption increases as the number of stages/effects increases
[b]*LTD* low temperature distillation

pressures and therefore lower specific power consumption. Furthermore more sophisticated energy recovery devices are being introduced in the market.

Innovative Desalination Technologies and Renewable Energy

In the past 5 years, there has been a strong drive towards the development of solar desalination. This trend has been moved forward by a generally more environmental and energy conscious approach to the power and desalination market in the region. The successful application of renewable energy in the region will depend on several factors, and primarily on the capacity of creating a strong platform of interest that involves both researchers and investors as well as the governments and includes a set of policies that can seriously promote this application further.

At the present time, combinations between renewable energy sources and desalination are implemented as pilot plant size applications or are still in the R&D phase. Therefore implemented capacities -with few exceptions- are relatively small and typically range between a few m^3 up to 100 m^3 per day.

Table 2 summarizes the innovative desalination technology development outlook for some of the most promising developing technologies.

The desalination industry is committed to a program to reduce the energy consumption in all major seawater desalination processes. The objectives are the development and implementation of new desalination technologies characterized by lower energy footprints that could be easily combined with renewable energy sources, and the implementation of more energy efficient solutions, retrofitted into existing desalination plants.

References

25th IDA/GWI Worldwide Desalting Plant Inventory.

Al Zahrani, S. G., Sommariva, C., & Ramachandran, V. (2004) Management issues and long term planning on large power and desalination plant: The Al Jubail example.

Sommariva, C. (2004, August) Desalination Management and Economics. Faversham House Group.

Sommariva, C. (2008, March). Utilisation of power plant waste heat steams to enhance efficiency in thermal desalination. *Desalination*, (1–3), 592–595.

Sommariva, C. (2010, April) Desalination and advanced water treatment. Economics and Financing ISBN 0–86689–069-6. Hopkinton: Miriam Balaban Desalination Publications.

Sommariva, C., & Awerbuch, L. (2005, September). Novel hybrid MED-MSF concept: Increasing efficiency in combined power and desalination plants. Sibngapore IDA conference.

Sommariva, C., Hogg, H., & Callister, K. (2001a). Forty years design life: the next target material selection and operating conditions in thermal desalination plants. *Desalination, 136*, 169–176. Amsterdam: Elsevier Science Publication.

Sommariva, C., Awerbuch, L., Hogg, H., Callister, K.. (2001b, October). Matching power generation and desalination by combining thermal and membrane process: the alternative to improve flexibility and performance. Accepted for publication in the proceedings of IDA international conference in Bahrain.

Sommariva, C., Hogg, H., & Callister, K. (2002, May 6–9) Maximum economic design life for desalination plant: the role of auxiliary equipment materials selection and specification in plant reliability. Sharm El Sheick, Conference on desalination strategies in south Mediterranean countries.

US Patent 3,133,132. High flow porous membranes for separating water from saline solutions.

Wade, N., Willis, J., & McSorley, J. (1999) The Taweelah A2 independent water and power project.

Part II
Energy

Sohail Murad

Hydrocarbon Fuels from Lignocellulose

John R. Regalbuto, Fahad Almalki, Qiuli Liu, Ritubarna Banerjee, Andrew Wong, and Jayson Keels

Abstract The main motivation for biofuels at present is to enable transportation fuels which do not contribute to global warming. Biofuels made from non-food biomass, collectively called lignocellulose, dramatically reduce the net carbon dioxide emissions from light and heavy duty vehicles. Lignocellulose consists of agricultural residue such as corn stover and sugarcane bagasse, waste from forest trimming, and energy crops such as switchgrass and short rotation poplar trees grown on marginally arable land with little irrigation or fertilizer.

In this chapter, catalytic and microbial routes for the conversion of lignocelluose into hydrocarbon biofuels will be reviewed, and the latest status of commercial development of the various routes will be reported. The biomass conversion routes and the order of the chapter is as follows: biomass gasification, pyrolysis, aqueous phase processing with inorganic catalysts and microbes, and biogas production via anaerobic digestion. There has not yet evolved a single dominant strategy for biomass conversion into fuel.

Keywords Biofuels • Lingo-cellulose • Hydrocarbons

Introduction

Biofuels can be made from non-food biomass, collectively called lignocellulose, consisting of agricultural residue such as corn stover and sugarcane bagasse, waste from forest trimming, and energy crops such as switchgrass and short rotation poplar trees grown on marginally arable land with little irrigation or fertilizer. In the past decade, research and development of lignocellulosic biofuels has evolved from efforts to produce "cellulosic ethanol", or ethanol derived from the cellulose fraction of lignocellulose (which also contains hemicellulose and lignin) to hydrocarbon biofuel production utilizing whole biomass (Regalbuto 2009, 2011).

This transition was sparked by a landmark paper in Chemical Reviews in 2006 titled "Synthesis of Transportation Fuels from Biomass: Chemistry, Catalysts and

J.R. Regalbuto (✉) • F. Almalki • Q. Liu • R. Banerjee • A. Wong • J. Keels
Department of Chemical Engineering, University of South Carolina, Columbia, SC, USA
e-mail: regalbuj@cec.sc.edu

S. Murad et al. (eds.), *Water, Energy & Food Sustainability in the Middle East*,
DOI 10.1007/978-3-319-48920-9_7

Engineering" (Huber et al. 2006a) which described a catalytic roadmap for the conversion of biomass into gasoline, diesel and jet fuels. This roadmap is depicted in Fig. 1. Feedstocks on the left are ordered relative to their abundance and cost; the most abundant and cheapest is lignocellulose shown in the upper left. Lesser abundant and more expensive feedstocks are the sugars and starches – corn and sugar cane – seen in the middle of the figure. The least abundant and most expensive lipid-based feedstocks from plant oil or animal fat, are shown at the bottom. Algae is included in this set of feedstock, and it is interesting to contrast the potential of algal and lignocellulosic feedstocks. Lipids derived from algae, like animal and plant lipids, are relatively easy to convert into fuels, but are hard to mass produce. On the other hand, lignocellulose very most abundant but, given its relative chemical dissimilarity (high oxygen content and polymeric structure), is more difficult to convert to hydrocarbon fuel. Both types of feedstocks will require novel chemical engineering solutions – mass production on the one hand and conversion on the other; algae is treated in a later chapter in this section and lignocellulose is the main focus of this one. Since it is not treated elsewhere in this text, a section on biogas conversion has also been included.

As seen in the upper section of Fig. 1, the three general routes for the conversion of lignocellulose to fuels are (1) gasification followed by Fischer-Tropsch synthesis, (2) pyrolysis, in various thermal or catalytic versions and a related process called hydrothermal liquefaction, and (3) aqueous phase processing. The latter are the newest routes of the roadmap as has been rapidly being advanced with inorganic catalysis, and via synthetic biology. Both of these routes require dissolved sugars, and with the appropriate catalysts or microbes, convert the sugar monomers into hydrocarbon fuel components. The advantage of gasification and pyrolysis compared to aqueous phase processing is the ability to process whole biomass including lignin, which is the most recalcitrant fraction of the biomass, but has the highest energy density. Processes utilizing lignin have a higher potential energy yield.

Biomass Gasification

Power generation based on fossil fuels is expected to produce 80% of the world's primary energy in 2040. The world's CO_2 emissions have increased by 44% from 2011 to 1993, which is increasing the earth's average temperature (http://biomassmagazine.com/articles/1674/solar-powered-biomass-gasification). The demand for energy is expected to grow by 20% in the next 25 years, especially in countries where the rising demand for energy comes from economic output and improved standard of living such as India and China (http://www.chevron.com/globalissues/energysupplydemand/).

This issue put great pressure on developing and improving energy produced from alternative fuels. Biomass is a potential and promising source of energy due to its nearly worldwide distribution and availability. If it is utilized at the same rate as of its growing, the produced energy would have zero CO_2 emission to the

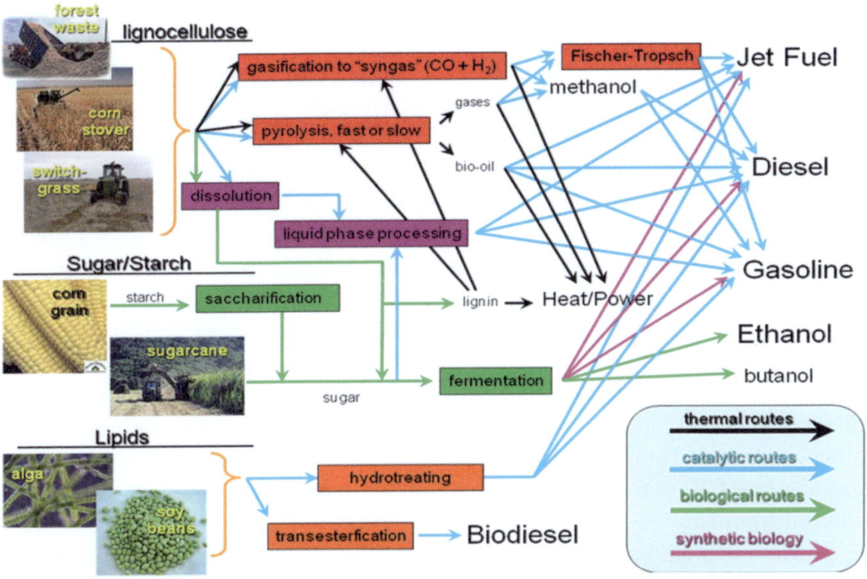

Fig. 1 Routes to lignocellulosic biofuels. Whole biomass is used for gasification and pyrolysis. New pathways include catalytic versions of gasification and pyrolysis, and aqueous phase processing of dissolved carbohydrates. Microbial routes (via synthetic biology) to hydrocarbons are also shown

environment. It is estimated that bioenergy could contribute up to a third of the world's primary energy supply by 2050. Typically, fossil fuels are mainly utilized to generate electricity and it is estimated that electricity will consume 93% of fossil fuels by 2040 (http://www.chevron.com/globalissues/energysupplydemand/). Due to this demand, great attention is focused on increasing the power generation using alternative fuels.

Biomass gasification (BG) is one of the most efficient thermochemical technologies compared to biomass combustion and pyrolysis, which converts biomass to a mixture of gases consisting of hydrogen (H_2), carbon monoxide (CO), carbon dioxide (CO_2), methane (CH_4) and water (Ausilio et al. 2009). In an ideal gasification process, the thermo-chemical conversion of biomass feedstock produces just CO and H_2 (syngas), but practically the producer gas consists of many gases which could be burned in boiler to produce heat, in internal combustion gas engine for combined heat and electricity (CHP) process, or in the case of mainly syngas compositions, it can be converted to synthetic natural gas (SNG), synthetic liquid fuels through Fischer-Tropsch synthesis (FTS), or green chemicals (Jared and John 2002).

BG is considered an extension of pyrolysis involving many heterogeneous and homogenous reactions.

$$C_{(S)}+O_2 \leftrightarrow CO_2 \qquad C_{(S)}+H_2O \leftrightarrow CO_2$$
$$C_{(S)}+CO_2 \leftrightarrow 2CO \qquad C_{(S)}+2H_2 \leftrightarrow CH_4$$
$$CO + H_2O \leftrightarrow CO_2 + CH_4 \qquad H_2O + CH_4 \leftrightarrow CO + H_2O$$
$$Tar \rightarrow CO + CH_4 + C$$

Basically BG takes place in four stages. First, is a drying process to reduce the moisture content. Second is a pyrolysis process where the tar and volatiles are driven off. Third, as the CO_2, H_2O, N_2 and the surplus of O_2 are produced from a complete combustion of biomass. Fourth, asvCO, H_2 and traces of light hydrocarbons are produced from reduction reaction (Younes and Mohammed 2013).

A biomass feedstock is mixed with gasifying agent at a temperature greater than 800 °C to produce syngas (producer gas) with varying degrees of caloric value (CV) (heating value) based on the gasifying agent (oxygen, air or steam), type of biomass feedstock, gasifier configuration, gasifying agent/biomass ratio and operating conditions. Typically, air is the most available and cheap agent, but producer gas has low CV, 4–6 MJ/Nm^3, due to dilution with excess amounts of nitrogen sourced from air.

A controlled amount of oxygen is a more efficient way than air to increase the producer gas CV to 10–18 MJ/Nm^3, which is more suitable for pipeline distribution and syngas conversion applications, but this option is more costly compared to air. A cheap and available gasifying agent such as steam can be used to increase the CV and varying steam/biomass ratio can affect the gas compositions in producer gas (Bocci et al. 2014).

For whichever gasifying agent is used, the amount of oxygen is controlled to maintain partial oxidation of biomass, unlike pyrolysis when the feed is heated in an inert environment, or total combustion when it is reacted with excess oxygen. In general, using local sources of biomass in gasification technologies could reduce CO_2 emission and reduce dependence on fossil fuels (Richard and Geoffrey 2008). Another feature of BG is the abundance of biomass material feedstocks available; these include forest and agriculture residues, municipal organic waste and animal waste (Nourredine et al. 2015).

There are many challenges in commercializing biomass-based power generation which currently limits its use at large scale. Supply chain logistics has many obstacles to sustain and optimize the supply of biomass associated with harvesting, collection, refining and transportation of biomass. The storage of biomass with moisture content higher than 50% is not feasible (Mohammad 2014).

In order to operate a turbine or internal combustion engine, the composition of producer gas has to be optimize by adjusting parameters such as the gasifier type; the fixed bed gasifier is a simple reactor which can be configured in two types, updraft gasifier (countercurrent) which is a downward motion of biomass where char moves down to be gasified and vapor moves up to pyrolyze in the hot gas; this results in low ash content and high tar concentration in producer gas. In a downdraft gasifier, biomass is introduced a lower throat where all reaction products in turbulently mixed to increase the residence time and consequently increase the conversion of char, but this configuration has low overall energy efficiency (http://www.eai.in/ref/ae/bio/bgt/type/fixed_bed_gasifier.html).

On the other hand, a fluidized bed gasifier using either an inert or catalytic materials to transport heat and mass, features even temperature throughout the reactor and provides high rates of heat transfer between materials, yielding low tar content in producer gas. However this type of gasifier has limited tolerance to biomass moisture content, may suffer from bed sintering especially with high ash content, and is more complex to operate (Marcin et al. 2011).

Another biomass gasifier configuration is entrained flow, where the size of biomass particles are very small to let more surface contact with gas in a short time. The conversion of this type is very high with low ash and tar content in product gas. The gasifying agent and biomass are fed co-currently at temperatures higher than 1000 °C (Review of technology for the gasification of biomass and wastes 2009).

Moisture content in biomass feedstock is one of the main challenges in biomass gasification technology. Biomass feed with moisture content greater than 40% decreases the thermal efficiency of the system due to the heat loss from heating and vaporing moisture. Therefore, drying biomass feed is important for high thermal efficiency of biomass gasification, but this is a costly and energy consuming pretreatment (Hitesh et al. 2014).

Based on biomass feedstock content, producer gas can have impurities such as aromatic hydrocarbon species (tar), particulate matter, sulfur compounds, ammonia, hydrochloric acids, etc. The producer gas can be used in many applications; for each there is a limit for the impurities acceptable for the operation. In this regard, many methods have been applied for gas cleaning. Different filtration methods are applied to remove tar from producer gas, but the main issues with this physical treatment are blocking and fouling from the sticky particulate tar (Mohammad 2013). In thermal treatment, of tar, high temperature is required for tar cracking which requires equipment constructed with expensive materials and an intense cooling system. Catalytic hot gas cleaning is one of the most attractive gas treatments and many catalysts have studied for tar and ammonia decomposition and that could meet the stringent requirements for many applications of using producer gas in biomass gasification system. Catalysts can decompose tar at the gasification operating temperature so no further cooling or heating is required for the catalytic gasification system. Noble metals such as Ru/CeO_2 showed higher activity to convert tar to CO and H_2. Less expensive nickel based catalysts have shown good activity for tar decomposition (Mohammad 2014).

The biomass gasification power plant in Vassa, Finland (Fig. 2) is considered the world's largest biomass gasification plant supplied by **Mesto** and was officially opened March, 11, 2013. Mesto's technology includes fuel handling, large scale dryer, circulating fluidized bed gasifier, DNA automation system and a coal boiler. Up to 40% of coal was replaced by biomass feed which is mainly obtained from forest residue wood. It was operated at a capacity of 140 MW as planned (http://www.metso.com/news/2013/3/metso-supplied-worlds-largest-biomass-gasification-plant-inaugurated-in-finland/).

Fig. 2 World's largest biomass gasification plant, Vaskiluodon Voima Oy in Vaasa, Finland (http://www.metso.com/news/2013/3/metso-supplied-worlds-largest-biomass-gasification-plant-inaugurated-in-finland/)

The **City of Covington Waste-to-Energy Gasification Plant** in Covington, Tennessee supplied by PHG Energy of Nashville was designed to convert wood trimmings and sewer sludge to electricity and started operation in September 2013. This unit is shown in Fig. 3. A downdraft gasifier with 12 ton per day capacity is used to produce gas with 6 M Btu per hour and consumes 10 tons of wood waste per day and 2 tons per day of sewer sledge, which previously cost the city \$30 per ton. The system includes wood chipping, material handling, gasification, thermal oxidizer and oil heater (http://www.phgenergy.com/case-study/covington-tenn).

In May 2011, a biomass gasification system started the operation with a capacity of 4.4 MW and it was design by Vancouver-based **Nesterra System Corporation** in University of Northern British Columbia (http://www.unbc.ca/green/energy/bioenergy-plant). The biomass gasification system (Fig. 4) consumes 4000 bone dry tonnes of hog fuels per year which is sawmill residue supplied from Lakeland Mills in Prince Gorge. The updraft gasifier heats water with gasified sawmill residue for campus buildings and was distributed through the existing hot water system. This has offset about 85% of fossil fuel which had been used to heat campus buildings.

The system includes the conveyer of biomass feed, gasifier with controlled amount of oxygen to covert the biomass to syngas, the oxidizer to combust syngas at a temperature 1200 °C, and the boiler to produce hot water. The UNBC gasification system successfully reduces the emission of the particulates by using Electrostatic Precipitator (ESP) and it was recognized by Third-Party testing of the Bioenergy Plant as among the lowest emissions of any biofuel plants in North America, producing emissions 18 times lower than typical bioenergy system. The

Fig. 3 City of Covington waste-to-Energy gasification plant in Covington, TN (http://www.phgenergy.com/case-study/covington-tenn)

Fig. 4 Bioenergy plant of University of Northern British Columbia, Canada (http://www.unbc.ca/green/energy/bioenergy-plant)

Fig. 5 Plasma biomass gasification plant in Wuhan, Hubei, China (http://www.westinghouse-plasma.com/projects/)

bioenergy plant won the Prestigious Canadian Green Building Award in June 2014, presented by Canadian Green Building Council.

Many researchers have been studied the promotion and integration of gasification process to improve the efficiency, increase the purity or reduce the cost of the technology. One of the advanced gasification system is Plasma Gasification. The high temperature 5000 K produced by ionization of gas molecules using electric discharge enables the decomposition of any organic compound to its elemental molecules. The high heating value of producer gas can be achieved by plasma gasification, but the main drawbacks of this technology are low efficiency and high cost due to energy consumption of the plasma (Steffen and Pier 2015).

150 tons per day of biomass is converting by **plasma gasification plant in Wuhan, Hubei**, China, supplied from Westinghouse Plasma Corporation and commissioned in first quarter of 2013 (Fig. 5). The plasma gasifier processes biomass to generate power and liquid chemicals through Fischer Tropsch (http://www.westinghouse-plasma.com/projects/).

Another advanced method is supercritical water gasification where the water at temperature and pressure above 372.12 °C and 22.12 MPa respectively has high solubility for organic materials and gases, therefore dissolves more of these materials and gases. As a result the main components of the producer gas are mainly methane and hydrogen. Consequently, supercritical water gasification is more tolerant to moisture content of biomass feedstock and has very low tar content in producer gas due to the high dissolving rate of organic materials (Steffen and Pier 2015).

Despites biomass gasification challenges, biomass resources are considered to be the fourth largest energy source in the world, able to provide up to 14% of the world's energy demand (World Energy Resources 2013). In addition, the gasification technology, as has been proved, is the most efficient and promising thermochemical process to convert biomass to heat, power, and chemicals. But more knowledge, improvements, and new ideas are necessary for this technology to become commercialized for cleaner and safer sources of energy.

Pyrolysis

Introduction

Pyrolysis is thermal decomposition occurring in the absence of oxygen (Huber et al. 2006b; Mohan et al. 2006). Under pyrolysis conditions, thermal energy cleaves chemical bonds of the original marco-polymeric cellulose, hemicellulose and lignin to produce mostly oxygenated molecular fragments of the starting biomass. These fragments have molecular weights ranging from 2 up to 300–400 (Huber et al. 2006b; Mohan et al. 2006). The lower molecular weight compounds remain as permanent gases at ambient temperature while the majority of compounds condense to a dark brown viscous liquid (bio-oil), which is a potential alternative of fossil fuels with no net release of carbon dioxide and low sulfur content (Huber et al. 2006b; Mohan et al. 2006; Czernik and Bridgwater 2004). Char is also formed in the pyrolysis process.

Biomass is a mixture of hemicellulose, cellulose, lignin which pyrolyse at different rates and by different mechanisms (Czernik and Bridgwater 2004). The rate and extent of decomposition of each of these components depends on the process parameters, such as temperature and heating rate and vapor residence time etc. As illustrated in Fig. 6, product distribution dramatically alters when different techniques and conditions are applied.

Lower process temperatures (<400 °C) and longer vapor residence times (20 s) favors the production of char, which is known as charcoal manufacture from biomass, while higher temperatures (>800 °C) and longer vapor residence times favors the gas formation. The highest liquid yield is obtained with fast or flash pyrolysis, where the biomass is heated up very fast (1000 °C/s) to a moderate temperature (500 °C) during very short vapor residence time (<2 s). Therefore fast pyrolysis is widely used to give high yield of liquid products in biomass conversion (Mohan et al. 2006).

For fast pyrolysis, a wide range of different reactor configurations have been investigated. The most important reactor types are bubbling fluidized bed, circulating fluidized bed, rotating cone, and ablative and entrained flow reactors. Circulating fluidized beds are commonly applied in biomass pyrolysis in both pilot plants and commercially; these are analogues of Fluid Catalytic Cracking, or FCC, a very common and mature process used in most refineries around the world.

Catalytic fast pyrolysis (CFP) of biomass has become dominant in both academics and industry. Significant attention has been paid to zeolites and zeolite-like materials for biomass CFP (Galadima and Muraza 2015; Carlson et al. 2008; Cheng

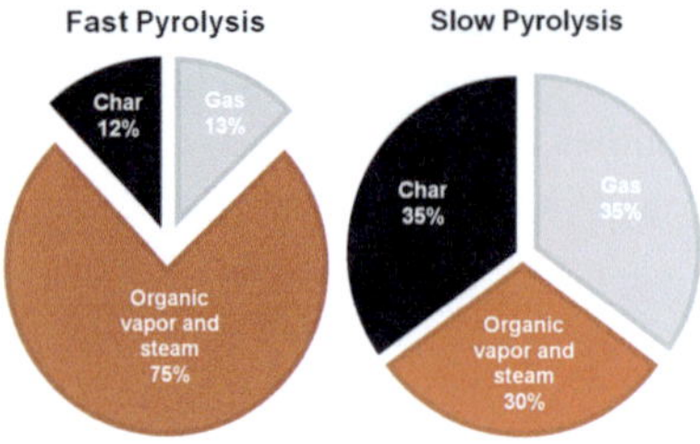

Reactor design	Organization/Company
Fluidized bed	Dynamotive, Canada
Ablative	Pytec, Germany
Vacuum	Pyrovac, Canada
Rotating cone	Biomass Technology Group, Netherlands
Circulating fluidized bed	Envergent (UOP and Ensyn), US and Canada
	Cool Planet, US
	Anellotech, US
	VTT, Finland
Auger transported bed	Bioliq at KIT

Fig. 6 Products distribution in fast and slow pyrolysis (*left*) and reactor configuration in pyrolysis technology (*right*) (Czernik and Bridgwater 2004)

et al. 2012). Highly acidic zeolites have proven to be quite effective in reducing oxygen levels in bio-oil (Galadima and Muraza 2015; Carlson et al. 2008; Cheng et al. 2012); however, the deoxygenation is accompanied by a large decrease in the amount of bio-oil produced and by increases in aromatic hydrocarbons and coke. In Ruan's work, SiO_2 were deposited onto the external surface of HZSM-5 by vapor deposition to decrease the acidity, and the the yield of coking decreased as the loading of SiO_2 increased (Zhang et al. 2015a). Supported noble metal catalysts are also applied in CFB. It is reported that 4-ethyl phenol (4-EP) was selectively produced in both ex-situ and in-situ CFP process of bagasse catalyzed by Pd/SBA-15 (Zhang et al. 2015b). Investigations on products distribution are also performed. Besides the feedstock sources and pyrolysis conditions (Huber et al. 2006b; Mohan et al. 2006), minerals content in biomass is another crucial factor (Collarda et al. 2015): iron- and nickel-impregnated cellulose were prepared to examine the influence of the metal on the yield and composition of fast pyrolysis products and results showed that metal impregnation led to a decrease of tar and CO yields balanced by an increase of char, H_2O and CO_2 yields. These findings open promising perspectives to optimize the production of fuels and chemicals from biomass.

Over the past a few decades, biomass pyrolysis have been investigated and explored by many companies to yield various products, from bio-oil to aromatics which have been commercialized, even though still at an early stage. It is often integrated with bio-oil upgrading to produce high value transportation oil which is the ultimate goal of biomass conversion.

Figure 7 shows a typical biomass pyrolysis process, patented by **Ensyn** and named as rapid thermal processing (RTP ™) (Ensyn 2012). In RTP pyrolysis, biomass feedstock (woodchips, agricultural residues etc.) are first pretreated in feed bin to remove moisture and grind to 2–6 mm to ensure rapid heat transfer. Then the prepared biomass is quickly broken down into pyrolytic vapor by a rapidly flowing stream of hot sand in the conversion unit. With controlled temperature and residence time, yield of solids and gaseous products are suppressed. Unconverted biomass becomes charcoal which will be separated from vapors in a cyclonic separator along with hot sands. The solid material is transferred to the sand reheater, charcoal is combusted in a bubbling bed to reheat the sand. After charcoal is burned

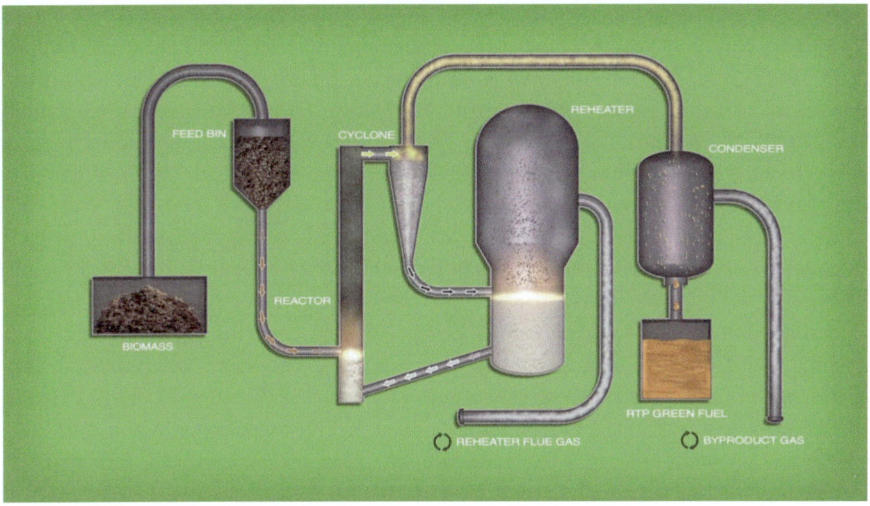

Fig. 7 Process flow of RTP technology at Ensyn (2012)

off, clean hot sand is transferred from the sand reheater back to the conversion unit. Meanwhile, pyrolytic vapor travels to the bio-oil collection system and then quickly quenched in the bio-oil collection system. The by-product gases are utilized to dry the raw materials during the pretreatment. The RTP process yields approximately 65–80% liquid, with 12–16% each of char and combustible gas. Instead of hot sand, catalysts can be circulated in catalytic biomass pyrolysis processes, and spent catalysts are regenerated in via the cyclone and recycled in the conversion unit.

In 2011, **UOP** began construction in Hawaii of a biofuels demonstration unit that will convert forest residuals, algae and other cellulosic biomass into green transportation fuels (Honeywell International Inc 2015). The project is part of the DOE's efforts to help spur the creation of the domestic biofuel industry, drive domestic job creation and reduce U.S. dependence on foreign oil. The project will also support the Hawaii Clean Energy Initiative goal to achieve 70% clean energy by the year 2030. The demonstration unit will utilize the RTP® rapid thermal processing technology to rapidly convert biomass into a liquid biofuel, followed by upgrading to green transportation fuels. RTP technology and equipment were offered by **Envergent Technologies LLC**, a joint venture business of UOP and Ensyn. An RTP unit integrated with product stabilization unit has been constructed in Kapolei and completed 30 days of processing with a capacity of 168 GDP pyoil, planning to operate 70 days in total by the end of this project; a 30 day commercial trial has been accomplished at a US refinery, which demonstrated a full pathway of biomass conversion to transportation fuels as well as its technology viability.

Renewable fuel oil (RFO), produced by RTP process, has been supplied in several locations in US. In 2015, Ensyn signed a 5 year contract of RFO supply (2,500,000 gallons of RFO per year) with Youngstown Thermal, Ohio, for district heating. Youngstown Thermal will use the RFO to displace up to 50% of its fuel needs in their natural gas fueled boiler. In 2014, Ensyn Fuels signed renewable long-term contracts with Memorial Hospital and Valley Regional Hospital, both of New Hampshire, for the supply of RFO for heating purposes. This contract allows Memorial to fully displace their petroleum heating fuels with Ensyn's renewable fuel, reducing Memorial's greenhouse gases from heating fuels by approximately 85%. In addition, adoption of Ensyn's RFO™ provides Memorial with substantial cost savings.

Ensyn is expanding its cooperation worldwide. A 20 million gallon/year project is under development in Aracruz, Brazil by Ensyn and Fibria Celulose SA, Ensyn's partner for RFO production in Brazil, and more in Southeast Asia are in progress.

Cool Planet's technology (Cool Planet Energy Systems 2015) combines a mechanical front-end that processes the biomass with heat and pressure into gases, with proprietary catalysts that convert those gases directly into Cool Planet's green fuels. Other than renewable fuels, biochar is also achieved which is currently produced commercially as a soil amendment. Fuel production has been demonstrated in pilot facilities in Camarillo, CA. In 2014, Cool Planet was awarded a $91 million loan from USDA, which will enable construction of the full facility, scheduled to come online in 2016 and be fully operational in 2017 at the Port of Alexandria, Louisiana. The commercial facility aims to process 90,000 tons of pine chips per year.

Anellotech's core technology, Thermo Catalytic Biomass Conversion (Bio-TCAT™) (Anellotech, Inc 2015), is based on Prof. George Huber's previous research at the University of Massachusetts-Amherst. By taking non-food biomass as feedstock, bio-based chemicals such as benzene, toluene and xylenes (bio-BTX) are produced in a clean and renewable way. BTX are widely used to make a wide array of industrial products, plastic bottles, clothing, automotive parts, electronic products, home and construction materials ect. After the start-up of its Pearl River Pilot Plant in 2014, Anellotech has allied with Johnson Matthey Process Technologies and IFP Energies. In November 2015, Anellotech received an equity investment of $7 million from a multinational corporate investor to co-develop and commercialize the new technology for the low cost production of bio-aromatics.

VTT, a Finnish company, provides a technology to convert biomaterials to liquid biofuels using fast pyrolysis (VTT Technical Research Center of Finland LTD 2015). The advantage of the VTT technology is that it is well suited for integration to existing power plants, and it also makes possible distributed production of high value products. It also offers a potential for phased construction of biofuel capacity. In such a case, upgrading of the primary liquid to transportation fuels and aromatics is carried out at a central refinery. VTT participated in the commercial development of the integrated pyrolysis with **Valmet, Fortum** and **UPM** during 2008–2014, in which a total of four fast pyrolysis pilot plants were operated. For Fortum Power and Heat Oy, VTT helped to develop the world's first integrated (pyrolysis) bio oil production plant. In 2013, bio-oil plant was integrated to Joensuu Combined Heat and Power (CHP), one of four CHPs Fortum has in Finland. The fast pyrolysis technology-based bio-oil plant is the first of its kind in

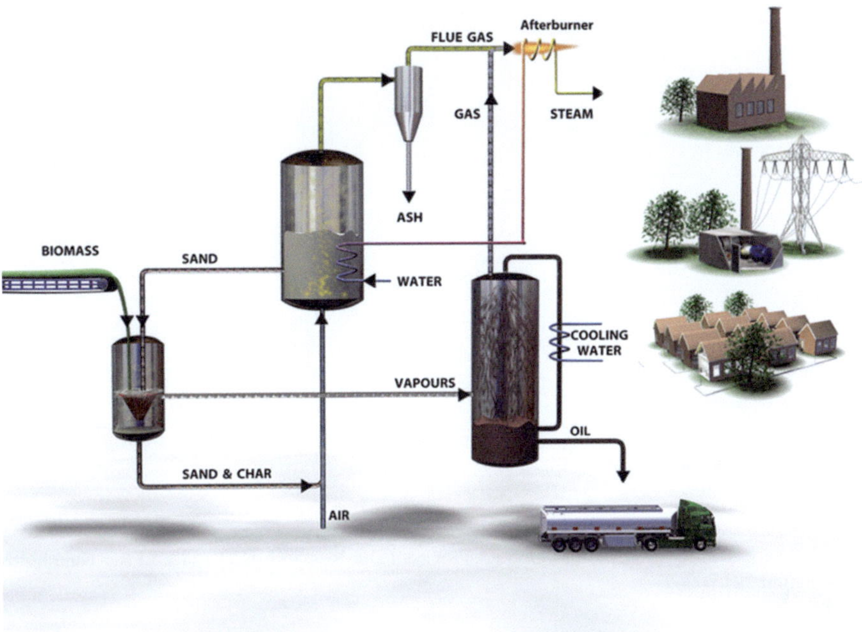

Fig. 8 RCR flash pyrolysis process flow diagram (BTG Biomass Technology Group and BlueBear 2015a)

the world on an industrial scale. The bio-oil plant was planned to be in full production phase in 2015.

BTG-BTL is a daughter company of **BTG**. Their technology, patented as RCR, is based on a Rotating Cone Reactor (BTG Biomass Technology Group and BlueBear 2015a) (see Fig. 8). In 2005, BTG installed a 2 t/h production plant in Malaysia. Up to now, over 1000 tons of oil have been produced and co-fired, replacing conventional diesel in a waste disposal system located 300 km from site.

In January 2014, **Empyro BV** has started construction of its pyrolysis oil production facility in Hengelo, the Netherlands (BTG Biomass Technology Group and BlueBear 2015b). Start-up of the installation has commenced early 2015 and production is gradually being increased which is considered as a success of the Empyro project as the first leap towards full-scale fast pyrolysis technology commercialization.

The **Karlsruhe Institute of Technology** (KIT) bioliq® process integrates fast pyrolysis of biomass with high-pressure entrained flow gasification, hot gas cleaning, and synthesis of synthetic fuels (The Research University in the Helmholtz Association 2015). A pilot plant funded by the federation, state, and EU (EUR 64 million) is in operation (Fig. 9).

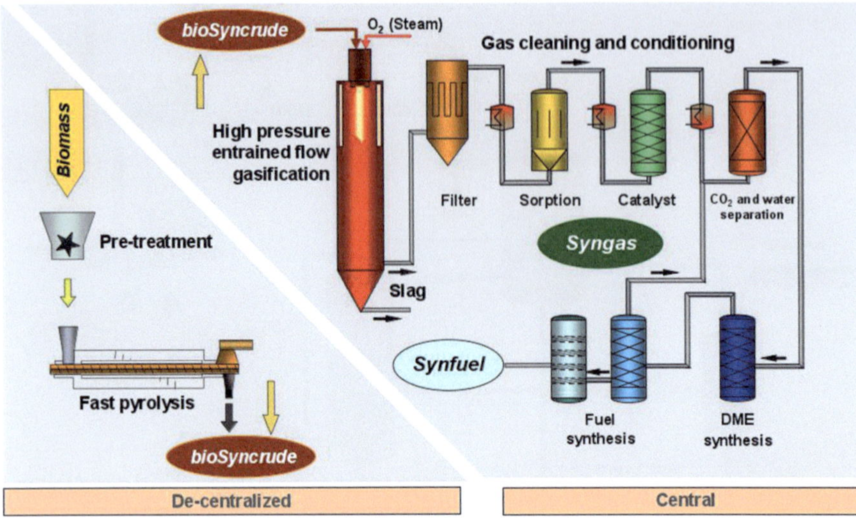

Fig. 9 Process Flow Chart of the bioliq® Process (The Research University in the Helmholtz Association 2015)

Aqueous Phase Processing – Inorganic Catalysis

Biomass is one of the most reliable energy sources at hand that promises minimum environmental consequences and has the potential to displace petroleum in the production of chemicals and liquid transportation fuels (Carlos Serrano-Ruiz et al. 2011). Hydrogen production from biomass-derived feedstock is being considered as a valuable and environment friendly alternative to fossil fuels owing to decreasing fossil fuel reserves, rising oil prices and a considerable decrease in greenhouse gas emissions. It currently accounts for about 7–11% of the world's energy needs (Demirbas et al. 2009; Balat 2008, 2009). However, the processing of biomass-derived feedstock requires unique reaction conditions and as such, there are a wide variety of methods catering to this need. Hydrogen technologies include gasification, fermentation, microbial technologies and aqueous phase processing. Gasification, fermentation and microbial synthesis have their own disadvantages that include reduced selectivity for hydrogen at high temperatures, high energy costs and other gaps in the research stages that need to be filled (Wei et al. 2014). However, biomass derived molecules possess a high oxygen content that gives them low volatility, high solubility in water, high reactivity and low thermal stability, properties that favor the processing of these resources by catalytic aqueous-phase technologies at moderate temperatures.

Aqueous Phase Reforming (APR) was first introduced by Dumesic and co-workers in 2002 (Huber and Dumesic 2006; Davda et al. 2005). These aqueous phase processes might be used in an integrated biorefinery to produce a range of fuels, as shown in Fig. 10. APR process involves chemical catalytic transformations

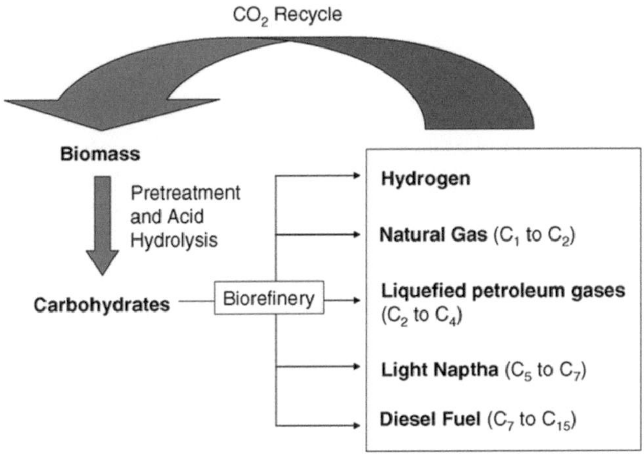

Fig. 10 An integrated biorefinery for conversion of carbohydrates to fuels by aqueous-phase processing (Huber and Dumesic 2006)

of biomass-derived oxygenated feedstocks (ranging from C1 to C15 sugars and sugar alcohols) in the liquid phase to chemicals and fuels. The key reactions involved in the processing of biomass are hydrolysis, dehydration, isomerization, aldol condensation, reforming, hydrogenation, and oxidation. Interests in APR has grown over time since it has been seen that producing hydrogen by APR of oxygenated organics provides several advantages. The main advantages are: (i) Reduction in energy consumption for the process due to elimination of the vaporization step (ii) Decrease in side/parallel reactions due to the low operation temperatures, typically about 500 K (iii) operational pressure, typically between 15 and 50 atm which allows easy hydrogen purification by membranes or other adsorption techniques (iv) can take place in a single chemical reactor system involving various reversible chemical reactions like water-gas shift (WGS) reaction, dehydration, carbon bond cleavage, etc.

The catalytic pathway for the production of H_2 and CO_2 by APR of oxygenated involves cleavage of C–C bonds as well as C–H and/or O–H bonds to form adsorbed species on the surface of the catalyst. Cleavage of these bonds occurs readily over Group VIII metals, such as Pt, Ru, Pd and Rh. However, the adsorbed CO species must be removed from the surface by the water-gas shift reaction to form CO_2 and H_2 because high surface coverages by CO lead to low catalytic activity (Davda et al. 2005).

The main barrier for the APR process is to reduce the cost for hydrogen production. Currently, 96% of hydrogen is produced by steam reforming of fossil fuels with a cost in the range of 2–4 dollars per kg at industrial hydrogen production (Holladay et al. 2009; Geantet and Guilhaume 2008). Moreover, it is necessary to come up with non-noble metals to reduce the reaction costs. As such, over the last few years, attention is being drawn towards Ni as a cheaper substitute for noble metals. However, the down side of using Ni is that it promotes a side reaction producing undesirable alkanes. So, bimetallic catalysts comprising of Ni-Sn, Ni-Cu are being currently researched to obtain higher hydrogen selectivity. Ni catalysts supported on carbon nanofibers have been the investigated under reducing and alkaline environments for catalytic activities (van Haasterecht et al. 2014). The aqueous-phase processing (APP) of the biomass-derived bio-oil model compounds, hydroxyacetone, propylene glycol, 1-propanol, 2-propanol, propanoic acid, and acetone, over Pt– Re/C has recently been explored (Fu et al. 2014). From the flux analysis, it was found that the hydrogenation/dehydration of hydroxyacetone was the most rapid reaction in the reaction system.

An optimal reactor design is essential for the production of hydrogen via aqueous phase reforming of ethylene glycol. It has been shown in recent findings that there is an inhibiting effect of hydrogen pressure on the reforming rate, while the yield to side products slightly increases with increasing hydrogen pressure. The addition of nitrogen as an inert gas results in a decrease of this inhibiting effect due to a decrease of the hydrogen pressure in the gas phase and to an increase in the liquid– gas mass-transfer rate (Neira D'Angelo et al. 2014).

Over the past couple of years, many companies and company-university consortia have been initiated that are trying to implement APR on an industrial scale although the process is still in the development phase. The companies below have been recognized among the top companies in the advanced bioeconomy in the annual rankings by Biofuels Digest, the world's most widely read biofuels and biobased chemicals daily journal: *40 Hottest Small Companies in the Advanced Bioeconomy for 2015–6 list (*Biofuels Digest 2015*).*

Virent, a company in Madison, WI is a world-leading company creating chemicals and fuels from a wide range of naturally occurring, renewable resources. Using patented catalytic chemistry, Virent converts soluble biomass-derived sugars into products molecularly identical to those made with petroleum, including gaso line, diesel, jet fuel, and c hemicals used for plastics and fibers. Virent's BioForming® platform is based on a novel combination of Aqueous Phase Reforming (APR) technology with modified conventional catalytic processing (Virent Inc 2015). The BioForming process converts aqueous carbohydrate solutions into mixtures of "drop-in" hydrocarbons. The process has been demonstrated with conventional sugars obtained from existing sugar sources (corn wet mills, sugarcane mills, etc.) as well as a wide variety of cellulosic biomass from nonfood sources. The process can accommodate a broad range of compounds derived from biomass, including C5/C6 sugars, polysaccharides, organic acids, furfurals and other degradation products generated from the deconstruction of biomass, as shown in Fig. 11.

Renmatix, based in Kennesaw, GA has patented the Plantrose® process which is a technology platform to convert non-food biomass into industrial sugars and novel polymers (Renmatix Inc 2015). Plantrose is a continuous process that deconstructs a range of plant material into renewable feedstocks to produce separate streams of sugars from hemicellulose and cellulose. After sugar extraction, remaining lignin solids are burned to supply heat energy required for the process (or utilized in higher value applications like adhesives or thermoplastics). Renmatix

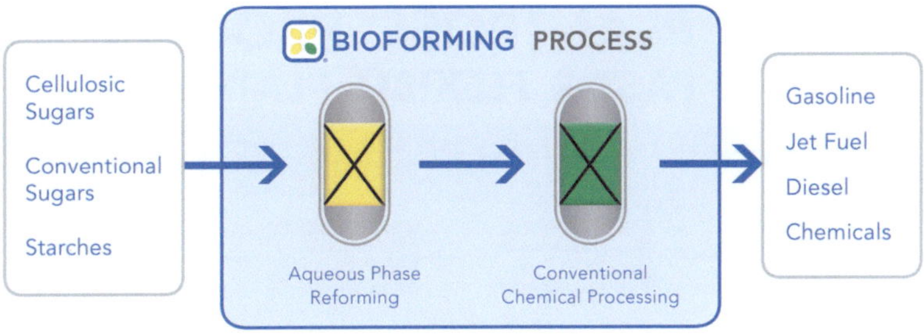

Fig. 11 Virent's BioForming Process (Virent Inc 2015)

uses supercritical water to breakdown biomass, specifically the cellulose in it, and convert it into cellulosic sugar contrary to most technologies that use enzymes, acids, flammable solvents etc. As compared to all the other methods, Renmatix claims that their process is biomass agnostic, cheap, does not require much consumables and provides very easy scaling-up in multiplex reactors (Fig. 12).

In 2006 **Rivertop™ Renewables'** founder, Dr. Don Kiely, and his research group developed an improved process that enabled the mass production of versatile, high-value renewable chemicals, including glucaric acid (Rivertop Renewables Inc 2015). The patented process does not discriminate between feedstocks — various sugars and sugar alcohols— making it capable of inexpensively producing C2, 3, 4, 5, 6 & 7 sugar acids. First targeting glucaric acid, Rivertop's oxidation platform utilizes all of the carbon atoms from the glucose feedstock and adds oxygen weight, resulting in a more pounds of product than feedstock input.

Founded in 2009, **Rennovia** Inc. is a specialty chemical company based in Santa Clara, CA focused on developing processes for the production of biobased glucaric acid, adipic acid, 1,6-hexanediol, hexamethylenediamine (HMD), and other important building blocks for a wide range of functional materials (Rennovia Inc 2015). Low cost sugar based feedstocks are processed through catalytic technology to produce bio-based chemicals.

The **Biomass Technology Group (BTG)** was officially founded in 1979 at the University of Twente in Enschede, The Netherlands (Biomass Technology Group 2015). BTG is one of the partners in the European Commission (EC) funded research project *SusFuelCat* (www.susfuelcat.eu), in full *'Sustainable fuel production by aqueous phase reforming – understanding catalysis and hydrothermal stability of carbon supported noble metals'*. Under the coordination of Prof. Dr. Bastian Etzold, nine further institutions are involved in SusFuelCat coming from Finland, United Kingdom, Italy, Netherlands, Germany, Russia, and Spain. The consortium is comprised of six universities and four companies, including three small and one large enterprise. Sustainable Fuel production by aqueous phase reforming – understanding Catalysis and hydrothermal stability of carbon supported noble metals. Within the project experts for multinational industry, SMEs and academia focus on the optimization of hydrothermally stable carbon supported catalysts for the APR to harness the potential of catalysts.

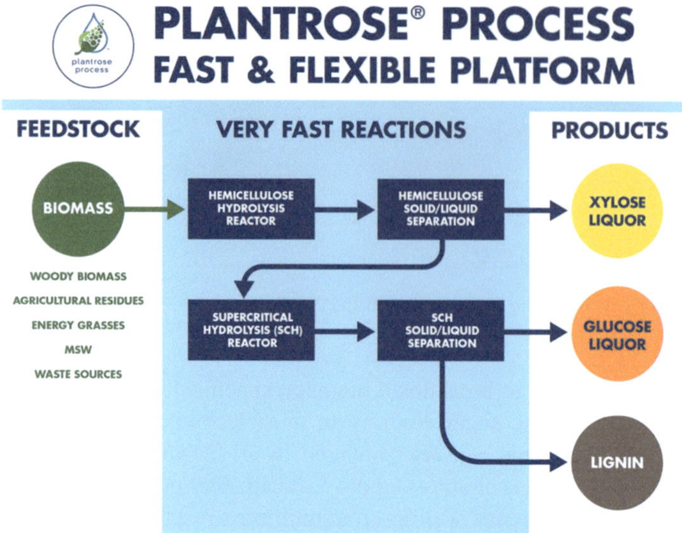

Fig. 12 Two-step continuous process that deconstructs a range of plant material into renewable feedstocks (Renmatix Inc 2015)

Aqueous Phase Processing – Microbial Fuels

The use of microorganisms for the production of fuel and commodity chemicals has shown some success in the recent decade, especially for corn and sugar cane ethanol production. Microbes use natural biological pathways to produce desirable products. Anaerobic microorganisms use either direct fermentation or indirect fermentation. Direct fermentation takes the biomass and breaks the biomass down into fermentable sugars to which can be converted to ethanol and other fuels by the microbes. Indirect fermentation requires converting the biomass into syngas (most often through pyrolysis) which can be later converted to ethanol by acetogenic bacteria (Elshahed 2010).

Microbial pathways for the production of fuels are of most interest due to the low temperature requirements, usually well below 100 °C. Various microbes can either perform saccharification of lignocellulosic materials or fermentation of sugars; however, no single microbe can efficiently perform saccharification and fermentation for cellulosic biomass at a high enough rate and titer (van Zyl et al. 2011). Other challenges of using microbes for fuel production include the careful control of temperature and pH, feedstock impurities, and selectivity (Manikkandan et al. 2009). Ways to improve the effectiveness of the microbial process include using a concoction of different microorganisms and using genetically modified microorganisms which are specifically developed for these special conditions, where fermentation of a broader range of sugars or increased titer is desired. Current microbes of interests include *Escherichia coli*, *Clostridium thermocellum*, *Clostridium sporogenes*, *Saccharomyces cerevisiae*, and *Zymomonas mobilis* (Boumba et al. 2012; He et al. 2014; Islam et al. 2015; Koeck et al. 2015). Genetically modified strains of similar microbes have been implemented by several companies for fuel production.

The production of 1st generation biofuels by microbial processes is being heavily utilized in Brazil and the central United States. The current annual production of ethanol in Brazil and the United States is 7.6 and 14.6 billion gallons (2015), respectively. These ethanol production facilities are close to the fertile land where sugar cane and corn can be easily grown for fuel production. Transportation of the biomass is also a concern due to the initial low energy density of the biomass which requires liquefying and increasing the bulk energy in the final product (Fischer et al. 2008). Other facilities across Europe, South America, and Africa are also producing ethanol. However, as the use of agricultural crops for fuel has been much debated over the last decade, a push for 2nd generation biofuels from lignocellulosic biomass is being further researched. In addition, characteristics of ethanol fail to make it an ideal renewable fuel. Ethanol can easily absorb water which is non-ideal for long-term storage and infrastructure compatibility. Moreover, the lower energy density of ethanol compared to other HC fuels make it unfavorable (Tollefson 2008).

The production of 2nd generation biofuels from agricultural waste products (corn stover and bagasse), woody biomass, perennial grasses, and other lignocellulosic biomasses is the current goal of the renewable energy sector. A consolidated bioprocess (CBP), also known as simultaneous saccharification and cofermentation (SSCF), where lignocellulosic biomass is broken down into sugars and then converted into fuel in a single operation unit is preferred in order to reduce the capital costs and process complexity; however, current enzyme technology lacks the rates, selectivity, and high titer of fuel production by CBP (van Zyl et al. 2011).

A multi-stage biomass process can be successfully implemented for the production of biofuels. The retrofitting of 1st generation ethanol production facilities with lignocellulosic pretreatment technologies or other bolt-ons for advance fuel production can be used for 2nd generation biofuels. This is often the cheapest route if the existing infrastructure exists and is being implemented by Gevo and Butamax to produce isobutanol and by POET-DSM to produce cellulosic ethanol. Other facilities are aimed at the production of biodiesel and commodity chemicals. Some of these companies will be detailed in the next section.

Figure 13 below shows the process pathways to convert lignocellulosic biomass into fuel.

The biomass is pretreated to facilitate the hydrolysis process by increasing the pore volume and area for which the acid or enzymatic hydrolysis to occur (Fischer et al. 2008). This step also aids in the removal of contaminations that would inhibit maximum yield from the enzymes. The hydrolysis process will break down the polysaccharides into monomeric sugars of glucose and xylose (Brown and Brown 2013). This mixture will then be fermented by select microbes, often *Saccharomyces cerevisiae* yeast strains, to ethanol or other intermediates for later hydrogenation (Elshahed 2010; Fischer et al. 2008). The mixture is finally separated to collect the fuel and lignin byproducts. Distillation is most commonly used for ethanol

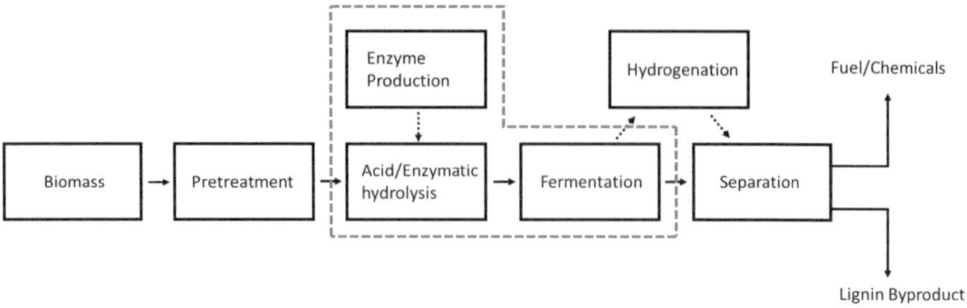

Fig. 13 Schematic of a lignocellulosic biomass feed into products (Fischer et al. 2008)

collection, and the lignin is burned for heat. A CBP or SSCF would combine the enzyme production, enzymatic hydrolysis and enzymatic fermentation into a single unit, highlighted in Fig. 13.

An increased demand for the microbial production of renewable fuels and chemicals is expected in the future (Elshahed 2010). Focus continues on selecting and optimizing the microbes for improved activity, selectivity, and titer. Combinations of microbes and environmental conditions (pH and temperature) are also being altered in order to increase the efficiency and selectivity. This includes single-stage CBPs and muti-staged pretreatments/fermentations. Professor Lee Lynd at Dartmouth has been a leader in pioneering CBP and SSCF with numerous current publications on microbial hydrolysis and fermentation (Lynd et al. 1989; Zhang et al. 2009; Olson et al. 2015; Izquierdo et al. 2014; Reed et al. 2014). The Lynd group primary focus is on *Clostridium thermocellum*. Other groups have focused on using *Escherichia coli* and *Saccharomyces cerevisiae* for use in the Melle-Boinot fermentation process (Yuan et al. 2014; Ha et al. 2011). While dozens of companies implement a microbial pathway for fuel and chemical production, only a selected few are highlighted below.

The recent completion of **DuPont's** cellulosic ethanol plant in Nevada, Iowa in October 2015 is the world's largest cellulosic ethanol plant (depicted in Fig. 14) (Dupont 2015). Production capacity is rated at 30 million gallons per year. Corn stover brought in by over 500 farmers in a 30 mile radius of the facility total up to 375,000 dry tons annually. The corn stover is initially ground up to small pieces that go through a pretreatment, where heat and chemicals sterilize the stover. Next, the pretreated stover is moved to the saccharification unit where the carbohydrates are subjected to proprietary enzymes to release sugars. These sugars go through a fermentation unit where more microbes convert these sugars into ethanol. Finally, a distillation unit is implemented to separate the final product. Whole stillage, a byproduct of the ethanol production is collected at the bottom of the distillation tower and can be sent to a biomass boiler for heat and steam.

LanzaTech's process involves the biological conversion of carbon to products through gas fermentation (Fig. 15) (Lanzatech 2015). These microbes use gases rather than sugars as in traditional fermentation. The LanzaTech process take a feed similar to syngas ($CO + H_2$) and uses microbes in a gas fermenter to produce valued

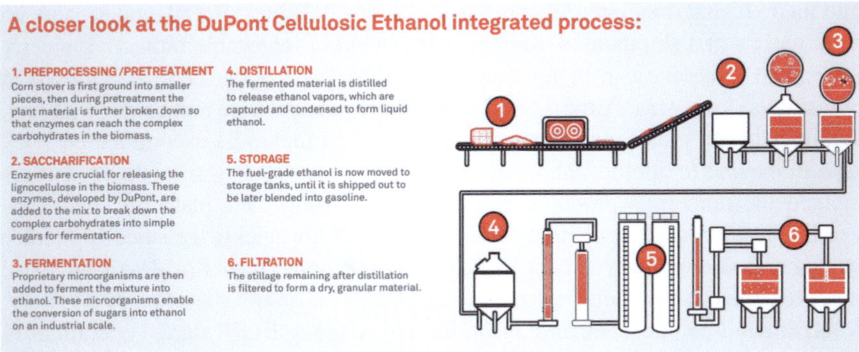

Fig. 14 DuPont cellulosic ethanol process used in Nevada, Iowa (Dupont 2015)

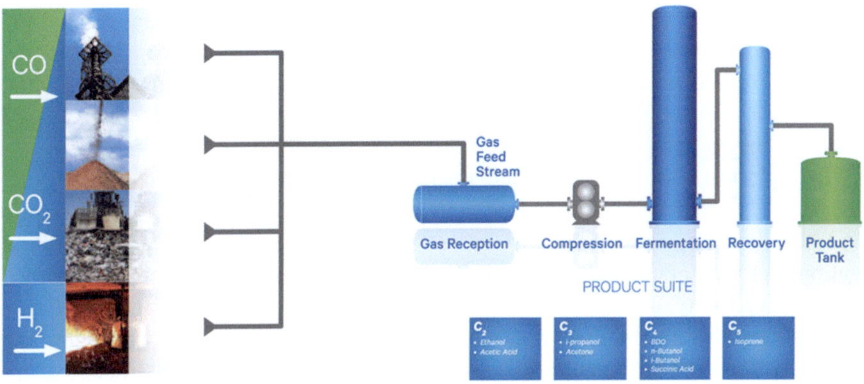

Fig. 15 LanzaTech Process of gas fermentation (Lanzatech 2015)

chemicals. The acetogenic *clostridium autoethanogeum* bacteria naturally produces ethanol. Advantages of the LanzaTech process include using gases and materials that would normally be considered wastes and pollutants. The steel industry produces CO rich exhausts as byproducts. LanzaTech has partnered with BaoSteel and Shougang steel mills in China to produce 300 mta ethanol at each location. Both facilities had a start-up date of 2012 and 2013 and are currently operational. In 2014, LanzaTech started a woody biomass syngas pilot plant in Georgia, USA for ethanol and commodity chemicals where 250 l/day is produced. Production of 2,3 butadiene is also possible and a pilot plant at BlueScope Steel in Glenbrook, New Zealand has been currently producing CO-based 2,3 BDO at a 15,000 gal/year. In addition to ethanol and 2,3-BDO, Lanza-Tech plans to make isoprene, isobutyl alcohol, methyl ethyl ketone, and isopropyl alcohol for customers in markets including diesel, jet fuel, olefins, and plastics. **Coskata** has a similar syngas fermentation process similar to LanzaTech but is not as widespread.

Amyris uses Brazilian sugarcane as their feedstock (Amyris 2015). Once the sugar is extracted, they use a proprietary enzymatic process to ferment the sugars into their HC intermediate. Amyris has operations in Brotas, Brazil, and its start-up plant had its first shipment of Biofene, their brand of renewable farnesene in early 2013. The farnesene can be upgraded to biodiesel and is currently being used in buses in the local area. Amyris and Total partnered to produce a renewable jet fuel, and in 2014, they started to produce commercial jet fuel to be used globally. They plan to expand to another intermediate molecule very soon. Amyris's advantages include the numerous products they can create using their platform chemicals, varying from cosmetics to fuels, and their ability to quickly genetically modify strains of bacteria. They expect to build two additional plants in Brazil by 2016; however, the end products may not be renewable fuels until the price of oil rises.

GranBio's cellulosic ethanol plant, located Alagoas, Brazil started operations in late 2014. This 2nd generation ethanol plant uses sugarcane bagasse as the feedstock (GranBio 2015). Enzyme technology was provided by Novozymes with other technologies aided by Beta Renewables and Chemtex. Production capacity at this facility is rated at 21.6 million gallons annually. GranBio pretreatment technology was gained from the American Process Company and utilizes a proprietary enzymatic hydrolysis with heat to convert the hemicellulose and cellulose into sugars. GranBio is expanding to a second plant that will start in 2016/2017 and several more by the end of 2020.

In September 2014, **POET-DSM** opened a commercial cellulosic ethanol plant that turned agriculture residues into renewable fuels (POET-DSM 2015). Located in Iowa, the plant was to take 770 tons of biomass per day and produce 20 million gallons per year of ethanol, later ramping up to 25 million gallons per year. Known as Project Liberty, this cellulosic ethanol plant incorporated an original 1st generation corn ethanol plant. The bolt-ons added cellulosic handling technologies and increased the production capacity. Fuel production is a multi-stage process where enzymatic hydrolysis is followed by POET-DSM's BPX fermentation process ("cold-cook"), which eliminates the need of heat during fermentation.

Global Bioenergies, founded in 2008, aims to turn renewable biomass into a drop-in fuel. Global Bioenergies partnership with Audi has focused on the fermentation of biomass into isobutene (Global Bioenergies 2015). The isobutene is a building block compound that can be blended with gasoline or upgraded into jet fuel. Global Bioenergies uses an advance fermentation process where the isobutene can evaporate from the fermentation broth, avoiding saturation of the broth and decreasing the separation costs. Since no microorganism naturally produces isobutene, Global Bioenergies have genetically modified microbes to produce isobutene. So far, they have been successful at the laboratory scale and initial scale-ups. A 5000 l isobutene demonstration plant is currently being built in Leuna, Germany with completion expected in mid-2016.

Located in Lebanon, NH, **Mascoma's** focus has been on finding a commercial pathway for CBP (Mascomam 2015). They recently sold their yeast business to Lallemand Inc., and their demonstration plant in Rome, NY was bought by Renmatrix. Mascoma uses a proprietary yeast called TransFerm which will produce enzymes to cost-effectively break down the complex sugars. They use the engineered yeast strains, TransFerm Yield+, in their fermentation process to break down the xylose and other sugars with broader capabilities and lower amounts of byproducts produced.

The microbial production of 1st generation fuels, especially ethanol, has been quite successful globally. Production of 1st generation ethanol was partially limited to the availability of local biomass and the debate of using edible crops for fuel. With the introduction of 2nd generation lignocellulosic biofuels, that debate is over. Many 1st generation ethanol plants are being retrofitted with lignocellulosic capabilities that enable a broader range of feedstocks. In the past 2 years, cellulosic ethanol production has started on the commercial scale, and the commercial outlook appears promising. These commercial plants have multi-stage bioprocessing units. The current technology for CBP or SSCF does not appear to be ready for commercial use. Most of these commercial and demonstration ethanol and hydrocarbon fuel facilities are located in the United States, South America, and Europe. Renewable fuels from microbial pathways in the Arab Middle East appears to less popular, especially behind wind and solar power. The key to more advance cellulosic fuel production in areas of high biomass availability is in finding the correct microorganism.

Biogas

Low oil and gas prices in recent times have provided stiff competition to the alternative energy industry; nevertheless, energy portfolio diversification is beneficial in order to provide sustainability and to compensate for future market fluctuations. Biogas utilization is one form of energy diversification that is gaining a lot of attention because it provides regional sustainability, utilizes waste, and creates a closed loop carbon cycle. Any biodegradable organic waste can be broken down to produce biogas along with liquid and solid byproducts. The biogas may ultimately be burned for energy which creates CO_2, while the byproducts make great fertilizer, all of which are used to make more biomass, thus closing the carbon cycle (Papadopoulou et al. 2012). Traditionally, biogas has been burned to provide energy for small applications in developing countries, but is now being incentivized for more expansive applications elsewhere in the world (Jorgensen 2009). For example, in 2014 the US EPA qualified biogas as a viable pathway to meet the cellulosic content requirements of the Renewable Fuels Standard (RFS) Program which increases its attractiveness to US fuel producers (Environmental Protection Agency 2014).

Biogas is produced when microorganisms biologically break down organic materials in the absence of oxygen. This process is known as anaerobic digestion (AD) and occurs in 4 main steps as shown in Fig. 16. First, the complex carbon compounds are hydrolyzed into smaller ones followed by acidogenesis in which fatty acids, CO_2, and H_2 are produced. The fatty acids are then converted to acetic acid with additional CO_2 and H_2. Finally, methanogens convert the fatty acids to CH_4 and CO_2 (Mao et al. 2015; Harvest Power 2015). Depending on the feedstock

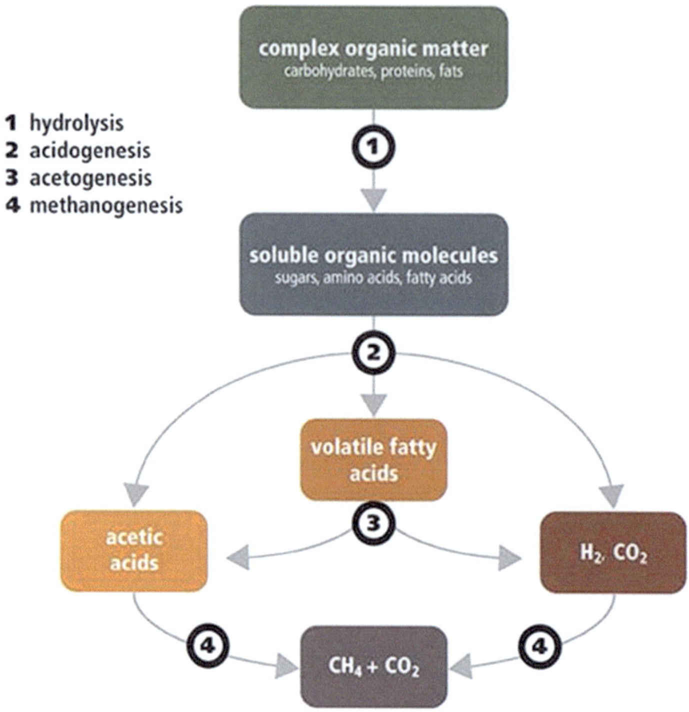

Fig. 16 Methane generation through anaerobic digestion (http://www.chevron.com/globalissues/energysupplydemand/)

and digestion conditions, the resulting biogas composition ranges from 50–70% CH_4 and 30–50% CO_2 with trace amounts of H_2S, N_2, and moisture (Papadopoulou et al. 2012; Harvest Power 2015). The H_2S often needs to be removed, as will be discussed later, since it is very corrosive and will damage equipment and machinery. One of the great advantages of biogas is the flexibility of the feedstock which may include: livestock manure, sewage, municipal solid waste, food waste, and even agro-industrial waste.

Anaerobic digestion is a long-standing, commercially proven technology with extensive research applied to improve the yield, production capacity, and composition tunability of the biogas (Papadopoulou et al. 2012; Mao et al. 2015; Carrere et al. 2016). Areas of research range from composition of feedstock and pretreatment to mechanical design and reaction conditions (Mao et al. 2015; Carrere et al. 2016; Ariunbaatar et al. 2014; Kikas et al. 2016). Reaction conditions that affect the product and efficiency include: temperature, pH, feed rate, retention time, C/N ratio, and moisture content (Jorgensen 2009; Carrere et al. 2016). Although any biodegradable material may be used, the organic make-up of the material greatly affects the AD process. When considering lingo-cellulosic materials, it is important to note that lignin content greatly reduces methane production, so pretreatment is necessary (Mao et al. 2015; Carrere et al. 2016; Ariunbaatar et al.

2014; Kikas et al. 2016). Hydrolysis is the rate determining step when dealing with any other complex organic substrates, therefore pre-hydrolyzing would be beneficial anyway in order to reduce the reaction time (Carrere et al. 2016; Ariunbaatar et al. 2014).

There are a number of different pretreatment techniques that have been implemented at both lab and full scale including: steam explosion, low temperature/biological treatment, sonication, centrifugation, pressurization, and electroporation (Kikas et al. 2016; Cambi 2015). Alkali and fungi additions have shown much promise towards delignification of lingo-cellulosic material, but neither have been implemented on full scale (Carrere et al. 2016). The most common and successful hydrolysis pretreatment is steam explosion which has been adopted by a handful of companies (Carrere et al. 2016). One such company, Cambi, has patented their Thermal Hydrolysis Process (CambiTHP™) which is aimed at pretreating leftover solids from wastewater treatment plants (WWTP) where solids are pressurized with steam before being sent to the digester (Cambi 2015). They are installing this process in over 50 WWTP around the world, with the largest one just being completed in 2014 in Washington DC, USA (Carrere et al. 2016; Cambi 2015). DC Water Blue Plains WWTP contains 32 CambiTHP™ reactors as shown in Fig. 17 and provides 135,000 t dry solids/yr. to the digesters which in turn produces enough biogas to generate 13 MW of electricity. The Cambi reactors not only enhance the digester residence time and methane yield, but they also improve the dewaterability of the solids and remove the pathogens so that they are fertilizer ready (Cambi 2015).

More recently, Aerothermal Group has announced great success with a demonstration of steam hydrolyzing "black bag" municipal waste. This process takes MSW as is, and steam pressurizes it in a revolving autoclave. This breaks down the organic material allowing easy separation from the non-biodegradable trash. The company claims that this improved AD feedstock results in 4 times higher gas yields (Aerothermal Group 2015).

Once the biogas is produced, a number of conversion processes exist (Fig. 18). To date, the most common utilization of biogas is to directly burn it (Yang et al. 2014; Itkulova et al. 2014; Kohn 2012). If the methane content is high enough, it may be economically feasible to extract work from its combustion through a heat engine, such as an internal combustion engine or turbine (Kohn 2012). This route is very common to produce combined heat and power (CHP) where the left-over heat from power generation is used to heat the anaerobic digester. If the methane content is not high enough for this, it can still be burned for direct heat, such as that for cooking or a water heater. When none of these are an option, the gas may be flared to the atmosphere which is still more environmentally friendly than not capturing it at all since CH_4 has about 25 times greater of a greenhouse effect than CO_2 (Kohn 2012).

A large scale example of a very recent project that utilizes biogas for CHP is **Harvest Power's** Energy Garden in Florida, USA. The company buys food waste from local amusement parks and hotels. The plant's digester processes about 130,000 t/yr. of biosolids, fats, oils, grease, and food waste. This results in about

Fig. 17 CambiTHP™ reactors (*left*) and 1 of 4 anaerobic digesters (*back*) at DC Water Blue Plains WWTP (Cambi 2015)

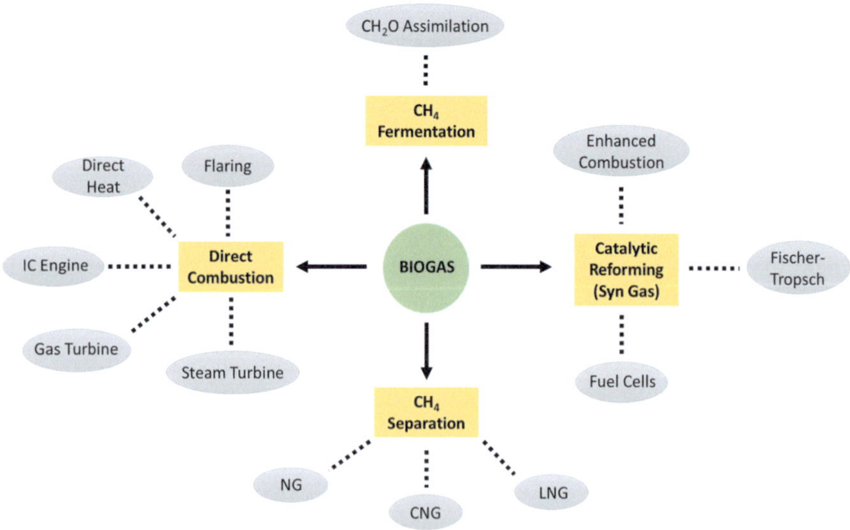

Fig. 18 Biogas utilization (Kohn 2012)

15,000 m^3/d of biogas which is fed to a 3.2 MW capacity generator. About 2.2 MW of heat is recovered from this and used to dry the digestate which is sold as 2 different products: class AA fertilizer and Struvite, a phosphorus fertilizer additive (Harvest Power 2015).

Another common way to use biogas is to separate out the methane and use it as an alternative to natural gas. The main components of the gas that need to be removed are the CO_2 and the H_2S. Once the impurities are removed, this renewable natural gas may be injected into natural gas pipelines, compressed, or liquefied for later use (Yang et al. 2014; Kohn 2012). Purification of biogas is essential if the end use is for natural gas vehicles. Worldwide, there are close to 250 commercial scale biogas purification plants which use techniques that are derived from conventional gas separation methods (Yang et al. 2014). Among those methods that have been commercialized are water scrubbing, pressure swing adsorption, chemical absorption, membrane separation, and cryogenic distillation (Yang et al. 2014).

Roeslein Alternative Energy is currently working on a project to cover manure/waste lagoons at a Smithfield Foods hog farm in Missouri, USA in order to capture and purify the biogas that is generated. This project is expected to produce 2.2 billion ft^3/yr. of renewable natural gas which will be injected into a natural gas pipeline and sold to nearby utility companies. In the future, they plan to supplement the manure with prairie grass (Roeslien Alternative Energy LLC 2015).

Schmack Carbotech is developing a similar project at one of the world's largest WWTP outside of Stockholm, Sweden. They are purifying the biogas coming from the plant's AD using pressure swing adsorption, and will sell it as biofuel for natural gas vehicles. The capacity of this plant is 106,000 ft^3/yr. of 97% CH_4, which is equivalent to 100 million kWh/yr. of energy (Schmack Carbotech 2014).

Today, natural gas is most commonly reformed using steam for the production of synthesis gas or H_2 (Papadopoulou et al. 2012). Biogas on the other hand already contains CO_2, which can act as the oxidant in the dry reforming of methane (DRM):

$$CO_2 + CH_4 \rightleftharpoons 2CO + 2H_2 \quad \Delta H = +247\ \text{kJ/mol}$$

This is a highly endothermic reaction so it will require a significant amount of heat input, and therefore high operating temperatures (Papadopoulou et al. 2012). This reaction also favors a CH_4/CO_2 ratio of 1, so if the biogas feed is methane rich, air may be added to the feed so that the CH_4 will partially combust, providing more CO_2 and heat to drive the reaction (Kohn 2012).

Catalytically reforming biogas into syngas is a very promising pathway for biogas utilization because it opens up many different options. Syngas is often used for enhanced combustion where it is burned in a gas turbine to generate electricity with cleaner combustion products (Kohn 2012). Another option for syngas is to convert it into a liquid fuel by using reactions such as Fischer-Tropsch or methanol synthesis. This route has a lot of potential to impact the transportation industry's energy input requirements. In addition there are also a few options for use in fuel cells such as separating the H_2 from syngas or just using the syngas in a fuel flexible solid oxide fuel cell (SOFC). Due to the high temperature nature of

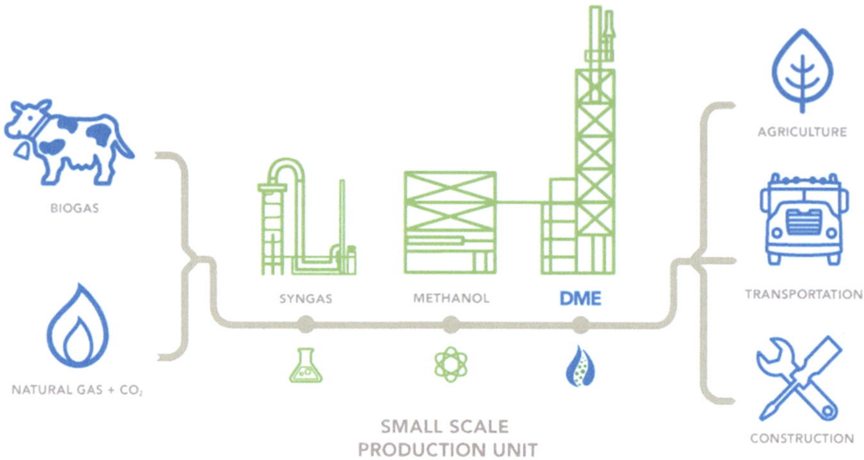

Fig. 19 Oberon Process: DME production through dry reforming of methane (Oberon Fuels 2015)

SOFC's, it has also been demonstrated to just feed the biogas directly into the cell where it can be reformed internally (Kirtley et al. 2014).

To date, there are very few companies that have employed a biogas reforming process at commercial scale. However, one such company, **Oberon Fuels**, has developed a skid mounted modular unit that converts biogas or natural gas and CO_2 into dimethyl ether (DME). DME is used as a clean alternative to diesel. These units have the capacity to produce 3000–10,000 gal DME/day. The first step to the Oberon Process, shown in Fig. 19 is the dry reforming of methane to syngas followed methanol synthesis by CO hydration. The final step is the dehydrogenation of methanol to produce DME. This process is aimed at smaller regional markets such as agriculture, trucking, construction, or stranded gas. An example of this application could be if a farmer has an AD system installed on his farm, he could tack on one of these Oberon units to make his own fuel for tractors and equipment (Oberon Fuels 2015).

A much newer technology is the use of methane oxidizing microbes called methanotrophs to convert biogas to liquid fuels. Methanotrophs are bacteria that utilize methane as their sole carbon source (Silverman Josh 2014; Hanson and Hanson 1996). They are able to metabolize methane and convert it to methanol followed by a conversion to formaldehyde (Fig. 20). The formaldehyde can then undergo a carbon assimilation cycle, such as the RuMP or Serine cycle, which is similar to photosynthesis (Hanson and Hanson 1996). This assimilation process ultimately forms lipids in way that is analogous to algae formation, except methane is used as the carbon source (Silverman Josh 2014). These lipids can then undergo traditional biomass conversion processes such as hydrotreatment to form gasoline, diesel, or jet fuel or transesterification to form biodiesel (Papadopoulou et al. 2012).

Calysta, based in California, USA, is one company that has commercialized this technology. Calysta develops Biological Gas-to-Liquids® processes that use

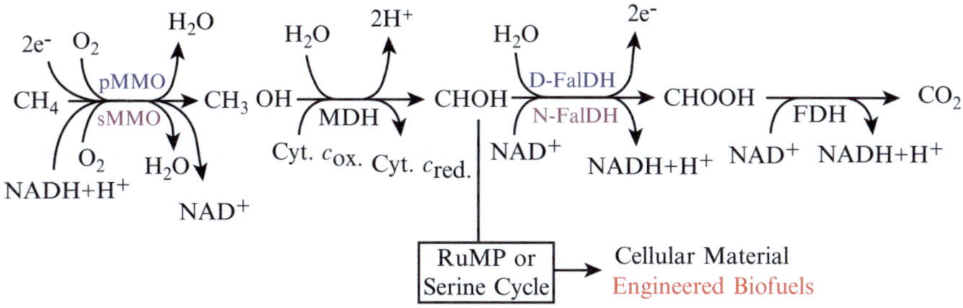

Fig. 20 Methane metabolization through methanotrophs (Silverman Josh 2014)

methanotrophs to convert methane to liquid hydrocarbons. They have also developed and patented a novel loop reactor that improves the dissolution of the gas. This reactor has been demonstrated at 10,000 ton/yr (Calysta Energy 2015).

References

Aerothermal Group. (2015). http://www.aerothermalgroup.com/latestnews.html. Accessed 7 Dec 2015.

Amyris. (2015). *Sustainability*. https://amyris.com/innovation/sustainability/. Accessed 01 Dec 2015.

Anellotech Inc. (2015). *New technology to produce clean, renewable chemicals currently made from petroleum*. http://anellotech.com/technology. Accessed 3 Dec 2015.

Ariunbaatar, J., Panico, A., Esposito, G., Pirozzi, F., & Lens, P. (2014). Pretreatment methods to enhance anaerobic digestion of organic solid waste. *Applied Energy, 123*, 143–156.

Ausilio B., Göran B., Martin, J., Marc L., & François V. (2009). *Bioenergy- A sustainable and reliable energy resource; A review of status and prospects* (Vol. 6). IEA Bioenergy: ExCo.

Balat, M. (2008). Mechanisms of thermochemical biomass conversion processes. Part 3: Reactions of liquefaction. *Energy Sources, Part A: Recovery, Utilization, and Environmental Effects, 30*, 649–659.

Balat, M. (2009). Development of worldwide green electricity in the past decade. *Energy Sources, Part B: Economics, Planning and Policy, 4*, 17–33.

Biofuels Digest. (2015). “40 Hottest Smaller Companies in the Advanced Bioeconomy” rankings for 2015–2016. http://www.biofuelsdigest.com/bdigest/tag/hot-40/. Accessed 8th November, 2015.

Biomass Technology Group. (2015). http://www.btgworld.com/en/project-development. Accessed 8 Nov 2015.

Bocci, E., Sisinni, M., Moneti, M., Vecchione, L., Di Carlo, A., & Villarini, M. (2014). State of art of small scale biomass gasification power systems: A review of the different typologies. *Energy Procedia, 45*, 247–256.

Boumba, V. A., Economou, V., Kourkoumelis, N., Gousia, P., Papadopoulou, C., & Vougiouklakis, T. (2012). Microbial ethanol production: Experimental study and multivariate evaluation. *Forensic Science International, 215*, 189–198.

Brown, T. R., & Brown, R. C. (2013). A review of cellulosic biofuel commercial-scale projects in the United States. *Biofuels, Bioproducts and Biorefining, 7*, 235–245.

BTG Biomass Technology Group & BlueBear. (2015a). *Fast pyrolysis*. http://www.btgworld.com/nl/rtd/technologies/fast-pyrolysis. Accessed 3 Dec 2015.

BTG Biomass Technology Group & BlueBear. (2015b). Official opening empyro. http://www.btg-btl.com/en/company/news/news/article?id=116. Accessed 3 Dec 2015.
Calysta Energy. (2015). *Gaseous fermentation* http://calystaenergy.com/technology/gaseous-fermentation/. Accessed 7 Dec 2015.
Cambi. (2015). http://www.cambi.com/Products/CambiTHP. Accessed 7 Dec 2015.
Carlos Serrano-Ruiz, J., Luque, R., & Sepulveda-Escribano, A. (2011). Transformations of biomass-derived platform molecules: From high added-value chemicals to fuels *via* aqueous-phase processing. *Chemical Society Reviews, 40*, 5266–5281.
Carlson, T. R., Vispute, T. P., & Huber, G. W. (2008). Green gasoline by catalytic fast pyrolysis of solid biomass derived compounds. *ChemSusChem, 1*, 397–400.
Carrere, H., Antonopoulou, G., Affes, R., Passos, F., Battimelli, A., Lyberatos, G., & Ferrer, I. (2016). Review of feedstock pretreatment strategies for improved anaerobic digestion: From lab-scale research to full-scale application. *Bioresource Technology, 199*, 386–397.
Cheng, Y. T., Wang, Z., Gilbert, C. J., Fan, W., & Huber, G. W. (2012). Production of p-xylene from biomass by catalytic fast pyrolysis using ZSM-5 catalysts with reduced pore openings. *Angewandte Chemie, 124*, 11259–11262.
Collarda, F.-X., Bensakhriab, A., Drobekc, M., & Volled GBlin, J. (2015). Influence of impregnated iron and nickel on the pyrolysis of cellulose. *Biomass and Bioenergy, 80*, 52–62.
Cool Planet Energy Systems. (2015). *USDA issues conditional commitment for $91MM loan guarantee to cool planet.* http://www.coolplanet.com/sites/default/files/docs/USDA_Issues_Conditional_Commitment_for_%2491MM_Loan_Guarantee_to_Cool_Planet.pdf. Accessed 3 Dec 2015.
Czernik, S., & Bridgwater, A. V. (2004). Overview of applications of biomass fast Pyrolysis Oil. *Energy & Fuels, 18*, 590–598.
Davda, R. R., Shabaker, J. W., Huber, G. W., Cortright, R. D., & Dumesic, J. A. (2005). *Applied Catalysis B: Environmental, 56*(2005), 171–186.
Demirbas, M. F., Balat, M., & Balat, H. (2009). Potential contribution of biomass to the sustainable energy development. *Energy Conversion and Management, 50*, 1746–1760.
Dupont. (2015). *Infographic: Reforming energy*. http://www.dupont.com/content/dam/assets/products-and-services/industrial-biotechnology/documents/DuPont_REFORM-ENERGY_infographic_151030.pdf. Accessed 1 Dec 2015.
Elshahed, M. S. (2010). Microbiological aspects of biofuel production: Current status and future directions. *Journal of Advanced Research, 1*, 103–111.
Ensyn. (2012). *RTP™ process.* http://www.ensyn.com/technology/overview/. Accessed 3 Dec 2015.
Environmental Protection Agency. (2014) Final Register Vol. 70 No. 138: 42127–42167.
Fischer, C. R., Klein-Marcuschamer, D., & Stephanopoulos, G. (2008). Selection and optimization of microbial hosts for biofuels production. *Metabolic Engineering, 10*, 295–304.
Fu, J., Sun, G., & Shanks, B.H. (2014). Aqueous-phase processing on multi-functional compounds over platinum−rhenium supported on carbon. Energy & Fuels, 28, 2123–2128. dx.doi.org/10.1021/ef500085v.
Galadima, A., & Muraza, O. (2015). In situ fast pyrolysis of biomass with zeolite catalysts for bioaromatics/gasoline production: A review. *Energy Conversion and Management, 105*, 338–354.
Geantet, C., & Guilhaume, N. (2008, October 22–24). Hydrogen production from biomass. In *Vanatrop*. Montpellier.
Global Bioenergies. (2015). *Isobutene process.* http://www.global-bioenergies.com/our-groupe/isobutene-process/?lang=en. Accessed 01 Dec 2015.
GranBio. (2015). *Bioflex I*. http://www.granbio.com.br/en/conteudos/biofuels/. Accessed 01 Dec 2015.
Ha, S.-J., Galazka, J. M., Rin Kim, S., Choi, J.-H., Yang, X., Seo, J.-H., et al. (2011). Engineered *Saccharomyces cerevisiae* capable of simultaneous cellobiose and xylose fermentation. *Proceedings of the National Academy of Sciences, 108*, 504–509.

Hanson, R., & Hanson, T. (1996). Methanotrophic bacteria. *Microbiological Reviews, 60*(2), 439–471.
Harvest Power. (2015). http://www.harvestpower.com/home/. Accessed 4 Dec 2015.
He, M. X., Wu, B., Qin, H., Ruan, Z. Y., Tan, F. R., Wang, J. L., et al. (2014). Zymomonas mobilis: A novel platform for future biorefineries. *Biotechnology for Biofuels, 7*, 1–15.
Hitesh, K., Prashant, B., Pankaj, A., & Soni, S. C. (2014). Effect of moisture content on gasification efficiency in down draft gasifier. *International Journal of Science, Engineering and Technology, 3*(4), 411–413.
Holladay, J. D., Hu, J., King, D. L., & Wang, Y. (2009). An overview of hydrogen production technologies. *Catalysis Today, 139*, 244–260.
Honeywell International Inc. (2015). *UOP breaks ground on facility to convert biomass to green transportation fuels*. http://www.uop.com/uop-breaks-ground-facility-convert-biomass-green-transportation-fuels/. Accessed 3 Dec 2015.
http://biomassmagazine.com/articles/1674/solar-powered-biomass-gasification
http://www.chevron.com/globalissues/energysupplydemand/
http://www.eai.in/ref/ae/bio/bgt/type/fixed_bed_gasifier.html
http://www.metso.com/news/2013/3/metso-supplied-worlds-largest-biomass-gasification-plant-inaugurated-in-finland/
http://www.phgenergy.com/case-study/covington-tenn
http://www.unbc.ca/green/energy/bioenergy-plant
http://www.westinghouse-plasma.com/projects/
Huber, G. W., & Dumesic, J. A. (2006). An overview of aqueous-phase catalytic processes for production of hydrogen and alkanes in a biorefinery. *Catalysis Today, 111*, 119–132.
Huber, G., Iborra, S., & Corma, A. (2006a). *Chemical Review*, 106, 4044.
Huber, G. W., Iborra, S., & Corma, A. (2006b). Synthesis of transportation fuels from biomass: Chemistry, catalysts, and engineering. *Chemical Reviews, 106*, 4044–4098.
Islam, R., Sparling, R., Cicek, N., & Levin, D. B. (2015). Optimization of influential nutrients during direct cellulose fermentation into hydrogen by clostridium thermocellum. *International Journal of Molecular Sciences, 16*, 3116–3132.
Itkulova, S., Zakumbaeva, G., Mukazhanova, A., & Nurmakanov, Y. (2014). Syngas production by biogas reforming over the Co-based multicomponent catalysts. *Central European Journal of Chemistry, 12*(12), 1255–1261.
Izquierdo, J. A., Pattathil, S., Guseva, A., Hahn, M. G., & Lynd, L. R. (2014). Comparative analysis of the ability of Clostridium clariflavum strains and Clostridium thermocellum to utilize hemicellulose and unpretreated plant material. *Biotechnology for Biofuels, 7*, 1–8.
Jared P. C., John J. M. (2002, June). *Benchmarking biomass gasification technologies*. U.S. Department of Energy, National Energy Laboratory.
Jorgensen, P. (2009). Biogas – Green energy. *PlanEnergi*, 1–36.
Kikas T, Tutt M, Raud M, Alaru M, , Lauk R, Olta J (2016) Basis of energy crop selection for biofuel production: Cellulose vs. lignin. International Journal of Green Energy 13(1): 49–54
Kirtley, J., Steinhurst, D., Owrutsky, J., Pomfret, M., & Walker, R. (2014). In situ optical studies of methane and simulated biogas oxidation on high temperature solid oxide fuel cell anodes. *Physical Chemistry Chemical Physics, 227*(16), 227–236.
Koeck, D. E., Koellmeier, T., Zverlov, V. V., Liebl, W., & Schwarz, W. H. (2015). Differences in biomass degradation between newly isolated environmental strains of Clostridium thermocellum and heterogeneity in the size of the cellulosomal scaffoldin. *Systematic and Applied Microbiology, 38*, 424–432.
Kohn MP (2012) *Catalytic reforming of biogas for syngas production*. PhD thesis, Columbia University.
Lanzatech. (2015). *Turning steelmaking off-gases into marketable commodities*. http://www.lanzatech.com/turning-steelmaking-off-gases-into-marketable-commodities/. Accessed 01 Dec 2015.

Lynd, L. R., Grethlein, H. E., & Wolkin, R. H. (1989). Fermentation of cellulosic substrates in batch and continuous culture by clostridium thermocellum. *Applied and Environmental Microbiology, 55*, 3131–3139.
Manikkandan, T., Dhanasekar, R., & Thirumavalavan, K. (2009). Microbial production of hydrogen from sugarcane bagasse using Bacillus Sp. *International Journal of ChemTech Research., 1*, 344–348.
Mao, C., Feng, Y., Wang, X., & Ren, G. (2015). Review on research achievements of biogas from anaerobic digestion. *Renewable and Sustainable Energy Reviews, 45*, 540–555.
Marcin, S., de Wiebren, J., & Adrian, H. M. V. (2011). Fluidized bed gasification as a mature and reliable technology for the production of bio-syngas and applied in the production of liquid transportation fuels—A review. *Energies, 4*, 389–434.
Mascomam, LLC. (2015) *Consolidated Bioprocessing (CBP) for High Efficiency Fermentation.* http://www.mascoma.com/technology/consolidated-bioprocessing/. Accessed 01 Dec 2015.
Mohammad, A. (2013). Technical challenges of utilizing biomass gasification gas for power generation: An overview. *Journal of Energy Technologies and Policy, 3*(11), 137–143.
Mohammad, A. (2014). Barriers of commercial power generation using biomass gasification gas: A review. *Renewable and Sustainable Energy Reviews, 29*, 201–215.
Mohan, D., Pittman, C. U., & Steele, P. H. (2006). Pyrolysis of wood/biomass for bio-oil: A critical review. *Energy and Fuels, 20*, 848–889.
Neira D'Angelo, M. F., Schouten, J. C. , van der Schaaf, J. & Nijhuis, T. A. (2014). Three-phase reactor model for the aqueous phase reforming of ethylene glycol. Industrial and Engineering Chemistry Research 53, 13892–13902. dx.doi.org/10.1021/ie5007382.
Nourredine, A., Sushil, A., Avanti, K., & Shyamsundar, C. (2015). A review on biomass gasification syngas cleanup. *Applied Energy, 155*, 294–307.
Oberon Fuels. (2015). http://www.oberonfuels.com/. Accessed 7 Dec 2015.
Olson, D. G., Sparling, R., & Lynd, L. R. (2015). Ethanol production by engineered thermophiles. *Current Opinion in Biotechnology, 33*, 130–141.
Papadopoulou, C., Matralis, H., & Verykios, X. (2012). Utilization of biogas as a renewable carbon source: dry reforming of methane. In L. Guczi & A. Erdohelyi (Eds.), *Catalysis for alternative energy generation* (pp. 57–127). New York: Springer.
POET-DSM. (2015). *First commercial-scale cellulosic ethanol plant in the U.S. opens for business.* http://poetdsm.com/pr/first-commercial-scale-cellulosic-plant. Accessed 01 Dec 2015.
Reed, P. T., Izquierdo, J. A., & Lynd, L. R. (2014). Cellulose fermentation by Clostridium thermocellum and a mixed consortium in an automated repetitive batch reactor. *Bioresource Technology, 155*, 50–56.
Regalbuto, J. R. (2009). Cellulosic biofuels: Got gasoline? *Science, 325*, 822.
Regalbuto, J. R. (2011). The sea change in US biofuels' funding: From cellulosic ethanol to green gasoline. *Biofuels, Bioproducts and Biorefining, 5*, 495.
Renmatix Inc. (2015). http://renmatix.com/technology/plantrose-technology/plantrose-process. Accessed 8 Nov 2015.
Rennovia Inc. (2015). http://www.rennovia.com/technology/. Accessed 8 Nov 2015.
Review of technology for the gasification of biomass and wastes. (2009, June). E4tech.
Richard, G. R., & Geoffrey, J. B. (2008). The value of technological advance in decarbonizing the U.S. economy. *Energy Economics, 30*(6), 2930–2946.
Rivertop Renewables Inc. (2015). http://www.rivertop.com/solutions/. Accessed 8 Nov 2015.
Roeslien Alternative Energy, LLC. (2015). *Key phase underway in $120 Million manure to energy project.* http://roesleinalternativeenergy.com/rae-smithfield-foods-project-progress-video/. Accessed 7 Dec 2015.
Schmack Carbotech. (2014). *Schmack Carbotech awarded the contract for gas processing in Stockholm.* http://www.carbotech.info/de/press/Schweden_Sofielund.html. Accessed 7 Dec 2015.

Silverman Josh. (2014) *BioGTL platform for the conversion of natural gas to fuels and chemicals.* Calysta Energy. http://calysta.com/pdfs/AIChE_final_33114.pdf. Accessed 7 Dec 2015.
Steffen, H., & Pier, U. F. (2015). New concept in biomass gasification. *Progress in Energy and Combustion Science, 46*, 72–95.
The Research University in the Helmholtz Association. (2015). *The bioliq® Process: Biomass-to-Liquids (BtL) – The bioliq® Process.* http://www.bioliq.de/english/55.php. Accessed 5 Dec 2015.
Tollefson, J. (2008). Energy: Not your father's biofuels. *Nature, 451*, 880–883.
van Haasterecht, T., Ludding, C. C. I., de Jong, K. P., & Bitter, J. H. (2014). Toward stable nickel catalysts for aqueous phase reforming of biomass-derived feedstock under reducing and alkaline conditions. *Journal of Catalysis, 319*, 27–35.
van Zyl, W. H., la Grange, D. C., & den Haan, R.(2011). *Developing organisms for consolidated bioprocessing of biomass to ethanol.* INTECH Open Access Publisher.
Virent Inc. (2015). http://www.virent.com/technology/bioforming/. Accessed 8 Nov 2015.
VTT Technical Research Center of Finland LTD. (2015). *Fast pyrolysis – conversion of biomass to bio-oil: Integrated production of fuels and chemicals.* http://www.vttresearch.com/services/bioeconomy/liquid-biofuels1/advanced-biodiesel2/fast-pyrolysis-%E2%80%93-conversion-of-biomass-to-bio-oil. Accessed 8 Dec 2015.
Wei, Y., Lei, H., Liu, Y., Wang, L., Zhu, L., Zhang, X., et al. (2014). Renewable hydrogen produced from different renewable feedstock by aqueous-phase reforming process. *Journal of Sustainable Bioenergy Systems, 4*, 113–127.
World Energy Resources. (2013). Bioenergy World Energy Council.
Yang, L., Ge, X., Wan, C., Yu, F., & Li, T. (2014). Progress and perspectives in converting biogas to transportation fuels. *Renewable and Sustainable Energy Reviews, 40*, 1133–1152.
Younes, C., & Mohammed, K. (2013). Thermal conversion of biomass, pyrolysis and gasification: A review. *International Journal of Energy and Science, 4*(3), 75–85.
Yuan, Y., Bi, C., Nicolaou, S., Zingaro, K., Ralston, M., & Papoutsakis, E. (2014). Overexpression of the 59 plantarum peptidoglycan biosynthesis murA2 gene increases the tolerance of *Escherichia coli* to alcohols and enhances ethanol production. *Applied Microbiology and Biotechnology, 98*, 8399–8411.
Zhang, J., Shao, X., & Lynd, L. R. (2009). Simultaneous saccharification and co-fermentation of paper sludge to ethanol by *Saccharomyces cerevisiae* RWB222. Part II: Investigation of discrepancies between predicted and observed performance at high solids concentration. *Biotechnology and Bioengineering, 104*, 932–938.
Zhang, B., Zhong, Z. P., Chen, P., & Ruan, R. (2015a). Microwave-assisted catalytic fast pyrolysis of biomass for bio-oil production using chemical vapor deposition modified HZSM-5 catalyst. *Bioresource Technology, 197*, 79–84.
Zhang, Z. B., Lu, Q., Ye, X. N., Li, W. T., Zhang, Y., & Dong, C. Q. (2015b). Selective production of 4-ethyl phenol from low-temperature catalytic fast pyrolysis of herbaceous biomass. *Journal of Analytical and Applied Pyrolysis, 115*, 307–315.

Energy Storage Systems for Smart Grid Applications

Said Al-Hallaj, Stephen Wilke, and Ben Schweitzer

Abstract Energy storage is a critical component of any initiative to make electric power and mobility more sustainable. As more solar and wind power generation are added to the electric grid, a mismatch between the periods of peak generation and peak demand necessitate some way to store energy and buffer transient fluctuations in the grid. Similarly, to transition from petroleum-based energy for transportation requires renewable technologies for storing energy with high energy density. This chapter addresses energy storage for smart grid systems, with a particular focus on the design aspects of electrical energy storage in lithium ion batteries.

Grid-tied energy storage projects can take many different forms with a variety of requirements. Commercially available technologies such as flywheel energy storage, pumped hydro, ice-based thermal energy storage, and lead acid or lithium ion batteries are already in widespread use. The energy storage industry is rapidly developing, introducing newer technologies such as compressed air energy storage and flow batteries in pilot project demonstrations. The appropriate selection of a particular technology depends on the system requirements for the type of energy to be stored/used, discharge rate, capacity, lifetime, and cost. Lithium ion batteries are a prominent candidate for smart grid applications due to their high specific energy and power, long cycle life, and recent reductions in cost.

Lithium ion system design is truly interdisciplinary. At a cell level, the specific type of Li-ion chemistry affects the feasible capacity, power, and longevity. Electrical, thermal, and mechanical engineering is required to ensure the battery system meets performance requirements while not exceeding safe operating temperatures. During normal operation, battery heat generation can increase temperatures and accelerate degradation mechanisms that shorten the usable lifespan. If heated above a certain threshold or mechanically damaged, the battery can enter a thermal runaway reaction leading to severe fires and safety hazards. Since these two thermal issues are some of the most significant for Li-ion batteries, a considerable portion of this chapter discusses the various thermal management strategies for avoiding high temperatures and preventing the propagation of thermal runaway.

S. Al-Hallaj (✉) • S. Wilke • B. Schweitzer
AllCell Technologies, LLC, 2321 W. 41st St, Chicago, IL 60609, USA
e-mail: SAlHallaj@allcelltech.com

S. Murad et al. (eds.), *Water, Energy & Food Sustainability in the Middle East*,
DOI 10.1007/978-3-319-48920-9_8

Lastly, this chapter provides a brief case study of a lithium ion battery to provide energy storage for a solar power farm, to buffer the grid when the farm goes on- or off-line. This example illustrates the many aspects involved in the cell selection, battery sizing, and thermal management.

Keywords Energy storage • Smart grids • Batteries

In discussing the situation of energy in the Middle East, central focus must be devoted to the development and implementation of sustainable energy sources and increased energy efficiency. National concerns over energy security and the global responsibility to respond to anthropogenic climate change both direct the energy conversation toward renewable technologies. It is appropriate, then, that many chapters in this book are devoted to exciting and promising advancements in the varied disciplines of renewable energy generation. However, generation is only one component of the energy equation, and in the case of most renewables another component carries equal importance: energy storage.

For technologies like wind and solar, their power generation is intrinsically periodic. The sun rises and sets each day, and in most cases the wind speed is unpredictably variable. If solar and wind power are to become dominant means of sustainable energy, what happens when the wind stops blowing or clouds obscure the sun? The need for a completely reliable energy infrastructure necessitates that renewable production be implemented in conjunction with technologies that help buffer these natural variations. In many instances, other production sources such as natural gas turbines or hydroelectricity can be ramped up and down to effectively compensate for the ebb-and-flow of renewable generation, but not as quickly as is required for fast demand spikes or rapid downturn of solar and wind. For these short transients, energy storage is necessary to provide the buffer (Denholm et al. 2010).

Figure 1 provides three examples of the role energy storage can play for grid-tied systems. The first, frequency regulation, addresses short-duration mismatches between power supply and demand due to turning on or off of large equipment loads. Variable generation ramping compensates for medium-duration changes in generation assets, often associated with changes in renewable power sources: a change in sunlight (i.e. – clouds) or wind speed. Lastly, peak shaving and load leveling involve day-long charge and discharge of an energy storage system to help reduce the peak generation/consumption, which is commonly the source of greatest expenses and infrastructural planning ("Peakers Cost Recovery Testimony" 2007). These three usage profiles span very different scales of time, energy, and power, and thus different storage technologies are necessary to meet each case's specific requirements.

Another key role of energy storage is in support of energy efficiency initiatives. Improving energy efficiency of buildings or industrial processes is often accomplished by taking advantage of "cheap" energy when it is available and storing it for later use when it is more difficult or expensive to meet peak demands.

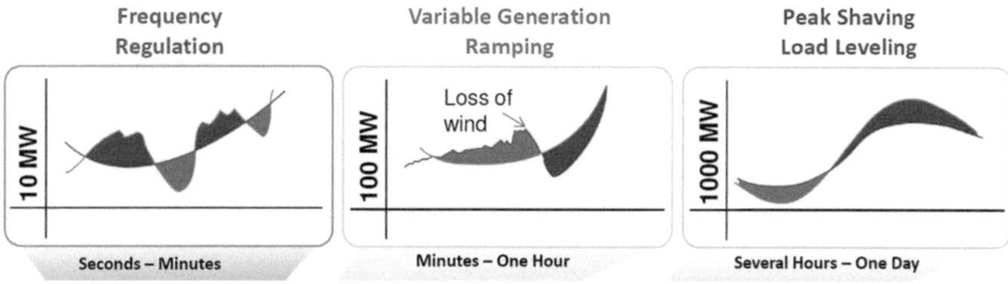

Fig. 1 Energy storage tied to the power grid can serve a variety of functions, from smoothing of momentary changes in load or renewable generation, to day-long shaving of the peak demand (Reproduced from Hearne 2014)

The nations of the Middle East have much to gain from evaluating and implementing appropriate energy storage technologies, especially in conjunction with renewable generation sources. The potential gains are numerous: increased energy security, better reliability of the power grid, and potentially lower energy pricing. This chapter is meant to provide a technical introduction to the broad field of energy storage technology, as a starting point for those unfamiliar with the topic or as a reference to those already engaged in these discussions. Toward this end, the chapter goals and its sections are threefold:

1. Provide a background on the broad categories of energy storage technologies, with particular attention given to batteries.
2. Discuss in detail the fundamentals and functioning of one particular battery technology, lithium ion. Li-ion batteries are currently the most promising battery chemistry for mobile applications and are increasing in use for stationary storage (grid and off-grid).
3. Present a specific case study to illustrate Li-ion batteries for grid applications, within the context of the Hashemite Kingdom of Jordan's energy economy.

Although the bulk of this chapter focuses on one particular energy storage technology, lithium ion batteries, references to significant texts are provided for the reader to consult regarding the other topics discussed herein.

Overview of Energy Storage Technologies

Numerous energy storage technologies are commercially viable and in various stages of implementation, of which the major categories are discussed here. From an introductory perspective, it is helpful to distinguish the different forms of energy storage based on a few key properties: the type of energy that is stored, the characteristic time to charge or discharge, the feasible scale of use, and the lifetime and cost of the integrated system. Countless other metrics can be used to classify these technologies – roundtrip energy efficiency, self-discharge rate, etc. – but these four provide a good starting point for discussion (Huggins 2010).

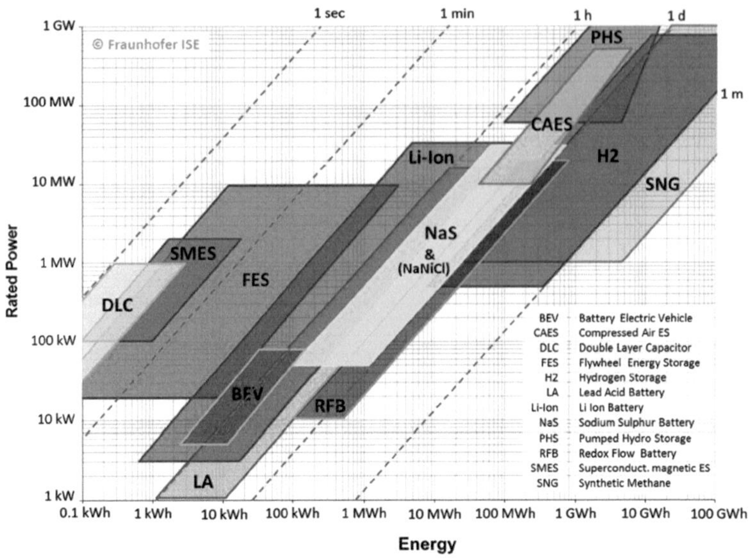

Fig. 2 Various energy storage technologies for grid applications, showing system power versus energy, with discharge time marked on the diagonals (Reproduced from Vetter 2013)

The type of energy stored – mechanical, thermal, or electrical – distinguishes how a particular technology can be utilized or converted, and so this section is subdivided into these basic categories. The characteristic charge or discharge time of a storage technology reflects its ratio of power to total energy storage, or P/E ratio. As mentioned in the chapter introduction, different grid use cases involve time scales ranging from minutes to days. For the very short time intervals, a higher P/E ratio (shorter discharge time) makes the most sense, but for a day-long discharge a much lower P/E may allow for cost savings by sacrificing power capability for a higher energy capacity. Figure 2 provides a general illustration of energy storage technologies currently in commercial use, with a comparison of feasible module sizes and discharge times.

A technology's feasible scale depends on its energy density, often expressed in gravimetric or volumetric units (kWh/kg or kWh/L), and cost normalized against its installed capacity ($/kWh or $/kW). For grid applications, energy density is not as crucial since weight and space are less often constrained than in a mobile application, such as an electric vehicle. However, the $/kWh or $/kW installation cost is one of the most critical design factors, which ties in with the fourth key metric, cost and lifetime. Given a product's expected lifetime and operating scenario, another helpful cost figure can be calculated to show the total cost of ownership normalized against the entire energy delivered throughout its lifetime. This different $/kWh figure reflects not only the initial purchase, but also the maintenance, total revenue generation, and planned frequency of system replacement. Knowledge of this lifetime $/kWh cost helps determine which different technologies will facilitate the lowest cost of electricity within the constraints and requirements for the application.

Mechanical Energy Storage

Mechanical energy storage systems rely on the translation of electrical or thermal energy directly to potential or kinetic energy. The varieties of such systems are numerous, but three architectures have proven to be reliable and mature enough to serve the needs of utilities in their efforts to balance electrical energy production and consumption: flywheel, compressed air, and pumped hydro energy storage systems.

Flywheel Energy Storage Systems

Flywheel energy storage systems (FESS) utilize the rotational kinetic energy of a rotor to store energy. A hub or disk rotor is coupled to an electric motor/generator, and the hub typically rotates in an evacuated chamber. The purpose of the evacuated chamber is to reduce the drag on the rotor from air, thereby increasing the "standby" time of the rotor when it is spinning but not actively absorbing or delivering energy to the motor/generator. Figure 3 shows a diagram of a typical FESS.

The FESS absorbs energy as the electric motor/generator spins the rotor, transforming electrical energy into rotational kinetic energy. When power is required from the FESS, the motor/generator absorbs power from the flywheel, generating electric current, and consequently the speed of the flywheel decreases. The amount of energy (E) that can be stored in a FESS is governed primarily by the speed of rotation (ω) and the mass of the rotor (m):

$$E \propto m\omega^2. \tag{1}$$

Equation 1 shows that the stored energy varies linearly with rotor mass, but increases with the square of the angular speed. This relationship manifests itself with designs that operate at high rotational velocity, in some cases exceeding 16,000 RPM ("Understanding Flywheel Energy Storage: Does High-Speed Really Imply a Better Design" 2008).

Compressed Air Energy Storage Systems

In contrast to FESS, compressed air energy storage systems (CAESS) rely on the potential energy of air when it is stored in a fixed volume under greater than atmospheric pressure. A typical grid-tied CAESS will consist of an electric motor

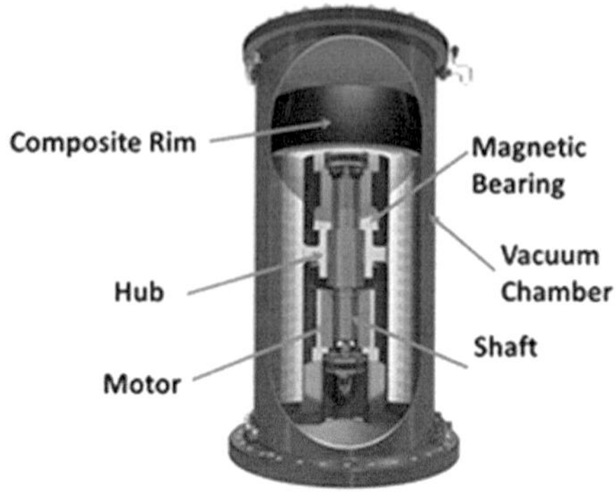

Fig. 3 A cutaway view of a flywheel energy storage system, showing the composite hub mechanically linked to an electric motor/generator, all housed in an evacuated case (Reproduced from "Carbon Fiber Flywheels" 2014)

coupled to an air compressor, and a turbo-expander coupled to a generator. When there is excess power available on the grid or when electricity rates are low, the compressor pumps air into an underground cavern, storing it between 1000–6500 PSI (McGrail et al. 2013). To generate electricity, the compressed air is routed to the turbo-expander/generator. A diagram of a CAESS plant is shown in Fig. 4.

Deployment of a CAESS plant is driven by two major factors: access to sufficient electrical grid infrastructure and availability of suitable geologic formations to serve as air reservoirs. As of 2013, there were only two operational facilities in the world, a 110 MW facility located in Alabama, USA, and a 290 MW facility located in Huntorf, Germany. Both plants utilize decommissioned salt mining caverns to store the compressed air. Analysis done by the Pacific Northwest National Laboratory shows that the Pacific Northwest region of the United States could be ideal for CAESS installations due to the porous geologic structures present there. Their estimated levelized cost of ownership for an installation ranges from 6.4–11.8 cents/kWhr, competitive with other sources of peak power generation (McGrail et al. 2013).

Pumped Hydro Energy Storage Systems

Pumped hydro energy storage systems (PHESS) operate in a similar fashion to CAESS, in that the potential energy of a working fluid is used to store or generate electrical energy. Since water is nearly incompressible, a PHESS utilizes the difference in height between two water reservoirs to store energy. A turbine is positioned in a tunnel between these two reservoirs, and when water flows from the higher reservoir to the lower one under the action of gravity, the turbine will spin a generator/motor, producing electrical current. To store excess energy from the grid, the turbine is powered by the motor/generator, pumping water to the higher reservoir.

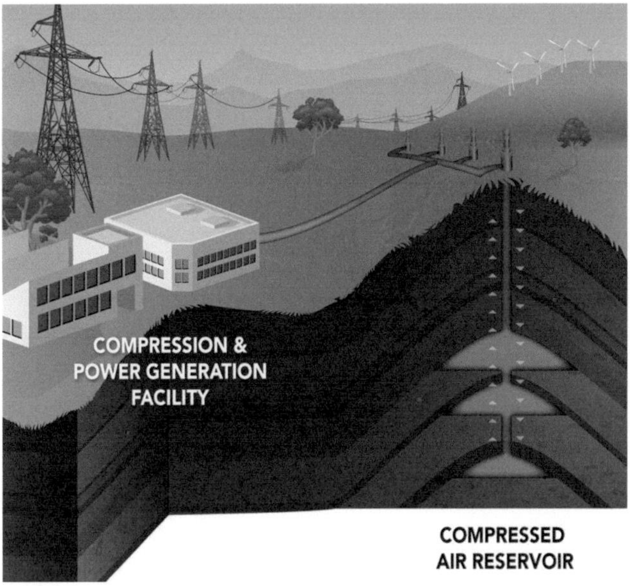

Fig. 4 Overview of a commercial compressed air energy storage plant, showing the compression/generation facility and underground cavern structure used to store the compressed air. (Image provided courtesy of the Pacific Northwest National Laboratory, operated by Battelle Memorial Institute for the U.S. Department of Energy, reproduced from McGrail et al. 2013)

As of 2015, there were 14 operational PHESS facilities in the USA with a total operational capacity of over 18 GW ("Pumped Storage Projects" 2015). Figure 5 shows a specific example of a PHESS facility, located in Ludington, MI, USA, that has a reservoir capacity of 27 billion US gallons. The peak power output of the plant is 1.87 GW and is primarily used to provide electricity during peak loads, such as during summer when air conditioning usage is high. As shown by the aerial photo, the Ludington plant consists of an elevated water reservoir with the turbines/generators in the middle of the picture.

Thermal Energy Storage

In thermal energy storage (TES), a reservoir of material is stored at a target temperature so that it can be used for on-demand heating or cooling. TES is increasing in usage for building heating, ventilation, and air conditioning (HVAC) systems. In the United States, 32–39% of total building energy consumption is expended for HVAC ("Buildings Energy Data Book" 2008), so an enormous amount of research and engineering design are being invested to improve efficiency of these systems and lower energy usage. Since these energy demands are thermal, direct storage as thermal energy is attractive because it can avoid the inefficiencies

Fig. 5 Ludington, MI, USA pumped hydro energy storage facility (Reproduced from "Pumped Storage Projects" 2015)

of conversion from electricity or other energy forms. Also, thermal energy is very easily captured from the sun, and the diurnal high-and-low cycle of outside temperatures offers the potential to store daytime solar heating and utilize it for nighttime (Dincer and Rosen 2010). TES generally uses abundant, cheap materials and has relatively simple system designs and operation. The main limitations are that the stored energy must be used for direct thermal applications, and such systems are beneficial only when the thermal energy demands are on regular cyclic periods.

One simple example of thermal storage is solar thermal water heating. Figure 6 shows a schematic drawing of how a rooftop solar collector can transfer heat to the building hot water supply through a dedicated fluid loop. Depending on the sizing of the water storage tank, daytime solar heat can effectively be stored in the water tank for usage even when the sun is not up. In this way, the building energy efficiency can be improved by taking advantage of the "free" solar energy to offset costs of heating the water from electricity or natural gas. Similar heat exchange schemes can be used to supply heat for the building air (through hot water radiative heaters) or swimming pools.

In the hot water example, the storage medium (water) operates by changing the temperature and using sensible, or specific heat. While effective, another interesting technological approach is to instead harness the thermal energy in the latent heat of a material's phase change. The primary benefit of latent heat is that much higher energy storage densities can be achieved. Using water, the phase change from solid to liquid can store 334 J/g of latent heat at one specific temperature; using sensible heat, this amount of energy would require changing the water temperature by 80 °C. Thus, latent heat can deliver much higher energy and power densities over narrow temperature ranges.

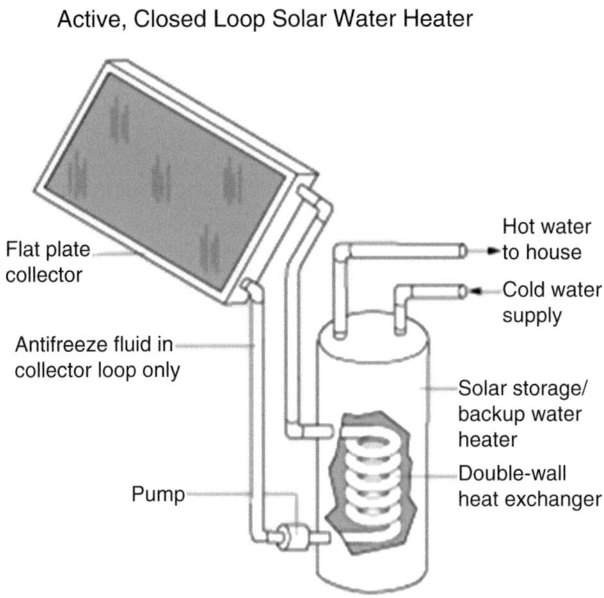

Fig. 6 Basic diagram of solar thermal hot water for domestic or commercial building use (Reproduced from "Solar Water Heaters")

In one embodiment of latent heat TES (LHTES) for building HVAC, the thermal reservoir can be charged at a cold temperature to aid in the cooling needs for the building. The energy is "stored" in the capacity of the LHTES unit to remove heat on demand. Unlike solar powered TES, cold storage does not take advantage of readily available or cheap energy resources. Rather, the cold TES market has grown due to utility energy pricing incentives. In many localities, time-of-use electricity pricing results in customers paying significantly more for electricity during peak periods, such as hot summer afternoons, than during off-peak times ("Business Rate Summary" 2008). This situation sometimes makes it cheaper for a building owner to install energy storage for peak shaving – storing off-peak energy for use during the day – despite the cost, conversion losses, and inefficiencies of the energy storage. One commercial example of this strategy is the Ice Bear LHTES unit manufactured by Ice Energy (St. John 2014), a USA-based company (Fig. 7).

Electrical Energy Storage

Electrical energy storage refers to the use of batteries and supercapacitors for accepting and delivering electric current. Electrical energy storage is, in fact, a bit of a misnomer since electricity cannot be stored. In batteries, the electrical charge and discharge is actually mediated through reversible electrochemical

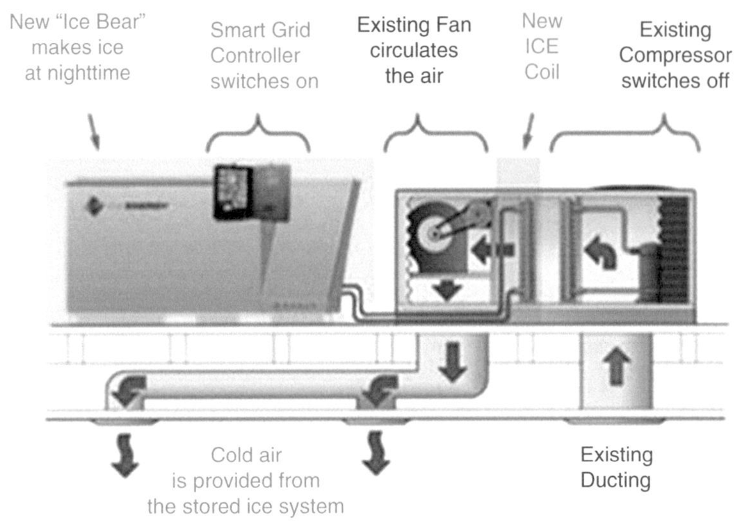

Fig. 7 Schematic illustration of Ice Energy's Ice Bear thermal energy storage system, which can shift the peak demand for building cooling by charging (i.e. – making ice) during off-peak hours and discharging during hot afternoons (Reproduced from "Ice Energy Project" 2008)

reactions – energy storage in chemical bonds – but from a practical standpoint the batteries are storing and providing electricity. Compared to mechanical and thermal energy storage, batteries offer the enormous benefit of direct electrical usage and grid connection without the need for conversion. This advantage, however, comes at the expense of much higher system cost and complexity, shorter lifetime, and, in some cases, more significant safety risks.

Although batteries can be made using a plethora of different chemistries, all batteries' basic components and mode of operation are more-or-less the same (Ehrlich 2002; Santhanagopalan et al. 2015). For the purposes of this chapter we consider only secondary, or rechargeable batteries, and not primary (single-use) batteries. Figure 8 shows the internal construction of a secondary battery and its operation during discharge. On the left-hand side of the battery is the positive electrode, or cathode, and on the right-hand side is the negative electrode, or anode. In between the electrodes is a separator, which electrically isolates the two electrodes, and an electrolyte that facilitates the movement of ions from one side to the other. In this example, the battery operates using positively charged transport ions. During discharge, the transport species is oxidized at the anode, which releases an electron, and the resulting transport ion travels through the electrolyte to the cathode. The electron travels through an external circuit and arrives at the cathode, where it reduces the transport ion. The flow of electrons from negative to positive electrodes results in current flow to an external load on the circuit. During charge, the battery operates in the same manner but in reverse direction.

A few standard definitions are helpful in understanding battery operation. State-of-charge (SOC) is the fraction of a battery's present stored capacity divided by the

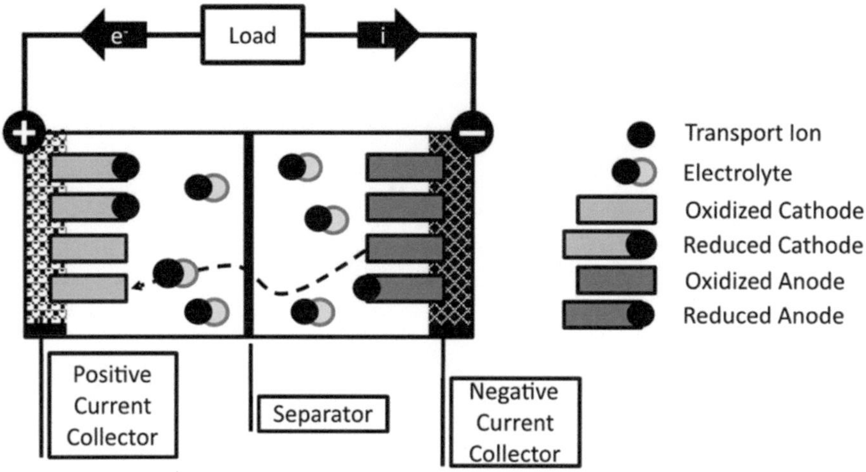

Fig. 8 Generalized battery operation during discharge. When the transport ion is positively charged, it flows from anode to cathode (*right to left*) through the electrolyte. (Reproduced by permission from Santhanagopalan et al. 2014. © 2014 by Artech House, Inc.)

maximum capacity it can hold when fully charged. The depth-of-discharge (DOD) is equal to 1 – SOC, which reflects what portion of the battery's capacity has already been discharged. The C-rate is the ratio of electrical current to the battery's nominal capacity (C-rate is a very similar concept to the P/E ratio). For example, a 10 Ah battery that is presently holding 6 Ah of charge and is discharging at 20A is at 60% SOC, or 40% DOD, discharging at a 2C rate. Depending on the chemistry, different battery types are capable of different DOD (sometimes restricted to less than 100%) and C-rates.

A few different battery chemistries have been successfully implemented for commercial power and grid applications: lead acid (PbA), nickel metal hydride (NiMH), sodium sulfur (NaS), lithium ion (Li-ion), and redox flow. Figure 9 provides a qualitative comparison of these chemistries on six key metrics for grid applications. Four of these criteria – specific energy, power, cycle life, and low cost – were already discussed in their relevance to grid applications. Tolerance of extreme hot or cold temperatures is another important factor for battery performance because normal operation can become impossible or unsafe outside of certain temperature ranges. The last criteria, efficiency, is pertinent because it reflects how much energy is lost in storage due to charge/discharge inefficiency.

Brief discussions for each of these battery types are provided in the subsections below, and references to significant texts and review papers are provided for further reading.

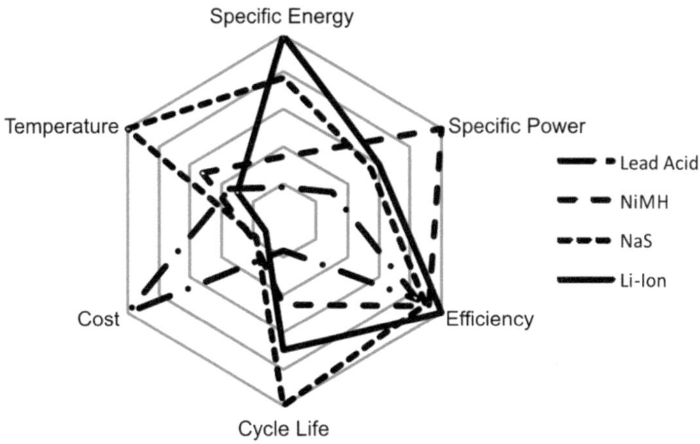

Fig. 9 A qualitative comparison of four battery chemistries, evaluated on criteria necessary for grid-based applications. Arbitrary units. (Reproduced by permission from Santhanagopalan et al. 2014. © 2014 by Artech House, Inc.)

Lead Acid

Lead acid (PbA) batteries have the longest history of use compared to other chemistries, and they are still extensively used, particularly for automotive and uninterruptible power supply applications. In a PbA battery, a lead dioxide cathode and lead anode are both submerged in a sulfuric acid electrolyte bath. During discharge, both electrodes are converted to lead sulfate; upon charging they revert back, and water in the electrolyte is broken down into gaseous hydrogen and oxygen. Two classes of lead acid batteries are distinguished by how they address this gas formation. Open systems release the gaseous products and therefore require periodic refill with water. Sealed systems, on the other hand, use pressure-regulating valves to contain the gases and promote their recombination to water. To further protect against gas buildup, sealed systems also make use of either a silica gel or an absorbed glass mat in the electrolyte.

The primary selling points for PbA are its very low cost and long history of reliability, which are why the chemistry continues in widespread use. However, lead acid batteries suffer from low specific energy (typically no more than 35 Wh/kg) and power density. Also, they normally only achieve a few hundred cycles before end-of-life, and they are not tolerant of deep DOD operation or high charge/discharge C-rates. The limited cycle life is further exacerbated by an acute sensitivity to temperature: battery life is cut roughly in half for every 8 °C increase in operating temperature (Pavlov 2011).

Nickel Metal Hydride

In a nickel metal hydride (NiMH) battery, a nickel oxyhydroxide cathode is separated from a hydrogen-containing metal alloy by a potassium hydroxide electrolyte. NiMH batteries are a newer technology than lead acid and offer substantial improvements in energy and power density. They are similar in performance to nickel cadmium (NiCd) batteries, but have displaced NiCd in almost all applications due to the undesirable toxicity and cost of cadmium. NiMH became the first battery technology to power hybrid electric vehicles at a large commercial scale, including cars such as the Toyota Prius and Ford Escape.

In addition to the benefit of high specific power, NiMH batteries accept high charge and discharge rates and offer a longer lifetime. However, their self-discharge rate is very high compared to other chemistries (sometimes as high as 30% per month), and they do not tolerate operation at low temperatures. One of the largest barriers to use is their high cost, due predominantly to the raw nickel material (Ma et al. 2011).

Sodium Sulfur

Sodium sulfur (NaS) batteries do not have nearly the same commercial scale of application compared to lead acid or NiMH, though they have been implemented for some grid projects. The primary feature that makes NaS desirable for grid use is the chemistry's long cycle life – over 5000 cycles of full DOD cycling. NaS also offers a substantial specific energy of 100 Wh/kg or more. The major challenges posed by this chemistry arise from its materials and operating temperature. A NaS battery comprises a liquid sulfur cathode, molten sodium anode, and a solid aluminum oxide electrolyte. To keep these components in the correct operating states, the battery must be maintained at a temperature above 300 °C. Such high temperature operation leads to increased corrosion and thermo-mechanical wear on the other battery components (Kim et al. 2013).

Lithium Ion

In most lithium ion cells, a lithium metal oxide cathode and a graphite anode are separated by an organic liquid electrolyte. Unlike the other battery chemistries discussed so far, the active materials in a Li-ion cell maintain their chemical crystal structure during charge and discharge. This means that the lithium ions that shuttle between the electrodes do not rearrange the positions of other atoms around them; the ions are simply inserted and removed, a process called intercalation (Fig. 10). For this reason, Li-ion materials are known as intercalation compounds.

Li-ion batteries have received the most research and commercialization attention over the past decade and have become the sole chemistry of choice for portable

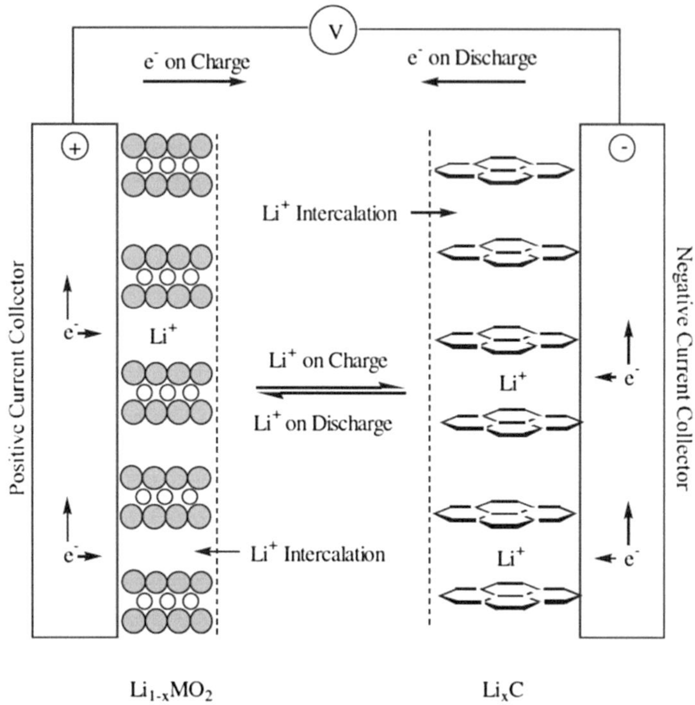

Fig. 10 Schematic drawing of charge and discharge inside a lithium ion battery, showing the intercalation of Li^+ ions in the atomic planar structure of the electrode materials. (Reproduced with permission from D. Linden and T. Reddy, *Handbook of Batteries, 3rd Edition*, McGraw-Hill, 2002.)

electronics and electric vehicle development. Boasting an unparalleled specific energy of 200 Wh/kg, Li-ion also offers high power density and a cycle life of several thousand deep DOD cycles. This combination of energy, power, and longevity make it the ideal chemistry for mobile applications that are weight sensitive (Santhanagopalan et al. 2015).

Cost is one of the main barriers to lithium ion technology, which remains the most expensive of the battery chemistries presented here. Li-ion also suffers from a temperature sensitivity similar to lead acid, though not quite as severe, which can drastically reduce the useful lifetime in hotter operating environments. Lastly, Li-ion batteries introduce a few particular safety issues, such as thermal runaway, which will be discussed more in a later section of this chapter.

Redox Flow

Redox flow batteries are an emerging technology that is only now being tested at a pilot scale for grid energy storage. Flow batteries are unique, in that their active

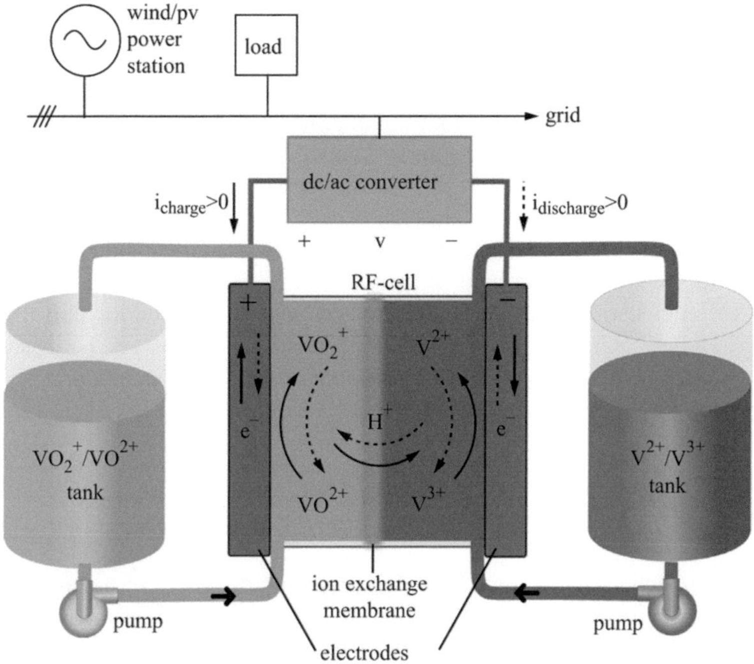

Fig. 11 Diagram of a vanadium redox flow battery connected to a power grid. (Reprinted from Alotto et al. 2014. With permission from Elsevier.)

materials are permanently dissolved in electrolyte solution. A diagram showing one of the more popular flow battery chemistries, vanadium, is shown in Fig. 11. The anode- and cathode-containing electrolyte solutions, called the anolyte and catholyte, are held in separate storage tanks. During charge and discharge, these solutions are pumped past a common ion exchange membrane where the electrochemical reactions take place.

One of the unique advantages of flow batteries is that their energy and power capabilities can be decoupled from a design perspective, since the size of storage tanks and the exchange membrane surface area can be adjusted independently. The actual size (energy) of the battery is also not strictly limited, allowing for simple scale-up for larger systems. However, due to the large solution volumes, flow batteries have rather low energy density, and the complexity of pump and control systems must be addressed prior to widespread implementation (Alotto et al. 2014; Weber et al. 2011).

Design and Optimization of Lithium Ion Batteries

In this section, major aspects of lithium ion battery design are discussed. The optimization of battery systems is a truly interdisciplinary field, which is reflected by the variety of topics here. Chemistry knowledge is required to identify which types of Li-ion chemistries are best suited for particular operating conditions. Mechanical design is needed to build the battery modules and integrated systems. Thermal analysis must be performed to understand and address battery heat generation, with special attention paid to the safety risks of thermal runaway and fire. Electrical engineering is necessary to connect individual batteries into the high voltage and capacity systems for grid-tied storage, and proper hardware and software must be developed to ensure their safe and reliable operation.

Cathode and Anode Materials

Lithium-ion technology is fundamentally driven by the attractive material properties of lithium metal, mainly its low density and low electrochemical reduction potential. These two properties enable Li-ion batteries to have a high specific capacity (mAh/g) and a high specific energy (Wh/g). To realize a completely rechargeable battery, the Li^+ ions must be able to intercalate into different host anode and cathode materials.

Common choices for Li-ion cathodes are shown in Table 1. These compounds have a high reduction potential relative to lithium, indicated by their average voltage. Most cathode materials fall into one of three crystal structures: layered, spinel, or olivine. Aside from the differences in capacity and electrochemical potential, the major differences are the pathways available for the Li^+ ions to move through the material. Layered structures offer a 1-dimensional pathway for Li^+ ions to move, while spinel and olivine structures offer a 2-D/3-D pathway for ions to move. In practice, this results in more facile movement of ions in the spinel and olivine structures, which leads to the battery being able to rapidly discharge during use – high maximum C-rate – as the ions move from the anode into the cathode. (The permissible charge C-rate is governed by the ability of the anode material to accept Li^+ ions.)

Different cathode materials also have different potential windows in which Li^+ ions can be reversibly inserted and extracted. For example, full de-lithiation for $LiFePO_4$ occurs at 3.7 V relative to Li/Li^+, while $LiCoO_2$ can tolerate higher voltages, up to 4.3–4.4 V relative to Li/Li^+ (Ehrlich 2002). Extracting more Li^+ ions past this voltage can result in degradation of the cathode crystal structure, heat generation, and permanent loss of battery capacity, so the normal operational voltage window is always specified by the cell manufacturer.

The voltage, C-rate capability, and specific capacity are the three main qualities that guide cathode selection based on an application's requirements. For example,

Table 1 Common Li-ion cathode materials (Nitta et al. 2015)

Compound	Acronym	Crystal structure	Specific capacity, theory/typ. (mAh/g)	Average voltage (V)
$LiCoO_2$	LCO	Layered	274/148	3.8
$LiNi_{0.33}Mn_{0.33}Co_{0.33}O_2$	NMC	Layered	280/160	3.7
$LiNi_{0.8}Co_{0.15}Al_{0.05}O_2$	NCA	Layered	279/199	3.7
$LiMn_2O_4$	LMO	Spinel	148/120	4.1
$LiFePO_4$	LFP	Olivine	170/165	3.4

power tools that require a high C-rate discharge typically utilize a $LiFePO_4$ cathode, which carries the tradeoff of a lower specific capacity. An electric car, however, prioritizes energy density over discharge rate, so $LiNi_{0.33}Mn_{0.33}Co_{0.33}O_2$ is the most commonly used cathode given its high specific capacity and voltage.

The other active component of a Li-ion battery is the anode material. Ideal attributes for an anode material are low reduction potential relative to lithium, high specific capacity, and a stable structure that does not degrade over time as Li^+ ions move in and out. The ideal anode material would be pure lithium metal, which is indeed used in primary lithium batteries. However, if a cell made with a lithium metal anode is recharged, the Li^+ ions do not re-incorporate neatly to the anode. Instead, they form dendritic structures that can pierce through the separator to touch the cathode, causing an electrical short circuit that can lead to cell self-discharge, or worse, thermal runaway. Given this risk, it is safer to intercalate the Li^+ ions into a host anode material, as is done with the cathode of the cell. The dominant anode material used in the current generation of Li-ion technology is graphite. Graphite exhibits good specific capacity (~372 mAh/g) and high reversibility of ion intercalation. Additionally, the material is easily sourced and relatively low in cost. Lithium titanate oxide (LTO) has been successfully commercialized as an anode material and displays exceptional cycle life capabilities. Compared with graphite, LTO can intercalate Li^+ ions much more easily, leading to very fast charge rates; however, LTO has a lower capacity (~175 mAh/g vs. 372 mAh/g) and a higher working potential (~1.55 V vs. Li/Li^+ compared to 0.1–0.3 V for graphite) (Nitta et al. 2015).

Other materials that can obtain higher specific capacities are actively being researched, most notably silicon. Silicon is highly attractive due to its incredible theoretical capacity (~3400 mAh/g), but suffers from massive volume expansion as the Li^+ ions intercalate into it. This large volume change (250%) results in fracturing of the Si particles, causing them to lose contact with each other and become electrochemically inactive (Nitta et al. 2015).

Cell Formats and Module Construction

Just as primary alkaline batteries come in a variety of sizes (AA, C, D, etc.), Li-ion rechargeable batteries come in different shapes and sizes, as well. Common cylindrical size Li-ion cells include the 18650 and 26650 families, where the first two digits indicate the cell diameter in mm (18 and 26 mm), and the last three digits represent the cell length in mm (65.0 mm). Common electrical capacities range from 2–4 Ah for cells in this size range. Other families of cylindrical cells exist as well, with some larger ones having capacities in the 40–50 Ah range. Other Li-ion cell formats include rectangular prismatic cans and soft pouch construction. Prismatic cans are advantageous in situations where it is convenient to package the cells together closely. Soft pouch construction results in a cell that can be made at lower cost compared to a prismatic can, due to the lack of rigid casing surrounding the cell internals. Figure 12 shows examples of these three cell formats.

Although the casing and shape of Li-ion can vary, they all share common internal construction to a large degree. The anode and cathode active materials are each mixed with a solvent to form a slurry, which is then coated onto rolls of thin copper sheet (for the anode foils) or aluminum sheet (for the cathode foils). In the case of a cylindrical cell, a separator sheet (typically made of polyethylene) is captured between the anode and cathode foils, and the three layers are wound to form what is known as a "jellyroll." The jellyroll is then inserted into the steel can, leads are bonded to the foils to electrically connect them to the cell can, and electrolyte is added. In the case of a pouch cell, the foils and separator are not wound, but stacked and then inserted into a soft metallized polypropylene case. As with the cylindrical cell, electrical tabs are welded to the foils, electrolyte is added, and the pouch is hermetically sealed.

Li-ion battery systems, such as those used in smart grid applications, are often assembled from thousands of individual Li-ion cells arranged in modules and packs. Using the 18650 format, a grid-tied battery can be constructed by electrically connecting 50–150 cells in parallel to obtain the required system capacity, typically 100–300 Ah. The electrical connections are made using a current collector, typically a laminated copper/nickel composite, spot welded to the individual cells, joining them in parallel to form a module. Several modules are then linked together in series to form packs, which when connected obtain the target system voltage, typically 600–900 V. The packs are usually connected using electrical contactors for safety, to limit the voltage present in the battery when the system is not in use and the contactors are de-energized ("AllCell Technologies Energy Storage Systems" 2015).

The assembly of this system highlights the tradeoffs of using cylindrical versus pouch cell formats. Cylindrical cells have been in commercial production for much longer and offer a lower cost per kWh than pouch cells. Pouch cells, however, can achieve much higher capacities per cell, sometimes as high as 75 Ah. One 75 Ah pouch cell is equivalent to 25 high-energy 18650-format cells (3 Ah each), so the module/pack assembly can be greatly simplified by using the pouch format, and

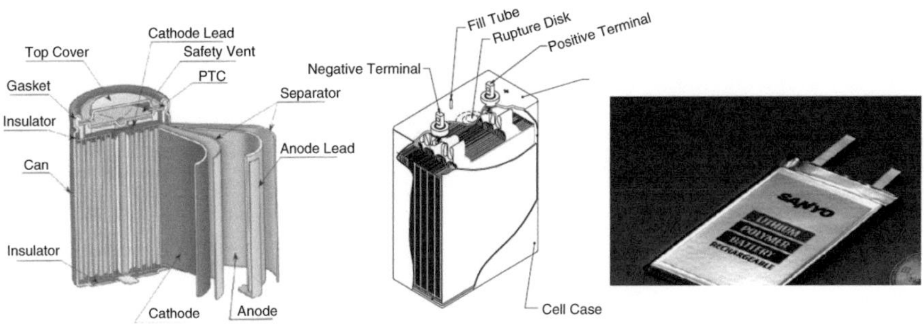

Fig. 12 Diagrams and images of cylindrical, prismatic, and soft pouch Li-ion cells. (*Leftmost* reprinted from Johnson and White (1998). With permission from Elsevier.) (*Center* and *righmost* reproduced with permission from D. Linden and T. Reddy, *Handbook of Batteries, 3rd Edition*, McGraw-Hill, 2002.)

these system simplifications can offer cost savings that compensate for the higher cell cost. Safety is another key differentiator for the different cell formats: if one cell is damaged and enters thermal runaway, a 75 Ah pouch will release 25 times more energy and flammable gases than a single 3 Ah cylindrical cell.

Thermal Issues and Management

No energy storage system returns all the stored energy back to the environment in a form that is suitable for work. The major form of inefficiency in a Li-ion battery is the generation of heat from several sources, including the cells themselves. The following sections outline the basic calculations for heat generation, hazards inherent with unregulated heat generation, and descriptions of thermal management systems to mitigate these hazards.

Heat Generation

All Li-ion cells generate heat as they are charged or discharged. The two major forms of heat generated are irreversible and reversible heat (Schweitzer et al. 2015; Bernardi et al. 1985). Irreversible heat generation occurs due to the ohmic heating effect, originating from the electrical current flowing through the anode and cathode foils, as well as from the flow of ions through the electrolyte. Reversible heat generation occurs due to the change in entropy of the anode and cathode materials as they undergo structural changes produced by the movement of Li^+ ions in and out of them.

The total amount of heat generated by a Li-ion cell, Q_{total}, can be expressed as the sum of the irreversible and reversible heat,

$$Q_{total} = Q_{irr} + Q_{rev} \tag{2}$$

which can be defined:

$$Q_{irr} = I^2 R_{ohmic} \tag{3}$$

$$Q_{rev} = IT\frac{dOCV}{dT} \tag{4}$$

Q_{total}, Q_{irr} , and Q_{rev} have units of watts, I is the total current passing through the cell in amps, R_{ohmic} is the equivalent cell resistance in Ohms, T is the temperature of the cell in Kelvin, and $\frac{dOCV}{dT}$ is the sensitivity of the cell open circuit voltage with temperature, which is determined experimentally. R_{ohmic} is a strong function of temperature and cell SOC, increasing with decreasing temperature and typically increasing as cell SOC decreases. To help illustrate the relative quantities of heat generated and the temperatures that Li-ion cells can reach during operation, Fig. 13 shows the expected temperature of two 18650 Li-ion cells with one discharged at a C/2 rate and the other at a faster 2C rate. Both cells were simulated in open air, with no forced air cooling (Schweitzer et al. 2015).

Thermal Runaway Hazards

In addition to the heat generated during normal operation, Li-ion cells can undergo a phenomenon known as thermal runaway. Thermal runaway can begin to occur when a Li-cell reaches an abnormally high temperature, usually in the range of 70–80 °C. At these temperatures, unwanted exothermic chemical reactions begin to occur inside the cell. If the reaction heat generation is not dissipated, the cell temperature continues to climb, initiating more exothermic chemical reactions. This feedback loop can quickly escalate, resulting in peak cell temperatures in excess of 700 °C and the expulsion of superheated gases and flames. Additionally, the heat released from the failing cell can raise the temperature of surrounding cells in a module high enough to force them into thermal runaway, as well, causing thermal runaway propagation.

For large grid-scale Li-ion systems, a single cell in thermal runaway could quickly turn into thousands of cells in thermal runaway, creating an emergency situation. Such situations are not hypothetical, as the battery fire that occurred in a Boeing 787 Dreamliner demonstrated in January 2013 (Williard et al. 2013). Thermal runaway has also been identified in consumer electronics fires, including 6 Dell laptops utilizing Sony Li-ion cells that caught fire between 2005 and 2006. These fires prompted Dell to recall 4.2 million laptops and brought the issue of thermal runaway into the public spotlight (Takahara 2006).

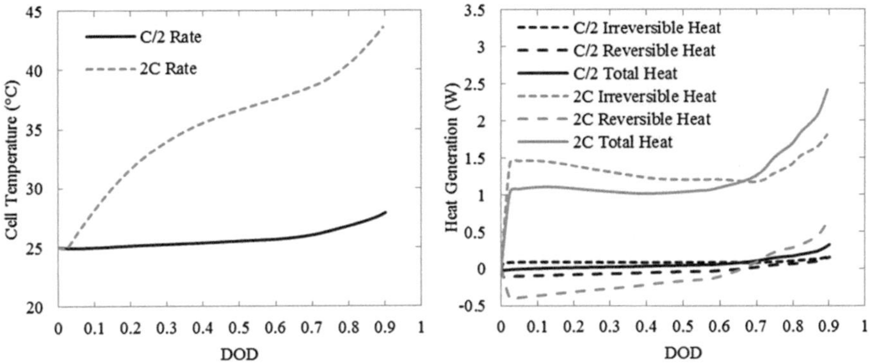

Fig. 13 Example of temperature rise and heat generation rate for an 18650 cell discharged at a C/2 and 2C rate in open air

Thermal Management Systems

Managing the temperature of a Li-ion battery system is the job of the thermal management system (TMS). A TMS should be designed so that the battery temperature can be kept low enough to reduce cell degradation (covered below) and prevent thermal runaway propagation. A Li-ion TMS can be classified as either active or passive in operation. An active system requires energy input to remove heat from the battery, whereas a passive system relies on natural convection to remove heat. A passive system may also simply rely on the heat capacity of the battery itself to limit temperature rise.

Active TMSs typically utilize pre-cooled air or liquid, which is then circulated around the battery cells, absorbing the heat generated. Although effective at controlling cell temperature, they require power input to pre-cool the fluid and pump it through the battery. Large grid scale energy storage systems that run continuously often employ an active cooling system to remove the constant heat load (e.g. – for frequency regulation services). The cooling system can either be integrated as part of the battery, or the battery can simply use fans to draw in conditioned air from the surroundings.

Unlike active thermal management, a passive TMS does not require additional energy input to control the temperature of the battery. The simplest form of passive TMS is simply the battery itself. If the cells and surrounding packaging have sufficient heat capacity, the total temperature rise of the battery may be limited to a safe temperature during operation. If the heat capacity of the battery is not enough, phase change material (PCM) may be thermally connected to the Li-ion cells. The PCM is capable of absorbing a large amount of thermal energy as it melts, up to 200–250 J/g ("PureTemp Technical Data Sheets" 2015). AllCell Technologies (Chicago, IL, USA) has pioneered the use of PCM for Li-ion batteries and has enhanced the PCM thermal conductivity by embedding it in a graphite matrix. The use of these PCM composites with Li-ion cells has been shown to be sufficient in managing the temperature rise of certain Li-ion batteries during transient operation

(Schweitzer et al. 2015). Graphite enhanced PCM materials have also been shown to arrest thermal runaway propagation in Li-ion battery packs (Kizilel et al. 2009). The thermal capacity gained from the melting of the PCM, coupled with the good heat conduction of the graphite matrix, results in the heat being conducted away from a failing cell that has entered thermal runaway. This operation is completely passive; no energy needs to be added to the systems to enable this propagation quenching behavior. In contrast, an air or liquid cooled TMS may not be able to arrest thermal runaway propagation if the system is not active.

To illustrate the effectiveness of PCM-graphite composites for preventing thermal runaway propagation, destructive testing studies have been performed on commercially produced e-bike batteries (Wilke et al. 2017). These battery packs incorporate the PCM composite encasing 40 cells connected with nickel current collectors, packaged inside an aluminum case with plastic endcaps (Fig. 14). To trigger an intentional thermal runaway event, a nail was inserted into one of the 18650 cells located in the pack corner (Cell 1). Thermocouples placed at various locations throughout the pack were used to measure the temperature response and corresponding danger to the surrounding cells. As shown in Fig. 14, upon nail penetration Cell 1 reached 180 °C, but the PCM material absorbed and distributed the heat such that the immediate cell neighbors did not exceed 109 °C. Within 1 min, all live cells returned to a safe temperature below 80 °C, preventing any risk of thermal runaway propagation.

Cell Lifetime and Performance Degradation

Another critical design concern for Li-ion systems, especially for grid applications, is the battery's useful lifetime. Due to various thermal, mechanical, and chemical processes within Li-ion cells, their capacity fades over time and their internal resistance increases, manifested as a reduction in achievable power. Figure 15 shows an example of cell capacity fade and resistance growth over the lifetime of a cell, cycled at various temperature and voltage conditions. As shown by the different curves, the decline in cell performance due to cycling is affected by both the temperature and peak charge voltage of the cell. Since the aging is a gradual process without a self-evident endpoint, the end-of-life (EOL) can be defined differently depending on the application, but is normally understood as when the cell declines to 80% of its beginning-of-life (BOL) capacity or rated power. During battery design, a battery is usually oversized for its initial use to ensure it can still meet the capacity and power requirements at EOL after the anticipated degradation. However, to keep battery cost low it must not be excessively overbuilt. To achieve both these goals, sufficient experimental testing and/or modeling is necessary to accurately predict degradation behavior.

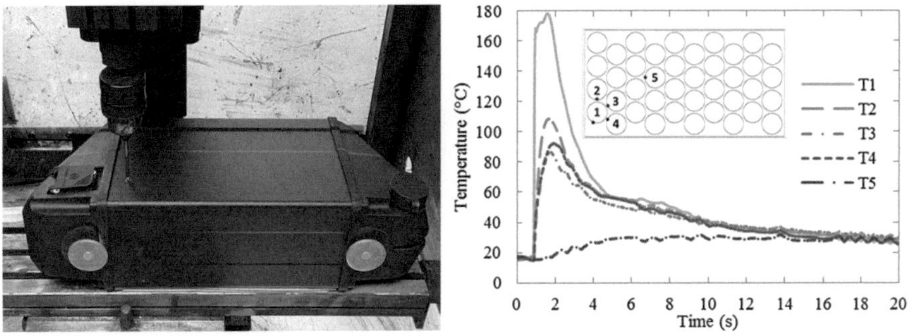

Fig. 14 Thermal runaway testing of an e-bike battery pack and temperature data of cells within the pack. After nail penetration of the corner cell, the temperatures of the surrounding cells show safe prevention of thermal runaway propagation

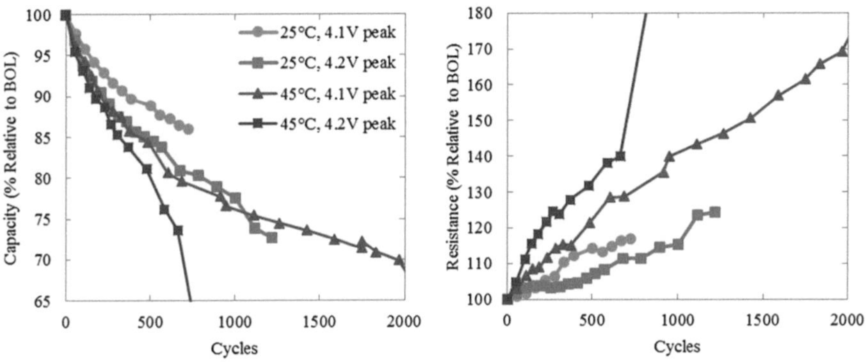

Fig. 15 Example of Li-ion capacity fade and resistance growth for a high-energy 18650 cell (2.9 Ah) with a graphite anode and NCA cathode chemistry. Data shows cycling results under various temperature and peak charge voltage conditions

Cell Degradation Mechanisms

The main mechanisms for cell aging can be traced to a few fundamental processes occurring within each cell, which are important to know because they guide the types of testing and modeling useful for life prediction. The degradation processes are classified by their effects on the cell's calendar and cycling aging. Calendar aging refers to those degradation effects that occur all the time, even when the battery is not in use. Cycling aging, conversely, deals with the degradation directly due to charge/discharge cycling of the cell. The following aging mechanisms, though not an exhaustive list, provide a basic understanding of cell degradation.

Calendar aging has one root cause: electrochemical instability of the cell's internal components. In a Li-ion cell, the graphite anode is unstable in the presence of the electrolyte at its normal operating potential. Similar to how steel develops a layer of rust in the presence of air, lithium inside the graphite reacts with the electrolyte to form a surface electrolyte interface layer, or SEI. Fortunately for Li-ion, the SEI is a passivating film, so its presence stabilizes the anode against further corrosion. While the SEI formation rate declines over time, the degradation process is always occurring whether the cell is used or not. From a practical viewpoint, the formation of SEI consumes some of the available lithium in the cell, resulting in capacity fade, and the film adds an additional barrier to the ion transport, causing increased internal resistance. The parasitic reactions that form the SEI are also highly sensitive to temperature, exhibiting an exponential dependence, so calendar aging is strongly affected by both the ambient environmental temperature and the self-generated heat during cycling. Due to this temperature linkage with cycling, if battery usage generates heat that is not effectively removed from the system, the cells' calendar aging rate will increase (Broussely et al. 2005).

Calendar aging is a key concept to understand because most battery purchasers only consider the cycle life requirement for their application. For hot ambient environments or for batteries under continuous usage (high, continuous heat generation), calendar aging can contribute more to the total battery degradation than cycling aging. For high-power batteries it is important to remember that the irreversible heating scales with the square of the electrical current (Eq. 3), so for a doubling in C-rate the heat generation will quadruple. The selection of thermal management – forced air, liquid, or passive cooling – must provide sufficient cooling capacity based on the battery's heat generation and target for operating temperature. Focusing solely on cycle life provides only half the picture of cell lifetime prediction.

Cycling aging is a result of thermal and mechanical strain on the cell components. The anode and cathode active materials experience a volume expansion and contraction whenever lithium is intercalated in and out of their crystal structures. This repeated volume change results in mechanical strain on the particles, which can fracture or delaminate from the electrode over time. Intuitively, smaller volume changes will result in less severe particle damage, which explains why cells exhibit longer cycle life when their DOD is reduced. The result of particle damage is seen in cell capacity fade, since there may be less useful material connected to the electrodes, and resistance growth, since the effective conductivity of the particles declines due to fracture (Broussely et al. 2005 Safari et al. 2010a, b). From a design perspective, cycling aging can be reduced by oversizing the battery so as to operate at a lower DOD, with the obvious tradeoff of a higher system cost.

Lifetime Performance Testing and Modeling

Since battery lifetime depends on numerous factors, significant testing and insightful modeling work is necessary in order to appropriately design a Li-ion system and be confident of its expected useful life. Such testing protocols must investigate and identify the effect of temperature, SOC, DOD, cycling frequency, charge and discharge rate, etc. on the battery calendar and cycling aging. These design considerations of lifetime are further complicated by the variety of Li-ion chemistries, manufacturers, and rapidly advancing materials developments. To gain a truly accurate picture of battery lifetime, extensive testing and modeling work must be performed on each new cell model. The enormity of this task has been addressed by different groups via three different approaches:

1. Perform exhaustive cell testing, as close to real-world application as possible, and use the test data to back up system design and lifetime targets. In most cases, accelerated testing is achieved by cycling continuously at elevated temperatures. This approach is often used by pack integrators, since it gives the highest confidence level for one particular use case, but it requires a long time of testing prior to product launch and is not necessarily applicable to other application scenarios.
2. Utilize sophisticated, complex software modeling tools to simulate the underlying physics of cell degradation ("AutoLion Software"; "Battery Design Studio Software"). These modeling packages require knowledge of the cell components: chemistry, mass loadings, and construction. This approach is only available to cell designers who have such knowledge of cell internals, and the life predictions are only useful from a design perspective (i.e. – Design A is better than Design B) and cannot accurately predict cell life.
3. Perform a limited set of standard cell testing, and use the data to fit semi-empirical models to cell degradation (Santhanagopalan et al. 2015; Schmalstieg et al. 2014). Though the test matrix is more manageable, the model is restricted to a few degradation mechanisms that are subjectively predefined. This approach is a useful compromise of the previous two, in that it delivers physics-based predictions for a variety of applications that are grounded by some limited amount of test data. However, since no testing is performed for exact usage scenarios, the confidence level is lower than provided by Approach 1.

Balance of System

Although the actual Li-ion cells make up the largest expense in battery pack cost, the several components that comprise the remaining balance of system also factor into the design (Fig. 16). The most important of these non-cell items is the battery management system (BMS), the electronics hardware and onboard software that monitor and control battery operation. Since Li-ion poses such significant safety risks if operated beyond the rated voltage, current, and temperature limits, the BMS acts as a smart controller and watchdog to prevent unsafe operation. If any of these unsafe conditions occur, the BMS will fault and isolate the battery to prevent damage. In many instances, the BMS also serves as the point of communication to other system hardware and the external operator (e.g. – a motor controller or grid demand signal). Other major components for grid-tied systems include thermal management, already discussed, and the AC-DC inverter, which is necessary to connect the DC battery with the grid.

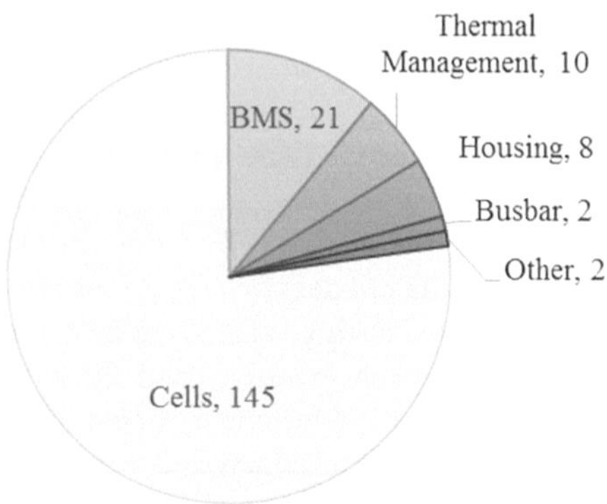

Fig. 16 Material costs of major components in a Li-ion battery pack for mass produced electric vehicles in 2020. Numbers shown are $/kWh (Reproduced from "Cost of Mass Produced EV Battery in 2020" 2015)

Case Study of Li-Ion Systems for Smart Grid Applications

In the last section of this chapter, a case study is presented to illustrate some of the basic design concepts applied to a lithium ion battery system to complement renewable power generation. Jordan's energy economy provides an illustrative context for where this project would be technologically and economically appropriate, though the design and analysis could apply at any location.

As of 2014, the Hashemite Kingdom of Jordan was importing 97% of its total energy. Having almost no conventional petroleum and little gas reserves, Jordan faces tremendous challenges in energy pricing and security. Since 2003, Jordan has received the majority of its gas supply from Egypt via the Arab Gas Pipeline, but when the pipeline became the target of repeated attacks in 2011–2014, the Jordanian government was forced to spend billions of dollars obtaining diesel and fuel oil from other sources. In recent years the Jordanian government has increased budgetary spending and development to improve the security and reliability of the nation's energy (Azzeh 2011). In addition to developments in shale oil and nuclear energy, Jordan has plans for 2000 MW of new solar and wind generation installations in 2014–2020. In 2015 alone, 500 MW of new production has increased the renewables' portion of domestic electricity generation from 1 to 14% (Almasri 2014; "'Unstable' Gas Supplies Highlight Potential Energy Crisis in Jordan" 2011).

With such a dramatic rise in renewable penetration to Jordan's domestic grid supply, the potential problems of frequency regulation and variable ramping (Fig. 1) may become significant. Appropriate regulatory requirements for energy storage installation can help mitigate these issues, as has been recommended in Texas, USA (Chang et al. 2014), or as has already been mandated in California, USA, where 1.35 GW of energy storage will be installed by 2020 ("Decision Adopting Energy Storage Procurement Framework and Design Program" 2013).

Requirements Definition

The case study will look at installation of energy storage to mitigate transient cloud cover effects on generation from a solar photovoltaic farm. The focus will be system sizing and optimization based on thermal management, product lifetime, and cost. In the real world, every application will have different requirements for energy, power, usage, and lifetime, all of which are often driven not only by technical requirements but also by market pricing and investment strategy. Here, the hypothetical system's requirements are based on similar battery use cases, guided by real-world input where possible.

To begin identifying the system requirements, energy storage mandates established in Puerto Rico will be used. Although the project context is in the Middle East and similar rules are being implemented elsewhere, Puerto Rico's adopted guidelines provide a quick, easy-to-understand requirement for a case study. In Puerto Rico, where growing penetration of renewables is contributing risk to grid stability, government rules have mandated that all new renewable generation projects include energy storage for grid balancing. For each new project, energy storage must be installed that can supply 30% of the project's nameplate power for 10 min of frequency regulation, and 45% of the nameplate power for 1 min of ramping (St. John 2013). Although the 1-min requirement is below the characteristic discharge time for Li-ion (see Fig. 2), a battery can be sized reasonably to meet the 10-min requirement.

The hypothetical solar installation will be 1 MW in size, so the battery for energy storage must be capable of 300 kW for 10 min, with a total delivery of 50 kWh. Selection of the battery chemistry, cell format, and module and system design could easily fill an entire book's content alone, so for the sake of brevity we will consider one particular cell (specifications in Table 2) that offers a good fit for this application. This cell uses a blended cathode chemistry, which provides the high specific capacity and voltage of NMC paired with the high rate capability of LMO. However, this cell is limited to a maximum discharge of 10A (4.5C rate) by the manufacturer specification, so in order to meet the 10-min discharge requirement (equivalent to 6C rate) the battery system must be oversized. 300 kW translates to a 66.7 kWh battery operating at 4.5C, so this is the minimum capacity for battery design. The hypothetical usage profile will be to cycle the battery twice daily (e.g. – once for sunrise and once for sunset).

Battery Sizing

Due to the degradation behavior of Li-ion batteries discussed earlier, the battery must be further oversized to ensure that it meets the power and capacity requirements after 10 years, a typical lifetime requirement for grid installations. To find what the BOL battery size should be, a semi-empirical life model developed for this

Table 2 Specifications for Li-ion cell used in solar farm energy storage case study

Property	Value
Form factor	18650
Capacity (Ah)	2.15
Voltage, nom. (V)	3.65
Impedance (mΩ)	30
Anode	Graphite
Cathode	NMC/LMO

cell will be used. In this model, calendar and cycling losses are treated as independent and additive (Santhanagopalan et al. 2015; Schmalstieg et al. 2014), so the battery capacity, C, is given by:

$$C = C_{BOL} - C_{cal} - C_{cyc} \tag{5}$$

Only the effect of capacity on energy delivery is considered here, though the real application would also need a lifetime analysis on resistance growth to guarantee the required power delivery.

In addition to the usage profile and life requirement already specified, this model requires the battery operating temperature (for calendar aging) in order to calculate the necessary DOD and corresponding initial battery size. However, the operating temperature is difficult to determine since it is interdependent with the choice of thermal management. Fortunately, the life model can be used to predict how DOD affects the maximum permissible operating temperature (necessary to reach the target lifetime). This relationship is shown in Fig. 17. As mentioned, at BOL the minimum battery size is 66.7 kWh. Due to cycling aging alone, the system must be oversized to an initial 78.4 kWh, so that it still provides the required energy and power at EOL. This value corresponds to no calendar aging, represented by the asymptotic far left end of the curve in Fig. 17. Since calendar aging increases exponentially with temperature, as the target operating temperature rises the calendar degradation becomes predominant. Above 40 °C, much larger initial battery sizes are required because they can tolerate the large calendar aging capacity loss and still manage to meet the EOL energy requirement.

Figure 17 also shows battery price, assuming a competitive $400/kWh. Here is where the balance of system and its cost comes into play. Consider two batteries sizes, shown in Table 3, taken from the model results. To keep initial costs low, a 91 kWh battery could be purchased to operate at 55% DOD, but its temperature must be maintained at an average of 20.1 °C in order to reach the 10-year lifetime. This temperature will require a substantial thermal management system, most likely liquid cooled, depending on the temperature of the ambient environment. In comparison, a larger 167 kWh battery can tolerate much hotter operation, at 39.8 °C. An even more dramatic difference between these two systems is the battery's heat generation: since the smaller battery operates at double the C-rate compared to the larger system, the former's heat generation will be four times that of the latter. Thus, the cooling system for the smaller battery not only needs to

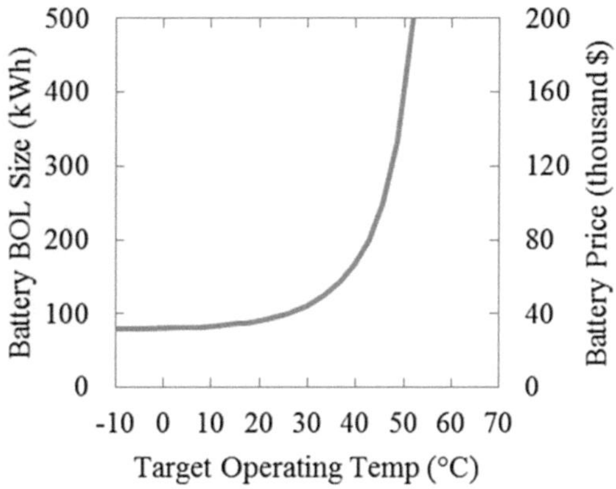

Fig. 17 The battery beginning-of-life size and price calculated using a semi-empirical lifetime model, as a function of the average operating temperature. Larger, oversized systems will operate at a lower DOD. Based on a target lifetime of 10 years

Table 3 Comparison of two battery sizes and their associated prices, operating temperatures, and C-rates during normal operation

Battery Size (kWh)	91	167
DOD (%)	55	30
Price ($)	36,400	66,800
Target Temp (°C)	20.1	39.8
C-rate	3.3	1.8

maintain a much low temperature; it must also remove four times the amount of heat. Further thermal analysis and module design work would be necessary to incorporate pricing of the cooling system and fully determine which option would be the most cost effective, but this section provides an illustration of the tradeoffs between battery size, cost, temperature, lifetime, and thermal management strategy.

System Integration

Once the battery size and thermal management are identified, the remaining system integration includes the BMS, AC-DC inverter, transformer, controller, and housing. An example of such system integration is shown in Fig. 18, a grid-tied energy storage system used by SunEdison that incorporates an AllCell Technologies battery and Dynapower inverter. Here, the battery uses phase change materials for passive thermal management in combination with forced air cooling. The battery cabinets and inverter are both located inside a standard sized shipping container, and auxiliary air conditioning removes the heat generated by the inverter and battery so the container stays within the target temperature range for the battery. For solar installations, a multiport inverter can be used to facilitate better integration of the solar arrays, battery, and AC grid connection.

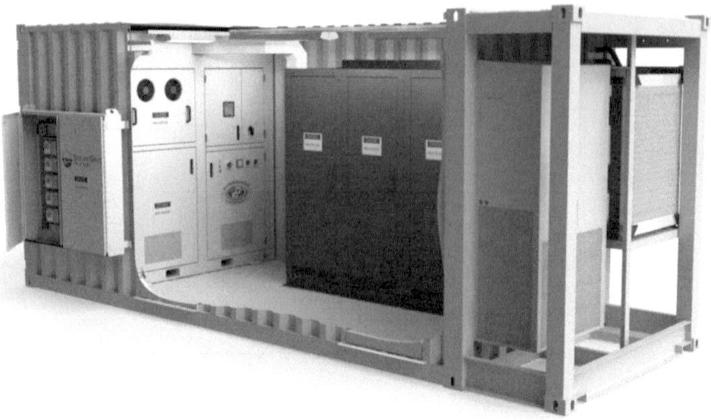

Fig. 18 An example energy storage system that integrates a lithium ion battery with AC-DC inverter and power electronics for grid-tied applications ("AllCell Technologies Energy Storage Systems" 2015)

Summary

The importance and implementation of energy storage technologies will continue to grow in coming years as renewable power production increases. Renewable sources such as wind and solar introduce particular challenges to grid stability, creating an energy storage need for frequency regulation and variable generation ramping. Lithium ion batteries are an excellent technological solution for these applications because of their high energy density, power capability, and cycle life. However, engineering design of these systems must carefully address the thermal and safety issues of Li-ion in order to make them commercially successful.

Energy storage not only provides a way to stabilize the grid, but also improves energy efficiency, advances national energy security, and lowers costs. This chapter has provided an introduction to the various types of energy storage technology and highlighted their benefits, challenges, and typical applications. By providing a detailed look at lithium ion batteries, specific illustrations have been made to show the aspects of engineering design and system integration for smart grid energy storage.

References

Almasri, I. (2014). Jordan to achieve 14% of electricity from renewable energy in 2015. *Jordan Economic News Analysis*, at http://www.iqtisadjordan.com/2014/11/jordan-will-achieve-14-of-electricity.html

Alotto, P., Guarnieri, M., & Moro, F. (2014). Redox flow batteries for the storage of renewable energy: A review. *Renewable and Sustainable Energy Reviews, 29*, 325–335.

Azzeh, L. (2011). Kingdom's first solar-powered electric vehicle charging station opens. *The Jordan Times*, at http://www.jordantimes.com/news/local/kingdom%E2%80%99s-first-solar-powered-electric-vehicle-charging-station-opens

Bernardi, D., Pawlikowski, E., & Newman, J. (1985). A General energy balance for battery systems. *Journal of the Electrochemical Society, 132*, 5–12.

Broussely, M., et al. (2005). Main aging mechanisms in Li Ion batteries. *Journal of Power Sources, 146*, 90–96.

Chang, J.; et al. (2014). *The value of distributed electricity storage in Texas*. The Brattle Group.

Denholm, P., Ela, E., Kirby, B., & Milligan, M. (2010). *The role of energy storage with renewable electricity generation*. Golden: National Renewable Energy Laboratory.

Dincer, I., & Rosen, M. A. (2010). *Thermal energy storage: Systems and applications* (2nd ed.). Hoboken: Wiley.

Ehrlich, G. M. (2002). Lithium-Ion batteries. In D. Linden & T. R. Reddy (Eds.), *Handbook of batteries* (3rd ed.). New York: McGraw-Hill.

Hearne, S. J. (2014). *Grid energy storage*. Sandia National Labs.

Huggins, R. A. (2010). *Energy storage*. New York: Springer.

St. John, J. (2013). Puerto Rico mandates energy storage in green power mix. at https://www.greentechmedia.com/articles/read/puerto-ricomandates-energy-storage-in-green-power-mix

St John, J. (2014). *SCE chooses ice energy for 25 MW of rooftop thermal energy storage*, at http://www.greentechmedia.com/articles/read/SCE-Chooses-Ice-Energy-for-25MW-of-Rooftop-Thermal-Energy-Storage

Kim, H., et al. (2013). Liquid metal batteries: Past, present, and future. *Chemical Reviews, 113*, 2075–2099.

Kizilel, R., Sabbah, R., Selman, J. R., & Al-Hallaj, S. (2009). An alternative cooling system to enhance the safety of Li-ion battery packs. *Journal of Power Sources, 194*, 1105–1112.

Ma, H., Cheng, F., & Chen, J. (2011). Nickel-Metal Hydride (Ni-MH) rechargeable batteries. In R.-S. Liu et al. (Eds.), *Electrochemical technologies for energy storage and conversion* (Vol. 1 & 2). Weinheim: Wiley.

McGrail, B. P.; et al. (2013). *Compressed air energy storage: Grid-scale technology for renewables integration in the pacific northwest*. Pacific Northwest National Laboratory.

Nitta, N., Wu, F., Lee, J. T., & Yushin, G. (2015). Li-ion battery materials: Present and future. *Materials Today, 18*, 252–264.

Pavlov, D. (2011). *Lead-acid batteries: Science and technology*. Amsterdam: Elsevier Science Ltd.

Safari, M., Morcrette, M., Teyssot, A., & Delacourt, C. (2010a). Life prediction methods for lithium-ion batteries derived from a fatigue approach, Part I. *Journal of the Electrochemical Society, 157*, A713–A720.

Safari, M., Morcrette, M., Teyssot, A., & Delacourt, C. (2010b). Life prediction methods for lithium-ion batteries derived from a fatigue approach, Part II. *Journal of the Electrochemical Society, 157*, A892–A898.

Santhanagopalan, S., Smith, K., Neubauer, J., Gi-heon, K., & Pesaran, A. (2015). *Design and analysis of large lithium-ion battery systems*. Boston: Artech House.

Schmalstieg, J., Käbitz, S., Ecker, M., & Sauer, D. U. (2014). A holistic aging model for Li (NiMnCo)O2 based 18650 lithium-ion batteries. *Journal of Power Sources, 257*, 325–334.

Schweitzer, B., Wilke, S., Khateeb, S., & Al-Hallaj, S. (2015). Experimental validation of a 0–D numerical model for phase change thermal management systems in lithium-ion batteries. *Journal of Power Sources, 287*, 211–219.

Takahara, K. (2006). Sony's battery fiasco a symptom of bigger woes at legendary firm. *McClatchy – Tribune Business News*.

Vetter, M. (2013). Overview of recent advances in battery systems for PV applications. *Ecobuild Conf.*

Weber, A. Z., et al. (2011). Redox flow batteries: A review. *Journal of Applied Electrochemistry, 41*, 1137–1164.

Wilke, S., Schweitzer, B., Khateeb, S., & Al-Hallaj, S. (2017). Preventing thermal runaway propagation in lithium ion battery packs using a phase change composite material: An experimental study. *J Power Sources, 340*, 51–59.

Williard, N., He, W., Hendricks, C., & Pecht, M. (2013). Lessons learned from the 787 dreamliner issue on lithium-ion battery reliability. *Energies, 6*, 4682–4695.

"AllCell Technologies Energy Storage Systems," AllCell Technologies, 2015.

"AutoLion Software," EC Power.

"Battery Design Studio Software," CD-adapco.

"Buildings Energy Data Book," U.S. Department of Energy, 2008.

"Business Rate Summary," Southern California Edison, 2008.

"Carbon Fiber Flywheels," Beacon Power, 2014.

"Cost of Mass Produced EV Battery in 2020," FEV, Global Clean Vehicle Summit, Changzhou, China, 2015.

"Decision Adopting Energy Storage Procurement Framework and Design Program (D.13–10-040)," Public Utilities Commission of the State of California, 2013.

"Ice Energy Project," Southern California Public Power Authority, 2008.

"Peakers Cost Recovery Testimony," Southern California Edison, 2007.

"Pumped Storage Projects," Federal Energy Regulatory Commission, 2015.

"PureTemp Technical Data Sheets," Entropy Solutions, 2015.

"Solar Water Heaters," U.S. Department of Energy, Office of Energy Efficiency & Renewable Energy at http://energy.gov/energysaver/solar-water-heaters.

"Understanding Flywheel Energy Storage: Does High-Speed Really Imply a Better Design," Active Power, 2008.

"'Unstable' Gas Supplies Highlight Potential Energy Crisis in Jordan," *The Jordan Times*, 2011 at http://www.albawaba.com/%E2%80%98unstable%E2%80%99-gas-supplies-highlight-potential-energy-crisis-402863.

Technologies and Options of Solar Energy Applications in the Middle East

Nasir El Bassam

Abstract The potential for solar energy in the Middle East is immense. It in general has the highest levels of solar input in terrestrial world. They also have cheap, plentiful space and the potential to generate solar power for electricity, heat, cooling and for water desalination. Continuously high solar radiation makes it ideal locations for solar installations, the potential reach of which is limited to deserts.

Direct Normal Radiation (DNR), the main measure of a region's suitability for solar thermal concentrated applications, ranges between 2050 and 2800 kilowatts-hour per square meter per year (kWh/m2/year) in the Arab region. This is equivalent to 1–2 barrel oil per square meter per year. These rates are among the best in the world making the region suitable for solar heating and cooling, Concentrated Solar Power (CSP) and Concentrated Photovoltaic (CPV) applications.

Despite these favorable conditions solar energies account in average to less than 0.2% of the region's total installed capacity for electricity, compared to currently ca. 7% (40,000 MW) in Germany and almost 33% Of German Electricity Came From Renewables in 2015. Costa Rica completes 2016 without having to burn a single fossil fuel for more than 250 days. 98.2 percent of Costa Rica's electricity came from renewable sources in 2016.

Read more: http://www.digitaljournal.com/news/environment/costa-rica-powered-by-renewable-energy-for-over-250-days-in-2016/article/482755#ixzz4Uin5wVeA

Keywords Solar energy • Renewable energy • Photovoltaics

Country all Renewable Energy Installed Capacity in Percent (%) of Total Installed Capacity in the Arab World: Algeria 0.25; Bahrain 0.14; Comoros 0.00; Djibouti 1.19; Egypt 2.00; Iraq 0.02; Jordan 0.29: Kuwait 0.01; Lebanon 0.04; Libya 0.06; Mauritania 7.71; Morocco 4.85; Oman 0.01; Palestine 0.98; Saudi Arabia 0.01; Somalia 0.00; Sudan 0.00; Syria 0.01; Tunisia 4.35, UAE 0.49 and Yemen 0.10. The contribution of solar energy in the region is less than 0.10%.

N. El Bassam (✉)
International Research Centre for Renewable Energy (IFEED), Lehrte, Germany
e-mail: info@ifeed.org; http://www.ifeed.org

S. Murad et al. (eds.), *Water, Energy & Food Sustainability in the Middle East*,
DOI 10.1007/978-3-319-48920-9_9

While the main focus of the region is on the deployment of RETs for electricity generation, solar water heating (SWH) has also been given due attention in many Arab countries. Its simple technology means a significant part of the system equipment is manufactured locally.

This leads to a direct effect of increasing the share of RETs and creating jobs in the region.

According to the Arab Forum for Environment and Development (AFED) report (2013), the total collector area of SWHs reached around 4.8 million square meters (m2); most SWHs are located in Jordan, Syria, Lebanon, Palestine, Egypt and Tunisia, although with different levels of penetration. In comparison: 14.2 million square meters exist in Germany.

We have the knowledge, skills and the technologies and all the theoretical, practical and technical underpinnings; the rich diversity of innovative projects and strategies to achieve the goals of sufficient supply with food, water and energy. Tackling poverty should be our priority. It is imperative that future initiatives for energy supply be based on renewable energy, for water desalination, combating desertification, development of rural regions and serving unserved population.

We have to face the urgent challenges rapidly mounting on our horizon. Among them, front and center, the devastation of poverty (food, water and energy poverty) that bears down on one-third of our Arab population The kind of poverty, that has become the root cause of hunger, malnutrition, economic dislocation and a host of related ills and imbalances. That degrades not just the every-day lives and dignity of those directly impacted; but also those who may feel they are far-removed from such conditions. Though in fact we are all victims of the crises brought on by unjust or ill-conceived policies and conflicts. Apart from technological feasibility, the adoption of solar energy as alternative to fossil fuels depends on regional as well as on national policy environments and concrete implementation strategies.

Preface

Various human societies have been established in the MENA region throughout history. And today deserts are important part of the world's economic, natural and cultural heritage. Among the greatest contribution of MENA cultures to the humanity are the three "Religions of the Book": Judaism, Christianity and Islam which have had tremendous impact far beyond their areas of origin. About 40–60% of minerals and fossil energy used globally is extracted from this part of the world (Oil and gas Belt) and it is privileged with the unique potential of solar energy "Sun Belt".

The potential for solar energy in the MENA region is immense It in general has the highest levels of solar input in terrestrial world. They also have cheap, plentiful space and the potential to generate solar power for electricity, heat and for water desalination. Continuously high solar radiation makes it ideal locations for solar installations, the potential reach of which is limited to deserts. The sun is supplying deserts (Sun Belt) with energy equivalent to 2–3 barrels of oil per square meter every year. In less than 6 h, deserts receive more energy than humankind uses in a whole year. In the desert, just across the Mediterranean Sea, Red Sea, the Gulf and Arabian Sea is a vast source of energy that holds the promise of a carbon-free and feedstock-free electrical future for the whole of Middle, Near East, Europe, if not the world. Apart from technological feasibility, the adoption of solar energy as alternative to fossil fuels depends on regional as well as on national policy environments and concrete implementation strategies.

Basics: Solar Energy and Technologies

Electromagnetic energy (solar radiation) transmitted by the sun (approximately one billionth of which reaches the earth) is the basis of all terrestrial life. "It amounts to about 420 trillion kilowatt-hours, and is several thousand times greater than all the energy used by all people. Solar energy is harnessed by capturing the sun's heat (through solar heaters) or light (through photovoltaic cells). It is estimated that one square kilometer (about0.4 square miles) of land area receives some 4000 kilowatts (4 megawatts) of solar energy every day, enough for the requirements of a medium-sized town" (Business Dictionary 2012). Solar technologies are broadly characterized as either passive solar or active solar depending on the way they capture, convert and distribute solar energy.

Active solar techniques include the use of photovoltaic panels and solar thermal collectors to harness the energy.

Passive solar techniques include orienting a building to the sun, selecting materials with favorable thermal mass or light dispersing properties, and designing spaces that naturally circulate air.

Solar energy's uses are limited only by human ingenuity. Solar applications includes also space heating and cooling through solar architecture, potable water via distillation and disinfection, daylighting, solar hot water, solar cooking, and high temperature process heat for industrial purposes. To harvest the solar energy, the most common method is to use solar panels (TheFreeDictionary.com 2012). Solar power is the conversion of sunlight into electricity. Sunlight can be converted directly into electricity using photovoltaics (PV), or indirectly with concentrated solar power (CSP), which normally focuses the sun's energy to boil water, which is then used to provide power. Other technologies also exist, such as Stirling engine dishes, which use a Stirling cycle engine to power a generator.

Fig. 1 Space Station. Solar panels powered the International powered by solar panels (NASA Science/Science News 2011)

Photovoltaics

Photovoltaics were initially used to power small- and medium-sized applications, from the calculator powered by a single solar cell to off-grid homes powered by a photovoltaic array.

Photovoltaics convert light into electric current using the photoelectric effect. A photovoltaic system (or PV system) is a system that uses one or more solar panels to convert sunlight into electricity. It consists of multiple components, including the photovoltaic modules, mechanical and electrical connections and mountings and means of regulating or modifying the electrical output (Fig. 1).

A solar cell, or photovoltaic cell (PV), is a device that converts light into electric current using the photoelectric effect. The first solar cell was constructed by Charles Fritts (Perlin, John, 1999) in the 1880s. In 1931 a German engineer, Dr. Bruno Lange (Popular Science 1931), developed a photo cell using silver selenide in place of copper oxide. Although the prototype selenium cells converted less than 1% of incident light into electricity, both Ernst Werner von Siemens and James Clerk Maxwell recognized the importance of this discovery. Following the work of Russell Ohl in the 1940s, researchers Gerald Pearson, Calvin Fuller and Daryl

Fig. 2 Solar panel as Sunshade and a power source for mobile phones (personal communication 2016, source unknown)

Chapin created the silicon solar cell in 1954. These early solar cells cost 286 USD/watt and reached efficiencies of 4.5–6% (Perlin 1999) (Fig. 2).

Different materials display different efficiencies and have different costs. Materials for efficient solar cells must have characteristics matched to the spectrum of available light. Some cells are designed to efficiently convert wavelengths of solar light that reach the Earth's surface. However, some solar cells are optimized for light absorption beyond Earth's atmosphere as well. Light-absorbing materials can often be used in multiple physical configurations to take advantage of different light absorption and charge separation mechanisms. Materials presently used for photovoltaic solar cells include monocrystalline silicon, polycrystalline silicon, amorphous silicon, cadmium telluride, and copper indium selenide/sulfide (Jacobson 2009) (Fig. 3).

Many currently available solar cells are made from bulk materials that are cut into wafers between 180 to 240 micrometers thick and are then processed like other semiconductors.

Other materials are made as thin-films layers, organic dyes, and organic polymers that are deposited on supporting substrates. A third group are made from nanocrystals and used as quantum dots (electron-confined nanoparticles). Silicon remains the only material that is well-researched in both bulk and thin-film forms.

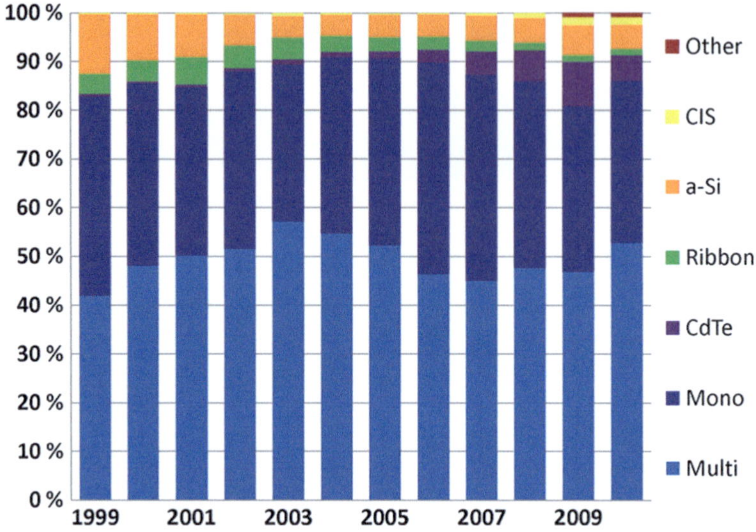

Fig. 3 The graph shows the market share of the different photovoltaic technologies from 1999 Until 2011. The *light* and *dark blue* are multi- and mono crystalline silicon, respectively; together they represented 87% of the market in 2010. http://en.wikipedia.org/wiki/File:PV_Technology.png.

As mentioned, solar cells produce direct current (DC) power, which fluctuates with the intensity of the irradiated light. This usually requires conversion to certain desired voltages or alternating current (AC), which requires the use of inverters. Multiple solar cells are connected inside the modules. Modules are wired together to form arrays, then tied to an inverter, which produces power with the desired voltage and frequency/phase (when it's AC).

Trackers and sensors to optimize performance are often seen as optional, but tracking systems can increase viable output by up to 100% (RISE 2010). PV arrays that approach or exceed one megawatt often use solar trackers. Accounting for clouds, and the fact that most of the world is not on the equator, and that the sun sets in the evening, the correct measure of solar power is insolation the average number of kilowatt-hours per square meter per day. For the weather and latitudes of the United States and Europe, typical insolation ranges from 4kWh/m2/day in northern climes to 6.5kWh/m2/day in the sunniest regions (Whitlock et al. 2000).

For large systems, the energy gained by using tracking systems outweighs the added complexity (trackers can increase efficiency by 30% or more) (Utility Scale Solar Power Plants (PDF) 2012).

Fig. 4 The PV power plant Girasol from Portugal is the second largest solar plant in the world, generating 62 MW of power from nearly 350,000 solar panels. (Urvish Dave, http://www.eai.in/club/users/narsi/blogs/23691#sthash.Nd3MXxet.dpuf)

Applications

Most commercially available solar panels are capable of producing electricity for at least 20 years (Zweibel 2010). The typical warranty given by panel manufacturers is over 90% of rated output for the first 10 years, and over 80% for the second 10 years. Panels are expected to function for a period of 30–35 years (Shenzhen JCN New Energy Technology CO). Many residential systems are connected to the grid wherever available, especially in the developed countries with large markets. In these grid connected PV systems, use of energy storages is optional (Mugnier, Daniel, 2014). In certain applications such as satellites, lighthouses, or in developing countries, batteries or additional power generators are often added as backups, which form stand-alone power systems (Fig. 4).

Between 1970 and 1983 photovoltaic installations grew rapidly, but falling oil prices in the early 1980s moderated the growth of PV from 1984 to 1996. Since 1997, PV development has accelerated due to supply issues with oil and natural gas, global warming concerns, and the improving economic position of PV relative to other energy technologies (Solar: photovoltaic 2009). Photovoltaic production growth has averaged 40% per year since 2000 and installed capacity reached 39.8 GW at the end of 2010, with 17.4 GW of that total in Germany (BP Statistical World Energy Review 2011). As of October 2011, the largest photovoltaic (PV) power plants in the world are the Sarnia Photovoltaic Power Plant (Canada, 97 MW), Montalto di Castro Photovoltaic Power Station (Italy, 84.2 MW) and Finsterwalde Solar Park (Germany, 80.7 MW) (PV Resources.com 2011) (Fig. 5).

Fig. 5 Charanka Solar Park, Gujarat, India, approx 300 MW (courtesy: Urvish Dave) – See more at: http://www.eai.in/club/users/narsi/blogs/23691#sthash.Nd3MXxet.dpuf

Fig. 6 Uruguay is now generating 95% of its electricity from renewable energy. Sarah Todd, December 2015. http://qz.com/566773/uruguay-is-now-generating-95-of-its-electricity-from-renewable-energy/

Concentrating Solar Thermal Power (CSP)

Concentrated solar power (CSP) systems use lenses or mirrors and tracking systems to focus a large area of sunlight into a small beam.

Commercial concentrated solar power plants were first developed in the 1980s. The 354 MW SEGS CSP installation is the largest solar power plant in the world, located in the Mojave Desert of California. Other large CSP plants include the Solnova solar power station (150 MW) and the Andasol solar power station (150 MW), both in Spain (Solar Millennium AG 2011) (Fig. 6).

Fig. 7 Ivanpah, the world's largest concentrating solar power plant, officially started generating energy for California's electric grid, can produce a whopping 392 megawatts of solar energy to power 140,000 California homes. Ethan Theo, 21 February 2014. Society & Environment

The 200 MW Golmud Solar Park in China is the world's largest photovoltaic plant (Wang, Ucilia 2011). Many power plants today use fossil fuels as a heat source to boil water. The steam from the boiling water spins a large turbine, which drives a generator to produce electricity. However, a new generation of power plants with concentrating solar power systems uses the sun as a heat source. The three main types of concentrating solar power systems are: linear concentrator, dish/engine, and power tower systems. Linear concentrator systems collect the sun's energy using long rectangular, curved (U-shaped) mirrors. The mirrors are tilted toward the sun, focusing sunlight on tubes (or receivers) that run the length of the mirrors (Fig. 7).

The reflected sunlight heats a fluid flowing through the tubes. The hot fluid then is used to boil water in a conventional steam-turbine generator to produce electricity. There are two major types of linear concentrator systems: parabolic trough systems, where receiver tubes are positioned along the focal line of each parabolic mirror; and linear Fresnel reflector systems, where one receiver tube is positioned above several mirrors to allow the mirrors greater mobility in tracking the sun. A dish/engine system uses a mirrored dish similar to a very large satellite dish. The dish-shaped surface directs and concentrates sunlight onto a thermal receiver, which absorbs and collects the heat and transfers it to the engine generator. The most common type of heat engine used today in dish/engine systems is the Stirling engine (Fig. 8).

Fig. 8 Prototype 150 kW dish/Stirling power plant at Sandia National Laboratory. U.S. DOE 2001. http://www1.eere.energy.gov/solar/pdfs/csp_water_study.pdf

Fig. 9 Evacuated tube. (Hot Water Now)

This system uses the fluid heated by the receiver to move pistons and create mechanical power. The mechanical power is then used to run a generator or alternator to produce electricity.

A power tower system uses a large field of flat, sun-tracking mirrors known as heliostats to focus and concentrate sunlight onto a receiver on the top of a tower. A heat-transfer fluid heated in the receiver is used to generate steam, which, in turn, is used in a conventional turbine generator to produce electricity. Some power towers use water/steam as the heat-transfer fluid. Other advanced designs are experimenting with molten nitrate salt because of its superior heat-transfer and energy-storage capabilities. The energy storage capability, or thermal storage, allows the system to continue to dispatch electricity during cloudy weather or at night (Fig. 9).

Solar Thermal Collectors

Solar thermal collectors transform solar radiation into heat and transfer that heat to a medium (water, solar fluid, or air). Solar water heating (SWH) or solar hot water (SHW) systems have been well established for many years, and are widely used throughout the world. In a "close-coupled" SWH system the storage tank is

Fig. 10 Flat-plate collectors. Solar Tribune 2012, http://solartribune.com/solar-thermal-power/#.T0_PvfU0iuI

horizontally mounted directly above the solar collectors on the roof. No pumping is required as the hot water naturally rises into the tank through passive heat exchange. In a "pump-circulated" system the storage tank is ground or floor mounted below the level of the collectors; a circulating pump moves the water or heat transfer fluid between the tank and the collectors. There are multiple types of solar thermal collectors:

- Evacuated tube collectors are the most efficient but most costly type of hot water solar collectors. These collectors have glass or metal tubes with a vacuum, allowing them to operate well in colder climates.
- Batch solar water heaters, also called integral collector-storage systems, have storage tanks or tubes inside an insulated box, the south side of which is glazed to capture the sun's energy.
- Flat plate collector, a box covered by glass or plastic with a metal absorber plate on the bottom. The glazing, or coating, on the absorber plate helps to better absorb and retain heat.
- Unglazed flat-plate collectors, typically made from rubber, are primarily used for heating pools.
- •Air collectors are used primarily for space heating in the home. Flat plate solar collectors durable, weatherproof boxes that contain a dark absorber plate located under a transparent coverd – are still the most common type of collector used for water heating in many countries despite being inferior to evacuated tube collectors in many ways (Fig. 10).

Evacuated heat pipe tubes are designed such that convection and heat loss are eliminated, whereas flat-plate solar panels contain an air gap between absorber and cover plate that allows heat loss to occur. Further, thermal heat pipe systems are capable of limiting the maximum working temperature, whereas flat-plate systems have no internal method of limiting heat build-up, which can cause system failure.

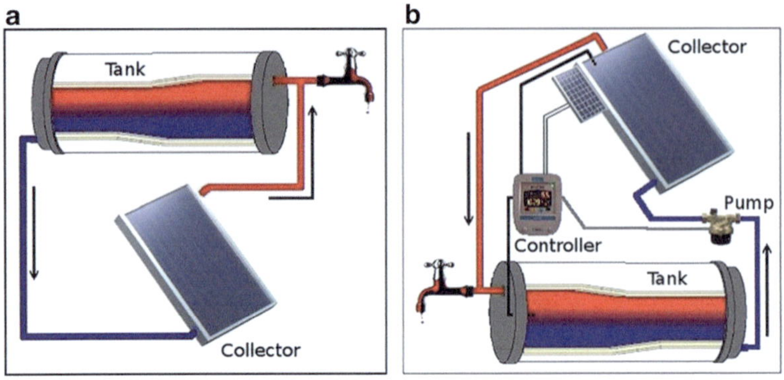

Fig. 11 Direct systems: (**a**) Passive CHS system with tank above collector. (**b**) Active system with pump and controller driven by a photovoltaic panel. Jwhferguson, selfpublished work 2010, accessed from URL http://www.solarcontact.com/solar-water/heater

Finally, evacuated heat pipe systems are lightweight, easy to install and require minimal maintenance. Flat-plate systems, on the other hand, are difficult to install and maintain, and must be completely replaced should one part of the system stop working (Fig. 11).

An Evacuated Tube Solar Collector is composed of hollow glass tubes. All the air is removed from the tubes to create a vacuum that acts as an excellent insulator. An absorber coating inside the tube absorbs the solar radiation. This energy is transferred to the fluid moving through the collector and then to the hot water storage tank. In cooler climates a heat exchanger is used to separate the potable water from the non-toxic antifreeze in the collector.

Batch Solar Water Heaters, also called integral collector-storage (ICS) systems, are made up of a water tank or tubes inside an insulated, glazed box. Cold water flows through the solar collector. The water is heated and then continues on to the backup water heating storage tank. Some water can be stored in the collector until it is needed. ICS systems are a type of direct solar water heating system, which circulates water to be heated, rather than using a heat transfer fluid to capture the solar radiation.

A Flat-Plate Solar Collector is an insulated box covered by glass or plastic with a metal absorber plate on the bottom. The weatherproofed collectors are usually glazed with a coating to better absorb and retain heat. Heat transfer fluid flows through metal tubes lying below the absorber plate. The fluid then flows through a heat exchanger before entering the storage tank.

Unglazed Flat-Plate Collectors (without insulation or absorber coatings) do not operate in cool or windy climates but are excellent for heating water in a pool (Solar Tribune 2012). Solar hot air collectors are mounted on south-facing vertical walls or roofs. Solar radiation reaching the collector heats the absorber plate. Air passing through the collector picks up heat from the absorber plate.

Freezing, overheating and leaks are less troublesome for solar air collectors than for liquid collectors. But since liquid is a better heat conductor, solar collectors using water or a heat transfer fluid are more suited to hot water heating for the home. A solar hot air collector is most often used for space heating. There are two types of air collectors: glazed and unglazed (Energy4You 2012).

SWH systems are designed to deliver hot water for most of the year. In colder climates a gas or electric booster may be needed as a backup to deliver sufficient hot water.

Solar Water Heater (SWH) Installed Capacity

While the main focus of the region is on the deployment of RETs for electricity generation, solar water heating (SWH) has also been given due attention in many Arab countries. Its simple technology means a significant part of the system equipment is manufactured locally.

This leads to a direct effect of increasing the share of RETs and creating jobs in the region.

According to the Arab Forum for Environment and Development (AFED) report (2013), the total collector area of SWHs reached around 4.8 million square meters (m2); most SWHs are located in Jordan, Syria, Lebanon, Palestine, Egypt and Tunisia, although with different levels of penetration. In comparison: 14.2 million square meters in Germany.

Solar Cookers and Solar Ovens

A solar cooker or solar oven is a device to generate heat from the energy of solar radiation. The radiation is concentrated in the focus of a concave mirror. The solar cookers are suitable for

- Boiling water for heating food
- Cooking, roasting, baking, grilling or frying
- Juice production
- Small food stalls in the open air
- Commercial use in dyeing
- Soap production
- Processing of natural fibers for weaving

Box types are constructed from metal or mirror reflectors and a glass plate. Aligned with the sun, the light is reflected by the reflector through the glass into the

Fig. 12 Bolivia Inti-Sud Soleil solar cooker construction workshop. SCI Net 2015 http://solarcooking.wikia.com/wiki/Bolivia

interior of the furnace. At about 150 _ C, water will boil to cook food. The parabolic types consist of mirrors focusing sunlight on a mostly matte black container in the focus area. The container absorbs solar radiation, so that its contents are heated (Fig. 12).

Because they use no fuel and cost nothing to operate, many non-profit organizations are promoting their use worldwide to help reduce fuel costs for low-income people, reduce air pollution and slow deforestation and desertification, caused by use of firewood for cooking. Many types of cookers exist. Simple solar cookers use the following basic principles:

- Concentrating sunlight: A reflective mirror of polished glass, metal or metallized film is used to concentrate light and heat from the sun into a small cooking area, making the energy more concentrated and increasing its heating power.
- Converting light to heat: A black or low reflectivity surface on a food container or the inside of a solar cooker will improve the effectiveness of turning light into heat. Light absorption converts the sun's visible light into heat, substantially improving the effectiveness of the cooker.
- Trapping heat: It is important to reduce convection by isolating the air inside the cooker from the air outside the cooker. A plastic bag or tightly sealed glass cover will trap the hot air inside. This makes it possible to reach similar temperatures on cold and windy days as on hot days.
- Greenhouse effect: Glass transmits visible light but blocks infrared thermal radiation from escaping. This amplifies the heat trapping effect (Solar Cooker Designs 2011) (Fig. 13).

Fig. 13 A camel transports cooling box for medicine by solar powered-refrigerators-for-mobile-health-clinics. Sarah Parsons 2009, Inhabitat. http://inhabitat.com/solar-powered-camel-clinics-carry-medicine-across-the-desert/

Solar Cooling

Solar cooling has been growing rapidly from around 60 systems in 2004 to more than 1000 systems installed in 2013 (IEA-SHC, 2014). Furthermore, 17 large-scale solar district systems are connected to cooling networks in Europe (IEA-SHC, 2014). However, compared to the potential of using solar energy to generate cooling, deployment levels are very low. Most deployment (80%) is in Europe and a number of companies like EAW, Invensor, Sortech, SolarNext (Germany), Pink (Austria) Broad (China), Thermax (India), Yazaki, Kawasaki, Mistubishi (Japan) and Jangsu Huineng (China) have released small, commercial solar cooling systems called "cooling kits" (Augsten 2014; Jakob 2014; REN21 2014).

In the Sunbelt region, where cooling demands are quickly rising, there are still only a few demonstration plants and studies available (Schwerdt and Ali 2014; Ssembatya et al. 2014), In those regions, rapid growth in the use of air conditioners for cooling is creating peaks in electricity demand.

Solar cooling technology can provide an effective solution to reduce the peak electricity consumption as it operates when the cooling demand is highest. Solar cooling can be complementary to solar heating, especially in regions that require cooling in the summer and heating in the winter. In this particular case, an increase in collector area to allow for cooling would simultaneously allow for a larger solar share of domestic water and space heating by a typical solar combi-system in winter. This approach avoids any long stagnation periods during summer or winter and, at the same time, increases economic efficiency. The increased efficiency of advanced flat-plate and evacuated vacuum tube collectors makes it possible to add a

Fig. 14 Parabolic trough collectors supplying a hotel in Turkey with process steam for solar cooling and the hotel laundry. (Courtesy: MAN Ferrostaal AG; de Riese) http://images.google.de/imgres?imgurl=http%3A%2F%2Fwww.solarserver.com%2Fuploads%2Fpics%2Fsolitem_anlage_tuerkei.jpg

thermally driven, adsorption or absorption cooling system to an existing solar combi-system, resulting in a so called "solar plus combi-system" (Faninger 2010).

There are three kinds of solar cooling systems: absorption chillers, adsorption chillers, and dessicant cooling systems. The technologies of solar space and water cooling are the same as those developed for gas cooling systems. In Europe, mostly ETC and FPC are used to provide heat, which can subsequently be converted into cooling using a thermally driven chiller or air-conditioning technologies. All told, 71% of these cooling systems run on absorption chillers, 13% on adsorption and 16% on desiccant systems (Jacob 2014).

(Solar Heating and Cooling for Residential Applications IEA-ETSAP and IRENA.Technology).

At Gebze High-Technology Institute near Istanbul, SOLITEM installed a solar energy system providing the renowned research institution with solar cooling for air-conditioning (Fig. 14).

Perspectives for the Arab World and MENA Region: "The Solar and DESERTEC Concept"

The DESERTEC concept was originated with Dr. Gerhard Knies, a German particle physicist and founder of the "Trans-Mediterranean Renewable Energy Cooperation (TREC)" network of researchers, to which the author (El Bassam) has also contributed.

The DESERTEC concept was developed further by TREC – an international network of scientists, experts and politicians from the field of renewable energies – founded in 2003 by the Club of Rome and the National Energy Research Center

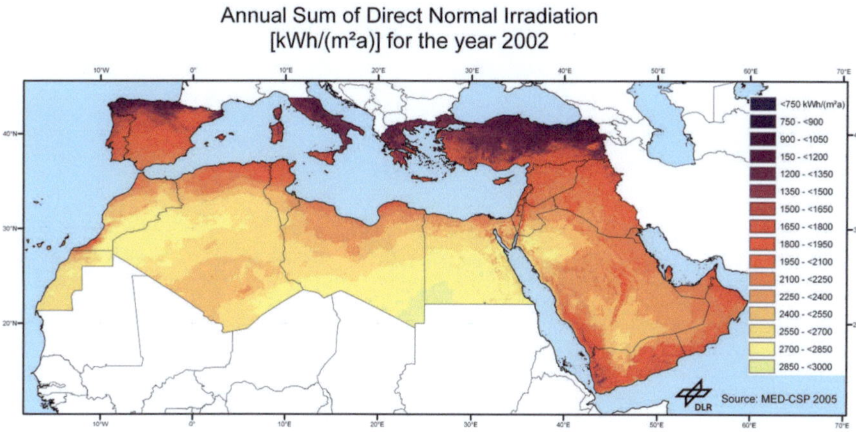

Fig. 15 Direct normal irradiation in the MENA region lies between 1500 and <3000 kWh/m^{2}a

Jordan. One of the most famous members was Prince Hassan bin Talal of Jordan. In 2009, TREC emerged to the non-profit DESERTEC Foundation.

Main scientific data has been delivered by an international network of scientists, experts and politicians from the field of renewable energies coordinated by Dr. Franz Trieb, German Aerospace Centre (DLR), conducted in three research studies to which IFEED has also substantially contributed (Fig. 15).

Basic Scientific Studies on Concentrating Solar Power for the Mediterranean Region

The DESERTEC Concept was developed by an international network of politicians, academics and economists, called TREC. The research institutes for renewable sources of the governments of Morocco (CDER), Algeria (NEAL), Libya (CSES), Egypt (NREA), Jordan (NERC) and Yemen (Universities of Sana'a and Aden) as well as the German Aerospace Center (DLR) made significant contributions towards the development of the DESERTEC Concept. The basic studies relating to DESERTEC were led by DLR scientist Dr. Franz Trieb working for the Institute for Technical Thermodynamics at the DLR.

The three studies were funded by the German Federal Ministry for the Environment, Nature Conservation, and Nuclear Safety (BMU). The studies, conducted between 2004 and 2007 evaluated the following as shown below:

Concentrating Solar Power for the Mediterranean Region (MED-CSP): Evaluated the potential for renewable energy in the Middle East and North Africa (MENA), and availability of resources and demand for energy in the region (2004–2005)

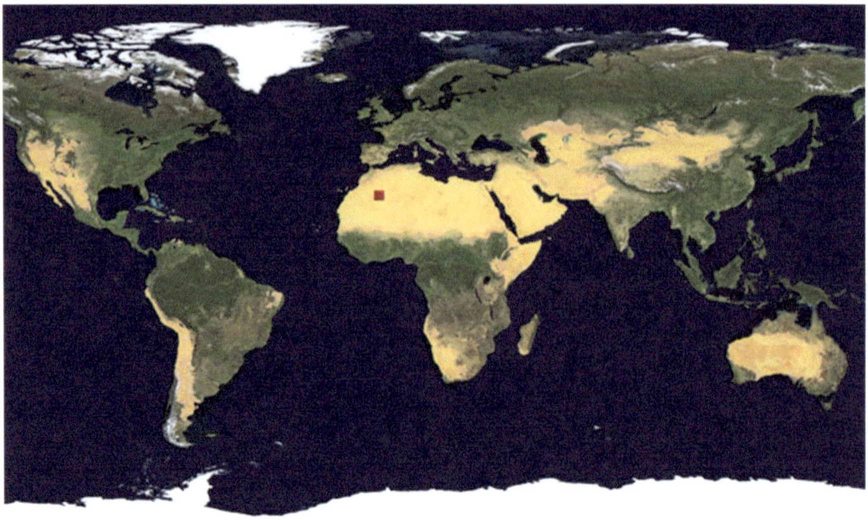

Fig. 16 One per cent of the surface area of the world's deserts would be enough to meet our current electricity needs by solar energy. Approximately 65% of world's deserts with maximum solar radiation exist in the Arab region

To keep global warming in a tolerable frame, the Scientific Council of the German Government for Global Environmental Change (WBGU) recommends in its latest study based on a scenario of the IPCC (Intergovernmental Panel for Climate Change) to reduce CO2 emissions on a global level by 30% until 2050. The advice of WBGU was to establish model projects to introduce renewable energies on a large scale as a strategic lever for a global change in energy policies. A strategic partnership between the European Union (EU), the Middle East (ME) and North Africa (NA) is a key element of such a policy for the benefit of both sides: MENA has vast resources of solar energy for its economic growth and as a valuable export product, while the EU can provide technologies and finances to activate those potentials and to cope with its national and international responsibility for climate protection, as documented in the Johannesburg agreement to considerably increase the global renewable energy share as a priority goal.

A special interest lies in the increasing demand for technically desalted water, which will require increasing energy input to the water supply sector. In order to establish appropriate instruments and strategies for the market introduction of renewables in the European and MENA countries, well founded information on demand and resources, technologies and applications is essential (Fig. 16).

The results of the **MED-CSP study** can be summarized in the following statements:

- Environmental, economic and social sustainability in the energy sector can only be achieved with renewable energies. Present measures are insufficient to achieve that goal.
- Renewable energy resources are plentiful and can cope with the growing demand of the EU-MENA region. The available resources are so vast that an additional supply of renewable energy to Central and Northern Europe is feasible.
- Renewable energies are the least-cost option for energy and water security in EU-MENA.
- Renewable energies are the key for socioeconomic development and for sustainable wealth in MENA, as they address both environmental and economic needs in a compatible way.
- An adequate set of policy instruments must be established immediately to accelerate renewable energy deployment in the EU and MENA (Trieb).

Trans- Concentrating Solar Power for the Mediterranean Region TRANS-CSP:

Evaluating the potential for an integrated electric power transmission grid connecting; the three regions – Europe, the Middle East, and North Africa; and the assessment of solar energy imports to Europe (2004–2006).

The results of the Trans- Concentrating Solar Power for the Mediterranean Region **TRANS-CSP study** can be summarized in the following statements:

- If initiated now, the change to a sustainable energy mix leads to less expensive power generation than a business as usual strategy in a timespan of about 15 years. Imported fuels with escalating cost will be increasingly substituted by renewable, mostly domestic energy sources. The negative socioeconomic impacts of fossil fuel price escalation can be reversed by 2020 if an adequate political and legal framework is established at time. Feed-in tariffs like the German or Spanish Renewable Energy Acts are very effective instruments for the market introduction of renewables. If tariff additions are subsequently reduced to zero, they can be considered a public investment rather than a subsidy• Solar electricity generated by concentrating solar thermal power stations in MENA and transferred to Europe via high voltage direct current transmission can provide firm capacity for base load, intermediate and peaking power, effectively complementing European electricity sources. Starting between 2020 and 2025 with a transfer of 60 TWh/y, solar electricity imports could subsequently be extended to 700 TWh/y in 2050. High solar irradiance in MENA and low transmission losses of 10–15% will yield a competitive import solar electricity cost of around 0.05 €/kWh.
- European support for MENA for the market introduction of renewables can attenuate the growing pressure on fossil fuel resources that would otherwise result from the economic growth of this region, thus helping indirectly to secure fossil fuel supply also in Europe. The necessary political process could be initiated by a renewable energy partnership and a common free trade area for

renewable energies in EUMENA and culminate in a Community for Energy, Water and Climate Security.

Concentrating Solar Power for Seawater Desalination (AQUA-CSP): Evaluated the anticipated water and power needs through 2050 in Europe, the Middle East, and North Africa; and the possibility to generate fresh water along with the electricity generation by the CSP (2004–2007).

Conventional large-scale desalination is perceived as expensive, energy consuming and limited to rich countries like those of the Arabian Gulf, especially in view of the quickly escalating cost of fossil fuels like oil, natural gas and coal.

None of the presently discussed strategies include concentrating solar power (CSP) for seawater desalination within their portfolio of possible alternatives. However, quickly growing population and water demand and quickly depleting groundwater resources in the arid regions of the world require solutions that are affordable, secure and compatible with the environment in one word: sustainable. Such solutions must also be able to cope with the magnitude of the demand and must be based on available or at least demonstrated technology, as strategies bound to uncertain technical breakthroughs, if not achieved in time, would seriously endanger the whole region. Its cost is today equivalent to about 50 US$ per barrel of fuel oil (8.8 US$/GJ), and coming down by 10–15% each time the worldwide installed capacity doubles. In the medium term by 2020, a cost equivalent to about 20 US$ per barrel (3.5 US$/GJ) will be achieved. In the long-term, it will become one of the cheapest sources of energy, at a level as low as 15 US$ per barrel of oil (2.5 US$/GJ). It can deliver energy "around the clock" for the continuous operation of desalination plants, and is therefore the "natural" resource for seawater desalination.

There are several good reasons for the implementation of large-scale concentrating solar powered desalination systems that have been identified within the AQUA-CSP study at hand:

- CSP desalination plants can be realized in very large units up to several 100,000 m3/day.
- Huge solar energy potentials of MENA can easily produce the energy necessary to avoid the threatening freshwater deficit that would otherwise grow from today 50 billion cubic meters per year to about 150 billion cubic meters per year by 2050.
- Within two decades, energy from solar thermal power plants will become the least cost option for electricity (below 4 ct/kWh) and desalted water (below 0.4 €/m3).
- Management and efficient use of water, enhanced distribution and irrigation systems, re-use of wastewater and better accountability are important measures for sustainability, but will only be able to avoid about 50% of the long-term deficit of the MENA region.

- Combining efficient use of water and large-scale solar desalination, overexploitation of groundwater in the MENA region can, and must, be ended around 2030.
- Advanced solar powered desalination with horizontal drain seabed intake and Nano-filtration will avoid most environmental impacts from desalination occurring today.

Electricity and one square kilometer of desert land using CSP Technology can harvest up to:

- 250 million kWh/year of electricity (250 liter oil/m2)
- 60 million cubic meters/year of desalted water (6000 liter water/m2)
- Produced electricity would be transmitted to European and African countries by a super grid of high-voltage direct current cables.

Implementation in the MENA Region

These studies have inspired companies, developers and governments in the region to deploy projects in their countries. First 2 CSP plants have been installed in Morocco and Abu Dhabi, UAE.

Noor 1 Solar Power Plant in Morocco

Morocco has inaugurated the Noor 1 project, a 160 MW solar CSP plant as first phase of Noor complex in February 2016, built at a location about 20 km from the city of Ouarzazate, approximately 200 KM south of Marrakesh. It covers 450 hectares (1112 acres) and is molten salt Thermal Energy Storage (Fig. 17).

Shams 1 Solar Power Plant in Abu Dhabi

Shams 1 solar power station is a concentrating solar power station near Madinat Zayed, Abu Dhabi, United Arab Emirates. It became operational on 17 March 2013. The first part, Shams 1, uses parabolic trough technology and has a capacity of 100 megawatts (MW), which makes it among the largest parabolic trough power stations in the world. Shams 1 have been commissioned in early 2013 and will be followed by Shams 2 and Shams 3 stations. Shams 1 solar power station consists of 768 solar collector assembly units and 27,648 absorber pipes (Fig. 18).

Fig. 17 Noor 1 Solar Power Plant in Morocco (Noor 1) VOA News, February 2016. http://www.voanews.com/content/morocco-dedicates-1st-phase-of-massive-solar-power-plant/3177855.html

Fig. 18 Solar Thermal Power Plant in Abu Dhabi (Shams 1) Francie Diep Posted March 19, 2013, *Popular Science* Massive Solar Power Plant Opens In Abu Dhabi. http://www.popsci.com/science/article/2013-03/largest-single-concentrated-solar-power-plant-opens-abu-dhabi

Fig. 19a CSP fields www.ifeed.org (El Bassam 2009)

Fig. 19b Pipeline for desalted water transport www.ifeed.org (El Bassam 2009)

Solar Oases Concept

The author has designed the concept of creating renewable energy "Solar Oases by using alternative energy sources and desalinated sea water for supplying the population in arid regions with food, energy and water and to establish a fundament for development and to ensure their survival in the future:

- Step one: Sea water desalination using concentrated solar thermal power and advanced desalination technologies (Fig. 19a).
- Step two: Transport of the water for implementation of farms (Fig. 19b).
- Step three: Establishing communities including:

Fig. 20 Projection of the oases including housing, lakes, food crops and greenhouse facilities www.ifeed.org (El Bassam 2009)

- Step four: Farms and food production facilities together with green houses, fish lakes, etc.
- Step five: Opportunities for job creation, education, handworks, workshops, and marketing.
- Step six: The vision for sustainable survival of the people in many regions.

The concept can be also applied for already existing villages and settlements. Specific modifications are then needed.

The mission is to develop a vision for this and future generations. The concept has been visualized in a video clip which can be seen at www.ifeed.org. (El Bassam 2009) (Fig. 20).

The UN Project: Integrated Renewable Energy Settlements (IRES)

The UN project of the Integrated Renewable Energy Settlements is based on the concept initiated by IFEED in 2004.

It provides a viable and sustainable development of rural areas in Europe and worldwide. The aim is to provide the population with energy, water and food and their decoupling from the consumption of fossil resources through the use of renewable domestic energy sources and in the same way, the creation and preservation of jobs through the integration of small and medium enterprises for the rural population. Overall, it includes environmental, socio-economic, energy and developmental elements. Parts of the concept have been implemented indifferent places. This is a groundbreaking plan for sustainable development of rural areas, poverty reduction, creation of decent livelihoods in Europe and particularly in developing countries, as well as avoidance of plight and drudgery. The approach stands for real-time, horizontal and vertical networking of people, nature, objects, equipment and systems for dynamic management of complex contexts (Fig. 21).

The anticipated main pillars of the IRES comprise the following Specifications:

- Decentralized, Autonomous and onsite production of energy for agriculture, enterprises and trade
- Combined use of different renewable energy sources (biomass, solar, wind, etc.)
- Suitability for remote areas: settlements, villages and islands
- It should include social, economic, ecological, education and job creation

This concept has been implemented in 10 locations, but only partially.

For the First Time and Worldwide the IREES has been fully implemented 2015 in the village of Wierthe in Lower Saxony, North Germany. It has been inaugurated officially on October 2015. The population of the village currently counts to a total of 378 people (Fig. 22).

More than 7000 photovoltaic modules have been installed and optimally aligned with the sun on approximately 4 hectares of land and on the roof tops of the buildings. The core of the project is a Photovoltaic Power Station, now with more than 1.6 MW peak power. The solar modules were built with the latest technology arrangement which calculated the annual electricity needs for about 460 households. The required heat (350 kW) is produced entirely from biomass (wood and pellets) on the grounds. The complex includes electric vehicles and electric bikes are also available, which are charged by solar power. In addition, hybrid vehicles are in use.

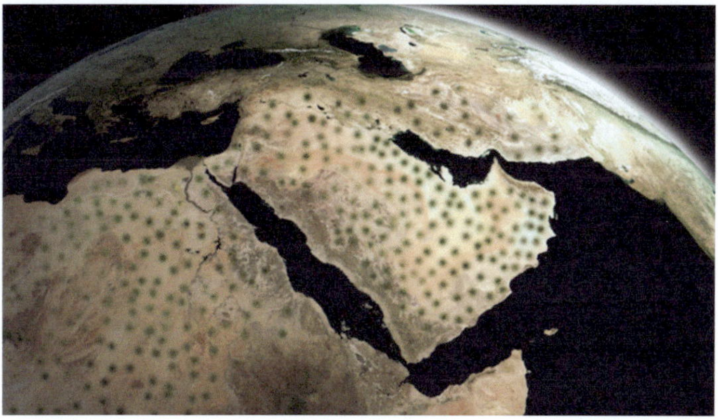

Fig. 21 The ultimate vision for distributed solar oases www.ifeed.org (El Bassam 2009)

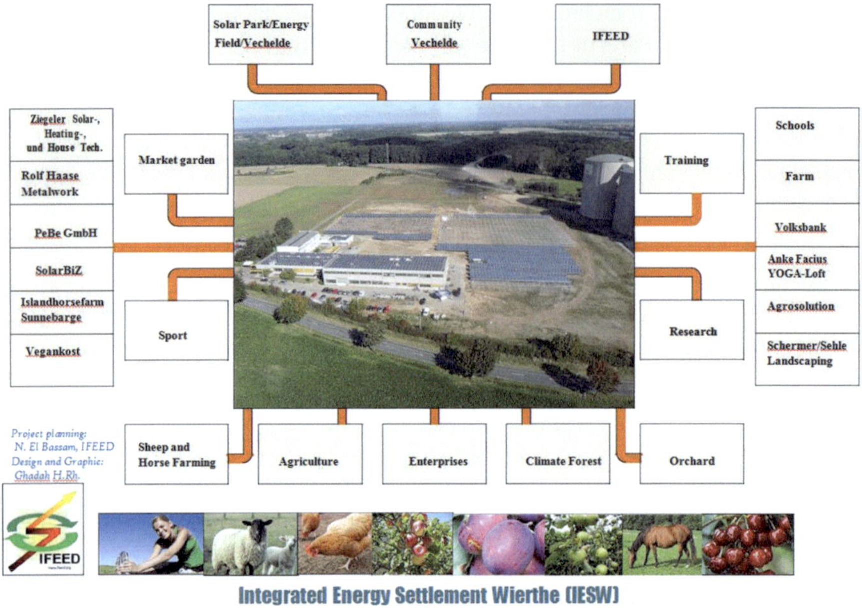

Fig. 22 Schematic representation of the whole project "Integrated Energy settlement Wierthe": production areas, the participants and the planned orchard (Design: El Bassam, Design: Ghada H. Rh, 2014)

A number of enterprises are located at the project site employing some 80 employees. Services of the commercial enterprises are not limited to the village but also are in Lower Saxony, as well as offered nationwide.

Approximately 4000 oak, beech, wild cherry, elderberry, hawthorn, etc. were planted by the pupils in a planting campaign. It also involved the schools of the community which planted part of the school forest. The fruit trees in the orchard and the numerous shrub species serve as food sources for bees and birds. They also offer nesting and breeding places and by doing so make a contribution to protecting our nature. The ground surface under and between solar systems is planted with grass and provides the resident sheep and ponies a food source. All these measures thus feature excellent ambient and soil conditions, which are considered optimal benefits for the purpose-built land (Fig. 23).

The Integrated Energy Wierthe is such an exemplary project which integrates most elements of sustainability of an ecosystem. It provides communities and settlements pragmatic ways to usher into sustainable development, poverty

Fig. 23 Field solar system latest technology arrangement which calculated the annual electricity needs for about 460 households

alleviation, livelihood and food security through education, training and application of science and technology for energy generation and energy utilization for holistic socio-economic development (Fig. 24). IFEED has developed more than 10 years ago for the United Nations the concept of the "Integrated Energy Settlements" which has been implemented in several counties. The concept comprises the following elements: Decentralized and onsite production of energy for households, small and medium-sized enterprises (SMEs), agriculture, water and waste water treatment, mobility, storage, trade etc.; it includes also social, economic, ecological, education and job creation components. This concept is a dynamic process. It has been modified and implemented on October 2015 in Wierthe, Lower Saxony, Germany. The project in Wierthe integrate 15 enterprises with around 80 employees and trainees, education, training and sport facilities, a forest, fruit and vegetable fields, bee, sheep and horse keeping, e-cars, solar park, bioenergy for heating, desalination and irrigation systems.

Wierthe Project is an innovative approach for maintaining the vital role of rural regions. It targets food and energy, nature and culture, diversity and dignity, economy and society. It combines strategic approaches of clean technologies and sustainable development to live in peace with our self and nature. We have the necessary knowledge, knowhow and the technologies to achieve these goals. We must act while we still have choices.

Fig. 24 Aerial view of the village Wierthe in Germany with the solar fields, bioenergy for heat supply, elecric cars for mobility and agricultural fields, forest and enterprises (Solar: photovoltaic 2009)

References

Augsten. (2014). Jakob, 2014; REN21, 2014, http://cleanleap.com/02-market-and-industry-trends/solar-thermal-heating-and-cooling

"BP Statistical World Energy Review 2011" (XLS). Retrieved 8 August 2011.

Business Dictionary. (2012). *Solar energy* [WWW], Available from: http://www.businessdictionary.com/definition/solar-energy.html#ixzz1lsoyXA8S

El Bassam. (2009). *Solar Oasis concept energy4You*, 2012. Solar hot air [WWW]. Available from: http://www.energy4you.net/hotair.htm. Energy Society of Canada Inc. and Solar, Oct. 21–24, 2000.

Faninger. (2010). https://www.researchgate.net/publication/268098993_High_Solar_Combi_Systems_in_Europe_OR-10-044

Hot Water Now. (n.d.). *Hot water adelaide, solar, gas & heat pump* [WWW]. Available from www.hotwaternow.com.au/colchester_solar_panel.jpg

Jacobson, M. Z. (2009). Review of solutions to global warming air pollution, and energy security. *Energy & Environmental Science, 4*, 148–173.

Mugnier, Daniel "State of the art for solar thermal or PV cooling and refrigeration", SHC. (2014). *Conference, Beijing (China)*. http://task48.iea-shc.org/Data/Sites/6/documents/events/shc2014/shc-2014-mugnier-tecsol-ac-keynote.pdf

NASA Science/Science News. (2011). *The edge of sunshine* [WWW]. Available from: http://science.nasa.gov/science-news/science-at-nasa/2002/08jan_sunshine/

Perlin, John. (1999). From space to earth (The story of solar electricity) (ISBN 0-674-01013-2, p. 147.). Harvard University Press: Ann Arbor.

Popular Science. (1931, June). *Magic Plates, Tap Sun For Power* (pp. 41, 134).

PV Resources.com. (2011). World's largest photovoltaic power plants.

Schwerdt & Ali. (2014). https://www.researchgate.net/publication/264436757_Solar_Cooling_Case_Study_of_Cooperation_between_Germany_and_Egypt_in_Assiut_University_Project

Shenzhen JCN New Energy Technology CO., Ltd. http://www.hktdc.com/manufacturers-suppliers/Shenzhen-JCN-New-Energy-Technology-Co-Ltd/en/1X08FPW1/

"Small Photovoltaic Arrays". Research Institute for Sustainable Energy (RISE), Murdoch University. Retrieved 5 February 2010. (Modified by El Bassam).

"Solar Cooker Designs". Solar Cookers World Network. http://solarcooking.wikia.com/wiki/Solar_Cookers_World_Network_%28Home%29

Solar Heating and Cooling for Residential Applications, IEA-ETSAP and IRENA.Technology Brief R12 – January 2015. www.etsap.org – www.irena.org

Solar Millennium AG. (2011). http://www.solarmillennium.de/english/technology/references_and_projects/andasol-spain/index.html. *Solar: photovoltaic: Lighting up the World*. Retrieved 19 May 2009.

Solar Tribune. (2012). *Guide to solar thermal energy* [WWW] Available from: http://solartribune.com/solar-thermal-power/#.T0_PvfU0iuI

Ssembatya, et al. (2014). *Simulation studies on performance of solar cooling system in UAE Conditions*. http://ac.els-cdn.com/S1876610214003774/1-s2.0-S1876610214003774-main.pdf?_tid=688ec742-eacd-11e5-89cd-00000aab0f02&acdnat=1458060544_2cdbe5f7717700a076dcb0e2196550e3

TheFreeDictionary.com. (2012). *Solar energy* [WWW]. Available from. http://encyclopedia.thefreedictionary.com/solarþenergy

"Trends in Photovoltaic Applications Survey report of selected IEA countries between 1992 and 2009, IEA-PVPS". Retrieved 8 November 2011.

Trieb, Dr Franz. *TRANS-CSP Trans-Mediterranean interconnection for Concentrating Solar Power*, www.dlr.de/tt/trans-csp

U.S. Department of Energy (DOE). (2001). *Concentrating solar power commercial application study: Reducing water consumption of concentrating solar power electricity generation A report to Congress*. Available from: http://www1.eere.energy.gov/solar/pdfs/csp_water_study.pdf

Utility Scale Solar Power Plants (PDF). (2012). IFC, Maruti Suzuki Building, 1 Nelson Mandela Road, Vasant Kunj, New Delhi – 110070 India.

Wang, Ucilia. (2011, 6, June). *The rise of concentrating solar thermal power*. Renewable Energy World.

Whitlock, C. E., et al.. (2000). *Release 3 NASA surface meteorology and solar energy data set for renewable energy industry use*.

Zweibel, K.. (2010). *Should solar photovoltaics be deployed sooner because of long operating life at low, predictable cost?*. Energy Policy. http://dx.doi.org/10.1016/j.enpol.2010.07.040

Reducing Energy Cost for Wastewater Treatment in the Middle East: A Physio-chemical Prospective

Digambara Patra

Abstract Middle East and North Africa (MENA) region having less than one percent of world's available water has 5% of world population putting additional stress on efficient water management and treatment in the region. Large amount of water used in agriculture, industry and municipality (household uses) in the region has deteriorated quality for further use, which can be treated and reused for efficient utilization of water resources. The principal challenges in wastewater treatments in MENA region derive from the different undesirable components, which vary depending on the sources due to variations in the discharged amounts of substances. Domestic/municipality wastewater has different source of pollutants in comparison to industrial or agriculture source wastewater. These impurities include metals such as Fe, Al, Zn, Pb, Cd, Hg, Cr, Ni, Cu, etc., nutrients like nitrate, phosphate, ammonia etc., inorganic materials like hydrogen sulphide, bases etc., biodegradable organic materials, organic materials like detergents, grease, coloring, solvents, fat, oil, solvents, pesticides, phenol, cyanide etc., microorganisms such as virus, worms egg and pathogenic bacteria, and suspended solids. Therefore, the challenges to treat wastewater by reducing energy cost are multi-folded and one scientific procedure is not sufficient to treat wastewater. It requires simultaneously many scientific methods to treat wastewater. In this chapter, various physical, chemical and biological processes involved in wastewater treatment, especially that can reduce energy cost, have been discussed. It briefly introduces to different scientific approaches that can be beneficial to treat wastewater coming from municipality, industries and agriculture sources in the MENA region. Study related to wastewater management and engineering is beyond the main objective in this chapter. Similarly, desalination techniques, which are not related to wastewater treatment, have been avoided. Focus has been particularly made for general readers on scientific procedural aspects without going much detailed into critical scientific development in this field.

Keywords Waste water • Purification • Energy efficiency

D. Patra (✉)
Department of Chemistry, American University of Beirut, Beirut, Lebanon
e-mail: dp03@aub.edu.lb

S. Murad et al. (eds.), *Water, Energy & Food Sustainability in the Middle East*,
DOI 10.1007/978-3-319-48920-9_10

Introduction

The population increase globally has put additional pressure in water demand. With increase in population growth, demand in water will increase. Especially population increase in developing countries will further need water resources for agriculture, industry, energy and particularly for domestic use. Annual renewable water resources (ANWR) should be at least 2000 m^3/capita. A country is likely to experience “absolute water scarcity” with ARWR less than 500 m^3/capita. Most of the Middle East and North Africa (MENA) region will have ARWR less than 1000 m^3/capita by the year of 2025. Only Iraq, Iran and Lebanon have ARWR more than 1000 m^3/capita. Kuwait, Libya, Qatar, Saudi Arabia, United Arab Emirates and Yemen have ARWR less than 100 m^3/capita. In 1975, the MENA region had an average ARWR 1325 m^3/capita, but it has declined to 525 m^3/capita and is expected to decline further to 414 m^3/capita in 2025 (Qadir et al. 2010). The International Panel on Climate Change foresees declining precipitation (−10% to −25%), declining soil moisture (−5% to −10%), declining runoff (−10% to 40%), increasing evaporation (+5% to +20%) and higher temperatures for MENA region. High temperature will further enhance evaporation, decrease soil moisture and increase evapotranspiration rates in vegetated areas leading to decrease run off and ground water recharge rates and increase crop water requirements in agriculture. These changes are expected to create a new and more difficult context for water management in MENA regions, which depends on seasonal rainfall. Water use in the MENA region can be divided into agriculture (86%), household activities (8%), and industrial sector (6%). Most of MENA region countries are agrarian although tourism and oil production are also important in many countries. Large difference exists between these countries in terms of water use in agriculture, domestic and industrial sectors. Treating municipal wastewater for agricultural purposes has been one of the key option as in most countries in the MENA region agriculture is primary water consumer, for example ~80% in Tunisia and ~90% in Syria. With rising requirement, different solutions and strategies have been being tried to get more reliable water resources.

Among the various approaches, one of the viable options is the possibility of treating wastewater coming out from various sources including industrial and agricultural activities, rural and municipal wastewater etc. In addition, compared to desalination and long distance water transportation, the cost effective and energy efficient alternative water source is reuse of wastewater after treatment. Therefore, treating wastewater is related to the standards and expectations. Recently advancement in scientific and technological innovation is producing recycled water for portable use by applying relatively energy efficient technology. Despite the fact that for wastewater treatment the multi-fold challenges come from economic, institutional, social, environmental, health, governmental, political and technical reasons, this chapter is mainly focused on problem and solutions related to scientific reasons, which can be generally a worldwide phenomena rather than limited to any specific region. The main scientific challenges in treatments come from the different unwanted components that include toxic components present in wastewater. These components vary depending on the sources due to variations in the discharged amounts of substances. For example, domestic wastewater and municipal wastewater may have different source of contaminants in comparison to industrial source wastewater. Both time and location also significantly influence composition of wastewater. Similarly, nature of industry may influence components present in the wastewater.

Normally, wastewater coming from municipality and house hold sources include: (i) metals such as Fe, Al, Zn, Pb, Cd, Hg, Cr, Ni, Cu, etc.; (ii) nutrients like nitrate, phosphate, ammonia etc.; (iii) inorganic materials like hydrogen sulphide, bases etc.; (iv) biodegradable organic materials; (v) organic materials like detergents, grease, colouring, solvents, fat, oil, solvents, pesticides, phenol, cyanide etc.; (v) microorganisms such as virus, worms egg and pathogenic bacteria; (vi) suspended solids. Beside these components other parameters like thermal effect (hot water), odour and taste (for instance, presence of H_2S) and radioactivity (toxic effect, accumulation) are also crucial. Therefore, treatment of wastewater necessitates removal of pathogenic bacteria, suspended solid particles, nutrients (phosphate, nitrate, etc.), biodegradable organics, metals etc.

Ancient Egyptians practiced " Basin Irrigation" based on natural rise and fall of the river that delivered ancient Egyptians with an efficient domestic wastewater treatment and disposal system. In this case the wastewater penetrates into the soil where anaerobic bacteria breaks down any nutrients before the wastewater reaches the water stream. When the flood in the Nile River occurs, the flood water gets mixed with the biologically degraded wastewater present in the upper soil layers and pushes it down as soon as the basin drain starts, thus, all salts and accumulated biomass is washed out and drained back to the main river stream. This practice is similar to the modern sludge activation wastewater treatment techniques. However, this ancient practice in Egypt was terminated after the construction of lake Nasser.

In the MENA region, treated municipality and industrial wastewater is largely used for irrigation, recreational/environmental use, ground water recharge, non-portable house hold use and portable reuses. Today Jordan, Lebanon, Kuwait, Syria, Egypt, Tunisia, etc. have been using wastewater after treatment for irrigation. Reuses of wastewater in the MENA region can be classified in three different groups. The first group comprises those countries where minimum or no treatment is carried out on waste effluents and subsequently used for irrigation of cropped land. In this case environment and health control are minimum. Countries like Yemen, Lebanon and West Bank (Palestine) come in this classification. The second group consists of countries like Jordan, Egypt, Iraq, Morocco and Syria where there is moderate regulations, thus, wastewater effluents are treated but a high percentage of the effluents are disposed to surface water bodies for later use in irrigation. The crops that can be irrigated during this process are specified. The last group is mainly GCC countries, which include UAE, Saudi Arabia, Oman, Qatar, Kuwait and Bahrain where advanced treatment is adopted to meet national or international standard. In this case after post treatment of wastewater, it is reused for irrigation and landscaping while reminder is disposed into the sea after many advanced treatment steps.

Although the procedure adopt may vary depending on the source of wastewater or industrial water, generally wastewater treatment plants in the MENA region can be classified in three different scientific steps. For large-scale wastewater treatment, following steps are generally practiced during removal process. (a) Pre-treatment process to alter microbial water quality at the entry point of treatment plant; (b) Coagaulation, flocculation and sedimentation processes in which small particles combine to form larger particles and settle down by gravity; (c) Ion exchange process that removes calcium, magnesium, radionucleotides etc.; (d) water is passed through granules materials after coagulation pre-treatment in granular filtration whereas in slow sand filtration water is passed slowly through a sand filter by gravity without the application of coagulation pretreatment. Most of the treatment plants in the region consist of primary and/or secondary steps, nevertheless, these processes may not be sufficient and wastewater might need chemical and other treatment to make it portable. Other steps like pre-chlorination, aeration, disinfection and ozone/Ultraviolet light treatments are often included.

Primary Step

The most simplest, energy and cost effective initial step is to identify and screen various solid and their sizes and then remove solid waste like wood, plastics, metal, cloth, paper, fecal solids, kitchen items, etc. The large pieces of debris are separated in the treatment plant for which wastewater is detained in a large sedimentation tank for several hours. This makes heavier and large solids to settle down to the bottom and form a sludge layer. In case of lighter solids such as wood, fats, oil, grease etc., they float on the surface of water and create a scum layer. Both the solids and scum are removed and wastewater is passed through for further treatment. In cases chemicals are added to the sedimentation tank, which help waste particles to form bond and settle out. Many of the small-scale plants carry out this process before releasing wastewater into the sea or river. Domestic wastewater management is one of the greatest headaches of Lebanese. Most of the wastewater treatment plants in Lebanon are small scale and primary in nature although many proposals are pending. Ghadir plant in Southern Beirut in Lebanon provides only preliminary treatment (i.e., grit and scum removal).

Morocco has about 100 small and medium size wastewater treatment plants, which treat about 13% of the total wastewater in Morocco. The technique employed in Morocco is generally basic and so far concentrated on stabilization pond technology, few of them are trickling filter plant. Therefore, the treated water is largely used for irrigation purposes. Only one plant is based in activated sludge method in Morocco. In 1980s a nation-wide water reuse program to increase the Tunisia's usable water resources was launched. Municipal wastewater receives secondary biological treatment and some tertiary treatment. The fact of reusing reclaimed water for irrigation is viewed as a method to increase water resources, provide supplemental nutrients and enhance wastewater treatment. Reclaimed water is used for agriculture irrigation of 8100 ha including citrus, olives, peaches, etc., for landscape irrigation such as golf courses and green areas. Tunisia has a long history since 1965 to irrigate the citrus orchards and olive trees (Bahri 2008). The general idea here is to remove large and small particles without giving emphasis on secondary and/or tertiary treatment. Large and small sized cities in Tunesia have sewage treatment facilities. About 110 plants are currently in operation in Tunesia, these plants are mostly located in urban coast to prevent marine pollution and Tunis contributes about 42% of the treated wastewater, which is primarily used for irrigation, golf courses irrigation, landscape irrigation and for artificial ground water recharge. The methods employed in these countries successfully reduce energy cost during wastewater treatment.

Removal of Larger Particles

Sedimentation is the technique used for gravity separation where the suspended solid particles settle under the influence of gravity. It is carried out before the conventional treatment process and is helpful for treating industrial effluents, especially wastewater from paper industries, refineries, etc. by reducing energy cost. In this process wastewater is moved slowly through a large basin, called as *sedimentation* or *settling basins*. Here, the water moves very slowly due to large size of the basins, which gives sufficient time for the waste particles to settle down at the bottom. Size and density of the solid particles influence the settling time. Similarly, velocity of the water during treatment is important for separation. Clay, turbid silt, natural organic matter and associated impurities such as manganese, iron, other toxic metals, synthetic chemicals, microbial contaminants, humic substances etc. can be removed in this process. Removal of humic substances is often desirable because during chlorine treatment these substances form disinfection byproducts with chlorine. Some of these byproducts like trihalomethanes are of great public health concern at high concentration. As an apprehension, humic substances are commonly found in wastewater as these are produced within natural water and sediments by chemical and biological process including decay of vegetation. Time to time alums can be added to supplement the sedimentation process. A hopper can be used to collect the waste particles using large rotating scrapers; sometime such hopper is used several times a day. Clear water at the top of the basin comes out of the sedimentation basin. This process can separate up to 60% of the suspended particles. However, water treated in this process can be used for industrial water supply and/or for further treatment like filtering processes including membrane/ion exchange processes. In particular, removal of solid particles in the sedimentation process helps the filter operation.

Iraq has a unique experience of treating wastewater using reed plants (al-Samawi 2000) that technically avoid any energy cost. Plants that grow naturally under harsh environmental conditions can offer a simple, energy efficient and cost effective method to treat wastewater. Iraq has favorable environmental condition to grow reed plants. In root-zone method, the treatment process does not need chemical materials or sophisticated mechanical/electrical equipments rather it makes use of a weed, namely reed. By this procedure average removal efficiency of BOD, suspended solids, chloride, phosphate and ammonia ranges between 82 and 92%. Chloride can be removed with 13% efficiency. Egypt has more than 350 wastewater treatment plants that treat about 10,100,000 m^3/day of wastewater, however, it is about 1/3 of the amount of wastewater released to Nile River. Gabal el Asfar plant is one of the largest wastewater treatment plants, which has a capacity of 2,800,000 m ^{3}/daily capacity. The plant gives a complete treatment of the carbonaceous contaminants along with a comprehensive sludge treatment, anaerobic digestion and, biogas production that generates 65% of the power supply adding to lowering energy cost. Mdinat Al-Jubail Alsinaiya is an industrial city located in Saudi Arabia. Figure 1 gives a flow chart of design for treatment of municipal wastewater including primary, secondary and tertiary process (Al-A'ama et al., 1995).

Although *sedimentation* is technically relatively an easy process, sometime it could be challenging as time to time dissolved and suspended solid particles do not settle down under the influence of gravity force. These dissolved particles arise from domestic discharges, land erosion, dissolution of minerals, etc. These materials include organic and inorganic materials in addition to bacterial, algae, viruses etc. These materials decompose and reduce clarity by causing turbidity and color that eventually carry pathogenic organisms and toxic compounds, thus, must be removed. For making these floating substances to get settle down chemicals known as *coagulants* are added, this process is known as *coagulation*. The main purpose of the coagulant is to stabilize the electrical double layer surrounding the very fine colloidal/solid suspension because the colloidal particles carry surface electrical charges that mutually repel each other. When *coagulant* is added, it supports to overcome the repulsive charges between colloidal suspensions and thus, van der Waal's force causes agglomeration of suspended particles into large particles, which settle down slowly under sedimentation. The process is relatively cost effective (also reduces energy cost) and simple but normally a pre-treatment, such as sedimentation process, is done before coagulation step. Chemicals that are commonly used for coagulation include alum, activated silica, aluminum salts, iron materials, etc. Besides inorganic materials mentioned above starch and synthetic polymers having positive, negative and neutral surface changes can be effectively used though it could be costlier than natural coagulants. The factors that could influence coagulation process are temperature, contact time and pH of the medium. Special organic coagulants from natural sources (Ramavandi 2014) are also used to remove microbes, bacteria, etc., and treat turbidity of wastewater.

In contrast *flotation* process is applied to remove by adhering suspended solid particles, oils, greases, biological solids, etc. with either air or gas. The solid particles get adhered to gas or air to form agglomerates; this accumulates over the time period at the waters' surfaces and eventually can easily be skimmed off. Passing of compressed air through the water boosts flotation process. Similarly,

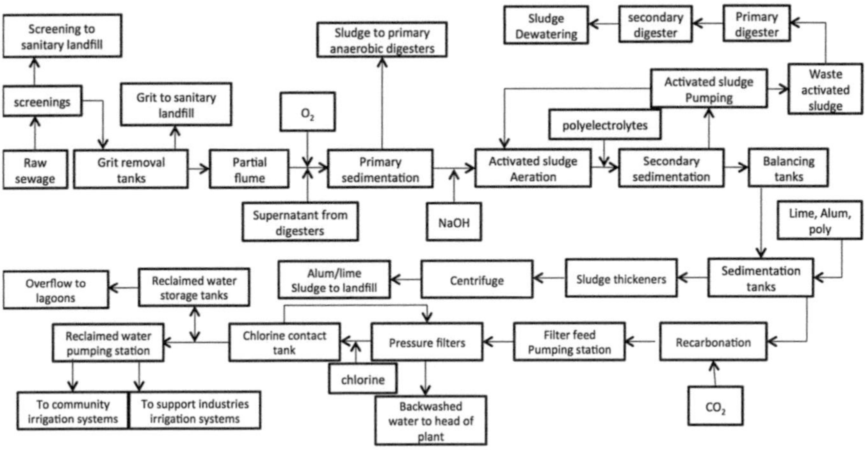

Fig. 1 Madinat Al-Jubail Alsinaiya municiapal wastewater treatment plant

alum, activated silica etc. also aids in the flotation process. Electro-flotation, also known as electro-flocculation, can be used as an effective process for water treatment and recycling purposes. In this process 75% of suspended solids and up to 99% of oil and grease can be removed, thus, flotation process has a great potential due to the high through put of modern equipment, low sludge generation and the high efficiency of the separation schemes already available. It is found that this method is quite efficient for wastewater treatment originated from paper industries and refineries. This can be incorporated as a technology in the minerals industry to treat these wastewaters and, when possible, to recycle process water and materials.

Removal of Smaller Particles

For removing particles, filtration process is applied where wastewater is passed through a medium having fine pores. Normally the filtration apparatus consists of a concrete tank containing sand, gravel and an under drain. Sand has pores and does the filtering process whereas gravel helps in keeping the sand from getting out and under drain stored the filtered water. However, over the time when sand becomes clogged with particles it must be backwashed. During backwash, water is flown in the reverse direction, thus, the sand and particles are suspended. Since particles are lighter than the sand, they rise up and can be flushed out from the system. After the backwash process is complete, the sand settles down onto the gravel, the process restarts again.

Sludge that is released from sedimentation/coagulation tanks can still have solids smaller suspended particles. To remove smaller suspended non-colloidal particles up to 1 μm, centrifugal process is applied. There are two kinds of centrifuge designed. The first kind is based on continuous operation, known as *"solid bowl centrifuge"*. Here wastewater sludge is continuously fed into the unit in a horizontally set up where a set of helical scrolls attached to the moving shaft push the solid waste towards the other side, far from the liquid moving towards another direction. Over the period solid content starts building up and the accumulated sludge becomes heavier to drop down to a collection bin. The second kind of centrifuge method works in a batch process, sometime referred as *basket type*, thus, needs operator to handle and remove accumulated sludge. In this case accumulation of sludge occurs on the vertical surface of the imperforate basket. The accumulation continues till it reaches the holding capacity before the bowl decelerates. A scraper is applied to eliminate the sludge. For better separation in both the methods polymer can be mixed into the wet sludge prior to feeding into the centrifugal unit. The speed of the centrifugal machine can influence removal of suspended particles. Similarly, efficiency and extend of separation of the suspended solid is directly related to their densities.

Secondary Step

Wastewater is treated and recycled to save energy (thus reduce energy cost) and raw materials following primary step known as "physico-chemical treatment" followed by secondary step (biological treatment) to remove most of the suspended solids and soluble organics. Secondary step is crucial, at least for industrial wastewater. The main objectives of the biological treatment is (i) to eliminate carbonaceous organic matter; (ii) to remove organic nitrogen compounds; (iii) to remove phosphorous; (iv) for nitrification and denitrification; and (v) for transformation of waste into harmless products. Such a treatment is required, may not be in all cases, because for several reasons. Suspended solids affect aquatic environment by leading to anaerobic condition and sludge deposits. Similarly, proteins, carbohydrates, fats etc. (biodegradable) organic matter reduce dissolved oxygen concentration in water and pathogens cause disease in humans and animals. Nitrogen, phosphorous etc. could cause eutrophication of lakes and reservoirs that helps to bloom algal. Some contaminants such as phenol, surfactants, pesticides, heavy metals etc. could cause cancer and/or are highly toxic. Floating materials and oil can make unsightly conditions. Removal of these contaminants can be done, depending on the origin of wastewater, by treating with microbes. In this procedure, a high concentration of microbes/microorganism is used. Microbes such as bacteria, fungal strains, etc. decompose organic matter present in wastewater. There are mainly two major processes involved depending on the presence and absence of oxygen/air.

Aerobic Biological Process

When decomposition occurs in the presence of available oxygen/air in wastewater, the procedure is known as *aerobic biological process*. In this case, efficiency and extend of process differs on the availability of oxygen, temperature, retention time, and biological activities of microbes. The rate of aerobic biological process can be further enhanced by addition of some chemicals that helps in microbial growth. Aerobic biological processes are applied for the transformation of harmful organic and inorganic substances (Barragan et al. 2007), thus, it is effective for the removal of chemical oxygen demand (COD), biological oxygen demand (BOD), volatile organics, nitrates, phosphates, dissolved and suspended organics, etc. (Abdel-Raoudi et al. 2012). Most common system for aerobic biological process are activated sludge systems and aerated stabilization basins. Aerated stabilization basins (ASBs) are often used in pulp and paper industries, municipalities and other industries. However, ASBs need regular maintenance with time. The untreated industrial influents have the most BOD. When untreated influents enter ASBs, it enters an oxygen rich environment where degradation/decomposition will start. There are different kinds of ASBs and widely used ones are surface aerators or diffuse aeration systems. In surface aeration, multiple units are applied to properly spaced and treat water whereas in diffused aeration air is delivered by blowers or compressors through a pipe, which is put under the surface for evenly release. Sometime pure oxygen is used instead of air. The bacteria (microbes) utilize oxygen as an electron acceptor in order to convert organic matters to carbon dioxide as per the reaction below.

$$\text{Organic Substrates} + O_2 \xrightarrow{\text{Bacteria}} CO_2 + H_2O + \text{other products}$$

This process continues and multiplies to create more bugs to break down more BOD. When water flows through the system, many changes occur and the amount of BOD in the system decreases. During this process amount of bacteria also falls. Decrease in oxygen demand makes the water acceptable for more advanced life forms like protozoa or metazoan, flagellates, free swimming ciliates, stalked ciliated, rotifers etc., which feed the dispersed bacteria and help in flocculating the bacteria that formed after decomposition. In *aerobic biological process* concentration of biodegradable organics can be reduced by up to 90%. The major limitation is the production of a large quantity of bio-solids in this process that might need another expensive treatment and management.

Anaerobic Biological Process

When decomposition occurs in the absence of available oxygen/air in wastewater, the procedure is known as *anaerobic biological process or putrefaction*. A*erobic* *biological process* is a single species phenomena but *anaerobic biological process* proceeds as chain process including several sequent organisms. It needs the synergetic action of the microorganisms that are involved. The important factors are partial pressure of hydrogen and the thermodynamics linked to it. During this process complex organic substrates are converted to simpler organic compounds containing nitrogen, carbon and sulphur. The gases formed in this process include N_2, NH_3, H_2S and CH_4. The chemical reaction takes place as follows and applied to decrease the biological load of wastewater (van der Zee and Villaverde 2005; Venkata Mohan et al. 2007).

$$\text{Organic Substrates} + O_2 \xrightarrow{\text{Bacteria}} CO_2 + H_2O + \text{other products}$$

For both the *aerobic* and *anaerobic biological processes* several class of organism are prime important such as pathogenic microorganisms, pollution indicator organisms, microorganisms for biological treatment etc. Among the biological treatment organism two major aspects are considered depending on the nutritional requirement of the microorganism and metabolism of microorganism based on need for oxygen. On the basic of cell structure these organism can be classified as eukaryotic and prokaryotic groups. Some important eukaryotes in wastewater treatment are fungi, protozoa, rotifers and algae, which need energy sources, carbon for the synthesis of new cellular material and inorganic elements like nitrogen, phosphorus, sulphur, potassium, calcium and magnesium and oxygen. Prokaryotic microorganisms (bacteria) are mainly needed for wastewater treatment and are crucial for the degradation of the organic material in the influent. Some of the examples of these bacteria involved in the activated-sludge process are Pseudomonas, Zoogloea, Achromobacter, Flavobacterium, Nocardia, Bdellovibrio, Nitrosomonas, Nitrobacter, Sphaerotilus, Beggiatoa, Thithrix, Lecicothrix, *E. coli*, Geotrichum, etc.

However, there are many other approached that can be considered to improve and enrich removal of organic and inorganic substrates. For example biological phosphorus can be removed by enriching with denitrifying phosphate accumulating organisms (DNPAOs) for simultaneous phosphate uptake and nitrate removal from wastewater, where the organisms utilize nitrate as electron acceptor instead of oxygen (Ahn et al. 2002). Further combination of anaerobic–anoxic–aerobic, anaerobic–aerobic–anoxic (Tsuneda et al. 2005), and aerobic–anoxic condition (Shi et al. 2014) can be used to treat wastewater from municipality sources. For example, textile industry consumes large quantities of water and chemicals for wet-processing units. Considering both volume discharge and wastewater composition, textile wastewater is considered as the most polluting among all industrial sectors. Wastewater from spent dye baths and dye rinsing operations in the textile industry contain unfixed dyes that may be highly colored. Through new environmental concerns and regulations, pressure is being placed on textile companies to reduce pollutant and reuse process water and chemicals. Neither simple chemical nor biological treatment has proved adequate in decolorization and in the depletion of dye/surfactant substances. Biological treatment is not very impressive in the color reduction. Moreover, color removal is not yet a serious concern to sanitary sewer (Koyuncu et al. 2001), the major objective in wastewater treatment for textile industry, especially in Turkey, has been removal of COD and BOD. Traditionally, treatment for wastewater from textile industries has been primarily by physio-chemical (Kim et al. 2004) or biological (Libra et al. 2004) methods. In one of the approach has been studied to reduce energy cost and evaluated for feasibility of activated sludge process as a pretreatment for nanofiltration of real denim waste-water for reclamation. In first step, effluent of a denim textile mill is treated in a bench scale activated sludge reactor (Sahinkaya et al. 2008). The treatment plant (flow chart of reactor shown in Fig. 2) located in the middle Anatolia region of Turkey, produces 20,000 tons of cotton fiber, 45 million meter of woven fabric, 15 million meter of denim per year and consumes about 3500–5000 tons/day of water. Thus, water reuse is inevitable in the plant. In this case the wastewater is diluted four and two time on day 0–26 and 26–56, respectively, to acclimate the microorganisms.

For mainly removing BOD and COD concentrations during wastewater treatment of Iranian industrial estates, integrated anaerobic and aerobic treatment systems are used (Piadeh et al. 2014). The limitation of aerobic treatment units, such as high surplus sludge production, high energy consumption and eliminating low organic loading rate etc., can be overcome by integrated anaerobic and aerobic treatment systems (Chan et al. 2009). Combination of up-flow anaerobic fixed bed reactor (UAFB) and integrated fixed-film activated sludge (IFAS) is the most chosen systems, about 26%, among all the processes for secondary wastewater treatment. Till recently about 690 industrial estate have been constructed in Iran and about 37% of total wastewater produced have been treated of which ~20% is reused for irrigation. The wastewater treatment process at Abbas-abad industrial estate in Iran has a capacity of 1700 m^3/day, which is connected to about 67% factories located in this region (Piadeh et al. 2014). While treating wastewater in industrial estates, biological treatment is generally applied of which 82% are a combination of aerobic and anaerobic units.

Jordan is one of the few countries where relevant wastewater policy framework and institutional structure exist. The country has extensive research on water reuse and the sage of reclaimed water in the total water supply is ~10%. Jordan has taken lead in terms of treating part of the domestic wastewater, generally known as greywater (washing, laundry, bathing). The projects on the treatment and use of greywater have implanted in north-east part and three southern governorates (Karak, Tafilah and Maan) in Jordan. The greywater reuse projects have revealed considerable potential of greywater reuse in irrigation at household level in poor communities by reducing energy cost. Greywater reuse has increased community participation in the national efforts to conserve limited water resources on-site and low-cost greywater treatment and reuse systems. The As-Samara wastewater treatment plant in Jordan has a primary settling tank, eight aeration tanks and eight settling tanks, found anaerobic sludge digesters, biogas and hydro-powered generators and an odour control system. Expansion of this plant includes construction of a new mechanical treatment plant equipped with advanced technology to treat

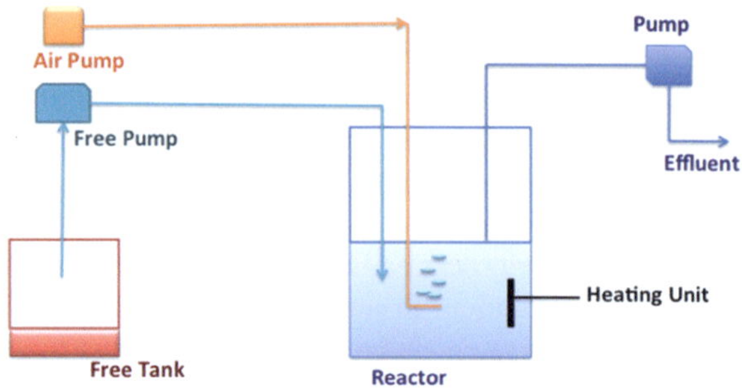

Fig. 2 Activated sludge reactor in Turkey

contaminants for agricultural reuse, produce fertilizer from sludge and generate clean energy.

Tertiary Steps

The second stage is the nanofiltration of biologically treated wastewater to improve the treated wastewater quality for reuse. Before nanofiltration, dead-end microfiltration is conducted with a nitrocellulose acetate membrane to prevent nanofiltration from fouling (Sahinkaya et al. 2008). Nano- or mic*ro*-filtration is done to remove particles of few nm to μm in size. Microfiltration is done for pore diameter 0.02–1 μm whereas nanofiltration or ultrafiltration is done with pore diameter 1–20 nm whereas reverse osmosis is carried out with pore diameter 0.1–1 nm. In case of micro/nano filtration the total suspended solids may not be more than 100 mg/l. Commonly used filers during micro−/nano-filtration are cotton, wool, rayon, cellulose, fiberglass, polypropylene, acrylics, nylon, asbestos and fluorated hydrocarbon polymers. Arrangement of these filters are crucial and are made in different shape such as spiral, plates, disc, tubular, hollow fibers etc. Depending upon the use and concentration of solid waste, the life span of the cartridge varies; this could be from few years to a decade. Pre-removing the suspended particles help in increasing the long life of the cartridge and filter.

Osmosis is a process in which solvent flows into the solution through a semi-permeable membrane. When an external pressure higher than osmotic pressure is applied, this process is reversed, known as reversed osmosis. This method has received greater attention in wastewater treatment because it can easily remove large, non-polar, ionic and toxic soluble contaminants. In this case semi-permeable membranes deliver a physical barrier for such contaminants based on their size making high level of automation by minimizing requirement of land and chemical and configuring flexible design. However, the major test for this process arises from the negotiation between permeability and selectivity. Beside this other important concern is requirement of high level of energy for generating pressure. Membrane fouling and complexity of the operation and process design further increases its energy need. Materials used as semi-permeable membrane is key to better performance. Therefore, there is great opportunity to enhance the performance by modifying functionality of the materials used as members. Such materials can boost permeability of the membrane, fouling resistance, mechanical and thermal stability and self-cleaning procedure, thus reducing the energy cost. Commercial semi-permeable membranes are asymmetric cellulose type such as cellulose acetate, triacetate, cellulose diacetate, etc. and thin film composite type. Thin film composite made up of aromatic polyamide barrier has superior water flux and salt rejection, resistance to pressure compaction, wider operating temperature range and pH range and higher stability to biological attack. However, membrane fouling has been a serious concern in reversed osmosis. Fouling could occur because of precipitation of metal hydroxide & carbonates, humic acid, colloidal suspended particles like silica and bacteria and fungi. These materials form a fouling layer on the membrane surface, thus, hydrophobicity, roughness, steric repulsion effect, and electrostatic charge of membrane surface affects membrane fouling. Increase in hydrophobicity, smoother surface and repulsive electrostatic force between membrane and foulant help in decreasing membrane fouling. In addition surface bound long hydrophilic molecules such as PEG can prevent adsorption of macromolecules, proteins etc. due to steric repulsion as a result of loss of configurational entropy, thus, reducing fouling. m-phenylenediamine and trmsoyl chloride have been commonly used as active monomers to form functional poly-amine layer in membranes to improve interfacial polymerization process. Mixing organic modifier with TMC solution or surface modification by simple adsorption can change the surface property and increase anti-fouling properties of the membrane.

Alternative physical processes such as distillation, evaporation and, crystallization can be applied to purify wastewater though few of these methods might need more energy, thus increase the cost. Distillation eliminates majority of the dissolved substances, except those substances whose boiling points are close to that of water, and also kills pathological pollutants. It also suffers as some of the chemicals like chlorine and its products can't be separated during this process. The major bottleneck is that it also consumes lot of energy to heat water above its boiling point. Evaporation is rather a natural process and cost effective. Vacuum evaporation has been a helpful tool for recycling wastewater. Crystallization process has been other cost and energy effective process to eliminate contaminants by increasing their concentration to a point to crystallize them out by evaporation or lowering the temperature. Mixing with other solvent is also helpful. High concentrations of TDS can be overcome during this process. Bicarbonate, ammonia, sulfite *etc.* can be broken down into various gases and pH can be regulated.

The highly concentrated civil wastewater in Sana's, Yemen is transported by road to the plant in cisterns filled directly from septic systems; thus, it has a very high organic load measured in terms of BOD. The procedure adopted here includes fine screening, oil and sand separation, pre-denitrification, biological oxidation-nitrification, final sedimentation, disinfection, sand pressure filtration, sludge recirculation, sludge thickening and sludge dewatering.

Algeria has about 166 wastewater treatment plants and is planning to increase to 272 in next 5 years. Oran in Algeria has one of the largest, second largest in Africa, wastewater treatment plant with a capacity of over 300,000 m^3/day that combines mechanical pre-treatment, biological treatment and disinfecting as the third stage. The technology used in this plant includes pre-treatment, nutrient removal, filtration equipment implementing disc filter technology, which is capable to capturing particles as small as 10 micrometer and disinfection using both UV and chlorine.

Tertiary treatment is needed if wastewater must be treated to very high levels, such as when it is released into fresh water bodies. Saudi Arabia has been focusing to shift the energy use in wastewater treatment from full dependence on external power supply source to self-sufficiency by tapping green power sources. Reuse of water will also reduce the cost and excessive consumption of energy during desalination, especially in countries like Saudi Arabia. Figure 3 shows a schematic diagram of the typical processes for industrial wastewater treatment including primary, secondary and tertiary steps. In this case first toxic materials like ammonia is removed passing the effluent through two successive aerated lagoons each with a volume of 40,000 m^3 and a detention time of 1.5 days. This is followed by second state. Tertiary water treatment technologies are very important in wastewater treatment strategy as these are used to obtain safe water for human consumption. Post treatment wastewater has been successfully used for landscape irrigation.

Umm Al-Hayman wastewater treatment plant in Kuwait city has been designed to received wastewater and design get reuse treated swage effluent for landscape irrigation that consists of five strategic reservoirs each with a capacity of 40,000 m^3. Many municipalities in United Arab Emirates have sewage treatment plants where sludge and effluents are treated. For example, Jebal Ali sewage treatment plant in Dubai municipality is planning to establish one of the largest of its kind to reuse wastewater. The sewage treatment plant has been designed especially for tanker reception and dedicated preliminary treatment, raw sewage preliminary treatment, primary treatment, secondary treatment by activated sludges, microflocculation, deep bed sand filtration, UV disinfection, treated effluent storage, treated effluent pumping, post chlorination and odour control by chemical and biological procedures. The wastewater treatment plant in Doha, Qatar treats water, sludge and odour. However, recently sewage treatment plant in Umm Salal Ali has applied advanced technique like ultrafiltration and UV technology to treat the sewage water and reduce energy cost.

Municipal wastewater treatment plant in Tobruk, Libya uses activated sludge process with simultaneous aerobic sludge digestion for biological treatment. The plant has a capacity of 18,000 m^3/day, which can be expanded to 36,000 m^3/day. To increase the space and energy efficiencies and cost, Oman has implemented a number of modern technology approaches to treat wastewater. The technologies include membrane bioreactor at the Al Ansab and Darsait wastewater treatment plants. Membrane bioreactors remarkably reduced the requirements of geographic footprint of the facility, especially to carry out wastewater treatment in the urban environment where land is limited. Ultafilter and sequencing batch reactor process are other methods adopted at the A'Seeb wastewater treatment plant in Oman. Ultra filter process is effective in producing treated contaminants of a particularly high quality whereas sequencing batch reactor is especially removes nutrients efficiently

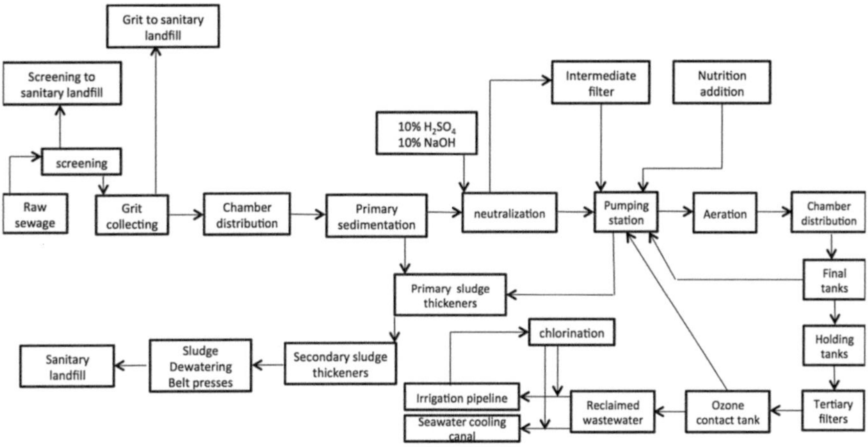

Fig. 3 Madinat Al-Jubail Alsinaiya industrial wastewater treatment plant

for irrigation purposes and produce sludge for fertilizers. Quriyat wastewater plant in Oman applied screw press dewatering technology, which is energy efficient in separating suspended solids from sewage.

Chemical Treatment Steps

Additional chemical processes can be further applied to increase the quality of the wastewater effluent. The advantage of chemical treatment is that most of the COD and TSS will be reduced during this process before reaching the secondary treatment, and the color of the effluent is decentralized (Irfan et al. 2013). This step further reduces the level of organic chemicals, nutrients, pathogens, and suspended solids in the treated effluent.

Adsorption Process

Adsorption is reducing energy cost process for wastewater treatment in which molecules, ions, or atoms adhere to a surface of another substance, known as adsorbent. Since adsorption is a surface based phenomena, it can occur between two different phases such as solid-liquid, solid-gas or liquid-gas. Therefore, this process is generally used as a refining procedure to eliminate inorganic and organic contaminants during wastewater treatment. Since it is a surface chemistry phenomenon, many different parameters could influence adsorption process. These parameters include temperature, pressure (in case of gas) and concentration of both contaminants. Beside these parameters efficiency of adsorbent depends on the surface area and/or active sites of the adsorbent. Selectivity of particular contaminant and adsorption kinetic (time of contact) are crucial for practical applicability. Often a pre-treatment procedure including pre-filtration and separation of suspended solid waste, oils etc. is needed to enhance the efficiency of the adsorption process. Reducing the size of the adsorbent to the nano-scale significantly increase the process due to their high specific area and associated sorption sites, short intra-particle diffusion distance, and tunable pore size and surface chemistry. In the large scale, columns and contractors filled with required adsorbents are used to remove contaminant from wastewater. In this case adsorption is carried out in a batch process by optimizing conditions, which are transformed to the column and contractor experiments. The feasibility of the procedure is tested by various adsorption isotherms models, such as Langmuir, BET, Freundlich, Hutchins, Reichenberg, etc.

The popular materials used as adsorbents are activated carbon, fly ashes, metal oxides, zeolites, moss, biomass, and geothites. Carbon based adsorbents are recently getting widely tested due to new carbon based materials such as carbon nanotubes (CNTs). CNTs have promised to have greater effectiveness than activated carbon while removing various organic contaminants due to its large surface area and multi-site interaction with contaminants (Yang and Xing 2010). CNTs are hydrophobic because of its graphitic surface and in water it aggregates that are of high affinity for organic pollutants due to presence of interstitial spaces and grooves of CNTs aggregates. Activated carbon indeed has similar surface area but a substantial number of its micropores are inaccessible to bulky organic molecules such a pharmaceutical products. Unlike activated carbon CNTs has the other advantages like adsorption of polar organic molecules through hydrogen bonding, π-π interaction, covalent bonding, etc. Heavy metal pollution exists in wastewater from many industries including metal plating facilities, mining operation and tanneries. Sometime ground water can also be source of heavy metals such as cadmium (Cd), chromium (Cr), lead (Pb) and mercury (Hg) and these are not biodegradable, therefore, these substances can accumulate and cause various diseases and disorders. Activated carbon is used to remove heavy metal. Bark, a by-product of timber industry, is rich in tannic and very effective in removing heavy metals like Cd, Hg, Pb etc. Different kinds of bark, sawdust, coffee, walnut shell, waste tea, lignin, chitosan, seafood processing wastes, dead biomass, rice hulls, seaweed, alginate, xanthene, clay, zeolite, fly ash, peat moss, CNTs, etc. have been used for removal of heavy metals. Metal oxide such as alumina, iron oxide, zinc oxide and titanium oxide are cost effective and efficient for eliminating heavy metal from wastewater. In this case sorption is guided by complexation between dissolved heavy metals and oxygen in metal oxides, where first fast adsorption of metal ions on the external surface occurs than rate-limiting intraparticle diffusion along the micropore walls follows. Metal oxide nanoparticle (Moussawi and Patra 2016a) has higher adsorption capacity, which can further be improved by conjugation/doping. Arsenic contamination in natural water renders the lives of millions of people around the world at risk. Arsenic in natural water can be the result of leaching from arsenic containing source rocks and sediments. Its presence is generally associated to geochemical environments such as basin-fill deposits of alluvial lacustrine origin, volcanic deposits, inputs from geothermal sources, mining wastes and land fills. ZnO acts as a good adsorbent, but so far it has its own limitations. Doping with organic system may influence adsorption of ZnO. This has been done through various methods such as surface modification by incorporation of organic materials (Moussawi and Patra 2016a; Qiu et al. 2008), fullerenes (Liao et al. 2013) and polymers (Li et al. 2003). Polymeric nano-adsorbents are other adsorbents that could remove both organic contaminants and heavy metals. Dendrimers can be tailored made where interior shell can be hydrophobic for attracting organic compounds while the exterior branches can be modified with hydroxyl or amine terminated groups for adsorbing heavy metals. Adsorption is primarily applied for source reduction, wastewater treatment and reclamation for potable, industrial and other purposes. The extent of the contaminants removal varies from 90 to 99%. However, the limitation associated with adsorption is the regeneration of columns and column life, which can be overcome by retreating the columns in cases. The management of the exhausted adsorbent is another concern.

Oxidation Process

During oxidation process, the organic compounds present in wastewater gets oxidized into water, carbon dioxide, and other easily biodegradable compounds such as alcohols, aldehydes, ketones, carboxylic acids, etc. The oxidation process is influenced by pH, temperature and presence of catalysis etc. and depends on the nature of contaminants and oxidants. The common oxidants used to carry out oxidation process included $KMnO_4$, Cl_2, O_3, H_2O_2, Fenton's reagents (H_2O_2 and Fe catalyst) and ClO_2 and the contaminants that are removed during this process include ammonia, phenols, dyes, hydrocarbons and other organic compounds. However, often a single oxidation process is not enough to decompose contaminants in wastewater. In that case advanced oxidation process (AOP) is carried out where more than one oxidation process is simultaneously involved. Such a method accelerates the process and produces highly reactive hydroxyl radicals. In this method commonly ultra violet (UV) photolysis, sonolysis, Fenton's reagent and other highly reactive species is used that can accomplish decomposition of organic contaminants at normal temperature and pressure. In Fenton's reaction, H_2O_2 decomposes by Fe^{2+} ions in aqueous phase resulting in the formation of OH*. The rational is to generate highly reactive transitory species such as OH*, O_2*-, O_3, H_2O_2, etc. for mineralization of organic compounds, pathogens and disinfection by-products (Pera-Titus et al. 2004). The procedures used to generate these reactive transitory are by using UV, Fenton reaction, photo-Fenton and catalytic wet air oxidation. Photo-catalysis is other form of AOP, which is efficient in degrading wide ranges of pollutants. Photo-catalysis can be done in homogenous medium where both catalyst and substrates are in the same solution medium or in heterogeneous medium where the catalysis is in solid phase and the substrates are in solution (Pignatello et al. 1999). Photo-Fenton reaction is one of the popular homogeneous photo-catalysis used for wastewater treatment. Fenton reaction takes place in the following steps.

$$Fe^{2+}(aq.) + H_2O_2 \rightarrow Fe^{3+}(aq.) + OH^- + HO^{\cdot}$$

Later on Fe^{3+} can be converted back to Fe^{2+} as:

$$Fe^{3+}(aq.) + H_2O_2 \rightarrow Fe^{2+}(aq.) + HO_2^{\cdot} + H^+$$
$$Fe^{3+}(aq.) + H_2O_2 \rightarrow Fe^{2+}(aq.) + HO_2^{\cdot} + H^+$$

Presence of light/photon accelerates the Fenton reaction in comparison to in dark condition due to regeneration of Fe^{2+}. Sometime this photo-Fenton reaction can be carried out along with heterogeneous catalysis.

Heterogeneous photo-catalysis using semi-conductors such as TiO_2, ZnO, ZnS, Fe_2O_3, CdS, GaP, etc. have been an efficient approach to degrade pollutants, which can use even use solar light and reduce energy cost. TiO_2 is among the most active photo-catalyst that has demonstrated competence in the 300 to 390 nm wavelength range (Gaya and Abdullah 2008; Furube et al., 2001). The idea is to give the catalyst with a photon having energy equivalent to band gap energy that excites the lone electron from valence band to conduction band in femtoseconds. The mechanism of electron-hole pair formation after excitation of TiO_2 is shown in Fig. 4. Excitation by light generates an empty unfilled valence band, which creates an electron (e^-)-hole (h^+) pair. The most widely used photocatalyst for water treatment is Degussa P-25 TiO_2 catalyst. This is because photo-catalysts like Degussa P-25 TiO_2 are also efficient in generative reactive oxygen species (ROS), which attacks with micro-organism. In this case photo-induced ROS binds to lipopolysaccharide layer of the external cell wall initially followed by the site attack on the peptidoglycan layer, peroxidation of the lipid membrane and the finally oxidation on the protein membrane. Subsequently, it will cause death of the cell.

In most of the cases photo-catalytic oxidation processes use semiconductor materials for the removal of the residual concentrations of contaminants, such as polycyclic aromatic hydrocarbons (PAHs) that are carcinogenic. Advantages of the photo-catalytic process over other techniques include its mild operating conditions and its reliance on sunlight as the power source, thus significantly reducing the operating costs of electric power required. Although many semiconductors such as TiO_2, WO_3, ZrO_2, SnO_2, Fe_2O_3, CdS, ZnS, WS_2, MoS_2 have been tested for the photo-catalytic degradation of various environmental contaminants (Lam et al. 2012), an advantage of ZnO is its low cost and that it absorbs over a larger portion of the solar spectrum (Sakthivel et al. 2003). For this reason, ZnO is considered the most suitable catalyst for photo-catalytic removal in the presence of sunlight. Since the minimum energy required for excitation of an electron for ZnO is 3.2 eV and solar light contains less than 4% of UV light, the application of ZnO is limited. To expand the use of ZnO photo-response to the visible region, surface modification and dye sensitization are effective methods. It has been shown that GaN:ZnO has

Fig. 4 Photo-catalyst mechanism of electron-hole pair in TiO_2 particle in the presence of contaminant (C) in wastewater

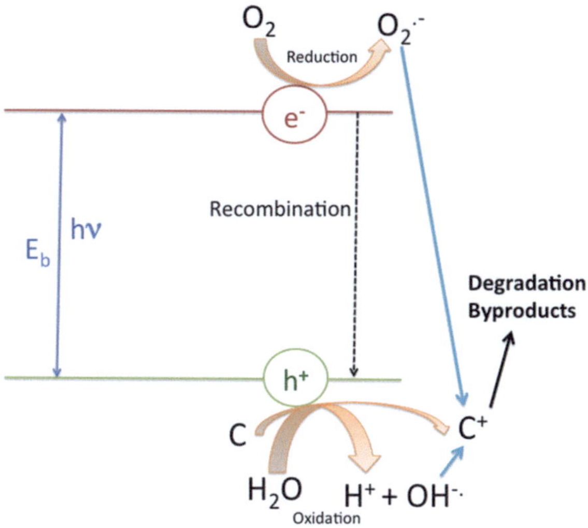

excellent properties for the photo-degradation of PAHs (Kou et al. 2010). Curcumin conjugated ZnO can remove to a far extent high concentrations of perylene, fluoranthene, and chrysene marginally faster than ZnO depending on the extent of curcumin conjugation in ZnO (Moussawi and Patra 2016b). The removal has been found to be faster for chrysene than perylene and extraordinary for flouranthene. The high rate of removal has been related to photo-degradation based on the following mechanism (Moussawi and Patra 2016b).

$$\text{Curcumin}/ZnO + \text{visible light} \rightarrow \text{curcumin}^{*}/ZnO$$

$$\text{curcumin}^{*}/ZnO \rightarrow \text{curcumin}^{*+}/ZnO + \text{e}^{-}CB$$

$$\text{e}^{-}CB + \text{O}_2 \rightarrow {}^{*}\text{O}_{2^-}$$

$$\text{H}_2\text{O}_2 + {}^{*}\text{O}_{2^-} \rightarrow {}^{*}OH + OH^{-} + \text{O}_2$$

$$2^{*}OOH \rightarrow \text{O}_2 + \text{H}_2\text{O}_2$$

$${}^{*}OOH + \text{H}_2\text{O} + \text{e}^{-}CB \rightarrow \text{H}_2\text{O}_2 + OH^{-}$$

$$\text{H}_2\text{O}_2 + \text{e}^{-}CB \rightarrow {}^{*}OH + OH^{-}$$

$$\text{H}_2\text{O}_2 + {}^{*}\text{O}_{2^-} \rightarrow {}^{*}OH + OH^{-} + \text{O}_2$$

$$\text{H}_2\text{O}_{\text{ads}} + \text{h}^{+} \rightarrow {}^{*}OH_{\text{ads}} + \text{H}^{+}$$

$$OH^{-} + \text{h}^{+} \rightarrow {}^{*}OH$$

$$*OH/{}^{*}\text{O}_{2^-}/\text{curcumin}^{*+} + \text{pollutants } (PAH_{\text{S}}) \rightarrow \text{degraded products}$$

Photo-catalysis can be divided into five different steps (Herrmann, 1999). Step 1 includes mass transfer of the organic pollutants in the liquid phase to the catalyst surfaces followed by adsorption of pollutants on the photo-activated catalyst surfaces in the second step. In 3rd step, photocatalysis reaction contunues on the catalyst surfaces for the adsorbed phase. In the 4th step desorption of the intermediate from the catalyst surfaces follows and finally in the last step mass transfer of the intermediate from the interface region to the bulk fluid occurs. Therefore, the operational condition plays important role during photo-catalysis. For example configuration of photocatalytic reactor, pH of the solution, temperature, dissolved oxygen, loading of contaminants, wavelength of light, light intensity, etc. are various driving force for efficient photo-catalytic reactions. Since solar light can be used as an activator for photocatalytic process, this process has greater potential to reduce energy cost while degrading contaminants in wastewater.

Conclusion

Wastewater treatment in large part of the MENA region is concetrated in primary and secondary stage treatments where reuse is carrided out largely for irrigation purposes. However, recently golf countries have taken serious interest where treated water can be portable by applying modern scientific approaches and technology. Treatment of wastewater has greater potential to reduce cost and energy consumption during desalination and transportation of water in the MENA region. Increase interest in nanotechnology, solar and renewable energy for secondary and tertiary treatments is going to infleunce wastewater treatment in the MENA region and reduce cost and energy consumption is going to play an important role to adopt such technology in other developing countries in the MENA region.

References

Abdel-Raoudi, N., Al-Homaidan, A. A., & Ibraheem, I. B. M. (2012). Microalgae and wastewater treatment. *Saudi Journal of Biological Science, 19*, 257–275.

Ahn, J., Daidou, T., Tsuneda, S., & Hirata, A. (2002). Characterization of denitrifying phosphate accumulating organisms cultivated under different electron acceptor conditions using polymerase chain reaction-denaturing gradient gel electrophoresis assay. *Water Research, 36*, 403–412.

Al-A'ama, M. S., & Nakhla, G. F. (1995). Wastewater reuse in Jubail, Saudia Arabia. *Water Research, 29*, 1579–1584.

al-Samawi, A. A. (2000). The use of reed plants for wastewater treatment: the Iraq experience. *Schriftenreihe des Vereins für Wasser-, Boden- und Lufthygiene, 105*, 311–317.

Bahri, A. (2008). Case studies in Middle Eastern and North African countries. In B. Jimenez & T. Asano (Eds.), *Water reuse: An international survey of current practice, issues and needs*. London: IWA Publishing.

Barragan, B. E., Costa, C., & Carmen Marquez, M. (2007). Biodegradation of azo dyes by bacteria inoculated on solid media. *Dyes and Pigments, 75*, 73–81.

Chan, Y. J., Ching, M. F., Law, C. L., & Hassell, D. G. (2009). A review on anaerobic-aerobic treatment of industrial and municipal wastewater. *Chemical Engineering Journal, 155*, 1–18.

Furube, A., Asahi, T., Masuhara, H., Yamashita, H., & Anpo, M. (2001). Direct observation of a picosecond charge separation process in photoexcited platinum-loaded TiO2 particlees by femtosecond diffuse reflectance spectroscopy. *Chemical Physics Letters, 336*, 424–430.

Gaya, U. I., & Abdullah, A. H. (2008). Heterogeneous photocatalytic degradation of organic contaminants over titanium dioxide: A review of fundamentals, progress and problems. *Journal of Photochemistry and Photobiology C Photochemistry Reviews, 9*, 1–12.

Herrmann, J. M. (1999). Heterogeneous photocatalysis: Fundamentals and applications to the removal of various types of aqueous pollutants. *Catalysis Today, 53*, 115–129.

Irfan, M., Butt, T., Imtiaz, N., Abbas, N., Khan, R., & Shafique, A. (2013). The removal of COD, TSS and colour of black liquor by coagulation-flocculation process at optimized pH, settling and dosing rate. *Arabian Journal of Chemistry, 48*. doi:10.1016/j.arabjc.2013.08.007.

Kim, T., Park, C., Yang, J., & Kim, S. (2004). Comparison of disperse and reactive dye removals by chemical coagulation and Fenton oxidation. *Journal of Hazardous Materials, 112*, 95–103.

Kou, J., Li, Z., Guo, J., Gao, J., Yang, M., & Zou, Z. (2010). Photocatalytic degradation of polycyclic aromatic hydrocarbons in GaN: ZnO solid solution-assisted process: Direct hole oxidation mechanism. *Journal of Molecular Catalysis A: Chemical, 325*, 48–54.

Koyuncu, I., Kural, E., & Topacik, D. (2001). Pilot scale nanofiltration membrane separation for waste management in textile industry. *Water Science and Technology, 43*, 233–240.

Lam, S.-M., Sin, J.-C., Abdullah, A. Z., & Mohamed, A. R. (2012). Degradation of wastewaters containing organic dyes photocatalysed by zinc oxide: A review. *Desalination and Water Treatment, 41*, 131–169.

Li, Z., Xiong, Y., & Xie, Y. (2003). Selected-control synthesis of ZnO nanowires and nanorods via a PEG-assisted route. *Inorganic Chemistry, 42*, 8105–8109.

Liao, S. H., Jhuo, H. J., Cheng, Y. S., & Chen, S. A. (2013). Fullerene derivative-doped zinc oxide nanofilm as the cathode of inverted polymer solar cells with low-bandgap polymer (PTB7-Th) for high performance. *Advanced Materials, 25*, 4766–4771.

Libra, J. A., Borchert, M., Vigelahn, L., & Storm, T. (2004). Two stage biological treatment of a diazo reactive textile dye and the fate of the dye metabolites. *Chemosphere, 56*, 167–180.

Moussawi, R. N., & Patra, D. (2016a). Modification of nanostructured ZnO surfaces with curcumin: A fluorescence-based sensing for arsenic and imporving arsenic removal by ZnO. *RSC Advances, 6*, 17256–17268.

Moussawi, R. N., & Patra, D. (2016b). Nanoparticle self-assembled grain like curcumin conjugated ZnO: Curcumin conjugated enhances removal of perylene, fluoranthene and chrysene by ZnO. *Scientific Reports, 6*, 24565. doi:10.1038/srep24565.

Pera-Titus, M., Garcı´a-Molina, V., Ban˜os, M. A., Gime´nez, J., & Esplugas, S. (2004). Degradation of chlorophenols by means of advanced oxidation processes: A general review. *Applied Catalysis B: Environmental, 47*, 219–256.

Piadeh, F., Moghaddam, M. R. A., & Mardan, S. (2014). Present situation of wastewater treatment in the Iranian industrial estates: Recycle and reuse as a solution for achieving goals of eco-industrial parks. *Resources, Conservation and Recycling, 92*, 172–178.

Pignatello, J. J., Liu, D., & Huston, P. (1999). Evidence of an additional oxidant in the photoassisted Fenton reaction. *Environmental Science & Technology, 33*, 1832–1839.

Qadir, M., Bahri, A., Sato, T., & Al-Karadsheh, E. (2010). Wastewater production, treatment, and irrigation in Middle East and North Africa. *Irrigation and Drainage Systems, 24*(1–2), 37–51.

Qiu, R., Zhang, D., Mo, Y., Song, L., Brewer, E., Huang, X., & Xiong, Y. (2008). Photocatalytic activity of polymer-modified ZnO under visible light irradiation. *Journal of Hazardous Materials, 156*, 80–85.

Ramavandi, B. (2014). Treatment of water turbidity and bacteria by using a coagulant extracted from *Plantago ovate*. *Water Research, 6*, 36–50.

Sahinkaya, E., Uzal, N., Yetis, U., & Dilek, F. B. (2008). Biological treatment and nanofiltration of denim textile wastewater for reuse. *Journal of Hazardous Materials, 153*, 1142–1148.

Sakthivel, S., Neppolian, B., Shankar, M. V., Palanichamy, M., & Murugesan, V. (2003). Solar photocatalytic degradation of azo dye: Comparison of photocatalytic efficiency of ZnO and TiO_2. *Solar Energy Materials and Solar Cells, 77*, 65–82.

Shi, J., Lu, X., Yu, R., Gu, Q., & Zho, Y. (2014). Influence of wastewater composition on nutrient removal behaviors in the new anaerobic–Anoxic/nitrifying/induced crystallizationprocess. *Saudi J. Biol. Sci., 21*, 71–80.

Tsuneda, S., Ohno, T., Soejima, K., & Hirata, A. (2005). Simultaneous nitrogen and phosphorous removal using denitrifying phosphate accumulating organisms in sequencing batch reactor. *Biochemical Engineering Journal, 27*, 191–196.

van der Zee, F. P., & Villaverde, S. (2005). Combined anaerobic-aerobic treatment of azo dyes– A short review of bioreactor studies. *Water Research, 39*, 1425–1440.

Venkata Mohan, S., Lalit Babu, V., & Sarma, P. N. (2007). Anaerobic biohydrogen production from dairy wastewater treatment in sequencing batch reactor (AnSBR): Effect of organic loading rate. *Enzyme and Microbial Technology, 41*, 506–515.

Yang, K., & Xing, B. S. (2010). Adsorption of organic compounds by carbon nanomaterials in aqueous phase: Polanyi theory and its application. *Chemical Reviews, 110*, 5989–6008.

Wind Power and Potential for its Exploitation in the Arab World

Muhammad R. Hajj

Abstract Wind power was used to propel boats along the Nile in 5000 B.C. and to pump water and for food production in the Middle East in 900 B.C. Today, only about 0.6% of the electricity demand in the Arab World is met by wind power generation. This ratio is much lower than corresponding ratios in Europe and the United States. Although one could point to specific reasons such as availability of wind, installation costs and local expertise, and availability of other energy resources, it is important to realize that generating wind power poses different challenges in different parts of the world. Like many places, issues such as data availability, grid capacity and transparent cost analysis are important for making decision regarding development of wind power. In the Arab world, the issues become more complicated when considering political uncertainty, which at a minimum impacts investment decision, and cultural values and attitudes towards power use and environmental impact.

Recognizing that it will be difficult to discuss or resolve many of the above issues, we discuss and review in this chapter technical details related to wind power generation including turbine blade aerodynamics, components and types of wind turbines and end with a discussion of potential benefits from wind power generation on the Arab world and societies and challenges to be met for reaping these benefits.

Keywords Wind Power • Turbulence • Aerodynamics

Introduction

Writing about the potential of using wind energy in the Arab world, one cannot but note that according to the US Department of Energy (http://energy.gov/eere/wind/history-wind-energy) wind power was used to propel boats along the Nile in 5000 B. C. and to pump water and grind grains in Persia in 900 B. C. From there, it spread to the rest of the Middle East where it was used for food production. Two thousands

M.R. Hajj (✉)
Center for Energy Harvesting Materials and Systems, Virginia Tech, Blacksburg, VA 24060, USA
e-mail: mhajj@vt.edu

S. Murad et al. (eds.), *Water, Energy & Food Sustainability in the Middle East*,
DOI 10.1007/978-3-319-48920-9_11

year later, around 1000 A. D., the crusaders carried the technology to Europe where it was used to drain lakes and marches in the Netherlands. In the late nineteenth century, Americans started using windmills to pump water for farms and ranches. During World War II, a 1.25-megawatt turbine fed electric power to the local utility network. Although wind turbines based on different technologies and of different sizes were developed during the twentieth century, their use remained limited, mostly because they were more expensive than other energy sources such as oil. The oil embargo in 1973 along with the needs to keep energy cost affordable and reduce the consumption of fossil fuels because of environmental concerns lead to significant investments in the research and development of wind turbine technologies in the last quarter of the last decade. The U.S. government worked with industry to advance the technology and enable the development and deployment of large commercial wind turbines. As of 2015, wind power in the US supplied about 65 GW, which is about 5% of the electricity demand (http://www.energy.gov/eere/wind/downloads/enabling-wind-power-nationwide). According to the same source, the potential power that can be generated from wind in the US is about 10 times the current total US electricity consumption. Clearly, wind power has been and will continue to be a solution that can address issues such as quality of life, job creation, economic advancement and environmental impact at the same time. The issues are in realizing effective means and reducing the cost to harvest this power.

The Arab world is very much behind the curve on renewable energy and on wind power production. In comparison to the 5% ratio of the electricity demand in the US being met by wind power generation, this ratio in the Arab world is about 0.6% (http://technologyreview.me/en/energy/the-potential-of-wind-power-in-the-arab-world/). Although one could point to specific reasons such as availability of wind, installation costs and local expertise, and availability of other energy resources, it is important to realize that generating wind power poses different challenges in different parts of the world. Like many places, issues such as data availability, grid capacity and transparent cost analysis are important for making decision regarding development of wind power. In the Arab world, the issues become more complicated when considering political uncertainty, which at a minimum impacts investment decision, and cultural values and attitudes towards power use and environmental impact. In summary, the issues to be considered when discussing wind power are complex and interrelated and decision making must be based on databases in a detailed analysis that assesses benefits and impacts from different perspectives. This article does not aim to perform this task. Rather it aims at explaining some technical issues related to wind power generation and provide a discussion of the benefits and challenges of using wind power in the Arab World.

Technical Background:

The Wind (Air in Motion)

Wind characteristics at a specific site, as determined through measurements, are very important for the assessment of potential sites for wind farms. Of particular concern is the intermittent aspect of winds, which results in fluctuations of the level of power generated by a wind farm. Such variations cause significant issues when trying to integrate power from a wind farm into a grid system.

To assess these variations, it is necessary to understand how wind is generated and what impacts its speed and direction. Wind is the result of spatial variations in atmospheric pressure, which cause air parcels that are initially at rest to move from a high-pressure region to a low-pressure one. Because of the earth rotation, the direction of the motion of an air parcel changes as soon as it starts to move. This deflection is caused by the Coriolis force, which acts to the right in the northern hemisphere and to the left in the southern hemisphere. As the wind speed increases, the deflection increases. So, as the air parcel increases its speed because of the pressure gradient, the deflection becomes larger. This continues until equilibrium is established between the pressure gradient and Coriolis forces. At that point, the wind starts to blow parallel to the isobars (contours of equal pressure) creating what is referred to as geostrophic wind.

Near the surface of the earth, the wind is subjected to a friction force that acts in the opposite direction to the wind direction (parallel to the isobars). This force slows the wind, which, in turn reduces the magnitude of the Coriolis force. As a result, this force cannot balance the force imposed by the pressure gradient. Consequently, the wind speed and direction changes near the earth surface. Specifically, its direction will assume a component across the isobars towards lower pressure. The effects of the friction force on the wind are diffused over a region above the earth that is referred to as the atmospheric boundary layer. Within this layer, the wind speed increases with the elevation. The difference in terrain conditions (rough or smooth) directly affects the magnitude of the friction force. This force is larger in complex terrains with hills and forests than in a smooth terrains (flat surfaces) or over water bodies. As such, the height of the atmospheric boundary layer can range from few hundred meters to several kilometers depending on the terrain (Simiu and Scanlan 1986). Clearly, the wind used to generate power is impacted by many factors including the climate, geographic location, elevation above ground level, roughness of the terrain, and the obstacles in the surroundings.

The characteristics of the geostrophic wind and their variations caused by the earth surface are best described by the van der Hoven spectrum (Van der Hoven 1957), which shows the energy distribution of the wind as a function of frequency. This spectrum usually exhibits two peaks that are separated by an energy gap. The low-frequency peak, which has a time scale of hours or few days, and corresponds to the geostrophic winds, is important for deciding on wind power sites. The high-frequency peak has time scales of the order of seconds or few minutes and is associated with local effects. Its characterization is useful for prediction of parameters related to the operation and control of wind turbines or farms.

In addition to determining the spectral distribution of the wind, it is important to determine the mean wind speed. This speed is obtained by averaging the measured wind speed over a time period that lies within the energy gap (10 to 20 mins.) of the van der Hoven spectrum. As such, the low-frequency variations will exhibit themselves as slow variations in the mean speed. That is why this speed is also referred to as quasi-steady wind speed. Subtracting the mean speed from the measured wind speed results in a measure of the local effects or level of atmospheric turbulence. Determining the quasi-steady wind speed is important for making a decision on the economic viability of a wind farm. This is best represented in terms of a probability distribution of the mean wind speed. A commonly used function is the Weibull distribution and is given by

$$f(v) = \left(\frac{k}{c}\right)\left(\frac{v}{c}\right)^{k-1} \exp^{-\left(\frac{v}{c}\right)^k}$$

Where v is the wind speed, k is the dimensionless shape parameter and c is the scale parameter that has the same units as the velocity. There has been several studies on wind distribution for different location in the Arab World. A recent study (Rehman et al. 1994) summarizes references earlier findings and provided new locations for high wind abundance in the Arabian Peninsula (Rehman et al. 1994; Ouarda et al. 2015; Darwish and Sayigh 1988; Hasan 1992; Bady 2012; Janajreh et al. 2013). This distribution allows for determining the most probable and average wind speeds as shown in Fig. 1.

It is important to note here that the peak wind energy occurs at wind speeds that are larger than the average wind speed. The reason is that wind power is proportional to the cube of its speed. Furthermore, The power at low speeds is very low and most of it will be required to overcome frictional losses. Consequently, power generation starts at wind speeds of at least 3 m/s to 4 m/s. This speed is referred to as cut-in speed. On the other hand, operating the turbine at very high wind speeds may cause serious damage to the whole turbine. To avoid these conditions, wind turbines are usually designed to cut out at specified wind speeds either by braking or feathering the rotor blades allowing the wind to spill over the blades. The highest speed at which a wind turbine can operate is referred to as the cut out speed. Realizing that the operation of a wind turbine is limited by cut in and cut out speeds, it is fair to assume that the wind turbine does not capture more than half of the available wind power.

It is important to also recognize that the mean wind speed is a function of the elevation above the surface with the wind speed increasing as the elevation is increased. This relation is represented usually by a wind profile that is written as (Simiu and Scanlan 1986):

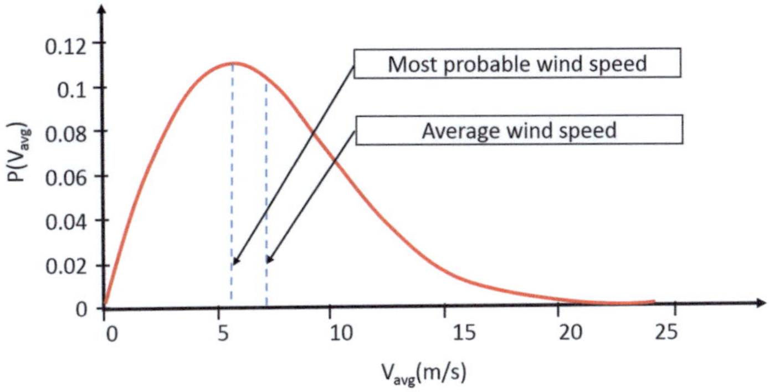

Fig. 1 Typical Weibull distribution of wind speed

$$u = u_r \left(\frac{z}{z_r} \right)^{\alpha}$$

where u is the wind speed at height z and u_r is the wind speed at a reference height z_r. The exponent α is an empirical constant that is a function of the terrain properties and stability of the atmosphere. We should note here that the density of air decreases with altitude. On the other hand, the wind speed is larger at higher elevations. Because the wind power is proportional to the cube of the wind speed and linearly proportional to the air density, the net effect is that the wind power is larger at higher elevations.

The percentage of nominal power generated by a wind turbine varies between 0–100% depending on the wind speed. This percentage is determined from World power velocity curves for specific turbines that relate the power generated (kW) to the hub-height wind speed. A typical power curve is shown in Fig. 2.

Given the Weibull distribution of wind speeds for a specific location and the power curves, one could calculate the number of hours per year that a wind turbine would operate at its rated power. This number of hours is referred to as the full load hours. In practical terms and for different reasons, such as the need to shut down the turbine for maintenance, the practical full load hours are about 80 to 90% of the theoretical full load hours. Furthermore, discrepancies between predicted and observed wind speed could lead to significant variations in full load hours. Table 1 shows the wind energy potential in the units of full load hours (kWh/y)/KW as determined from a study performed by UNEP/ROWA sponsored by the German Federal Ministry of Environment, Nature Conservation, and Nuclear Safety (Al-Karaghouli et al. 2007). The number of full load hours vary from a low of about 1200 for the UAE and Lebanon to values above 2500 in Egypt, Morocco and Oman. Given that annual full load hours of 1400 h/y are considered as long-term economic potential, one could conclude that many Arab countries have good wind energy potential. One should note here that because of the variation in wind power potential across these countries, the viability of wind power could

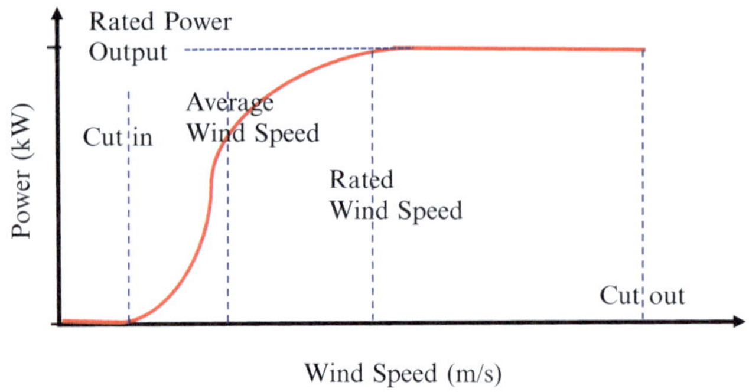

Fig. 2 Typical power-velocity curve of a wind turbine

Table 1 Wind energy potential (full load hours) for different countries in the Arab World (Al-Karaghouli et al. 2007)

Country	Full Load Hours per year (h/y)
Lebanon	1176
UAE	1176
Bahrain	1360
Qatar	1421
Yemen	1483
Jordan	1483
Kuwait	1605
Iraq	1789
Syria	1789
Saudi Arabia	1789
Algeria	1789
Tunisia	1789
Libya	1912
Oman	2463
Morocco	2708
Egypt	3015

become more relevant and better assessed if better cooperation and resources sharing are established.

The turbulence in the wind, which includes all fluctuations that are below the energy gap (seconds to few minutes), has no impact on the average level of energy generation over a long time scale (above few months or one year). On the other hand, turbulence impacts the aerodynamic loads and fluctuations in power generation. It also impacts the power velocity curve differently over different wind speed regimes and the extent of the wake of the turbine. Turbulent fluctuations are usually represented in terms of a power spectrum that exhibits the energy contribution of different frequencies or wavenumbers to the total energy of the fluctuations. Theoretical representations of the power spectrum of turbulence in the atmospheric boundary layer are given by spectral functions such as the von Karman spectrum or the Kaimal spectrum. These functions are related to the integral length (or time), which measures the correlation length of turbulence and the turbulence intensity, defined as the ratio of the standard deviation of the wind speed to the mean wind speed. Both of these quantities are specific to the terrain and are evaluated from experimental measurements. The correlation length can vary between 100 m and 350 m and the turbulence intensity is around 20% but can go to much larger values over short measurements periods. We should note here that turbulence intensity is larger in complex terrains. Also, it decreases with elevation. The readers are referred to the following references for further information about evaluation, modeling and representation of atmospheric turbulence (Simiu and Scanlan 1986; Hajj et al. 2000).

Types and Components of Wind Turbines

Wind turbines are classified according to the direction of their axis of rotation. If the axis of rotation of the wind turbine is parallel to the ground, it is referred to as a horizontal-axis wind turbine (HAWT). In cases where the axis of rotation is perpendicular to the ground, the wind turbine is referred to as a vertical axis wind turbine (VAWT). Both types have their advantages and disadvantages. Still, most commercially available wind turbines are of the horizontal axis type. The reason is that HWAT are more efficient at higher wind speeds and elevations above the ground than VWAT. Below, we will discuss HAWT.

A wind turbine consists of the following major components:

- A nacelle that is placed on top of the tower and contains key components of the wind turbine including the gearbox, electrical generator, controller, and brake.
- A tower that supports the structure of the turbine and carries the nacelle and rotor. The tower is usually made from tubular steel, concrete, or steel lattice. It is, in general, advantageous to have a higher tower because wind speeds increase with height.
- Blades that rotate when wind is blown over them and, as such, transfer the wind power to a rotor hub. Most turbines have either two or three blades.
- A rotor hub to which the rotor blades are attached. Cast iron or cast steel is used. This hub transfers the energy from the blades to the generator. This is achieved through the gearbox or directly to the ring generator.
- A generator that converts the mechanical energy of the rotating shaft to electrical energy

Other components of a wind turbine include an anemometer to measure the wind speed and direction, a brake to stop the rotor when needed, controllers to either shut down the turbine at high wind speeds to avoid damage or to pitch the blades out of the wind to control the rotor speed, and a yaw drive (in case of upwind turbines) to make sure that the wind turbine is always facing the wind.

Energy Conversion and Efficiency

Wind power is the power generated by converting the kinetic energy of the wind into useful electrical or mechanical power. The kinetic energy of the wind per unit mass is dependent on its mean speed and is defined as $½V^2$. Per unit volume, the kinetic energy is defined as $½ \rho V^2$. For large systems, the most efficient way to convert this energy to electric power is by using a wind turbine that consists of rotating blades, a shaft and a generator. As the wind passes by the blades, it forces them to rotate. Through a rotor, the rotation of the blades is used to rotate a shaft. The shaft rotation is then converted to electrical power by a generator. Power electronic controls are then used to convert the electricity into the correct frequency and voltage that feeds into the power grid.

Figure 3 shows a schematic for the aerodynamic performance of a wind turbine when placed in an air stream, which is governed by fundamental laws that define the operational parameters for wind turbines. Defining by A the surface area of the wind turbine and V the velocity of the wind as it hits the turbine blades, we note that the wind speed behind the turbine, defined by V_2, must be smaller than V because energy was extracted from the wind. On the other hand, the wind speed upstream of the blades, defined by V_1, must be larger than V because the blades have slowed it down. If one defines the mass flow rate as the mass of wind that hits the blades of the turbine per unit time, e.g. one second, one can express the mass flow rate as $½\rho AV^2$ (the units are in kg/s and that is why it is called mass flow rate) where A is the surface area of the blades. One fundamental law is that the mass must be conserved, which implies that

$$½\rho A_1 V_1^2 = ½\rho AV^2 = ½\rho A_2 V_2^2$$

Consequently, the same air mass that hits the blades of the turbine per unit time must occupy a smaller area in front of it and a larger area right after it, i.e. $A_2 > A > A_1$ as shown in the schematic.

A second fundamental law that governs the aerodynamic performance of wind turbines is the principle of linear momentum. Momentum is a measure of the tendency of an object to keep moving once it is set in motion and the linear momentum of a specific wind mass passing by the blade is defined to be its mass times velocity. The principle (or balance) of linear momentum states that the rate of change of momentum is equal to the sum of applied forces (Newton's second law). Considering the schematic of the wind turbine shown below and the fact that the wind speeds ahead and behind the blades are different, we write the change of linear momentum as $\rho AV(V_1-V_2)$ (units are kg.m/s^2 or Newton (N)). This change is equal to the force F applied by the blades on the wind. This force can then be used to determine the power delivered from the wind to the turbine, which is given by the product of the force and velocity, i.e. $P=FV= \rho AV^2(V_1-V_2)$ (units are N.m/s or Watts). Alternatively, the power is related to the rate of drop of kinetic energy as the

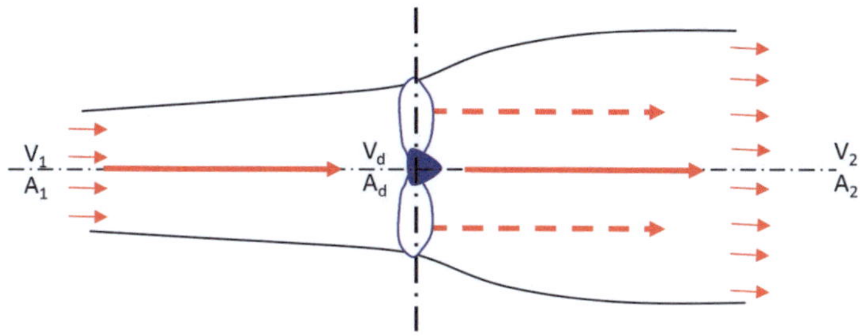

Fig. 3 Schematic of air flow through a wind turbine

wind moves from a position upstream of the blades to a position downstream of the blade. This rate is given by ½ $\rho AV(V_1^2-V_2^2)$. Equating these quantities one obtains a relation between the wind speed at the turbine and the speeds upstream and downstream of the turbine, i.e. $V=½ (V_1+V_2)$. The power is then written as ¼ $\rho A(V_1+V_2)(V_1^2-V_2^2)$ or ¼ $\rho AV_1^3(1-b^2)$ $(1+b)$ where $b=V_1/V_2$. Next, we define the available wind power by the upstream wind speed, V_1, and the surface area of the turbine, A, and write $P_{av}=$ ½ ρAV_1^3. The ratio of the power taken by the turbine to the available power is then defined as the aerodynamic efficiency and is given by $C_p = ½(1-b^2)\ (1+b)$. This efficiency is maximum when b=1/3, which yields a maximum aerodynamic efficiency of 59.26%. This value, known as the Betz limit, is the theoretical power fraction that can be extracted from the available wind power. The percentage of available wind power that can be extracted depends on other losses, which amount to about 10% in the drive train and the generator and another 10% in the inverter and cabling. Additional losses can be incurred when the wind speed exceeds the rated speed. Accounting for all of these losses, a wind turbine will convert about 35% of the available wind energy into electrical energy.

Turbine Blade Aerodynamics

The blades of wind turbines are designed generate the maximum power from the wind. They do so by generating lift in a similar manner to the wing of an airplane. When air moves past an airfoil section of the wing, it generates lift and drag forces on the wing as shown in Fig. 4. The lift is perpendicular to the direction of the wind and is the force that causes the blade to rotate. The drag is a force that is parallel to the wind flow. For optimal operation, the ratio of the lift to drag forces should be maximized. This ratio is a function of the airfoil shape and the angle of attack defined as the angle between the vector representing the relative velocity of the wind with respect to the airfoil and the chord line of an airfoil. At low angles of attack, the lift increases as the angle of attack is increased. At a specific angle of attack, the flow separates from the top surface, which results in a significant drop in the magnitude of the lift force. This condition is referred to as stall.

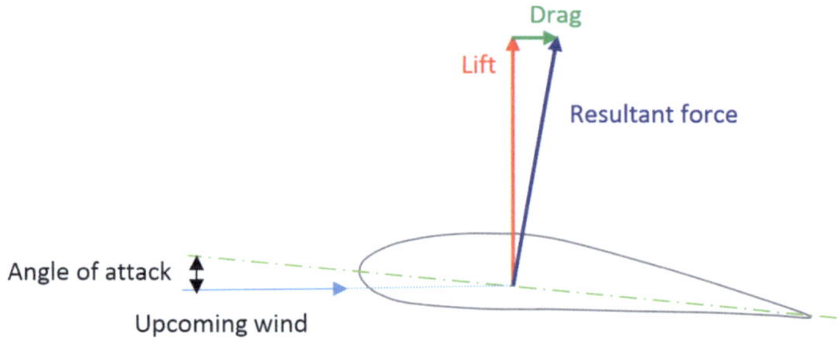

Fig. 4 Lift and drag directions on an airfoil section

The wind turbine moves in the airflow at a relative speed. At each point along the blade, the airfoil has a relative velocity, $\mathbf{V_{rel}}$, that is the vector sum of the upstream wind speed **V** and the tangential velocity vector of the airfoil that is perpendicular to **V** and has a magnitude $r\Omega_r$. Note that the velocity vector is dependent on the radial position of the airfoil section, i.e. the magnitude of this velocity is progressively larger along the blade as one moves from the blade root towards its tip. Consequently, and to maintain an effective lift to drag ratio along the blade, the blade must be twisted as shown in Fig. 5.

The drag and lift forces on this airfoil section per unit length are respectively given by:

$$F_D = C_D \tfrac{1}{2}\rho V^2 c \, \text{and} \, F_L = C_L \tfrac{1}{2}\rho V^2 c$$

where c is the chord length of airfoil, ρ is the air density, and C_L and C_D are respectively the lift and drag coefficients. Both of these forces contribute to the axial thrust force, F_T, and to the tangential force, F_r, that produces the rotational torque. The lift force gives a positive contribution to F_r. On the other hand, the drag force contributes negatively to F_r. For this reason, a high C_L/C_D ratio is desirable.

One important parameter that is related to the operation of wind turbines is the tip speed ratio (TSR) defined as the ratio of the rotational speed of the tip of the blade calculated as ΩR over the incident mean speed V. Although a higher TSR implies a higher rotational speed and better power generation, operating at a high TSR has its own shortcomings, which include noise generation, potential vibrations, reduced efficiency to increased drag, large breaking systems and potential erosion from particles carried by the wind and hitting the blade at high speeds. On the other hand, a low TSR (slow rotation) implies that too much wind is passing through the blades without contributing to the energy extraction. That is why the TSR depends on the pitch angle and the airfoil and planform geometries of the blades and their number. Typically, three-bladed wind turbines operate at TSR values between 7 and 10.

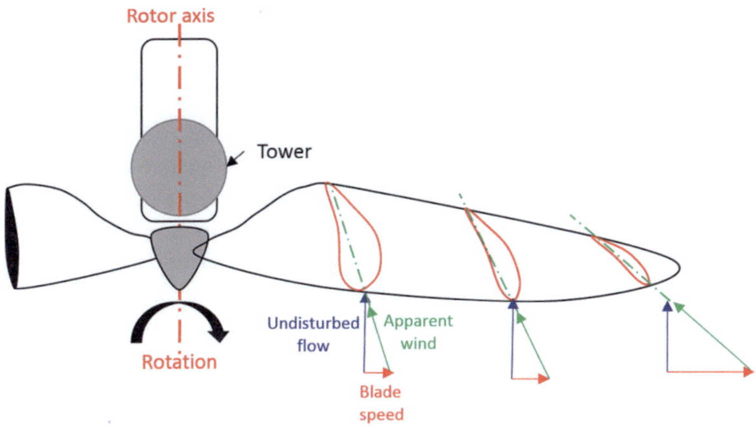

Fig. 5 Variation of the relative speed (in *green*) along a rotating blade

Operating Limits of Wind Turbines

For safety and efficiency considerations, wind turbines are usually subjected to operating limits that are identified by the following terms:

- Start-up Speed – This is the speed at which the rotor and blade assembly begins to rotate.
- Cut-in Speed – It is the minimum wind speed at which the wind turbine will generate usable power. This wind speed is typically between 3 and 4 m/s.
- Rated Speed – This speed, between 12 and 17 m/s, is the minimum wind speed at which the wind turbine will generate its designated rated power. Between the cut-in and rated speeds, the generated power increases as the wind speed is increased. Above the rated speed, the generated power is kept constant as the speed is increased till the cut-out speed is reached in order to protect the generator. Reducing the power output beyond the rated speed is achieved through pitch control where the pitch angle is varied to reduce the lift force and slow down the blade rotation, or through stall control where the blades are actively pitched to produce stall or the twist in the blades is angled so that wind above a certain speed will cause stall and slows them down.
- Cut-out Speed – This is the speed, between 25 and 35 m/s, at which the wind turbine will shut down. It is also referred to as the furling speed. The shut down is a safety feature that aims at protecting the wind turbine from damage. The shut down is implemented through an automatic brake that is activated by a wind speed sensor, by pitching the blades, by using drag flaps mounted on the blades or the hub that are activated at high wind speeds, or by turning the turbine sideways to the wind stream.

Considerations for Benefiting from Wind Energy in the Arab World

When considering wind energy as a potential source for electric power, one has to consider opportunities and challenges in addition to positive and negative impacts. According to the US department of Energy (Wind Energy Benefits (Fact Sheet) 2015), wind energy is beneficial from different perspectives. Wind energy (1) is cost effective (Wiser 2015), (2) creates jobs (American Wind Energy Association 2015), (3) diversifies the national energy portfolio, which stabilizes electricity costs and reduces vulnerability to supply disruptions, (4) is plentiful and readily available (American Wind Energy Association 2016; Wiser et al. 2015), (5) produces electric power without consuming water which results in significant savings in water consumption, (6) is clean and its use means less smog, less acid rain and less greenhouse gas emissions, and (7) requires low-operating costs. In summary, wind energy is a good option for power generation with relatively low costs that provides economical and environmental benefits. On the other hand, wind energy has its own disadvantages, negative impacts and shortcomings. For instance, sites with high potential for wind energy may be far from cities where electricity is needed. Building the transmission lines and integrating the wind power in an existing grid may be costly. Also, the development of a wind farm may not be the most effective use of the land. Concerns have also been expressed at the noise produced by rotating wind turbine blades, aesthetic pollution and potential damage to local wildlife.

Challenges to implement wind power generation at a large scale in the Arab world are many. First, efforts to collect reliable, site-specific wind and hydrological resource data in the Arab world have been modest. The lack of these data makes it difficult to estimate potential gains from wind energy within a big picture that considers other energy sources or renewables. Without these data and a clear projection of benefits it is very hard to estimate economic gains from using wind power. Second, electricity generated by wind is dependent on the wind speed, which is time-dependent. As such, wind energy does not eliminate the need for other sources to supply reliable energy. Rather, the wind energy, when made available to the grid, should be used to reduce energy production from other power plants. This reduction may itself come at a price because it may impact the efficiency of the operation of other power plants. Simply put, connecting a wind farm to a grid and quantifying gains by wind energy in terms of reducing fossil fuel emissions are not at all straightforward. Finally, wind turbines are often grouped together to form what is usually referred to as a wind farm and generate bulk electrical power. The generated power from these turbines is fed into a utility grid and distributed to customers. In the addition to new transmission lines required to connect the wind farm, the existing grid may have to be upgraded to handle occasional surges of wind-generated power. As such, feeding wind power into the grid can be costly.

In addition to the above challenges, the drivers for using renewable energy may be different in different parts of the world. Accordingly, the challenges and opportunities are different. One has to consider all aspects that would impact a large-scale project such as the development of wind farms. These aspects include political, environmental, cultural, social, and financial complications and challenges. The interrelations of these challenges must also be addressed. For instance, political uncertainty and crises and the non-existence of clear legislation and commitments adversely impact investor confidence and foreign participation at the required levels for developing wind farms.

Conclusions

The objective of this chapter has been to explain the technical operation of large scale wind turbines to a general public and provide some data about potential use of this resource in the Arab world. The benefits and challenges faced when using or making decision to use this resource are outlined. In the opinion of this author, the issues to be considered when discussing the potential use of wind power in the Arab world are complex and interrelated and decision making must be based on databases and a detailed analysis that assesses benefits and impacts from different perspectives. A good start would be to develop education programs that broaden students and future generations perspective beyond the scientific and technological knowledge to include analysis of complex and interrelated databases, development of business models and assessment of social, aesthetic and economic impacts.

References

Al-Karaghouli, A., Al-Yousfi, B.., & Zitzewitz, E. (2007). *Current status of renewable energies in the Middle East-North African region*. UNEP/ROWA.

American Wind Energy Association. (2015). *US wind industry annual market report year ending 2014*. American Wind Energy Association.

American Wind Energy Association. (2016). *US wind industry: Fourth quarter 2015 market report*. Washington, DC: American Wind Energy Association.

Bady, M. (2012). Assessment of wind conditions in Assiut City, Egypt using long-term measurements. *Journal of Engineering Sciences, Assiut University, 40*(6), 1679–1694.

Darwish, A. S. K., & Sayigh, A. A. M. (1988). Wind energy potential in Iraq. *Solar & Wind Technology, 5*(3), 215–222.

Hajj, M. R., Tieleman, H. W., & Tian, L. (2000). Wind tunnel simulation of time variations of turbulence and effects on pressure on surface-mounted prisms. *Journal of Wind Engineering and Industrial Aerodynamics, 88*(2), 197–212.

Hasan, A. (1992). Wind energy in West Bank and Gaza Strip. *Renewable Energy, 2*(6), 637–639.

http://energy.gov/eere/wind/history-wind-energy

http://www.energy.gov/eere/wind/downloads/enabling-wind-power-nationwide

Janajreh, I., Su, L., & Alan, F. (2013). Wind energy assessment: Masdar city case study. *Renewable Energy, 52*, 8–15.

Ouarda, T. B. M. J., Charron, C., Shin, J-Y., Marpu, P. R., Al-Mandoos, A. H., Al-Tamimi, M. H., Ghedira, H., & Al Hosary, T. N. (2015). Probability distributions of wind speed in the UAE. *Energy Conversion and Management, 93*, 414–434.

Rehman, S., Halawani, T. O., & Husain, T. (1994). Weibull parameters for wind speed distribution in Saudi Arabia. *Solar Energy, 53*(6), 473–479.

Simiu, E., & Scanlan, R. H. (1986). *Wind effects on structures* (2 nd ed.). New York: Wiley.

Van der Hoven, I. (1957). Power spectrum of horizontal wind speed in the frequency range from 0.0007 to 900 cycles per hour. *Journal of meteorology, 14*(2), 160–164.

Wind Energy Benefits (Fact Sheet). (2015). *United States.* http://www.osti.gov/scitech/servlets/purl/1169772.

Wiser, R. (2015). *2014 wind technologies market report.* Washington, DC: United States. Dept. of Energy. Office of Energy Efficiency and Renewable Energy.

Wiser, R., Lantz, E., Mai, T., Zayas, J., DeMeo, E., Eugeni, E., Lin-Powers, J., & Tusing, R.. 2015. Wind vision: A new era for wind power in the United States. *The Electricity Journal, 28*(9), 120–132.

Part III
Food

Nuhad Daghir

Food Security in an Insecure Future

John R. Hillman and Elias Baydoun

Abstract Food security in the Middle East is directly affected by a challenging combination of ongoing destructive conflicts, a global economic downturn, widespread poverty, high population growth, corruption, intolerance, and the potentially damaging consequences of climate change. Many Arab countries demonstrate nearly all the features of those countries classified as poor, less developed, or failing to achieve the eight Millennium Goals. Even the economies of the richer oil-exporting countries in the Region have been seriously damaged by the downturn in oil and gas prices as new sources come on stream elsewhere and demand falls as a result of renewable sources of energy becoming available.

Keywords Sustainable energy • Food security • Water management

Climate change predictions for the Region are worrisome even if most of the environmental commitments made by countries at the UN Climate Change Conference in Paris at the end of 2015 are delivered. Large parts of the Region already suffer from periods of extreme temperatures and shortages of fresh water, so harsher conditions of higher temperatures and weather extremes pose special problems for inhabitants and policymakers, with heightened geopolitical risks and uncertainties. Agriculture is especially vulnerable.

Modern agricultural technologies with improved cultivars and livestock breeds must displace poor agricultural and pastoral practices. Larger-scale, better-capitalised production units are required with policy shifts to remove market-distorting subsidies, tariffs, and other regulatory impedances. Agricultural "roadmaps" are needed to encourage the most appropriate crops and maintain livestock in the most appropriate places. The trading of meat, dairy products and many types of crops to water-poor regions can be regarded as an efficient way to

J.R. Hillman (✉)
James Hutton Institute, Invergowrie, Dundee, UK
e-mail: jrhillman@tiscali.co.uk

E. Baydoun
Department of Biology, American University of Beirut, Beirut, Lebanon
e-mail: eliasbay@aub.edu.lb

S. Murad et al. (eds.), *Water, Energy & Food Sustainability in the Middle East*,
DOI 10.1007/978-3-319-48920-9_12

redistribute water. In unsettled times, food security requires protection of food-producing, food-processing, and food-distribution industries are protected along with research and development facilities, and the maintenance of skilled personnel.

There are doubts as to whether agencies of the United Nations, other international agencies, donor countries and charities are able to provide adequate humanitarian aid and humanitarian intervention to enable safeguarding of displaced peoples and ensuring the resumption of normal peaceful conditions essential for food security to be guaranteed. Promises of aid funds must become reality, followed by the establishment of functional supply and transport networks, and rapid transition to self-reliance thereafter. Food security and social stability are inextricably linked.

The solution to the food-security problems in the Region ultimately lies in the actions of governments, businesses and individuals. Governments need embedded scientific and technological expertise, strong civil-society institutions, integrity, and a particular focus on high-quality education.

Introduction

In a previous article (Hillman and Baydoun 2012), we considered definitions of food security in the modern era of rising global populations, discussing how food security might be attained in terms of security of water and fossil-fuel-derived energy supplies, climate change, rapid urbanisation, changing dietary trends, and modification of the natural environment leading to depleted natural resources, increasing environmental pollution, and the need to introduce modern technologies. The concepts of sustainable agriculture and uncertainty were also addressed, notably in respect of fresh thinking about key components of agricultural systems. These included (Babu and Blom 2014) vertical and horizontal integration of farming-related businesses to allow adequate capitalisation for enhanced efficiency measures; (Bardshaw and Brook 2014) policy shifts to remove market-distorting subsidies, tariffs, import and export bans, and excessive bureaucracy; (Baydoun and Hillman 2012) improved crop and livestock breeding, including entirely new species; (Bolukbasi et al. 2015) automation in agriculture and horticulture; (Breisinger et al. 2014) protected cropping; (Cong et al. 2013) new-generation agrochemicals; (Elasha 2010) new agronomic practices; (Fan et al. 2014) novel foodstuffs; (FAO 2011) habitat reconstruction and land renovation; (FAO 2015) biofuels and biodiesel; (Garland et al. 2014) periurban and urban agriculture; (Grebner et al. 2014) industrial biotechnology; (Grivetti and Ogle 2000) farming the seas and oceans; (Hillman and Baydoun 2012) long-term carbon storage; and (Hillman and Baydoun 2015) new ways of thinking about carbon trading. More recently, we reviewed mitigation and adaptation processes and strategies to address the impacts of climate change on food, water, and energy security in the Arab Middle East (Hillman and Baydoun 2015).

Here, we consider potential adaptations to an insecure global future generally, and to the concerns in the Arab Middle East specifically, in the light of the economic realities of wide disparities in wealth, competition for resources, and widespread poverty in many parts of the globe, coupled to a relatively high population growth, on-going conflicts, attempted cultural genocides, potential conflicts, endemic corruption and nepotism, and epidemics of infectious diseases. Most Arab countries are classified as poor, less developed, or failing to achieve the eight Millennium Goals of the United Nations, and these Arab countries share several undesirable features (Table 1). Even the much richer oil-exporting Arab nations are under pressure. After a decade of relatively high oil prices, these nations have accumulated more \$2.5 trillion in sovereign assets reinforced by substantial infrastructural investments supplemented by high levels of spending on imported military hardware. Now, however, oil prices are under pressure as global oil and gas prices slumped in 2014 and remain depressed, possibly for the medium to long term as new sources of oil and gas come from hydraulic fracturing (fracking) and from Iranian exports as the economic blockade on Iran is being relaxed, coupled to greater energy efficiency in industry and new sources of renewable energy. This price depression has exposed the degree to which the economies of these Arab oil exporters are dependent on oil and their failure in most instances to diversify their economies as their populations continue to expand alongside public expectations of continuing governmental largesse. Nearly all of the immediate adaptations Arab countries must undertake in order to adjust to a raft of severe insecurity issues require strategic planning and value-for-money infrastructural and civil-society improvements, and any preparatory changes in rural and urban areas will differ in scale and design. Longer-term adaptations will be reliant on more stable conditions and a stepwise improvement of educational standards and attitudes. At the time of writing, no Arab country is deemed to have acceptable levels of budget transparency according to the latest 2015 Open Budget Index released by the International Budget Partnership (see www.internationalbudget.org).

The future is especially insecure because of the persistence in the Region of a combination of incompatible political and economic ideologies, religious and ethnic groupings overtly intolerant of others, introvert nationalism and disrespect of others, disconnection from democratic principles, profound cultural divides, ignorance – some wanton, inability to adapt to modernity, and malevolent community and national leaders. From a noble history of toleration, hospitality, and learning, Arab society is fragmenting, defiled by the actions of relatively few. Arabs are killing Arabs either directly or indirectly; Arabs are inflicting as-yet-untold horrific crimes on other Arabs either directly or indirectly. Attacks on Arab countries by their neighbours might be used as an excuse to divert attention from their own failing donor-dependent economies or social structures, or most often to steal resources and ensure that the neighbouring country is suppressed from developing normally. At the global level, many would say that future conflicts and insecurity in much of the world are inevitable, simply because of the impacts of expanding global populations and the obvious competition for limited resources. The intensity of this competition must be analysed in context of the alarming

Table 1 Fourteen features of countries classified as poor, less developed, or failing to achieve the eight united nations millennium goals. The listed features are closely interrelated

1. Poverty
Common in both the urban and especially the rural poor, poverty is sometimes concentrated in regions, often in marginalised ethnic or religious groups, and may relate primarily to girls and women. Many of the urban poor operate in the unofficial economy. The rural poor tend to be land-constrained, dependent on rain-fed, low-yield subsistence agriculture with little or no access to modern technology (modern cultivars and livestock breeds, fertilisers, pesticides, automation, agronomy advice, veterinary support), and usually do not own their land. The rural poor encounter barriers to trade (e.g. transport, storage) and are unable to meet quality assurance standards. With no or limited access to social benefits (primarily pensions, child support, education, training, and healthcare), the poor may be hungry, thirsty, suffer ill health and low standards of accommodation, and die early. The poor are susceptible to exploitation.
2. Hunger and thirst
Access to food and potable water may involve substantial travel on foot, and the basic requirements may only be met wholly or in part by humanitarian assistance. Food quality and safety are usually low, and cooking is often dependent on wood for fuel. Symptoms of malnutrition are prevalent.
3. Disease
Slum dwellings, insanitary conditions, poverty, and hunger lead to a vulnerability to pandemics, made worse by poor or no public health provision. High maternal and child mortality and low general life expectancy characterise poverty and there is a reliance on traditional and/or herbal medicine. Crops and livestock are subject to catastrophic attacks by pests and diseases.
4. Poor environmental management
Poor countries suffer a depletion of their natural resources, including freshwater supplies and native flora and fauna. Mineral and fossil-fuel resources are extracted to be exploited by industries in other more-developed countries. Land, water, and the atmosphere may be polluted with few or no remediation efforts. National, regional, and international environmental regulations are not properly implemented. Agricultural soils tend to be subject to erosion, salination, solarisation, desertification, and nutrient depletion. Even without climate-change predictions, the general anthropogenic environmental degradation currently taking place increases the vulnerability to flooding, sand storms, ill health, and displacement of peoples. Most poor countries are enduring adverse climate trends, and climate-change predictions point to even harsher conditions (erratic rainfall with rising temperatures, heat waves, hot extremes, and storms; rising sea levels and acidification of seas and oceans; increased desertification with effects on agriculture as well as the natural flora and fauna; socio-economic and health implications).
5. Poor infrastructure
Quantity and quality of the built infrastructure tends to be low, or even absent (*e.g.* roads, ports, airports, telecommunications and access to the internet, hospitals and clinics, reliable power and fuel supplies, potable water sources, sanitation systems and sewage disposal, protected natural environments, cold and pest-free storage of agricultural and horticultural produce). The cost of living is worsened by relatively high overland transport costs. Facilities may be inadequate to meet demand and maintenance neglected. Rapid urbanisation and conflicts exacerbate infrastructural deficiencies.
6. Corruption
Political autocracy coupled to a lack of transparency in government and public services, a weak judiciary, lack of consultation on major issues, and suppressed media enable a climate of corruption and nepotism to permeate all areas of society, including schools, colleges, and universities. A low regard for human rights, democratic processes, health and safety measures, and international law, is made worse by an overburdensome bureaucracy (staffed by poorly paid civil servants and police) dependent on bribery to function. Corruption allows laws to be broken with impunity, and protests ignored. In most areas of society, a lack of altruistic leadership and

(continued)

Table 1 (continued)

charity is commonplace. Unsurprisingly, there is widespread malcontentedness at corruption and its effect on reinforcing inequality and impeding meritocratic advancement. Little effort is made to eliminate corruption without external pressures from donor countries.
7. Low GDP *per Capita*
Poor countries manufacture few value-added products, offer little or no advanced training, and lack participation in the global knowledge economy. There is little or no foreign direct investment and free trade is constrained. Financial assistance from donor countries is increasingly being audited to ensure compliance with attempts to prevent unauthorised expenditure. Many poor countries have not managed to have constructive relationships with potential donor countries, and some try to align themselves with one of the main international political power blocs. Unrelieved debt burdens have led to high interest rates on loans and difficulty of obtaining credit. Private savings are minimal and the country may be subject to periodic flights of capital. Their economies are grossly imbalanced with little spent on healthcare, education, and other social benefits compared with defence and vanity projects. The economy may be damaged by previous and/or ongoing conflicts, and may in any case have limited absorptive capacity properly to manage additional resource inflows and outflows. Agriculture, unofficial transactions, and remittances from those working abroad contribute disproportionately to the real economy.
8. Poor education
Limited or no access to free schooling accounts for a general low level of literacy and numeracy, especially of females. This is reflected in the absence of high-grade internationally competitive universities, colleges, and research institutes, despite a high level of parental financial sacrifice to secure a supposedly good education for their children. Science and technology tend to be poorly taught and there is undue influence of religions with regressive attitudes to modernity. A lack of investment in research and development (R&D) and a lack of a critical mass of scientists, engineers, and technologists impede industrial development, and prevent those that remain from joining international consortia, participate in learned societies, access essential literature and training programmes, and have the opportunity to use state-of-the-art instrumentation, software, and laboratory consumables. The best educated and the most talented and entrepreneurial usually emigrate, leading to a brain drain. Intellectual property rights are often ignored and little benefit is derived from traditional knowledge and its products tend to be exploited in other countries. Crucial demands and needs of less-developed countries are rarely the targets of major international R&D projects.
9. Gender inequality
Low educational attainment in most girls and women is reflected in female political, economic, and social representation and participation failing to match their proportion of the general population. Cultural and religious influences often lead to females being regarded as inferior to males. Family planning is limited, and in the absence of social security and reasonable incomes, large families are the norm.
10. High population growth
This is a feature of the poorest social groups, and correlates with low life expectancy and gender inequality. In some countries high population growth may exceed the economic capacity of the country to feed itself, leading to a propensity to generate refugees, displaced persons, and terrorists.
11. Vulnerability to transnational terrorism
Terrorism often relates to a combination of one or more of the following: poverty, hunger, poor educational attainment, disconnection from democratic principles, susceptibility to indoctrination by intolerant religious ideologies, criminal activities, and psychiatric disorders. Poor countries are usually unable to defend themselves from terrorists, and groups of terrorists may receive covert support from other countries and agencies. Poor countries may form the battlefield for fighting between different groups of terrorists. Recovery from terrorism and conflicts in general may take several generations, and lead to psychological and physiological after-effects in the survivors and their progeny.

(continued)

Table 1 (continued)

12. Weak public sector
Low salaries, complex and inefficient bureaucratic processes, poor educational attainment, incompetence, widespread acceptance of corruption through bribery and political interference, and the lack of an investigative free media, account for the justifiable lack of confidence in the public sector. A low regard for human rights, legal processes, and justice undermines societal advancement. There is also poor custodianship of cultural heritage and essential infrastructure.
13. Neighbouring countries
Many poor countries have reliance on and vulnerability to neighbouring countries for access to water, transport and communication networks, foodstuffs, energy, control of environmental issues (*e.g.* desertification, flooding, biodiversity protection, pollution *etc.*), and security. Neighbouring countries may (a) create security problems by aiding terrorists and insurgents, (b) provide an uncontrolled source of refugees, (c) invade, (d) steal water or other natural resources, and/or (e) issue mendacious media releases or operate diplomatically to undermine the confidence of donor countries, aid agencies, and investors. The more-developed countries rarely understand the sheer difficulty of managing a poor country facing inter-ethnic conflict, and terrorism and poverty.
14. International agencies, non-government organisations, international media and the United Nations
Poor countries are monitored by a plethora of international and national bodies, and extensive reports generated, but the necessary actions – political and military – to solve the main issues and problems are rarely carried out. Unwarranted aggression inflicted on other countries – poor or not, or the suppression of their populations or specific parts of their populations by nation states eventually must be counteracted by military intervention. Sadly, poor nations are oftentimes regarded as pawns in international power struggles, and remain either exploited for their resources, or ignored. Moreover, economic downturns in donor countries reduce the level of aid and absorption of refugees, made worse by certain countries cynically failing to meet their initially well-publicised aid pledges.

projected changes and options for adaptation as well as mitigation (curbing emissions) detailed in the latest 2015 Fifth Assessment Report of the International Panel on Climate Change (IPCC, www.ipcc.ch/). The immediacy of food security during social instability in the Arab Region forms the backdrop to this article, rather than the longer-term infrastructural and social transformations needed to mitigate and adapt to climate-changing emissions, transformations that demand political stability and sophistication.

Challenges, Assistance, and Needs

The global population is estimated to be around 7.2 billion at present and there is an 80% probability that it will increase to between 9.6 billion and 12.3 billion in 2100 (Garland et al. 2014), so stabilisation of the population is highly unlikely this century. Moreover, human population reduction is not a quick fix for environmental problems, and even a catastrophic event that killed billions of people would have relatively little effect on the overall impact of humans on the environment

(Bardshaw and Brook 2014). Population increases could undermine attempts to ameliorate attempts to reduce climate-modifying emissions.

Climate-change predictions for the Arab Region are deeply concerning. Most scientists support the conclusions of the latest Fifth Assessment Report of the IPCC. Formed in 1998 by the United Nations Environment Programme (UNEP) and the World Meteorological Organization (WMO), and relating closely to the UN Framework Convention on Climate Change – the main multilateral forum for addressing climate change, AR5 comprises three Working Group Reports and a Synthesis Report with its Summary for Policymakers. The main issues are covered in detail: observed changes and their causes; future climate changes and their risks and impacts; future pathways for adaptation, mitigation and sustainable development; and a more detailed analysis of adaptation and mitigation including policy options, technology and finance. In the Arab Human Development report authored by Balgis Osman Elasha, (2010), the impacts of climate-change projections in the Arab Region are given in stark detail. In the coming decades, Arab societies and their industries will be profoundly and adversely affected by projected temperature increases in excess of 4 °C and severely reduced rainfall, threats of increasing frequencies of impacts originating from El Nino events, changes in the seasonal distribution and predictability of rainfall, depleted aquifers, reduced river flows, rising sea levels, flash floods, in addition to increased numbers of dust storms and hurricanes. As the Arab Region itself is a relatively small direct contributor to global greenhouse-gas emissions, although its fossil-fuel exports and its importation of goods that took energy to create are substantial contributors to them, adaptation must be a crucial factor in policy developments in more settled times that will themselves be dependent on stable and peaceful Arab countries.

Questions arise as to whether organisations such as those overseen by the United Nations, multilateral groups such as the European Union, individual nations and several charities will be capable of initiating and then maintaining peaceful conditions and food security. Food and other forms of aid are likely to face increasing demands at a time when the economies of many donor countries are enduring continuing austerity and recessionary conditions, and when there is growing but unjustified cynicism about the effectiveness of these aid organisations. Irrespective of the many estimates that total global food production can readily meet the needs of the present global population, political reality is that all countries and people are patently not equal and are unlikely to be so for the foreseeable future. Aid can assist in partially rebalancing the inequalities, not least where several countries in the Arab Region are currently in turmoil and others are deeply troubled. Some regard most of the Arab world as regressing, out of synchrony with, and lacking sympathy from, the world at large. Food costs and poverty are primary concerns, exacerbated by insecurity of energy and water supplies as well as rampaging insurgents and those wishing to impose unacceptable regimes and suppression of minorities, denying their citizens proper democratic freedoms.

Wars and conflicts are all too easily incited in the absence of strong democratic civil society involvement and usually quickly bankrupt countries; destroy nearly all parts of their economies; ruin infrastructure including homes, businesses, transport and communication networks; disrupt family life and social interactions; generate displaced people and refugees; and attract foreign interference, including active participants. War crimes are commonplace. Psychological aftereffects are noted in civilians and combatants long after fighting has died down. Populations in the Arab Middle East have endured asymmetric warfare between countries with vastly different military capabilities, provoking guerrilla tactics, chemical warfare, civil war, and even unconventional warfare through acquiescence, capitulation, and clandestine support for long-term insurgencies. Wherever they occur, warfare and conflicts are always associated with the participants having distorted views of history – some drawing on ancient history – with widely diverse concepts of ethics, and complicated by ethnic and profound religious differences. Little consensus exists on triggering factors, theories, and outcomes. Prodigious sums of money have been expended in the Arab Middle East on armaments and defence forces to the exclusion of adequate investments in social welfare, education, and research and development.

Political instability for whatever reason usually leads to food insecurity and conflict. During social disturbances and the onset of widespread conflict, the normal mechanisms underpinning food production, importation, storage, processing, transport and retailing are profoundly disrupted, creating conditions that promote the formation of ghettos, and further stimulate corruption, robbery, and the black market. Normal policing and social order collapse, social behaviour degrades, and criminal activities dominate. As the economy collapses, the terms of trade are transformed. Disease epidemics become manifest. The restoration of domestic and business normality takes years and may never be achieved within one or more generations.

Food insecurity itself may lead to political instability, not least in a world of global intercommunications when citizens of a poor country or region can view with understandable envy the lifestyles of those in rich countries or communities. It is therefore a basic duty of political and community leaders to ensure that food security is a foremost priority for those people within their sphere of responsibility. Insecurity of food supplies can be created by adverse weather conditions; the depredations of pests, weeds, and diseases; and salt contamination that can be caused by poor irrigation and agronomic practices. Around 62 million hectares of hitherto fertile land on earth have been damaged by salt. According to *Economics of Salt-induced Land Degradation and Restoration* (Qadir et al. 2014), only tree planting, deep ploughing, and growing salt-tolerant crops coupled to digging drains and dykes around the affected area can address the problem. Neglecting the health of Africa's soils, many of which suffer almost irreversible degradation and nutrient deficiency, will lock the continent into cycles of food insecurity for generations to come, according to the 2014 Montpellier Panel report. Indeed, 2015 was designated the 'International Year of Soils' by the 68th UN General Assembly. Since the 1920s, there are particular issues relating to selection pressures on destructive pests, weeds, and diseases in the vast monocultural single-cultivar agricultural systems that are also the present-day main sources of global food aid. Many of these main producing areas are experiencing irregular rainfall patterns and failing irrigation arrangements. Crop failures in these areas have quick knock-on effects on volatility of the global agricultural commodity markets; this is a situation likely to get worse as the population inexorably increases and demands more food. Moreover, the world has yet to experience the sort of dramatic harvest failures that occurred in the 1930s and before. Another aspect of the agriculture sector (generally accepted to include crops, livestock, fishing, and forestry) in the developing world is that it absorbs *circa* 22% of the economic impacts caused by medium- and large-scale natural hazards and disasters. Between 2003 and 2013, these events in developing countries affected more than 1.9 billion people and caused more than $494 billion, but agriculture only received about 3.4% of all humanitarian aid (FAO 2015 www.fao.org/emergencies/how-we-work/resilience/en/). Delegates from FAO at the conference announced the launch of a facility that will focus on bringing together technical expertise and financial resources with the aim of building greater resilience of agriculture to natural extreme weather events.

The combination of advanced crop and livestock breeding, agrochemicals, automation, better agronomy and livestock husbandry, increases in the land area farmed, and efficient larger-scale better-capitalised production units have collectively prevented Malthusian disasters. Since 1950, food production has more than quadrupled, using less than 1% more cultivated land, allowing civilisation to proceed and expand. Food security is no longer an issue in many countries, and the global economy, human health, and societal development have been, for the most part, positively influenced by agricultural advancement. In the period 1900–2010, when the global population increased from around 1.5 billion to around 7 billion, average agricultural commodity prices decreased by an average of 0.9% p.a. because supplies rose faster than demand (unpublished presentation by Prof. Ingo Pies in 2014 to biennial development meeting of Pottinger). Nonetheless, feeding any substantial increases in the global population will only be possible by technological innovations because of the ecological limitations on increased water and fertilizer supplies and increasing the area farmed. Likewise, various climate-change predictions amplify justifiable concerns about global agricultural productivity. Supply and demand market dynamics are complex and ultimately resilient to political interference, although many countries and trading blocs try to manipulate production by tariffs, export bans, subsidies, inhibiting technological developments and market processes. The spikes in prices of traded wheat, rice, maize, and soya in 2008 and 2011 were initially blamed on speculation, especially the index tracking funds and derivatives markets. Yet this type of speculation does not trade physical goods but price risks, and is therefore a form of insurance market. In addition, speculation would be expected to be associated with high stocks as farmers opt for storage rather than sales. In 2008 and 2011, however, stocks were very low and caused the price rises. Even so, government policy failures, including protectionist export bans and inadequate promotion of agricultural efficiency, contributed to panic buying, as exporters reduced their offer and importers increased their demand in response. Calls by civil society groups to ban speculation by index-tracking funds and derivatives markets were always and continue to be profoundly misguided. All countries should have policies to sustain and constantly review food production and supply, especially if there is a significant dependence on food imports. Reasonably substantial reserve food stocks are essential cushions to prevent price bubbles and food shortages, but excessive stocks can distort markets such as when they are released in large quantities as general food aid and undermine the operation of normal agriculture markets in developing countries. Some price volatility is an essential component of healthy competitive markets, driving adaptation, risk taking and innovation.

Debates about the environmental costs of different kinds of agriculture, not least in terms of water and energy security in respect of the Arab Middle East, and in terms of the destruction of natural habitats and loss of biodiversity, as well as cultural and other changes, have stimulated possible strategies to address these concerns. One approach is for each country to have a "roadmap" for its agricultural development, and these roadmaps might be aggregated into a regional roadmap. The US report: *A Science Roadmap for Agriculture* – cited as *Task Force on Building a Science Roadmap for Agriculture*, National Association of State Universities and Land-Grant Colleges, Experiment Station Committee on Organisation and Policy, "A Science Roadmap for the Future", November 2001 (www.nasulgc.org/comm_food.htm) http://agsci.oregonstate.edu/files/main/roadmap2.pdf pioneered a way to define the needs of agriculture and help shape the future direction of the various strands of agriculturally relevant science. This impressive US-specific study followed a conceptual framework of needs to (a) be competitive in a global economy; (b) add value in future harvests; (c) adjust agriculture to a changing climate; (d) be good stewards of the natural environment and natural resources; (e) make agricultural enterprises profitable; (f) make families and communities strong; and (g) modify foods for improved health and safety. One important relevant outcome of this work is the obvious requirement to grow crops and practice livestock husbandry in the most appropriate environments, enabling different environmental zones to have a critical mass of expertise and facilities. For the water-constrained Arab Middle East, this would mean increased reliance on food imports that could also be regarded as a form of water importation. But this would be possible only if there were formal agreements between food-producing and food-recipient countries, and these agreements were economically and politically stress-resistant. To this caveat would be questions of how to pay for imports, and acquiring resources required to redirect the agricultural workforce into other wealth-creating activities. Introduction of a logical, science-led agricultural roadmap in the Region may be impossible at present but needs to be initiated. On a global level, food commodity and non-food agricultural commodities prices have declined by 17–18% in the period mid-September 2014 to mid-September 2015 (see *The Economist* Commodity–Price Index (2015) in www.Economist.com/indicators), reflecting relatively clement weather conditions in most of the producing areas, greater production efficiency, balance sheets strong enough to bear losses, competition to gain market share, and continuing investments. Importing countries, however, are affected by the strength of their currencies amongst other factors, such as social upheaval. Unlike other commodities whose prices direct reflect industrial demand, prices of foodstuffs reflect the effects of weather, pests and diseases, the demands of a rising global population, and many kinds cannot be readily stockpiled. A further factor operating at the global level is the domination of farm supplies by six international companies: Monsanto (the largest seed producer), Syngenta (the largest agrochemical producer), Bayer, BASF, Dow Chemical, and Du Pont). All of these companies have been active in acquiring other companies and patents, thus reducing competition, and potentially reducing innovation as they robustly defend their intellectual property. In this era of low commodity prices and developing resistance to older-type herbicides, farmers are constrained by input costs of fertilisers, seeds, agrochemicals and veterinary medicines that have steadily risen in over the past decade, only partially alleviated by the recent reduction in fuel and lubricant costs.

Pandemics – epidemics of infectious or contagious disease that have spread through populations across a large region, crossing international boundaries – drastically curtail food production and distribution, aggravating poverty in both the rural and urban poor. Pandemic-causing diseases include the ever-present cholera, influenza such as the 1918 and 2009 H1N1 outbreaks, typhus, smallpox, measles, tuberculosis, plague (*Yersinia pestis*), leprosy, malaria, human immunodeficiency virus infection and acquired immune deficiency syndrome (HIV/AIDS), viral haemorrhagic fevers (ebola, Marburg, Crimean-Congo, Lassa, Rift Valley, dengue, yellow fever, *etc.*), and now there are new diseases such as severe acute respiratory syndrome (SARS). Vaccine development is necessarily slow, and treatment of bacterial diseases is hampered by the rapid development of antibiotic resistance. The propensity of diseases to mutate, acquire new vectoring capabilities, have reservoirs in wild animals, and even persist in spore form, mean that there must be constant vigilance.

As one of the major global food-producing bloc of nations, and as one of the major food importers and donors of food and other forms of humanitarian aid, the European Union (EU) bears crucial responsibility for the deleterious effects of its complex and highly bureaucratic common agricultural policy. Its massive subsidy regimes impact adversely on global markets and its well-meaning but often poorly thought-through environmental regulatory decisions are not based on sound scientific evidence. So-called “greening” policies are being introduced that may be deemed to enhance the environment but are likely to decrease profitable production. Social measures to support small-scale inefficient producers also distort the global marketplace. Likewise, the series of restrictions being introduced on the use of a wide range of agrochemicals within the EU and for imported commodities, without carrying out proper impact assessments and fast-tracking alternatives, imperil production. In the medium to long term, a more serious issue is the virtual ban in genetically modified (GM)) crops, inhibiting their uptake in countries intending to export to the EU as well as suppressing state-of-the-art research and development and associated investments in EU countries. Of particular relevance in this regard is the recent and largest statistically rigorous review of the agronomic and economic effects of the current range of commercially available GM crops on farming (Klümper and Qaim 2014). In examining publications between 1995 and March 2014, it is therefore a near-complete survey. In essence, the two main types of GM crop – resistance to insect pests and tolerance to the wide-spectrum weedkiller glyphosate – conferred considerable yield improvements and much higher profits than conventional crops. GM crops and related products in the development pipeline were not considered, and these promise great advances in nutrition enhancement, environmental clean-up, new medicines, new products for manufacturing industries, and improved crop and forest species. Moreover, existing GM crops have greater impacts in poorer countries than in richer countries because their insect pests and pernicious weeds are more difficult to control. By including non-peer-reviewed papers (book chapters, working papers, conference papers *etc*.) as well as peer-reviewed papers for the meta-analysis, it was possible both to correct for academic bias in focusing just on the most dramatic effects, and include data for many ancillary effects, such as the effects of fertilisers. The EU ban on GM crops is therefore denying poor exporting countries from reaping the full benefits in yield, profitability, and commodity quality, in addition to reducing potential EU food-aid exports. One positive feature of the EU is the bureaucratic system to improve and monitor the quality and safety of foodstuffs, from raw ingredients through to ready-meals, and GM crops have been found to be safe. That for many years most of the feed protein in the EU comprises imported GM maize and soya bean, the issue of GM crops should not be ignored for much longer. Opposition to the processes of modern genetic modification and ownership of the processes (often deeming them to be "unnatural"), and disregarding the quality, safety, and value-for-money of the product, actually condemns conventional agricultural practice, and demonstrates ignorance of naturally occurring mutations and horizontal gene transfer. Even so, in early November 2014, the newly elected President of the European Commission, J-P Juncker, in what is widely regarded as a blatant act of appeasement to Greenpeace and other so-called "environment" pressure groups, sacked the EU's chief scientific advisor, Professor Anne Glover, for her support of GM technology, compounding this regressive stance with abolition of the post. All countries and trading blocs should have influential and competent teams of chief scientific advisors.

Research and development in gene identification, construction, insertion, editing, and expression, coupled to high-throughput phenotyping are collectively revolutionising agricultural, horticultural, and forestry sciences. GM technology is not simply the insertion of genes using various technologies from a similar or different species into a recipient organism. It includes the concept of gene silencing – the prevention the reduction of expression of certain genes – a process that can take place at either the transcription or translation cellular processes. It is not the equivalent of gene knockout but is essentially gene knockdown because the methods to silence genes do not completely eliminate the expression of a specified gene. The methods to silence genes include RNA interference (RNAi or post-transcriptional gene silencing), small interfering double-stranded RNA (siRNA), and CRISPR. Of special interest is the CRISPR (clustered, regularly interspaced, short palindromic repeats) toolkit that is derived from research on prokaryotic antiviral systems and currently involving the Cas9 and Cpr1 endonucleases (Jinek et al. 2012; Cong et al. 2013; Bolukbasi et al. 2015). As viruses constantly evolve to escape from these antiviral systems, bacteria probably evolve new systems.

CRISPR technology is able to recruit heterologous domains that can regulate endogenous gene expression as well as label specific genomic loci in cells, so that is feasible to engineer germ lines and thus the path of evolution. This technique is replacing methods using mutagenic agents, virus vectors, zinc-finger nucleases and transcription activator-like effector nucleases (TALENs). Its relative simplicity and evolutionary significance for all life forms, including humans, means that internationally agreed regulatory frameworks are essential. The technology does not involve implanting genes from one organism into another, and is not therefore creating transgenic organisms; it is gene editing.

There is now realistic expectation of new perennial cereals; incorporation of C-4 photosynthetic characteristics in existing C-3 crops; enhancement of nitrogen fixation by free-living soil microorganisms in the vicinity of crop roots; tolerance and resistance to biotic and abiotic stresses; and modification of lignification, texture and endogenous components (such as vitamin content, acrylamide in potatoes, antinutritionals, toxins, proteins, oils, and carbohydrates) of a wide range of existing and potential crop species (see www.isaaa.org/kc/cropbiotechupdate). Besides the present-day generation of improved livestock species and new forms of husbandry, the use of balanced diets based on competitively priced synthetic amino-acid and fatty-acid products will lessen the need for large-scale soya and maize production. Ancillary advances are taking place in mechanisation; diagnostics; predictive modelling and decision-support systems; remote sensing; protected-cropping systems; and weed, pest, and disease control. Diminution of abiotic and biotic stresses in the field and under protective cropping is now the focus of major research initiatives.

Of the *circa* 250,000 species of angiosperms, according to the Food and Agriculture Organisation of the United Nations (FAO 2004), only 30 species provide 95% of human energy needs and only four species (rice, wheat, maize, and potato) account for 60% of energy intake, and 75% of crop diversity was lost in the last century. Around 80% of human calorie intake comes from 12 crop species (Grivetti and Ogle 2000) and 50% comes from just three grasses – wheat, maize and rice (see www.knowledgebank.irri.org/ericeproduction/importanceofRice.htm) dependency on such a narrow genetic base is a threat to food security and is only partially alleviated by investments in *in* situ and *ex situ* plant gene/seed banks, germplasm collections, and DNA libraries. Just 7000 species of angiosperms and gymnosperms have been cultivated for human consumption in human history, with around 80,000 angiosperm species yet to be discovered. In theory, most angiosperms should be capable of being biotechnologically modified for food and non-food uses. Will scientists in the Arab Middle East fully participate in these exciting developments? Valuable collections in the Arab Region are inadequately respected for their worth under peaceful conditions but are now extremely vulnerable during this period of war-like conditions and enduring financial pressures.

Coming into force in 2004, the International Treaty on Plant Genetic Resources for Food and Agriculture (International Seed Treaty) was designed to complement the 1993 Convention on Biological Diversity (CBD) and was designed to guarantee food security by (a) conservation, exchange, and sustainable use of all types of plant genetic resources; (b) offering fair and equitable benefit-sharing; and (c) recognition of farmers' rights. Critics of the International Seed Treaty point to great variability across countries of access to collections and interpretation and implementation of farmers' rights. Moreover, in adopting the CBD's outlawing of biopiracy – the uncompensated commercialisation and profiteering of seeds, propagules, growing plants, and their products from source areas – has severely inhibited the acquisition and exchange arrangements in collections until better processes come into force.

Gap analyses are methods to identify gaps in *ex situ* collections of wild-plant relatives of agriculturally relevant species as a means to guide efficient and effective collection strategies (Villegas et al. 2010). A gap analysis by the International Center for Tropical Agriculture managed by the Global Crop Diversity Trust and the Millennium Seed Bank in Kew examined 29 priority gene pools of the globally most important food crops and their wild relatives. Most at risk were eggplant, potato, apple, sunflower, carrot, sorghum, and finger millet (see www.cwrdiversity.org/conservation-gaps/). Ongoing conflicts and disorder in the Arab Region justify an independent review of a regional Red List Index based on the List of Threatened Species released by the International Union for the Conservation of Nature (www.iucn.org) in order to evaluate the extinction risk of species and subspecies of the natural flora and fauna.

Certain individual countries in the EU are important donors of humanitarian aid in their own right. According to a recent report (November 6, 2014) from the Organisation for Economic Co-operation and Development, economic stagnation especially in the Eurozone portion of the EU poses a major risk to world growth. If the stagnation were to continue or even get worse, humanitarian assistance would inevitably be reduced, and countries that hitherto were willing donors would become increasingly introverted. As evidenced by growing problems of graft, corruption and authoritarian government in certain (but not all) members of the EU that were formerly dictatorships or in the sphere of influence of the former Soviet Union, democratic norms (human rights, respect for minorities, tolerance, free press, independent judiciary and rule of law, active civil-society groups, transparency of accounting for taxpayers' money *etc.*) take time to become bedded into the fabric of society. In considering the medium-to-long-term future of the EU, it is sobering to note that throughout European history, confederations between its diverse nations and subsets have rarely persisted unless full political, legal, monetary, and more profoundly, cultural fusion had taken place.

All governments worthy of the title must ensure that there are relief mechanisms to enable the provision of basic food supplies and fresh water together with functioning standby electricity-generating equipment in unsettled times caused by natural or man-made disasters. Surely governments have the ultimate responsibility to attend to the needs of their people and not themselves. Rarely observed, governments need genuine food, nutrition and agricultural experts as an integral part of the decision-making hierarchy. Such experts must have a proper understanding of the pre-conflict or pre-disaster food, fresh-water, and energy supplies and their distribution systems, and how they can be safeguarded, modified and employed to proper effect, and how alternative mechanisms can be deployed. Sadly, this aspect seems to be neglected at the present time in the chaotic condition of certain countries in the Middle East. Much can be learned from countries in Europe during the wars that raged in the nineteenth and twentieth centuries. Simply standing by and watching the population adapt slowly to acquiring barely adequate water and food supplies inflicts untold misery on innocent people. All governments should have readily accessible emergency supplies (reserve stocks) and transport systems, and be willing to introduce rationing if need be. Special protection measures are needed for water supplies and farms to make sure production can continue no matter the degree of impairment. In more settled times, each government should establish a group of experts to construct interactive databases as the foundation of an agri-informatics and metrics organisation. This would collate information on supply and demand changes, supply-chain details, imports, crop and livestock genetics, commodity production levels, labour-force composition, pricing, inputs, availability of decision-support systems, advisors and research bodies, grant funding, biotic and abiotic stress factors, natural resource constraints, predictive modelling of shocks to the agricultural system and disruptive events, *etc*. Other research organisations would interact with this organisation to ensure best practice and enhance agricultural resilience, demonstrate efficient use of inputs, exploit wastes, optimise the use of mechanisation, and foster skills. Arab countries still have to utilise fully the international capabilities and potential of (International Centre for Agricultural Research in the Dry Areas – currently based in Beirut given the conflicts in Syria) and other members of the CGIAR consortium (formerly the Consultative Group on International Agricultural Research).

The plethora of aid agencies offering humanitarian and development aid encompass those that are organised by a single government, multilateral donors, non-governmental organisations, philanthropic and charitable organisations, businesses, and individuals. ReliefWeb (www.reliefweb.int) provides a relatively comprehensive Directory of Humanitarian Organisations. Fragmentation of the total aid effort is becoming a worrisome issue. The International Committee of the Red Cross, part of the International Red Cross and Red Crescent Movement along with International Federation of Red Cross and Red Crescent Societies and 189 National Societies, is mandated internationally to uphold the four treaties and three additional protocols of the Geneva Conventions. These Conventions are rules that apply in times of armed conflict both within and between countries, and define the rights of civil and military prisoners and protections for wounded people and for civilians. Weapons of war are dealt with by The Hague Conventions and the biochemical warfare Geneva Protocol. Enforcement of the Conventions is through the UN Security Council but is rarely invoked, primarily because of profound ideological differences about democracy and human rights between the five permanent members of the Security Council with veto powers, so there tends to be diplomatic reliance on regional treaties and national laws. Parenthetically, there are ten non-permanent members of the Security Council without veto powers that are elected by the General Assembly of the United Nations for a two-year period. On 19 November 2014, the UN Security Council pledged to counter the global terrorist threat and increase cooperation to address the perils posed by foreign terrorist fighters such as those that are a notable feature of conflicts in the Arab Region. The UN Office on Drugs and Crime is also involved in this initiative. Other international related treaties include the United Nations multilateral treaty referred to as the Geneva Protocol or Convention relating to the Status of Refugees as well as the Declaration on the Protection of Women and Children in Emergency and Armed Conflict adopted by the United Nations in 1979.

Humanitarian aid is distinguished from humanitarian intervention, which involves armed forces protecting civilians from violence or genocide. The United Nations Office for the Coordination of Humanitarian Affairs is mandated to coordinate humanitarian responses, usually in concert with the International Committee of the Red Cross. Valuable reference material can be found in (a) the postings from the Global Policy Group; (b) Humanitarian Policy Group Policy Note 34 *Providing Aid in Insecure Environments*; (c) the global resource for non-governmental organisations People in Aid Code of Good Practice, which is an internationally recognised management tool; (d) postings of the Humanitarian Accountability Partnership International, notably the HAP 2007 *Standard on Humanitarian Accountability and Quality Management;* and (e) the Food Assistance Convention Treaty that entered force in January 2013 and is aimed at meeting the food and nutritional needs of the most vulnerable people, complementing the Food Aid Convention which unlike the Assistance Convention is focused on a limited number of food items expressed as 'wheat equivalent tons'. Both have their legal depository in the Secretary General of the United Nations. Three other sources of crucial information are (a) the FAO 2014 (b) *Food and Nutritional Needs in Emergencies* (www.who.int/nutrition/publication/en/nut_needs_emergencies_text.pdf), consisting of guidelines to estimate and monitor food and nutritional needs of populations in emergencies, developed jointly by the United Nations High Commissioner for Refugees (UNHCR), UNICEF, World Food Programme, and World Health Organisation; and (c) publications of the International Food Policy Research Institute (IFPRI), especially *Global Nutrition Report 2014*, *Resilience for Food and Nutrition Security* (2014), and *How to Build Resilience to Conflict: The Role of Food Security 2014*. According to the World Bank, IFPRI, and FAO, the number of undernourished people in the world, based on calorific intake, now stands at about 805 million, despite real global growth rates of 3.6% over the period 1990–2014. This figure increases to *circa* two billion suffering from micronutrient (minerals, vitamins *etc.*) deficiency. According to the IFPRI *2014 Global Hunger Index* by Grebner et al. (2014), the state of hunger in developing countries as a group has improved by 39% since 1990. Even so, the level of hunger is still serious with an estimate of 805 million people continuing to go hungry. The highest levels are south of the Sahara and south Asia.

In the IFPRI *Global Nutrition Report 2014,* evidence is summarised to show that improvements in nutrition status will make large contributions to sustainable development goals, namely poverty, food, health, education, gender, and employment Investment in nutrition has a highest benefit ratio. Projections from the World health organization (WHO) and UNICEF demonstrate that the world is not on track to meet any of the six World Health Assembly (WHA) nutrition targets (reducing child stunting, reducing anaemia in women of reproductive age, reducing low birth weight, reducing the number of overweight children, increasing exclusive breast feeding, and reducing child wasting), although many countries are making good progress in meeting nutrition outcomes. The manifestation of malnutrition is changing as countries are now facing complex, overlapping, and connected malnutrition burdens. Three of the chapters in the IFPRI publication Resilience for Food and Nutrition Security (Fan et al. 2014) are germane to this article. Breisinger et al. (2014) briefly mention the Arab Spring and uses Egypt, Somalia, Sudan, and Yemen as case studies of conflict-affected countries. Mabiso et al. (2014) have specific reference to the Syrian refugee crisis, and take a global overview of the complex relationships between refugees and host countries. Babu and Blom (2014) introduce a model that seeks to delineate the key capacity components of a resilient food system, considering a country's capacity to create, manage, and utilise human resources for a resilient food system.

Significant challenges to aid provision include (a) harnessing the necessary stream of funding when grandstanding promises by countries are often never met; (b) establishing and coordinating the basic support network; (c) ensuring the logistics arrangements are effective, including communication networks; (d) prevention of resource misappropriation; (e) protection for officials and support workers on the ground; (f) protecting the vulnerable people needing aid; (g) operating with transparency and integrity; and (h) laying the structural and procedural foundations for self-reliance.

Effective lines of communication with donors and international agencies and charities are pivotal so that emergency arrangements can be established without delay and hindrance. These bodies need to deal with those individuals in the recipient countries truly knowledgeable about the capacity and specific problems facing food and water security, and fast-moving internal developments. The experts in the recipient countries must have the authority to be able to (a) quantify the levels of demand, (b) direct supplies, (c) recommend the siting of depots and distribution centres, and (d) highlight points of accessibility and vulnerability. In poorly governed countries, experts must be prepared to deal directly with these donors, agencies, and charities, difficult as that might be. The complexities of globalisation extend beyond food and water security (Lerche 1998).

When people are deliberately persecuted, and honest law enforcement collapses, then non-partisan protection must be afforded, usually with outside security forces. Unfortunately, ideological differences mean that the international community has often been shown to be ineffective in bringing about rapid termination of conflicts by imposing observers or armed forces, although thanks to relatively few major international donors, humanitarian relief has been forthcoming, albeit frequently late and inadequately funded. Dealing with refugees and displaced people requires expertise and sympathetic support. Housing provision together with monitoring and combatting infectious diseases, usually run in parallel with the issuance of food supplies. Governments that prepare for worst-case scenarios are to be commended. Even the distribution of authoritative guidance for populations in stress would represent a small step in the right direction, as would reinforcing the institutions that bind civil society, such as voluntary rescue and care organisations. Networks of low-temperature clean and secure depots with associated distribution centres should be set up at the outset of disasters and conflicts. Even in peaceful times, a marked cut in food waste helps food security. According to M. M. Rutten (Rutten 2013) around a third of the food for humans produced annually (about 1.3 billion tonnes) is either lost or wasted, and in developing economies the situation tends to be worse, with in excess of 40% lost during harvesting, processing and storage (FAO. 2011. Global food losses and food waste – Extent, causes and prevention (See www.go.nature.com/um7vga).

Basic needs of refugees, as recommended by the UNHCR and related organisations, are modest but are directly applicable to those displaced or besieged in their own country. UNHCR recommends each refugee receive more than 2100 calories per day, recognising that a lack of food variety and inadequate supply of fruit and vegetables lead to deficiencies in essential vitamins and minerals. Calorific intake can be reduced if the provided foodstuffs do not conform to traditional diets, or if the rations are traded to acquire other non-food goods and services. Encouragement is needed to set up temporary gardens. Fresh-water provision is of primary importance, with a minimum of 20 litres per person. A greater volume is needed, though, to prevent public-health problems of diarrhoea, cholera, and even polio. Thus, clean-water sources and pumps are required along with taps within walking distance. Vessels are needed for transfer and storage of water. Water-purification tablets should be provided. Sanitation systems are essential for hand washing and the safe disposal of urine, faeces, sanitary towels, wound dressings, infected and contaminated materials, and for the disposal of dead bodies. Monitoring of faecal contamination is recommended. Housing refugees and displaced people at short notice demands special expertise to avoid overcrowding and give adequate protection against inclement conditions. Overlaying the fundamental needs for food, water, and shelter are meeting basic medical needs, particularly of the young, women, and the old and frail. In addition, within a short time, children require to be educated.

Host communities and host countries sometimes resist integration of forcibly deracinated people and can grow resentful at the costs incurred, especially if the host economy is weak. Most financial assistance from donor countries is given to aid agencies rather than host countries. Large-scale influxes of refugees can soon overwhelm the host country's infrastructural resources (chiefly fresh water, energy, housing, hospitals and healthcare systems, education, and waste disposal). Other problems arise from combatants embedded in refugee cohorts, spreading the conflict and increasing policing costs. Cultural incompatibilities between refugees and the host population create hostilities. Refugees can suffer the dire consequences of being rendered stateless. In general, it is fair to say that humanitarian care is not able to sustain basic needs in the medium to long term. As a consequence of a funding crisis for humanitarian aid in the Arab Middle East, the World Food Programme was forced to suspend its desperately needed food-aid-voucher scheme for more than 1.6 million Syrian refugees at the beginning of December 2014, the onset of winter. This suspension meant that refugees were less welcome in host countries and border closures are already being implemented in the immediate area as well as in the European Union.

Axiomatically, just as responsible governments must be alert to and prepared for civil and other forms of unrest, they should always promote food production and remove any impediments to the uptake of improved technologies so that their economies have inbuilt resilience to dreadful events. Likewise, governments should have in their ranks, or instantly available for consultation, competent scientists, technologists, and engineers able to advise on food, water, and energy resource distribution and allocation. Over the past few decades, public-sector agricultural research and development in virtually all countries have suffered financial reductions and financial resources have been switched to activities regarded as more exciting and with greater wealth-creating potential; history shows this to be monumentally misguided. The urban disregard for agriculture is likely to continue as urbanisation increases, until the point food security threatens social stability. Active or benign neglect of food-producing, food-processing, and food–distribution industries as well as of the scientists, technologists, and engineers underpinning its productivity, improvement and efficiency reveal incompetent governance. As an aside, the dearth of scientists, technologists, and engineers in active politics accounts for numerous policy failures. Graduates in the arts (such as history and politics) and social sciences dominate politics and the upper echelons of the machinery of government (civil service) worldwide, people with little understanding or appreciation of business let alone of the "hard" sciences and engineering and their essential utility (and limitations) for mankind. Perhaps this explains the growing dissatisfaction with the prevailing political classes. The scientific approach is that of the quest for knowledge by constantly questioning, developing and testing hypotheses by experimentation so that opinions change as "facts" change, oftentimes undermining policies that are not evidence-based, whereas many political parties are founded on inflexible belief systems, as are almost all religions.

One aspect of food security in times of conflict and community disharmony has been the remarkable resilience of researchers to continue their studies or just maintain libraries, databases, records, laboratories, and genetic resources under the most trying conditions. The pursuit of knowledge is a fundamental feature of humans, as is the search for improvement. When there is blatant disregard of national constitutions as well as United Nations treaties, protocols and conventions, and universities, colleges, schools, and research institutes become targets of malevolent forces, then the rest of the world must have no other option than to intervene, regardless of diplomatic niceties, in order to restore at least the vestiges of societal normality. As a first step, food security and the provision of fresh water for the besieged people must be a priority.

If and when particularly large, heavily populated countries become embroiled in conflicts and/or major natural disasters, the existing international support efforts are likely to fail. This, in turn, may lead to a series of related conflicts, as opposing ideological pressures culminate in outright wars, invasions, and suffering on a huge scale. Throughout the world, history has shown that unless they are relatively rich (and that may not be enough), smaller or militarily weak larger countries are influenced, for good or ill, by their more powerful and sometimes aggressive neighbours. As recent events demonstrate, conflicts in smaller countries rarely bring about rapid corrective measures from the international community, and adverse and damaging propaganda actively promoted in donor countries can prolong the suffering. Ultimately though, food and fresh-water security are a prerequisite and eventually underpin stability, peaceful and thriving economies.

Conclusions

Today, much of the Arab world is poorly governed and insecure for its citizens; they urgently deserve a better life. Many of the most talented Arabs seek a better life elsewhere. The warfare must be ended forthwith. Grossly and unfairly misunderstood by much of the rest of the world, Arabs demonstrate admirable resilience and stoicism yet retain their sense of humour tempered by understandable cynicism and justifiable suspicion of conspiracies. Enemies of the Arabs subject them to a tirade of insults and demeaning innuendos, often designed to deny them basic rights and international support. Nonetheless, Arabs must not be the continuing authors of their own misfortune, and a first step would be an end to internal conflicts followed by an effective Region-wide clampdown on corruption at all levels. Remember – it is the victor who determines the writing and shape of history. If the level of insecurity in the Region were to get worse, then not only the Arabs but also the rest of the world would pay a high price, so it is in everybody's interest to help restore peace. Bluntly, the solution to their problems lies in the actions of the Arabs themselves.

In fully grasping the opportunities available through top-quality education (Baydoun and Hillman 2012), high standards of integrity and tolerance can be demanded from those in leadership roles in communities, organisations, businesses, and local and national government. Ignorance can be reduced, even if not eliminated. Essential components of democracy can be established, including independent and diverse news media, an autonomous judiciary operating to high standards of justice and unaffected by pressure groups and politicians, freedom of speech, and dynamic humanities and artistic sectors. Wealth, and security of food, water, and energy, can and must be assured through the knowledge economy. Harmony can be restored to communities suffering deep-seated divisions. Furthermore, countries in the Arab Middle East will then be in a position to interact much more effectively and comprehensively in the international arena so that, if needed, external support and assistance can be fully and timeously harnessed. Despite all the odds, this transition must be accelerated from the current dangerous condition to a much more enlightened and prosperous existence. Education throughout society has proved to be a slow process, and can be resisted by regressive forces and indolence, so responsible leadership is a prerequisite. Arab scientists, technologists, and engineers must contribute actively to this transition, thereby securing a safe, healthy, and buoyant future for all Arabs. Research and development priorities must be reassessed in the light of worsening nexus of water, food, and energy insecurity, and the desperate need to return to peaceful conditions. In legal jargon: time is of the essence. Finally, a buoyant growth potential for the Arab Middle East is dependent on the fundamentals of demography, education, access to capital, technology, careful custodianship of its natural resources and environment, and social stability; all are threatened by this insecurity nexus.

References

Babu, S. C., & Blom, S. (2014). Building capacity for resilient food systems. In: S. Fan, R. Pandya-Lorch, & S. Yosef (Eds.), *Resilience for food and nutrition security* (pp. 119–126). International Food Policy Research Institute. https://www.ifpri.org/publication/resilience-food-and-nutrition-security

Bardshaw, C. J. A., & Brook, B. W. (2014). Human population reduction is not a quick fix for environmental problems. *Proceedings of the National Academy of Sciences, 111*(46), 16610–16615. doi: 10.1073/pnas.1410465111. www.pnas.org/lookup/supp/doi:10.1073/pnas.14110465111

Baydoun, E., & Hillman, J. R. (2012). The future of universities in the Arab World. Arab Academy of Science. www.arabacademyofsciences.org

Bolukbasi, M. F., Gupta, A., & Wolfe, S. A. (2015). Creating and evaluating accurate CRISPR-Cas9 scalpels for genomic surgery. *Nature Methods, 13*(1), 41–50. doi: 10.1038/nmeth.3684. http://www.nature.com/nmeth/journal/v13/n1/full/nmeth.3684.html

Breisinger, C., Ecker, O., Maystadt, J.-F., Trinh Tan, J.-F., Al-Riffai, P., Bouzar, K., Sma, A., & Abdelgidir, M. (2014). Food security policies for building resilience to conflict. In: S. Fan, R. Pandya-Lorch, & S. Yosef (Eds.), *Resilience for food and nutrition security* (pp 37–44). International Food Policy Research Institute. https://www.ifpri.org/publication/resilience-food-and-nutrition-security

Cong, L., Ran, F. A., Cox, D., Lin, S., Barretto, R., Habib, N., Hsu, P. D., Wu, X., Jian, W., Marraffini, L. A., & Zhang, F. (2013). Multiplex genome engineering Using CRIPR/Cas Systems. *Science, 339*(6121), 819–823. doi:10.1126/science 1231143.

Elasha, B. O. (2010). *Mapping of climate change threats and human development impacts in the Arab region* (51 pp.). United Nations Development Programme, Regional Bureau for Arab Affairs. http://www.arabstates.undp.org/content/rbas/en/home/library/huma_development/mapping-of-climate-change-threats-and-human-development-impacts-.html

Fan, S., Pandya-Lorch, R., & Yosef, S. (2014). *Resilience for food and nutrition security* (211 pp.). International Food Policy Research Institute. https://www.ifpri.org/publication/resilience-food-and-nutrition-security

Food and Agricultural Organisation of the United Nations. (2004). *What is happening to agrobiodiversity?* http://www.fao.org/docrep/007/y5609e/y5609e02.htm

FAO. (2011). *Global food losses and food waste – Extent, causes and prevention.* Rome. www.go.nature.com/um7vga

FAO. (2015). *The impact of natural hazards and disasters on agriculture and food and nutrition security. Action to build resilient livelihoods.* Prepared by the Food and Agricultural Organization of the United Nations (FAO) for the World Conference on Disaster Risk Reduction. *Sendai, Japan,* 14–16 March 2015. http://www.fao.org/3/a-i4434e.pdf

Garland, P., Raftery, A. E., Ševčikovà, H., Li, N., Gu, D., Spoorenberg, T., Alkema, L., Fosdick, B. K., Chunn, J., Lalic, N., Bay, G., Buettner, T., Heilig, G. H., & Wilmoth, J.. (2014). World stabilization population unlikely this century. *Science, 346*(6206), 234–237. https://www.sciencemag.org/content/346/6206/234?related-urls=yes&legid=sci;346/6206/234

Grebner, K., Saltzman, A., Birol, E., Wiesmann, D., Prasai, N., Yin, S., Yohannes, Y., Menon, P., Thomson, J., & Sonntag, A. (2014). *Global hunger index: the challenge of hidden hunger* (56 pp.).Washington, DC: International Food Policy Research Institution. http://www.ifpri.org/publication/2014-global-hunger-index

Grivetti, L. E., & Ogle, B. M. (2000). Value of traditional foods in meeting macro- and micronutrient needs: The wild plant connection. *Nutrition Research Reviews, 13*, 31–46. http://www.ncbi.nlm.nih.gov/pubmed/19087432

Hillman, J. R., & Baydoun, E. (2012). Overview of the roles of energy and water in addressing global food security. *International Journal of Thermal & Environmental Engineering, 4*(2), 149–156. http://iasks.org/sites/default/files/ijtee20120402149156.pdf

Hillman, J. R., & Baydoun, E. (2015). An overview of mitigation and adaptation processes and strategies to address the impacts of climate change on food, water, and energy security in the arab middle east. In: International Conference on Climate Change and Water-Energy-Food Nexus in the Arab Middle East. Amman, Jordan. 5-6 December 2015, pp. 15–35

Jinek, M., Chylinski, K., Fonfara, I., Hauer, M., Doudna, J. A., & Charpentier, E. (2012). A programmable dual-RNA-guided DNA endonucleases in adaptive bacterial immunity. *Science, 337*(6096), 816–821. doi:10.1126/science.1225829.

Klümper, W., & Qaim, M. (2014). A meta-analysis of the impacts of genetically modified crops. *PloS One, 9*(11), e111629. doi:10.1371/journal.pone.0111629. http://journals.plos.org/plosone/article?id=10.1371/journal.pone.0111629

Lerche III, C. O. (1998). The conflicts of globalization. *The International Journal of Peace Studies, 3*(1). ISSN 1085–7494. http://www.gmu.edu/programs/icar/ijps/vol3_1/learch.htm

Mabiso, A., Maystadt, J.-F., Vandercasteelen, J., & Hirvonen, K. (2014). Resilience for food security in refugee-hosting communities. In S. Fan, R. Pandya-Lorch, & S. Yosef (Eds.), *Resilience for food and nutrition security* (pp. 45–52). Washington, DC: International Food Policy Research Institute.

Open Budget Index. (2015). *International budget partnership*. www.internationalbudget.org

Organisation of Economic Co-operation and Development. (2014, November 6). http://www.oecd.org/newsroom/comprehensive-action-needed-to-shift-the-global-economy-into-higher-gear.htm

Qadir, M., Quillerou, E., Nangia, V, Murtaza, G, Singh, M. Thomas R. J., Dreshsel, P., & Noble, A. D. (2014). Economics of salt-induced land degradation and restoration. *United Nations Sustainable Development Journal Natural Resources Forum, 38*(4), 282–295. https://sustainabledevelopment.un.org/index.php?menu=1517

Rutten, M. M. (2013). What economic theory tells us about the impacts of reducing food losses and/or waste: Implications for research, policy and practice. *Agriculture and Food Security, 2*, 13. doi: 10.1186/2048-7010-2-13. http://www.agricultureandfoodsecurity.com/content/2/1/13

Task Force on Building a Science Roadmap for Agriculture. (2001, November). *National association of state universities and land-grant colleges, experiment station committee on organization and policy, A science roadmap for the future*. www.nasulgc.org/comm_food.htm

The Economist Commodity-Price Index. (2015). *Economic and financial indicators*. http://www.economist.com/news/economic-and-financial-indicators/21659718-economist-commodity-price-index

Villegas, J. R., Khoury, J., Jarvis, A., Debouck, D. G., & Guarino, L. (2010). A gap analysis methodologhy for collecting crop genepools: A case study with *Phaseolus* beans. *PloS One, 5* (10), e13497. doi: 10.1371/journal.pone.0013497. http://journals.plos.org/plosone/article?id=10.1371/journal.pone.0013497

Soils and Food Security in the Arab World

Isam Bashour

Abstract The area of the Arab region is about 14 million Km^2 ($1.4 * 10^9$ ha) of which about 90% lies in the arid and semi-arid harsh environments and fragile ecosystems. During the past three decades, limited efforts were put to combat soil degradation, and desertification and soil salinization are still the major environmental problems in the Arab region. The annual rapid increase in the population, in addition to the change in life style and the increase in food consumption, led to the widening of the void between production and consumption of food in the region.

Arable land per capita is one of the lowest in the world because only less than 10% of the total area is cultivated and agricultural technology is still basic with relatively low productivity.

Investment in agricultural and water projects in the region should be encouraged to increase production and reduce dependence on imported products. Improving the conservation agriculture (CA) technique by conducting scientific research and field trials on farmers' land, along with the dissemination of results and experiences is one way to reduce soil degradation and improve food security in this region. The promotion and spread of CA is crucial, yet challenging, for scientists and decision makers in the Arab world.

Keywords Soil salinity • Food security • Desertification

Introduction

Historians place the beginning of agriculture in Mesopotamia to 7000 years ago. About 3000 years ago, the Phoenicians were cutting the cedars of Lebanon, leaving eroded rocky barren mountains. In the eighteenth century B.C. Hammurabi of Babylon, left description of brick-lined canals of a system encompassing 25,000 Km^2, which supported 15–20 million people, silt deposits, plugged the canals which were eventually abandoned. In ancient Egypt, agriculture and civilization

I. Bashour (✉)
Department of Agricultural Sciences, Faculty of Agricultural and Food Sciences, American University of Beirut, Beirut, Lebanon
e-mail: ib02@aub.edu.lb

S. Murad et al. (eds.), *Water, Energy & Food Sustainability in the Middle East*,
DOI 10.1007/978-3-319-48920-9_13

flourished along the Nile. The flooding of the Nile was a blessing with its new soil and washing the residual salt from the soil away. In the first century A.D. plains in Syria, Lebanon and northern Africa were the bread basket of the Roman Empire. By the middle of the third century the flow of nutrients out of the soil and the decline of soil fertility set the course for environmental and economic instability in the exploited regions (Gardiner and Miller 2008).

One can't overstate the value soil and the many purposes it serves for humans and nature, especially in a region like the Middle East where it takes about 1000 years for the formation of 2–3 cm of soil.

The world population is approaching 7.0 billion. Ninety-five percent of the growth occurring in less developed regions with a fertility rate average about 5 children per woman (1.4 in Europe and 2 children in North America). In the Arab countries population is growing at a fast rate, more than 2.3% per year.

This mushrooming human population in the Arab region imposes an increasing pressure on farmers to produce more food each year. The quantity and quality of food available for the human diet are determined by the area of arable (farmable) land available per person and by the productivity of the land. In the Arab world two countries were able to approach self-sufficiency in food production, Syria and Iraq, but due to the present instability in both countries their food production has dropped drastically and is now similar to the other countries in the region, dependent on the importation of the majority of their food from outside.

Major Soil Types in Arab Countries

Soils are formed from interaction of geology, relief, climate, human activity and time. In the Middle East, soils can be grouped in two main types: arid (desert soils) and semi-arid (mountain/valley soils).

The international soil taxonomy system divides soils of the world into 12 orders, 60 suborders, 294 great groups and 2366 subgroups and many more divisions into families and series. Among the 12 soil orders, 6 orders are the most common in the Arab world: Aridisols, Inceptisols, Alfisols, Entisols, Vertisols, and Mollisols. Following is a map showing major soil orders in the Arab World, followed by a short description and a photo of each soil order.

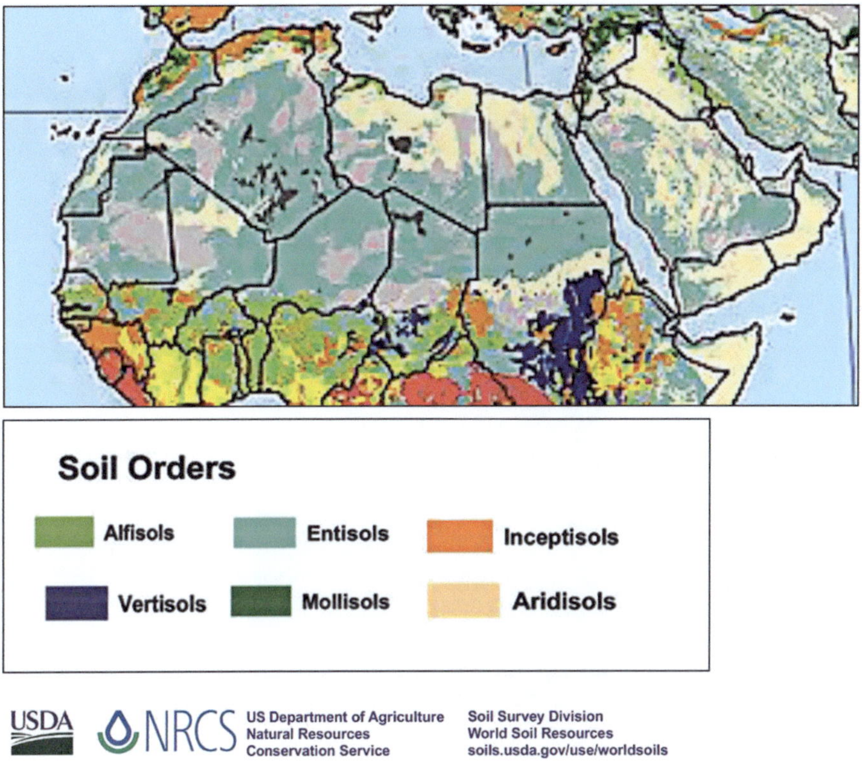

Soil map showing major soil orders in Arab countries

1. Aridisols:

- Aridisols are common in the deserts of the world. They often accumulate gypsum, salt, calcium carbonate and other materials that are easily leached from soils in more humid environments.
- Soils that are too dry for mesophytic plants to grow. The lack of moisture greatly restricts the intensity of weathering processes and limits most soil development processes to the upper part of the soils.

Image courtesy of USDA-NRCS

2. Inceptisols:

- Soils of semiarid to humid that generally exhibit only moderate degrees of soil weathering and development.
- Inceptisols have altered horizons that have lost bases or iron and aluminum but retain some weatherable minerals. They do not have an illuvial horizon enriched with either silicate clay or with an amorphous mixture of aluminum and organic carbon.

Image courtesy of USDA-NRCS

3. Alfisols:

- Soils result from weathering processes that leach clay minerals and other constituents out of the surface layer and into the subsoil, where they can hold and supply moisture and nutrients to plants. They are formed primarily under forest or mixed vegetative cover and are productive for most crops.
- Alfisols are common in semiarid to moist areas.

Image courtesy of USDA-NRCS

4. Entisols:

- Soils that have little or no evidence of development of pedogenic horizons.
- They occur in areas of recently deposited parent materials or in areas where erosion or deposition rates are faster than the rate of soil development; such as dunes, steep slopes and flood plains.

Image courtesy of USDA-NRCS

5. Vertisols:

- Soils that have a high content of expending clay minerals. They undergo pronounced changes in volume with changes in moisture. They have cracks that open & close periodically, and show evidence of soil movement in the profile.
- Because they swell when wet, vertisols transmit water very slowly with little leaching. They tend to be fairly high in natural fertility.

Image courtesy of USDA-NRCS

6. Mollisols:

- Soils that have a dark colored surface horizon relatively high in their Organic Matter content.
- Mollisols form under grass in climates that have a moderate to pronounced seasonal moisture deficit.

Image courtesy of USDA-NRCS

Soil Degradation in the Arab Region

Soil degradation affects large areas in the Arab World. It is described by FAO as "any contamination, loss or alteration that makes the soil less productive". Soil erosion, soil salinization, and desertification are major soil degradation factors in the Arab region.

Soil Erosion

The long history of cultivation in this region makes it difficult to construct the original pattern of vegetation. In the mountains of the Eastern Mediterranean, forests probably were dominant, while on the foothills and low land forests gave away to grassland. Remnants of the forests are still found in some slopes. Since farming was first established, grazing and human activities have changed the vegetation cover, leading to soil erosion and desertification. Much of the natural vegetation that exists today in the arable land of the Arab World is only a degraded remnant of what it once was. The vegetation degradation that has been taking place for a very long time led to soil degradation, formation of dust storms and serious wind erosion in major parts of Arab lands.

Several airports in the Arabian Gulf region, Sudan, Egypt, Libya, and Iraq have to close for a day or more when the *Shamal* wind blows. Most of the desert houses in these areas have no windows in their northern walls to reduce dust from entering during a dust storm. In Kuwait dust storms that reduce visibility to below one kilometer are recorded on average 27 days/year and in north- eastern parts of Syria there are dust storms on average 15 days/year. Egypt and other Arab countries in North Africa are exposed to strong dust storms for several days per year, too. The size of the wind eroded soil particles varies from clay (less than 2 microns) to silt (0.2–0.02 mm) and fine sand (0.02–0. 1 mm). The fine sand is blown along, bouncing up to a few centimeters above soil surface. Silt can be carried up for many meters in the air and the fine clay particles can be carried in the air for several days and be transported for very far distances, reaching Lebanon from the African Desert or the Arabian Peninsula. Dust storms remove valuable soil from agricultural land and young crops can be destroyed when blasting by wind or covered by deposited sand. The disturbance of land surface, mainly by plowing, in dry areas increases the severity of wind and water erosion. Plowing helps breaking the soil structure and buries vegetation (natural soil cover) that would otherwise protect the top soil from wind and water erosion. These detrimental effects are particularly marked in dry farming areas in Lebanon, Syria, Jordan and Iraq. The already high rates of soil erosion in this area are expected to increase in the future due to land cover degradation and precipitation changes. Soil erosion removes the most fertile portion of the soil. Knowing that more than 95% of our food originates from the land, we must respect and protect this resource and avoid degrading it. As Winston Churchill said, "To build may have to be the slow and laborious task of years. To destroy can be the thoughtless act of a single day."

Desertification

FAO defines desertification as "land degradation in arid, semi-arid and sub-humid areas resulting from various factors, including climatic variations and human activities." According to the Arab Center for the Study of Arid Zone and Dry Land (ACSAD), most of the Arab region falls within the boundaries of arid land; in fact degradation of dry lands affects about 70% of land in the Arab World; around 49% of land area in Mashreq, 29% in the Nile valley, 17% of north Africa and 9% of Arabian Peninsula is endangered on account of desertification. The countries facing the greater danger are Libya, Egypt, Jordan, Bahrain, Kuwait, Qatar and the UAE.

Desertification is a dynamic process in various levels of intensity that is affected by various environmental factors including drought, land degradation, natural resources depletion, population growth, increased temperature, decline in precipitation, scarcity of irrigation water, soil erosion and salinization.

Actions dealing with desertification in the Arab world are mostly monitoring exercises concentrating on evaluating the damages of desertification. The efforts and resources devoted to combating desertification in the Arab world and the Middle East, in general, are less than that required to tackle this growing problem. (Al-Zubi and Keough 2013).

The economic costs of environmental degradation are high in the Middle East. According to the World Bank, they vary from 2.1% of gross domestic product (GDP) in Tunisia, to 7.1 of GDP in Iran.

Programs should be developed for agricultural and range lands. These programs should encompass the collection of biological and physical data, providing information on domestic and wild animal, human population, vegetation production, plant cover, climatic data, and soil characteristics and degradation. The ecological monitoring should also provide information on socioeconomic practices and human population and behavior which affect the ecosystem and productivity.

Soil Salinization

Soil and water salinization has been identified as a major process of degradation in most Arab countries. The detrimental effect of soil salinization is prominent in the Arabian Peninsula, Egypt, Iraq, Syria, Libya and Jordan. To improve production in the Arab world, large areas of arid land was put under irrigation in regions where water contains moderate to relatively high concentrations of soluble salts. The extension of irrigation to arid land led to an increase in the area affected by shallow water table and the extension of the hazard of soil salinity. This is because irrigation water brings in additional salts and releases immobilized salts in the soil through mineral dissolution. Losing water through evaporation leads to the concentration of salts in soil solution. Sea water intrusion at coastal areas is another source of salinization in coastal zones such as the Delta in Egypt, the narrow coastal strip of land in Lebanon, and the coastal agricultural land along the Mediterranean Sea in North Africa.

High levels of salts in soil reduce plant growth and at certain levels it may restrict it entirely. However, salt affected soils can be reclaimed and high crop productivity can be attained if the soil salinity problem correctly diagnosed and appropriate reclamations and management practices are adopted.

Investment in Agricultural Projects in Sudan and Egypt

The past 10 years have seen a surge of investment in agriculture in Sudan and Egypt by investors, mainly, from the Gulf States. Chinese, European and North American companies are also exploring opportunities in several African countries, motivated by the potentially high returns on investment (Hallam 2009). Due to the lack of water that is available for agriculture, in most of the Arab countries, increasing food production to reach self-sufficiency is not a fully attainable option. Therefore, Arab countries have been depending on food imports for more than 70% of their consumption. For the rich Arab countries, the concern is not so much the price of imported food as its availability, where as in times of crises, major exporters may put restrictions on food export (Hallam 2011).

The lack of investment in sustainable agriculture in the Arab World over decades has led to the low production in many Arab countries, mainly Egypt and Sudan. The civil wars and instability in Iraq and Syria have led to a marked reduction in food and feed production in both countries, forcing them to depend on importing the majority of their needs from outside the Arab countries.

The large private investment in Sudan and Egypt are important developments in these two countries and could make significant contribution to their production and stability. Unfortunately, no detailed information is yet available on these investments and their impact on the investors, local people, and food security in the region. The challenge is what can be done to optimize the benefits for all involved and encourage the establishment of additional mutually benefitting investments.

From the limited published literature (Hallam 2011) and the information that is available through my providing technical assistance in the field of soil science and plant production, for a few investors in Sudan, Egypt and Ethiopia, the following points may be mentioned:

1. The number of agricultural projects/investments has markedly increased during the past 15 years; major investors are from the Gulf States and China. However, the total area of leased land is still minimal relative to total of cultivated land. (It is less than 1% of the cultivated land.)
2. Land acquisition is mostly contracted on large areas on a long-term lease for 99 years. The leased land areas may differ in size from 2000 ha to more than 50,000 ha.
3. Contracts are drawn between investors and governments. Time consumed in negotiations and finalization of agreement between investors and government is often longer than it should be.
4. Delays to start the actual operations vary among investors and can be long in many projects. Some projects move within a few years and others are delayed for than a decade.
5. The idea that investing of Gulf States in agriculture in other Arab countries like Sudan and Egypt, where the land and water constrains are not present, is a good strategy that may lead to less dependence on food importation.

6. Major benefits of these investments to the local people are: employment opportunities, technology transfer, and structural developments. The realization of these benefits depends on the positive cooperation between investors, government and the local people.
7. The Arab region is one of the most arid regions in the world. The area of arable land is estimated to be about 550 million hectares, but only some 13% is cultivated with a production of about 65% of the world level. These constrains could be reduced by increasing investments in agriculture, applying up-to-date technological production methods, and improving the system of governess and cooperation between the Arab countries.
8. Local people in host countries often question, "Why foreign investment? Why not local investors or local government development projects?" They often consider the investment an excuse for "land grab". Alternatives to land acquisition, is controversial and carries a number of inherent risks. Other forms of investment such as joint ventures or contract farming can, in principle, offer just as much security to investors.
9. Regardless of the type of legal arrangement between investors and host countries, long-term leases of governmental lands are one option to increase land cultivation and production in the Arab world that should be encouraged and protected by new laws that are fair to all involved parties, especially considering the rights of the local population that are requested to move from a region to another or at least lose the right of grazing on a large piece of land that has been available to them for generations is now being negotiated for investment.
10. Investors should provide educational, health, and training programs in addition to employment of the local population.

Conservation Agriculture (CA)

Conservation agriculture (CA) is often referred to by several names: no-till, direct drilling or direct seeding. It is defined as a concept for resource-saving agricultural crop production that strives to achieve acceptable profits, high sustained production levels, while conserving the environment. CA is characterized by three principles, which are linked to each other.

1. Minimum soils disturbance
2. Keeping a permanent soil cover
3. Applying suitable crop rotation

During the last 15 years CA has been gaining popularity all over the world. It is now applied on more than 150 million hectares. The FAO and other international organizations have been introducing and promoting CA in Africa and Asia. About 90 years ago agricultural research in the United States showed that keeping plant residues on soil surface (no-till) reduces run-off and evaporation. In the Arab world Tunisia was among the first countries to research direct drilling in 1999 (Balloume

2007). In Syria, Lebanon and Jordan this technique was introduced by German Corporation for International Cooperation (GIZ) in a joint program with The Arab Center for the Studies of Arid Zones and Dry Lands (ACSAD) in 2007 and the American University of Beirut (AUB) was the technical partner in Lebanon. Several field plots were conducted in various locations in Lebanon in cooperation with 64 farmers and the obtained results were encouraging (Bashour 2007). In 2014 a promotional program in cooperation with the extension department at the Ministry of Agriculture was initiated. This practice was one of the 18 winners out of 749 programs submitted from 140 countries to the Feeding Knowledge Program on Best Sustainable Development Practices on Food Security in Milano Expo 2015.

The climate in the Arab countries is hot and dry in the summer and cold and wet in the winter. The rain is irregular, sudden and intensive. The soil types are arid and semi-arid in more than 90% of the entire area of about 14 million $Km^2 = 1.4*10^9$ ha. This makes the area among the most fragile and unstable ecosystem in the world.

The rapid population growth (2.4% per annum) increases the pressure on the natural resources (AOAD 2010) and consequently the degree of degradation of land and water resources to reach alarming levels. This situation implies the need for a compromise between sustainable agricultural productions that conserves the environment and provides good income to farmers at an acceptable productivity with a feasible cost. Conservation agriculture could be the solution for agricultural production in the Arab world without degrading the natural resources.

The statistics on the use of water resources in the Arab region indicate a volume of about 195 billion cubic meter. Agriculture accounts to about 85% of the use. Despite this high use of water for agriculture, more than 60% of agricultural products are imported from outside the Arab region. The efficiency of water use is about 40% (the ratio between crop water requirement and total water supplied to crop), this is mainly due to surface irrigation and the low level of farmers' education. The increase in water use efficiency in agriculture is of a first priority.

The area of arable land in the Arab countries is about 197 million hectares with 70 million hectares are cultivated (constitutes about 36.5% of the arable area); about 15% of which irrigated (AOAD 2010). The results of research conducted in Tunisia, Syria, Lebanon and other countries in the middle East and Mediterranean basin show that no-till increases soil moisture, decreases soil erosion, increases variety of life in soil and it may increase herbicide usage, especially in weed infested soil (Nardali 2007 and Bashour 2007). The results show also that CA is an effective management technique which led to the improvement of soil quality and fertility as well as yield stability in the dry Mediterranean region, Morocco (Mrabet 2003), Tunisia (Ben Moussa et al. 2010), Iraq (Piggin et al. 2011) and Syria (Al-Ouda 2013).

The above information indicate that CA could be a solution for agriculture without degrading the natural resources. The dissemination of CA in the Arab World needs a lot of work and effort. It requires governmental support, participative research programs to be conducted on farmers land in order to deepen the knowledge about this new system and convince farmers of its benefit to them by increasing their net profits.

Conclusion

The rising food demand in the Arab World against the increase in cost of energy, production input, land degradation, and climate change calls for an increase in agricultural production in a sustainable way. This involves producing more from the same area of land while reducing the negative environmental impact.

Improving the conservation agriculture technique by scientific research and practice, along with the dissemination of results and experiences are crucial and challenges for scientists and decision makers in the Arab world.

Investment in agricultural activity and water resources should be encouraged to improve production and reduce dependence on imported products. These large investments may drive herdsmen from their land. It is the responsibility of government, to treat these people fairly and provide an alternative source of income, for these poor people. Investors should also comply with the ethical responsibility of providing employment and training to improve their life.

Land acquisition and development in Egypt, Sudan, along with Syria and Iraq, in the future, should be encouraged to increase production, but protected by suitable laws to ensure the rights of the local population, striving to move this region closer to food security.

References

AL-Ouda, S. A. (2013). Effect of tillage systems on wheat productivity and precipitation use efficiency under dry farming system in the Northern East of Syria. *The Arab Journal for Arid Environments, 6*(2), 3–11.

Al-Zubi, M., & Keough D. (2013). *Responding to climate change impacts: Water, energy, and food nexus in Arab cities*. Poster paper, Faculty of Environmental Design, University of Calgary, Canada.

AOAD. (2010, April 29–30). *An overview of agricultural development and food security in Arab Countries* (pp. 4–8). Vienna: Joint Technical Meeting Arab Organization Group Institute and the OECD Development Assistance Committee.

Bashour, I. (2007). Impact of conservation agriculture on soil fertility in dry regions. In *Proceedings, international workshop on conservation land management to improve the livelihood of people in dry regions* (pp. 111–119), ACSAD, Damascus, Syria.

Belloume, A. (2007, May 7–9). Conservation agriculture in the Arab region between concept and application. In B. Stewart, A. Fares Asfary, A. Belloum, K. Steiner, T. Friedrich (Eds.), *The Proceedings of the international workshop on conservation agriculture for sustainable land management to improve the livelihood of people in dry areas* (pp. 11–24), ACSAD & GTZ, Damascus, Syria.

Ben Moussa-Machraoui, S., Errouissi, F., Ben-Hammouda, M., & Nouira, S. (2010). Comparative effects of conventional and no-tillage management on some soil properties under Mediterranean semi-arid conditions in north-western Tunisia. *Soil and Tillage Research, 106*, 247–253.
Gardiner D. & Miller R. (2008) *Soils in our environment* (7th ed., pp. 11–15). Columbus: PEARSON.
Hallam, D. (2009). *International investments in agricultural production.* Paper presented at Expert Meeting on "How to Feed the World in 2050," Food and Agriculture Organization of the United Nations, Rome, Italy. Retrieved April 9, 2010, from ftp://ftp.fao.org/docrep/fao/012/ak976e/ak976e00.pdf
Haggett, P. (2001). *World encyclopedia of world geography. Physical geography, soils and vegetation.* New York: Marshall Cavendish, p 2061.
Hallam, D. (2011). International investment in developing country agriculture—issues and challenges. *Food Security Journal, 3* (Suppl. 1): S91–S98. In P. Liu (Ed.), *Impacts of foreign agricultural investment on developing countries: Evidence from case studies.* FAO, Rome, 2014.
Marbet, R. (2003). No-tillage as technology to prevent desertification. In *Proceeding II world congress on conservation agriculture* (pp. 182–193). Iguassu Falls, Paranama-Brazil.
Nardali E.T. (2007). *No- till farming, effect on soil, pros and cons and potential.* Agriculture issues and policies series. New York: Nova Science publishers, Inc. Online at: http://www.fao.org/ag/ca/6c.html
Piggin, C., Haddad, A., & Khalil, Y. (2011, September 26–29). Development and promotion of zero-tillage in Iraq and Syria. In *Proceedings of the 5th world congress on conservation agriculture* (pp. 304–305), Brisbane, Australia.

Rainfed Agriculture and Food Security in Dry Areas

Kamil Shideed

Abstract The world's ability to produce enough food to feed the growing population is further constrained by water scarcity, particularly in dry areas. Water is an increasingly scarce resource and the FAO estimates that nearly 1.8 billion people will be living in countries or regions with absolute water- scarcity by 2025.

The problem faced by people and countries in dry areas amounts to more than resource scarcity. It is a combination of resource limitations, land and water degradation, and the low efficiency of resource use. Under conditions of resource limitations in dry areas – particularly water – future increases in productivity and production for improving food security and ensuring environmental quality, need to come from enhancing the efficiency of resource-use – rather than using more inputs or increasing the food production area.

The challenges of feeding the growing population under the conditions of climate changes, shortages of water for irrigated agriculture and degradation of arable land are increasing the demand to improve grain production from rainfed areas. The contribution of rainfed farming to food security in dryland countries can be substantially enhanced through increased adoption of currently available technologies supported by enabling policy and institutional environments. Rainfed farming can contribute more significantly to achieve new targets of food security if desired investment levels are realized. On-farm results show the huge potential for improving land and water productivity and profitability of smallholder rainfed agriculture. Yield gap in rainfed crops remain large enough to suggest considerable scope for increasing achievable yields.

Applied agricultural research-for-development suggests the following strategies for producing more food with less resources, particularly in rainfed areas of developing countries which are characterized by resource-poor small-holder farming systems: closing the yield gap of rainfed crops, enhancing adoption of improved technologies, promoting sustainable intensification and diversification of production systems, strengthening innovation systems, reducing vulnerability and managing risk, encouraging the use of water saving technologies, informing policy development, and increasing investment in agricultural research and development.

K. Shideed (✉)
ADG-ICC, International Center for Agricultural Research in the Dry Areas (ICARDA), 950764, Amman 11195, Jordan
e-mail: k.shideed@cgiar.org

S. Murad et al. (eds.), *Water, Energy & Food Sustainability in the Middle East*,
DOI 10.1007/978-3-319-48920-9_14

Advances in science to produce improved and higher-performing crops and livestock hold exciting prospects for making dryland food production systems more efficient, and more resistant to climatic pressures and new pests and diseases. This chapter illustrates the huge potential of technological innovation to improve food security, but also the need for supportive policies and institutions to encourage farmers to adopt these innovations.

Keywords Climate change • Water shortage • Rainfed farming • Food security • Resource use efficiency

Introduction

Despite remarkable growth in agriculture in the past 50 years, the world still faces a critical challenge: how to feed a population expected to reach nine billion by the year 2050. The goal for the agricultural sector is no longer simply to maximize productivity, but to optimize across a complex landscape of production, rural development, environmental, social justice and food consumption outcomes (Pretty et al. 2010). In dry areas, where water is the most limiting factor, achieving this goal will require coordinated action on several issues. First, to optimize the use of scarce water resources while continuing to increase food production. Second, to develop national, regional and international policies that support the development of more productive and sustainable food production systems. Third, to narrow the (widening) gap between potential and actual yields, i.e. what is possible using available technology versus what farmers are actually harvesting? Fourth, to diversify farming and systems, with intensive but sustainable production methods.

A recent study on the "water footprint of humanity" shows that water used by the agricultural sector accounts for nearly 92% of annual global freshwater consumption, the largest proportion of which is the green water footprint (FP) accounting for 74% (Hoekstra and Mekonnen 2012). In the agricultural sector, 19% of the total WF relates to production for export. Industrial production contributes 4.4% to the total WF and domestic water supply 3.6%. The global annual average WF related to all sectors (agricultural, industrial and domestic supply) is 74% green, 11% blue and 15% gray.

Farmers generally over-irrigate as a result of their perceptions of water requirements and their expectations of rainfall and market values. Studies in Syria, Iraq, Jordan and Egypt – by ICARDA and the UN Economic and Social Commission for West Asia – showed that farmers over-irrigated wheat by 20–60% (Shideed et al. 2005). Producers perceive water as a fixed input in the short run, but allocable among competing crops on the farm. As water prices were highly subsidized, they did not have a major impact on water allocation.

Water use efficiency is low in many areas. For example, it is 40–60% for irrigated agriculture in Syria (MunlaHasan 2007). This low figure is due to the widespread use of traditional surface irrigation methods with their low efficiency, high seepage and evaporation losses and uneven field coverage. A recent ICARDA study (Yigezu et al. 2011b) measured the economic and environmental impacts of wheat farmers in three provinces in Syria if they would shift from traditional supplemental irrigation to improved supplemental irrigation (ISI). At the current adoption level (22.3%), ISI helps save at least 120 million m^3 of water per year. Introducing a water use charge of US$ 0.20 for every cubic meter applied in excess of the recommended application rate can lead to near-universal adoption and conserve an additional 46 million m^3 of water per year (Yigezu et al. 2011a). Such a policy would also increase total farm profits by US$ 16.14 million per year, and generate a total yearly impact of US$ 36–90 million. However, introduction of water pricing in dry areas remains a sensitive political issue.

Climate change is amplifying food security challenges, as it impacts all aspects of food security. It affects crop yields, the availability and distribution of freshwater and rainfall, and food prices (Vermeulen 2014).

In summary, the problem faced by people and countries in dry areas amounts to more than resource scarcity. It is a combination of resource limitations, land and water degradation, and the low efficiency of resource use. Under conditions of resource limitations in dry areas – particularly water – future increases in productivity and production for improving food security and ensuring environmental quality, need to come from enhancing the efficiency of resource-use – rather than using more inputs or increasing the food production area.

A wide range of technology solutions is available. Enabling policies that encourage wider adoption and higher resource-use efficiency are critical to achieve productivity growth and food security targets.

As a result of rainfall variability, temperature fluctuations and frequent drought, farming in rainfed areas is highly risky and unpredictable, implying that food production in NENA countries is insecure. This is evident in FAO food security indicators (FAO 2013). Two important indicators of food security vulnerability are the cereal import dependency ratio and the value of food imports in total merchandise exports. These data indicate that the cereal import dependency ratio of all NENA countries is 73.9% as an average of 1990–2011 period, which is the highest globally. The world cereal import dependency ratio is 15.7% and that of developing regions is 15.5%.

These data clearly indicate that the NENA countries are the most vulnerable in terms of their food security globally. As such, these countries are the most vulnerable to shortages of food supply and fluctuations of food prices in the international markets. This situation is further complicated by the fact that the value of food imports accounts for nearly one-third of total merchandise exports, while food imports represent only 5.6% of the world total merchandise exports.

Some 80% of the world's cultivated area is rainfed and produces 62% of the staple food (Haddad et al. 2011). Likewise, NENA countries depend mainly on rainfed agriculture where most of cereal production is taking place. However, investments in rainfed agriculture do not match the high and increasing importance of rained farming currently contributing to food production. Rainfed agriculture has been neglected in favor of irrigated agriculture over the past five decades.

The contribution of rainfed farming to food security in NENA countries can be substantially enhanced through increased adoption of currently available technologies supported by enabling policy and institutional environments (Khouri, Shideed and Kherallah 2011). Rainfed farming can contribute more significantly to achieve new targets of food security if desired investment levels are realized. On-farm results show the huge potential for improving land and water productivity and profitability of smallholder rainfed agriculture.

Previous results show increasing TFP growth at the world level as a result of improved productivity performance in developing countries. The wide variation in agricultural TFP growth rates among countries is explained by differences in national capacities in agricultural and industrial research. National capacity for agricultural R&D is the most important factor explaining the sustainable long-term agricultural productivity growth in some countries and not in others. Countries with national research systems capable of continuously producing new technologies adaptable to local farming systems generally achieve higher growth rates in agricultural TFP. In addition, actively collaborating with international research institutions facilitates spillover impacts of technology and thus significantly raises returns to national agricultural research investments. The presence of an enabling environment that encourages the uptake and adoption of new technologies and practices is another factor that explains cross-country differences in agricultural TFP.

Global agricultural output growth peaked at 2.7% per year in the 1960s and maintained its annual growth at 2.1 and 2.5% for every decade in the last 40 years. An important observation associated with this growth is the shift in the source of output growth from being primarily input driven to productivity driven (Fuglie 2012). Growth in total inputs declined from 2.5% in the 1960s to 0.7% in the 2000s, whereas annual TFP growth rose from 0.2 in the 1960s to about 1.7% since 1990 (Fuglie 2012).

The steady growth in total yield of 2.1% per year during the past five decades has driven the trend in the annual output growth. The growth rate of cereal yield, however, has slowed down after 1990. Its annual growth rate decreased from 2.5% in the 1970s and 1980s to 1.3% in the 1991–2009 period. But this decline in cereal yield does not significantly affect the growth of agriculture as it has been offset by productivity improvement in other commodity groups.

> The challenges of feeding the growing population under the conditions of climate changes, shortages of water for irrigated agriculture and degradation of arable land are increasing the demand to improve grain production from rainfed areas (Anderson, Johansen and Siddique 2015).

The Context: The State of Food and Water Security Globally, Dry Areas, and the Arab Middle East Countries

This section highlights major trends and future projections and draw policy, development and research implications. Human population is expected to grow from its present level of nearly 7 billion to more than 9 billion by 2050. Meanwhile, global food insecurity is intensified, leading to the "double burden" of both obesity and malnutrition due to the fact that high-calorie food is cheaper and more readily available than the nutritious food required for healthy growth (http://www.independent.co.uk/news/world/). The challenges ahead of us are a combination of increased instability in food supplies and changes in the diversity and types of available food. Food affordability and quality are the two main reasons behind the "double burden' as poor families do not afford buying nutritious food such as dairy, eggs and fruit and are mainly dependent on oil and cereal-based carbohydrates which are rich in calorie.

Because the climate change will affect different crops differently in different parts of the world, food insecurity cannot be precisely projected. But there is a consensus that food production under current levels of productivity growth will not meet the growing demands for food. This imbalance between food supply and demand will result in increased food insecurity. However, science and technology can reverse this trend if appropriate investments in agricultural research and development be made available by all concerned. Therefore, the real challenge is no longer on the production of enough food, it is rather on the type and accessibility of the food that is produced.

The world's ability to produce enough food to feed the growing population is further constrained by water scarcity, particularly in dry areas. Water is an increasingly scarce resource and the FAO estimates that nearly 1.8 billion people will be living in countries or regions with absolute water- scarcity by 2025.

Water is also the subject of increasing competition between different sectors, with each seeking to exploit this "precious resource" for its own specific interest and gain. Rapidly increasing urban areas require more water for domestic use; agriculture sector demands more water to increase food production for growing population; and industrial sector not only creates an additional demand for freshwater limited supplies, but also depletes available resources through pollution. This increased competition on limited water supplies necessitates the need for integrated water management strategy that addresses availability, quality, and access to water for food production and processing, drinking and human use, industrial use, and sanitation in the context of an ecosystem approach (ICARDA 2015). This means dealing with the currently unacceptable levels of water waste, particularly in the agricultural sector- which is the largest consumer of water worldwide, and removing inefficient practices. Introduction of supplemental and deficit irrigation measures represent feasible options for unlocking rainfed agriculture to achieve higher water productivity. Rainwater harvesting is another proven option- exploiting indigenous knowledge to construct small, cost-effective, and practical storage structures.

However, while these measures will help improve access to water, they alone cannot solve the crisis that now confronts the world's smallholder farmers, producers of more than 70% of the world's food. Beyond these technical solutions, there is a need for a "rights-based" approach to water management at the international and national level, which commits states to equitable provision and the reliable supply of water for food production, with clear prioritization to the needs of small farmers.

Fortunately, the United Nations have already recognized two opportunities: the right to safe drinking water and sanitation, and the right to food. The first entities everyone, without discrimination, access to sufficient, safe, and affordable water for drinking and sanitation; the second realizes the right to adequate food in the context of national food security. Both are interrelated as safe drinking water and sanitation are crucial for health and nutrition, and access to water is indispensable for food production. Yet neither right considers the productive uses of water. By combining these two commitments and building on their complementarities, we can address this shortcoming, and promote a right to water for food security and nutrition.

Adopting a "rights-based" approach helps ensure that smallholder farmers receive their fair share of water, protecting their interests against more powerful organizations and sectors. This will mean a reliable supply of food for rural communities and families depending on them for their food consumption, and ultimately stronger systems for food and nutritional security.

Challenges of the Twenty-First Century Faced by the Dry Areas, and Particularly Arab Middle East Countries as Related to Food and Water Security

Global economic growth is forecasted to be reasonably robust over the next 10+ years. Global trade continues to grow in importance and income gains in fast-developing countries have important linkages to agriculture in other developing countries (Alwang 2016). As incomes and trade grows, important changes are re-shaping global food markets. Demands are shifting toward higher-quality foods and animal proteins continue to replace staples in consumer diets. Liberalized trade has been accompanied by a growing concern for food safety and phytosanitary restrictions on trade (especially with Europe and the U.S.) are replacing tariffs as barriers. Agricultural producers need to be aware of these trends and adjust their practices to meet the demands of maturing global markets.

Cereals are the most important source of food consumption in developing countries (providing about 53% of total calories) and the world as a whole (49%). Per capita food use of cereals has declined gradually (Fig. 1) since its peak in the mid-1990s, but this decline was mainly attributable to income growth

	1969/1971	1979/1981	1989/1991	2005/2007	2030	2050	Comparison 1999/2001	
Kg / person / year							New	Old
World								
Cereals, food	144	153	161	158	160	160	158	165
Cereals, all uses	304	325	321	314	329	330	309	309
Roots and tubers	84	74	66	68	73	77	69	69
Sugar and sugar crops (raw sugar eq.)	22	23	22	22	24	25	23	24
Pulses, dry	7.6	6.5	6.2	6.1	6.6	7.0	6	5.9
Vegetable oils, oilseeds and products (oil eq.)	7	8	10	12	14	16	11	12
Meat (carcass weight)	26	30	33	39	45	49	37	37
Milk and dairy, excl. butter (fresh milk eq.)	76	77	77	83	92	99	78	78
Other food (kcal/person/ day)	194	206	239	294	313	325	285	289
Total food (kcal/person/day)	**7,373**	**2,497**	**2,633**	**2,722**	**2,960**	**3,070**	**2,719**	**2,789**
Developing countries								
Cereals, food	140	152	160	155	159	158	157	166
Cereals, all uses	193	219	229	242	254	262	239	238
Roots and tubers	79	70	62	66	73	78	67	67
(Developing minus China)	62	59	58	64	74	81	62	63
Sugar and sugar crops (raw sugar eq.)	15	17	18	19	22	24	19	21
Pulses, dry	9.3	7.8	7.3	7	7.4	7.7	6.8	6.7
Vegetable oils, oilseeds and products (oil eq.)	4.9	6.4	8.4	10.1	13.1	15.4	9.4	10.4
Meat (carcass weight)	11	14	18	28	36	42	26	27
(Developing minus China and Brazil)	11	12	13	17	23	30	15.4	16
Milk and dairy, excl. butter (fresh milk eq.)	29	34	38	52	66	76	45	45
Other food (kcal/person/ day)	115	130	177	253	279	239	242	242
Total food (kcal/person/day)	**2,056**	**2,236**	**2,429**	**2,619**	**2,860**	**3,000**	**2,572**	**2,654**

Fig. 1 Changes in the commodity composition of food demand (Source: Alwang (2016))

and consumption pattern changes in China and India (Alwang 2016). Wheat food consumption per capita is not likely to grow for the world as a whole, but it will continue to rise in sub-Saharan Africa, South Asia, and, to a lesser degree, in Latin America. It will level off and eventually decline in the Near East/North Africa, and East Asia (Alwang 2016). Growth in wheat consumption in some developing regions will be translated into increases in imports.

The relatively high income elasticity of demand for livestock products (particularly poultry) means that income growth has resulted in substantial increase in livestock consumption globally. The main consequence of income growth in lower-income countries has been diversification of diets. This diversification is most visible as increased consumption of livestock products and vegetable oils (Alwang 2016). This increase in demand for animal proteins puts pressure on feedgrain markets. As a result, global demand for feedstocks is expected to increase dramatically through 2050 (Fig. 2).

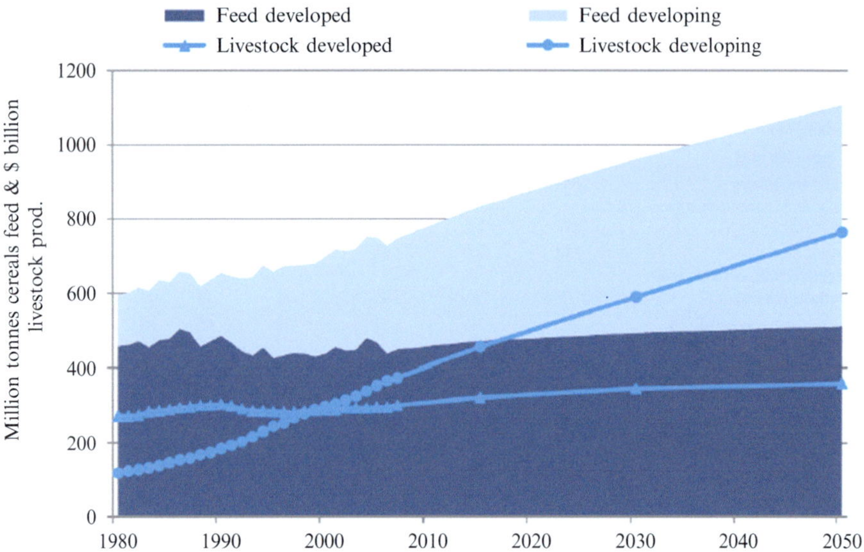

Fig. 2 Demand for livestock feed (Source: Alwang (2016))

Resource constraints for agricultural production have become increasingly binding while rates of growth in cereal yields is slowing. This is a primary reason why people express fears about risks that world food production may not be enough to feed a growing population and ensure food security for all. To understand these concerns, it is important to understand the interplay between supply and demand. Global demand for food will increase as a result of population growth, increased incomes, and use of crops for biofuel production. The supply side is more complicated.

There are three main sources of growth in crop production: (i) expanding the land area under production (horizontal expansion); (ii) increasing the frequency with which land is cropped (which could be through irrigation or introducing crops into fallow periods); and (iii) increasing yields on existing farmland (vertical expansion). According to Alwang (2016), land in agriculture grew by about 1.8% per year in developing countries during 2002–2011; during the same period, factor productivity grew at about the same rate, and increased intensification accounted for about 0.5% per year in agricultural output growth (Global Harvest Initiative 2015). Land expansion into rainfed areas is now limited and irrigation potential is becoming an increasingly binding constraint to production over time (Alwang 2016).

Any country using more than 20% of its renewable resources for irrigation is considered as crossing the threshold of impending water scarcity (Alwang 2016). Already 22 countries (developing but including some in the Central Asia region) have crossed this threshold, 13 of them in the critical over 40% category. It is estimated that four countries (Libya, Saudi Arabia, Yemen and Egypt) use volumes of water for irrigation larger than their annual renewable resources. For these and many other countries, the scope for maintaining irrigated production depends crucially on exploiting whatever margins there exist for using irrigation water more efficiently. This can provide some limited relief in the water scarce regions, particularly in the region that needs it most, the Near East/North Africa. (FAO 2013).

Yield increases thus will come through technical change – resulting from investments in agricultural research. Research outputs include improved crop varieties, means for expanded use of modern inputs, and increased efficiency in use of existing inputs. Of the three sources of productivity growth mentioned above, area expansion is least likely to contribute to output growth in the future. The strength of the other two will depend on investments in research. Irrigation expansion has potential in some areas, but other areas are acutely water stressed and current uses of irrigation are no longer sustainable. If productivity growth does not keep pace, increased land will have to be brought into production to meet global needs. Recent estimates show that this growth has slowed. In fact, the recent gap analysis conducted by the Global Harvest Initiative (2015) states: "TFP growth is not accelerating fast enough to meet the expected agricultural output needs of 2050".

Developing countries account for 56% of world cereal production and this share is expected to increase to 60% in 2050 (Alexandratos and Bruinsma 2012). However, and in response to rising prices, there has been an increase in the rate of growth of world cereal production of 2.3% during 2003–2010, up from −0.1% during 1996–2003.

For CWANA region, major trends are summarized below (Alwang 2016):

Economic Growth Recent patterns of economic growth in the CWANA region have varied. In MENA, the macroeconomic performance of seven countries--Egypt, Tunisia, Iran, Lebanon, Jordan, Yemen and Libya–stands out. Each of these countries experienced rapid economic growth during 2000–2010, with minor setbacks during the 2008–2009 global recession. They then suffered a sharp economic slowdown following 2011, as a result of multiple economic and political factors (World Bank 2015). Looking forward, the macro-economic outlook is mixed and forecasts rely on assumptions about the resolution or continuation of conflict. Conflict in Libya and Yemen is likely to stunt growth in the medium term, but forecasters are more optimistic about the growth recovery in Egypt, Tunisia, Jordan and Iran in 2015. Moreover, the sustainability of growth remains uncertain in MENA as these economies have been long suffering from structural problems which have restrained them from moving to a higher and sustainable growth path (World Bank 2014).

Growth in Central Asia has similarly been strong, hitting a peak of 8.1% in 2011. Since then, growth has fallen somewhat (to 5.5% in 2014 and down to a forecasted 3.1% for 2015). The fall is mainly due to uncertainty about regional tensions. Looking forward and assuming geopolitical tensions are reduced, the economic outlook is reasonably strong. Stronger trade linkages with the EU and the rest of the world are partly responsible for the improved economic outlook. Turkey's growth also peaked in 2011 (at 8.8%) and recently has fallen dramatically, mainly due to conflict in the Middle East and domestic political uncertainty. Future prospects are uncertain, due to continued instability in the region.

Demographic Trends Human development indices in CWANA lag behind what would be expected based on economic indicators (Fig. 3). While regional economic growth rates are very high, this growth is unevenly distributed across and within

Country	Total Population in 1000 (2005)	Rural Population (%)	GDP PPP in CI$ (2003)	Human poverty index (2002)[a]	Human development index (2004)[b]
Afghanistan	25,971	75	n/a	n/a	n/a
Algeria	32,877	40	6,248	21.9	0.728
Armenia	3,043	36	3,607	n/a	0.768
Azerbaijan	8,527	50	3,606	n/a	0.736
Bahrain	755	10	n/a	n/a	0.859
Djibouti	721	15	2,144	34.3	0.494
Egypt	76,117	58	3,960	30.9	0.702
Iran	70,675	32	7,145	16.4	0.746
Iraq	26,555	33	n/a	n/a	n/a
Jordan	5,750	21	4,319	7.2	0.76
Kazakhstan	15,365	44	6,556	n/a	0.774
Kuwait	2,671	4	n/a	n/a	0.871
Kyrgyzstan	5,278	66	1,714	n/a	0.705
Lebanon	3,777	12	n/a	9.5	0.774
Libya	5,631	11	n/a	n/a	0.798
Mauritania	3,068	36	1,896	48.3	0.486
Morocco	32,209	40	4,012	34.5	0.64
Oman	3,020	21	n/a	31.5	0.81
Pakistan	161,151	65	1,971	41.9	0.539
Palestine (West Bank, Gaza Strip)	3,901	35	n/a	n/a	0.736
Qatar	627	8	n/a	n/a	0.844
Saudi Arabia	25,795	12	n/a	15.8	0.777
Somalia	10,742	64	n/a	n/a	n/a
Sudan	35,040	59	2,046	31.6	0.516
Syria	18,651	50	3,575	13.7	0.716
Tajikistan	6,356	76	1,119	n/a	0.652
Tunisia	10,042	36	7,083	19.2	0.76
Turkey	73,301	33	6,749	12.0	0.757
Turkmenistan	5,014	54	5,884	n/a	0.724
United Arab Emirates	3,107	15	n/a	n/a	0.839
Uzbekistan	26,410	63	1,737	n/a	0.696
Yemen	21,481	74	889	40.0	0.492
Total	723,628				

PPP = purchasing power parity; CIS = constant USD; n/a = not available
[a]scale of 1 to 100
[b]scale of 0 to 1
Source: World Bank, 2006.

Fig. 3 Population growth and human development indicators in CWANA (Source: Alwang (2016))

countries and due to long-standing under-investments in several social sectors, other social indicators are far below the expected.

Population growth rates vary among CWANA sub-regions and countries. During the twentieth century, in an era of rapid population growth, the populations of Egypt, Iran and Turkey more than quadrupled. Robust economic growth at the end of the twentieth century and through 2011 has reduced these rates of population growth. Population growth rates among countries and sub-regions range from less than 2% in Algeria, Azerbaijan, Caucasus, Iran, Kuwait, Kyrgyzstan, Morocco, Tajikistan, Tunisia, Turkey, Turkmenistan and United Arab Emirates; to 2.9% in the Nile Valley; 3.3% in Jordan, Libya and Saudi Arabia; 3.7% in Yemen and 4.2% in Oman (IAASTD 2009). These growth rates are forecast to fall as incomes grow in the region, but they are still likely to exceed global averages. These growth rates, coupled with increasing incomes, imply that demand for foods will likely grow across the region through 2025.

Poverty in MENA is relatively low. However, this overall picture masks other important dimensions. First, within the region, there are stark differences. Second, the region faces a problem of high (and growing) inequality and this inequality is so severe that there is evidence that it reduces economic growth (Ncube, Anyanwu and Hausken 2013). Third, a high percentage of the population is near the poverty line and vulnerable to falling below it. Large segments live in vulnerable situations and are exposed to external shocks. About 14% of the population (11 million) in Egypt lived on \$2.00 – 2.50 a day in 2010. Moving the poverty line from \$2.50 to \$4.00 a day will imply that the overall poverty rate in Egypt is about 40%. In Yemen, with the highest poverty rate in the MENA region, about one third of the population lived on under \$1.25–\$2.50 a day in 2010. Estimates show that more than 60% of Yemen's population (about 14 million) is currently living below \$2.50 a day and almost all of the population lives under \$4.00 a day.

The original stated purpose of many of the subsidy programs commonly found in the region was to protect the poor. However, these programs have been shown to be wildly inefficient, and many are being scaled down or eliminated over time. MENA lags far behind the rest of the developing world (except Africa and it is comparable to Africa) in terms of access to social safety nets (74% of households receive no transfers at all) (IFPRI 2015).

Political Instability Political instability has characterized the MENA region for quite some time, but the period from 2003–2015 was exceptional. There is no indication that this instability will be abated in the near future. Ongoing conflict in Iraq, Syria, Libya and Yemen are worrying signs.

Outmigration and Remittances Partly as a result of the ongoing political turmoil, regional outmigration has exploded. The most recent estimate from the United Nations High Commission for Refugees, for example, notes that from January-June 2015, approximately 137,000 people crossed the Mediterranean from Africa. This represents an 83% increase from the same time period in 2014 and is testament to the desperation in MENA.

Youth Unemployment Unemployment is a huge problem in the MENA region, especially among youth and women. Along with slow growth, unemployment rates have remained stubbornly high particularly among youth (15–24 years) with an average rate of 22% for young males and 39% for young females in the MENA countries. Some estimates show that the youth unemployment rate is as high as 40% in Tunisia and even higher in inland governorates. A recent analysis of youth unemployment in MENA shows, surprisingly, that many unemployed youth are well-educated.

This situation has implications for agriculture and its role in development. As a part of the regular pattern of development, agricultural productivity growth tends to lead to farm consolidations as mechanization substitutes for hand labor in many activities. This pattern in most instances is associated with agriculture-led growth, but with fewer people employed in the sector. In the face of large-scale youth unemployment and underemployment coupled with rapid urbanization, important questions arise. First, can agriculture be a source of opportunity for youth? Large-scale commercialized agriculture is not likely to be a major source of employment, but efforts to stimulate productivity on smaller-scale farms may open some opportunities. Land tenure rigidity, limited access to credit, and the rigidity of public supporting institutions may represent obstacles for young farmers. Research to develop mechanization for small-scale farmers, however, may have the effect of ameliorating some of the more onerous on-farm tasks and making farming more attractive to young people.

Second, what can be the role of the agricultural value chain in absorbing unemployed youth? ICARDA has gained substantial experience in evaluating agricultural value chains and perhaps might focus its future research opportunities for youth. Globally, agricultural value chains are quite diverse, and many have multiple nodes and actors ranging from quite small-scale to very large. Opportunities for small-scale business development should be explored, but as noted elsewhere bureaucracies, lack of credit, and weak rural institutions in much of the CWANA region make business start-ups difficult. Policy analysis is needed to explore avenues to reduce obstacles to small-scale business development.

Crop Production

There are two important sources of growth in output of the major crops in the region: area expansion and yield increases. For wheat, there have been significant increases in area under production in CWANA countries between 2000 and 2012 in Afghanistan, Egypt, India, Iran, Kazakhstan and Morocco (area for Morocco is, however, highly variable). Since area expansion into specific crops may be accomplished by switching out of other crops, expansion does not necessarily mean an aggregate increase in land under agricultural production. Thus, substantial gains in wheat area could have occurred due to crop switching (beyond the scope of this analysis). Substantial increases were observed in several CWANA countries wheat area increased by more than 40% each in Kazakhstan, Afghanistan, Egypt and Iran. In India and Pakistan, area planted to wheat grew by 20 and 14%, respectively. In Algeria, Iraq and Sudan, wheat area increased, but with so much variability around the trend that the increase was not statistically significant.

For countries with expanding areas of wheat, wheat yields generally increased over 2000–2012. Increased yields are statistically significant for Afghanistan, Algeria, India, Pakistan and Uzbekistan. Yield growth is especially pronounced in Afghanistan, Algeria and Morocco, where reported wheat yields almost doubled between 2000 and 2012. Pakistan's yield growth of about 22% is also impressive, especially given the large size of the wheat area and its relatively high base yields. Wheat clearly remains an important strategic crop in this country.

While Egyptian wheat yields did not increase over the period, its high base productivity combined with the large area increase indicates that area expansion was not achieved by bringing unproductive lands into wheat cultivation.

For the CWANA countries experiencing a reduction in wheat areas during 2000–2012, Turkey, Tunisia, Tajikistan, and Iraq all experienced increased wheat yields that were statistically significant. Yield growth in Turkey (about 35%), Tajikistan (more than doubling) and Iran are especially impressive.

The overall message from the wheat story is a good one. As wheat is highly traded on international markets, and demand for wheat is expanding in the CWANA countries it would be expected that as prices increase and over time, wheat area would expand. This has largely happened; of major producers in the region, only Turkey and Syria did not increase their land planted to wheat (and Turkey's overall production was given a boost by the yield increase). The yield picture shows that area expansion was not achieved by bringing marginal lands into production – most countries that increased their land also experienced yield increases.

Importance of Rainfed Agriculture in Food and Water Security in Dry Areas and Arab Middle East Countries: Opportunities Provided by Applied Agricultural Research-for-Development for Producing More Food with Less Resources

Given limited opportunities to bring new land to agriculture production, the grains required to feed the growing population by 2050 must rely on intensification of available crop lands. However, there is often a surplus of food grains in developed countries and a deficit in developing countries. Given that consumers in developing countries may not afford to pay the high prices of food produced in developed countries, the problem of food security becomes one of distribution rather than production. This leads to the conclusion that research on yield improvement will contribute more to food security if it is focused on food deficit regions, such as

MENA and SSA regions (for wheat), South Asia (for grain legumes) where local production cannot meet the growing demand, and thus imports (mainly from developed countries) are required (Anderson, Johansen and Siddique 2015). Applied agricultural research-for-development suggests the following strategies for producing more food with less resources, particularly in rainfed areas of developing countries which are characterized by resource-poor small-holder farming systems:

Closing the Yield Gap of Rainfed Crops

The yield gap is the difference between potential and actual yield achieved on-farm. Actual yields are of the order of 60% of potential, as documented in different farming systems: irrigated rice in south-east Asia, rainfed wheat in Central Asia, rainfed cereals in Argentina and Brazil (Godfray et al. 2010). Yield gaps are due to a number of factors that may be technical, economic (relating to markets or investment capacity), information-related etc. The yield gap is not static. Both potential and actual yields will change with time, as new technologies or new constraints develop.

Facing the challenges of meeting the food requirements of growing populations under the conditions of water scarcity, land degradation and climate change calls for more role to rainfed agriculture to feed the world in general and Arab Middle East countries in particular. A recent review in yield gap in rainfed crops provided the following new insights on the potential to increase the crop yield in rainfed areas (Anderson, Johansen and Siddique 2016):

- The gap between actual (or average) yields of cereal and legume crops and the potential yields in the "rainfall-limited" areas indicates that there is still substantial room to increase the average yields of crops in rainfed cropping systems in both developed and developing countries. The size of the gap between actual and potential yields varies according to the agro-ecological zone and the available technology from about 0.5 to over 5.0 t/ha, suggesting substantial scope for yield increase.
- Less information is available regarding the spatial and temporal variability of the yield gap, the sources for the gap and the possible methods to close the gap, particularly at the farm or field level.
- There is limited information on the feasibility and profitability of applying various approaches to bridge the yield gap, including management practices (both cultural and agronomic practices as well as soil improvement measures) and plant breeding.
- The evidence of the impact of conservation agriculture on crop yields in a wide range of agro-ecologies supports the adoption of zero tillage and crop rotation but is less clear in support of residue retention.

- Farmer-participatory varietal selection, including breeding for specific adaptation can contribute substantially to close the yield gap under different environments.
- Sustainable yield increase requires the use of different packages (a range of components) that are appropriate to specific agro-ecological conditions. Previous interventions based on single inputs, cultural and agronomic practices or genotypes, can only be partial solutions.

Strategies designed to close the yield gap in the poorest countries face some particular challenges relating to small landholding size, the dominant role of women in the agricultural workforce, lack of secure land tenure, remote location etc. Closing the yield gap involves not just transferring known technologies and practices to farmers, but "putting in place the institutional structure -- especially well-functioning input and output markets, access to finance, and ways to manage risk -- that farmers need to adopt the technology" (Keating et al. 2010).

Promoting Adoption of Improved Technologies

Agriculture continues to be the main source of the livelihoods of farming communities, the largest source of employment and contributes substantially to national income. However, agricultural productivity is very low in many of the developing countries in dry areas. This suggests that increasing agricultural productivity is critical to economic growth and development in developing countries.

One important way to increase productivity, as experienced by both developed and developing countries, is the introduction and promotion of improved agricultural technologies and management practices. However, there are many constraints limiting the adoption and wide dissemination of improved technologies and management practices. This is in fact the main challenge in developing countries as the technology adoption is a prerequisite for increasing agricultural productivity.

Previous studies have identified several reasons to explain the low or lack of adoption of improved technologies by farmers (Doss 2006; Shideed 2008):

- Farmers are simply not aware of new technologies or not aware of their benefits to them. This may be attributed to weak or lack of extensions and services delivery systems
- Improved technologies are not available, or not available at the time that farmers would need them. Again this is related to the lack of local or national capacities to produce the improved technologies in quantity, quality and timely manners that suit farmers' needs and skills.
- Improved technologies are not profitable, given the complex sets of decisions that farmers are making about how to allocate their land and labor across agricultural and non-agricultural activities.
- Improved technologies are not affordable. Affordability is a key factor in technology adoption by small-holder farmers. Given cash limitations and lack

of access to easy micro credits, small-holder farmers have clear preference to cost-reducing technologies that can be purchased with minimum cost possible to match their financial conditions.
- Improved technologies are not risk reducing. Faced with weather and climate variability accompanied with biotic and a biotic stress, farmers need risk-reducing technologies that minimize their losses and stabilize their yields, and thus income. Drought- tolerant and disease- resistant varieties as well as water-saving technologies and management practices are very much needed by small-holder farmers in dry areas.

Sustainable Intensification of Production Systems

How can more food be produced sustainably? Opportunities for bringing more land under cultivation are extremely limited. In the future, production growth must come mainly from productivity growth. For example, it is estimated that 80% of the required increase in food production between 2015 and 2030 has to come from intensification (FAO 2011). We must work to increase potential yield (maximum yield with the best varieties and agronomy, with constraints removed) and simultaneously to close the yield gap (the difference between potential and actual yields).

The FAO defines sustainable intensification of production systems as "producing more from the same area of land while reducing negative environmental impacts and increasing contributions to natural capital and the flow of environmental services". This approach is emerging as a major priority for policy makers, and international development partners. The Asian Development bank (ADB 2010) identified three main constraints to agricultural growth: land and water shortages; lack of access to rural finance, infrastructure, technology, markets and distribution networks, and nonfarm income opportunities; and threat of climate change and price volatility.

Producing more food per unit of land, water and other resources is essentially the concept of 'Eco-efficiency' which encompasses both the ecological and economic dimensions of sustainable agriculture (Keating et al. 2010). It opts for more agricultural output (both quantity and quality) with less input of land, water, nutrient, energy, labor or capital. This is a major challenge for agriculture in future in achieving such eco-efficiency while addressing the risk and variability.

> The agricultural revolution over the next 40 years has to be the eco-efficiency revolution, with 50–100% increases in the efficiency with which resources of land, water, nutrients, and energy are used. Importantly, this greater output and efficiency has to be achieved without further greenhouse gas emissions while maintaining or restoring land, water, biodiversity, and agro-ecosystems (Keating et al. 2010).

Agricultural research is the main source for developing the technologies and practices that can shift "eco-efficiency" while keeping risks low by shifting the efficiency frontier. Social and economic systems and institutions are the necessary conditions whereby improved technologies and practices can be embedded with constant or declining production risks. Food security involves both closing the yield gap and breakthrough technologies and practices to raise yield potential and increase efficiency of resource use (Keating et al. 2010).

Integrated Systems Approaches

Dryland farming systems are complex, with inter-dependent and interacting components – including crops, livestock, forestry and fish, with livelihoods supported by various, and seasonally changing, farm and off-farm sources of income. Past experience has shown that narrowly focused (e.g. commodity based) research is unlikely to develop successful technologies for these systems. A new multi-partner program, led by ICARDA, used an integrated, multi-disciplinary approach that fully takes these complexities into account.

The CGIAR Research Program on Dryland Systems (Integrated Agricultural Production Systems for the Poor and Vulnerable in Dry Areas) aimed to develop technology, policy and institutional innovations to improve rural livelihoods. Research and capacity development activities encompass crop, livestock and natural resources technologies, markets, policy, and gender and equity issues. The program targeted five dryland regions: West African Sahel and dry savannas, East and Southern Africa, North Africa and West Asia, Central Asia, and South Asia. In each of these regions, research was structured under four themes:

- Strengthening innovation systems, building stakeholder innovation capacity, and linking knowledge to policy action
- Reducing vulnerability and managing risk (technologies, institutional options)
- Sustainable intensification for more productive, profitable and diversified dryland agriculture with well-established linkages to markets
- Measuring impacts and cross-regional synthesis. The program design clearly identifies two types of farming systems that may require different technology 'mixes: vulnerable systems (e.g. subsistence farming with severe resource limitations) where the primary objective is to reduce risk and increase resilience; and better favored systems (smallholder systems with potential for intensification and market access) where the goal is to increase productivity.

The program brings together the full range of stakeholders: farming communities, national research and extension systems, policy makers, international and regional organizations, advanced research institutes, NGOs, the private sector, and development agencies. Research was conducted at pilot sites in each target region, with emphasis on community participation in design, implementation and monitoring of interventions.

The aims are to prioritize key agricultural systems for impact, identify key researchable issues within target agro-ecosystems, increase the efficiency and sustainability of natural resource use, develop more resilient agricultural systems to manage risk and production variability, promote conservation and sustainable use of dryland agrobiodiversity, improve the productivity and profitability of agricultural systems through sustainable intensification, diversification, value-added products and market linkages, identify niches of importance to the most vulnerable livelihoods (even if they appear to have low marketing potential), address constraints faced by the most marginal farmers, and develop new partnerships and models of working together.

Water Saving Technologies

Water-efficient technologies such as improved irrigation methods are crucial for both rainfed and irrigated areas; this chapter focuses mainly on rainfed areas. There is a wealth of literature on available, proven technologies, and on the sources of inefficiency in irrigation water use.

Given severe water scarcity and limitation in dry areas, future food security and productivity enhancement strategies should give due focus to improving "on-farm water use efficiency" in the production of various commodities under rainfed conditions through the use and promotion of water-saving technologies and practices.

Water-use efficiency can be measured in different ways such as irrigation efficiency (amount of water from the main water source which can be effectively supplied to the root zone), or crop water-use efficiency (the fraction of water transpired by the crop to that stored in the root zone). But technical measures of efficiency are not sufficient to assess the economic use of water. The economic efficiency of water use depends on the relative prices of water and other inputs, the marginal products and prices of the inputs, and the amounts of other inputs, including rainfall. At farm level, this is referred to as "on-farm water-use efficiency (FWUE)", defined for the purpose of this analysis as "the ratio of the required amount of water for a target production level to the actual amount of water used" (Shideed et al. 2005). The resulting indicators of FWUE are very useful in guiding policies toward improving irrigation efficiency.

Irrigation accounts for 80–90% of water consumption in the WANA region. Thus, small improvements in FWUE can substantially improve water availability. Low irrigation efficiency is associated with poor timing and lack of uniformity of water applications, leaving parts of the field over- or under-irrigated relative to crop needs. Moreover, operators of irrigation systems do not have an incentive to supply farmers with a timely and reliable delivery of water that would be optimal FWUE. Farmers generally over-irrigate as a result of their perceptions of water requirements and their expectations of rainfall and market values.

Efficiency Frontiers and Pathways for Technology Intervention

Innovations such as conservation agriculture are helping to address the challenge of productivity growth in the face of resource scarcity. While conservation agriculture has been widely adopted in some regions (e.g. USA, Australia, Latin America) and some crops, adoption has been slow in many developing countries – and particularly in West Asia and North Africa, where land and water resources are particularly scarce. This section discusses some of the theoretical aspects, and highlights the lessons learnt from the Australian success, and some of the results achieved by ICARDA and its partners.

An important point here is to understand the "efficiency frontiers for conservation agriculture". The return-risk framework suggests that "efficiency frontiers" exist at which the return from existing technology is maximized for different risk levels. The curves in Fig. 4 describe such frontiers: the lower, solid line represents the frontier for currently adopted technologies and practices; the higher, dashed line represents the frontier for "yet-to-be" adopted technologies which create new opportunities and return-risk dimensions. Using this framework, pathways for technology intervention can include (Carberry et al. 2010; Keating et al. 2010):

- Moving from B to D by adopting current best practices to remove system inefficiencies with no increased exposure to risk
- Moving along the efficiency frontier using existing technologies (D to A) but with an associated increase in inputs and risk
- Adopting breakthrough practices or technologies to move to a new efficiency frontier (dashed line) that enables,
  - Maintaining output through increased efficiency of resources use (reduced level of resources use) while reducing exposure to risk (D to C)
  - Increasing output with the same exposure to risk (D to F) or (same level of resources)

In reality, most farmers choose acceptable risk investments which return closer to 60–80% of the potential (point D). To increase returns with little added risk, the only real option for these producers is to move from the current efficiency frontier (solid line) by adopting breakthrough technologies (D to F). Figure 4 illustrates how Australian farmers have adopted technology and information systems over the past 30 years to improve their economic performance and deal with the high risk of farming in a variable climate.

The widespread adoption of conservation agriculture in Australia is a significant management practice which has raised cropping productivity over the past 30 years. Conservation tillage practices improve rainfall infiltration and increase soil water storage, which commonly results in increased water use efficiency of crops in Australian dryland farming (Carberry et al. 2010).

As water supplies for irrigation become less secure under climate change, conventional irrigated agricultural systems may be replaced by dryland agronomic

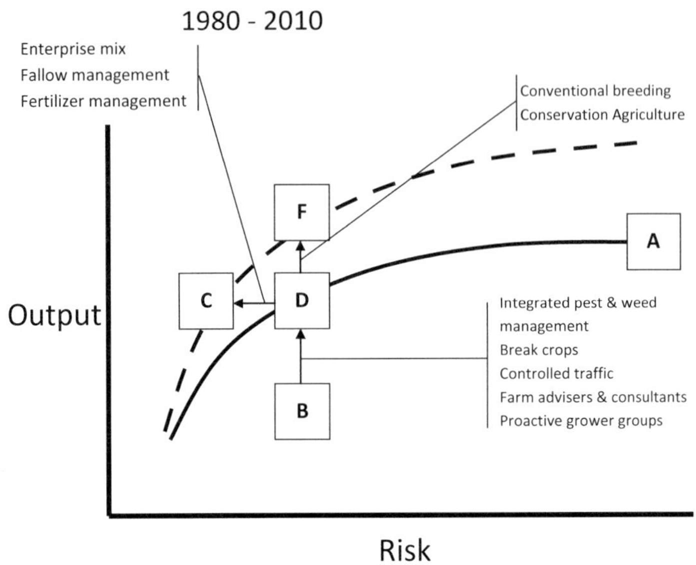

Fig. 4 Conceptual framework for efficiency frontiers of conservation agriculture (After Carberry et al. (2010) and Keating et al. (2010))

practices with supplementary irrigation to increase the efficiency of water use. Opportunities to modify government policies that impede efficiency gains could be addressed. New technology packages would include conservation agriculture techniques, together with other resource-efficient technologies such as supplemental irrigation. In practice, productivity increases can be realized through combinations of these strategies as managers respond to market signals, availability (and affordability) of technologies and immediate drivers such as drought (and price volatility as well as policy changes).

An investment approach to create new return-risk frontiers is to obtain productivity gains through innovation (research, development and adoption of new technologies). Productivity can be increased through three intervention pathways:

- Develop and implement technologies or interventions that remove system inefficiencies, i.e. move from point B to pint D in Fig. 4
- Invest in breakthrough technologies that increase the efficiency of resource use whilst reducing risk (point D to point C, Fig. 4)
- Invest in breakthrough technologies that offer greater returns for the same level of risk (point D to point F).

The efficiency frontier framework can also be applied to other technologies. Keating et al. (2010) apply the framework of Fig. 4 to the case of water use efficiency (WUE) for grain production, defined as the ratio between grain yield and evapotranspiration. Empirical studies demonstrate opportunities to improve

WUE using two pathways of the return-risk framework, by closing the yield gap and move crop yield up onto the efficiency frontier (point B to D) by better controlling biotic stresses. The second pathway is to move along the efficiency frontier and increase WUE with a yield-maximizing strategy (point A). This would require farmers to increase their investment risk either by increasing inputs or the chance of crop failure (Keating et al. 2010). The only pathways for farmers to increase production are either to move along the efficiency frontier (and so increase their investment risk) or to adopt new technologies that generate a new efficiency frontier (Keating et al. 2010).

Informing Policy Development

Food security and productivity enhancement targets cannot be realistically achieved without enabling policy and institutional environment.

Food security concerns have led to extensive policy debate, which has highlighted several issues (Solh 2011). First, current agricultural policies in developing dryland countries are inadequate in general, and particularly ineffective in protecting the fragile natural resource base. Second, land degradation and water scarcity are occurring rapidly, both in marginal environments and in irrigated areas. Third, it is extremely hard to protect and conserve communally owned natural resources such as rangeland and water. Technology alone is not sufficient but must be supported by policy, community-led action and continuous monitoring. This underscores the crucial role of researchers in informing policy development. However, it must be noted that direct use of research results to fundamentally change policy is rare. Rather, the primary pathways by which decisions are shifted are often indirect, and involve improving general understanding of the context in which decisions are taken through "conceptual influence". As a result, such influence is primarily indirect, and may involve a large number of intermediate adoption and diffusion events before new understanding contributes to a shift in policy (CGIAR 2006).

Political-economic factors can be crucial, particularly for management of natural resources. Policy reform on natural resources will not happen unless researchers have fully taken these factors into account. Of course, providing policy makers with new research information is necessary, but not sufficient to foster adoption of such recommendations by politicians (Zilberman and Waibel 2007).

In measuring the contribution of research to changes in policy, establishing causality from research to implementation of the policy is critical (Norton and Alwang 1998). To establish policy influence, Ryan (1999) and Shideed et al. (2008) interviewed partners, stakeholders, and policy makers about their perception of how policy changes had taken place and the role of different institutions and information in the change.

Research that successfully influences policy can generate large impacts (e.g. Ryan 1999, 2002). For example, ICARDA's research on barley fertilization in rainfed areas in Syria led to a fundamental change in government policy on fertilizer allocation. Although some other policy distortions remained, the returns to research on barley fertilization were still substantial and consistent with an estimated rate of return to investments in research and dissemination of 70%, and a benefit-cost ratio of 35–41 (Ahmed et al. 2010; Shideed et al. 2008).

The Syria case also provided important lessons. Policy makers may only partially adopt research results or recommendations. It is vital that national partner institutions are involved in any policy-oriented research. Building mechanisms for leveraging policy influence in project design and implementation also helps address policy makers' concerns, and increase the chances of success. Thus, understanding the impact pathway and identifying key partners along the pathway is a prerequisite for successful research-for-development.

The Need for Research Investment

Research has played a major role in improving food security through higher output and especially higher productivity, and thus increasing agricultural production growth. Most of the growth has been achieved through productivity increase. Science and technological innovation are critical for various reasons, as summarized by Austin (2010, quoting The World Bank World Development Report, as reported by Austin 2010): to meet growing demand; to maintain market competitiveness; to address poverty and to tailor technologies to growing heterogeneity among farmers and gender groups; and for adapting to and mitigating environmental externalities such as climate change.

Despite the very high returns to investment, estimated at 65%, agricultural research-for-development has experienced significant under-investment (Austin 2010). International investments in the CGIAR, for example, have stagnated during the past decade. The CGIAR's Strategy and Results Framework (2011) notes that:

> To achieve a food-secure world by 2025, an annual increase in agricultural productivity of 0.5% across all regions until that year is required . . . This equates to a massive expansion of investment in agricultural research for development . . . from US$ 5.1 billion per year today to US$ 16.4 billion per year by 2025 . . .. This includes the investment needed in national as well as international public sector research. Investment in international public goods research is currently about 10% of total public R&D spending (slightly over US$ 500 million in 2009) . . . Extrapolating to 2025, a CGIAR budget of US$ 1.6 billion (10% of US$ 16.4 billion) by 2025 is required.

One of the causes of the slowdown in productivity growth was a fall in the growth of public research investments. Increased investment is needed to stimulate sustainable growth in agriculture, which is the engine for the economic growth and poverty reduction, as well as a major job market. Meeting the global challenge to increase food production by 70% by 2050 depends largely on increasing investments in research, development and extension; promoting risk management

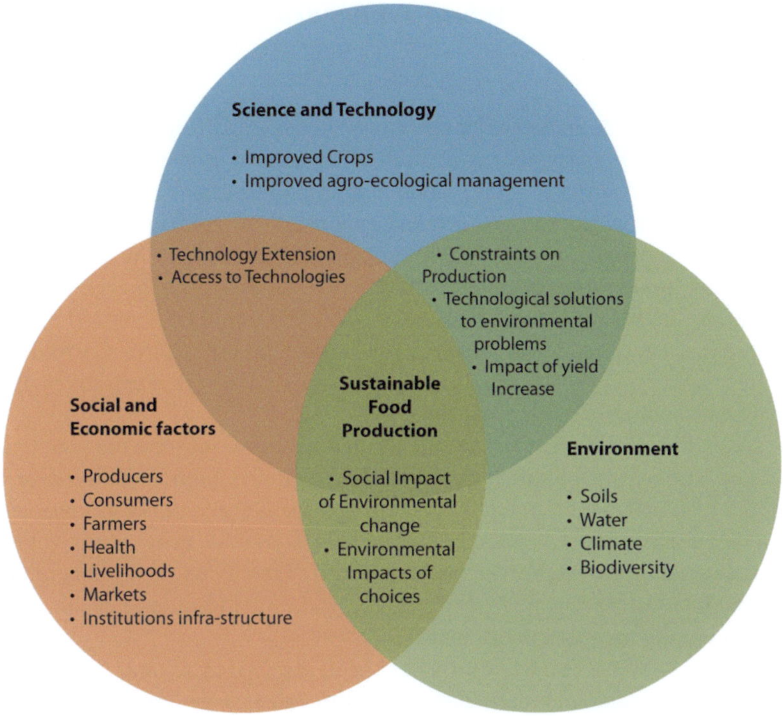

Fig. 5 Role of science and technology in sustainable food production systems (After Austin 2010)

systems; improving farmers' skill and human capital; and developing policies that encourage efficiency gains.

Science is essential, but not sufficient to ensure productivity growth and food security. Broader socio-economic and environmental factors are influential (Fig. 5). Multi-disciplinary research, together with supportive policies, is required to develop and promote innovation, and ultimately to apply innovation to ensure food security while protecting natural resources.

Prospects and Opportunities for Enhancing the Role of Dry Areas in Enhancing Food Security Globally and in the Arab Middle East Countries

This section will provide empirical examples and success stories on a large scale and elaborate on their scalability.

Evidence on Increasing Wheat Productivity and Production at Scale

The Arab counties are facing major challenges in meeting the increased demand for food. The region is classified as a food deficit region, and thus predominantly dependent on food imports. Total amount of imported grains reached 66 million tons in 2010, and with increased demand as a result of population growth, grains imports are expected to reach 73 million tons by 2020. This huge food gap is explained by limited local production due to low yield levels achieved by farmers compared to the higher levels of potential yield that could be achieved under the same levels of input use. There are many reasons behind the huge yield gaps in Arab countries, among which is limited capacity in transferring the research results and improved technologies to farmers' fields.

In spite of all the challenges and the limited natural resources, the potential for increasing agriculture productivity in MENA region is high since average crop yields are well below their potential. Restoring agricultural productive capacity can have a broad impact on economic growth as well as food security, rural livelihoods and employment. Sustainable intensification and diversification of the agricultural sector and investment in added value products and rural industries can contribute to providing employment and more income.

According to above factors, efforts have been invested by several Arab countries and ICARDA in implementing a pilot project to enhance food security to increase wheat productivity and production in Arab countries, while conserving and optimizing the use of scarce water resources. The project was funded in its first phase (2011–2014) by AFESD, KFAED, IsDB and OFID and implemented in Egypt, Sudan, Morocco, Tunisia, Jordan, Yemen, Syria, Palestine, Iraq and Algeria. The project was focused on improving wheat productivity under irrigates, rainfed, and supplemental irrigation conditions in Arab countries to reduce the food gap.

Considerable success has been achieved of the first phase of the project (2011–2014). Partner countries and ICARDA have successfully demonstrated and promoted improved technologies that have allowed improving wheat yield and improving the on-farm water management practices in several Arab countries.

Adoption and Impact

Results obtained show a positive impact in wheat productivity in the project sites. This is illustrated in the following country cases.

Egypt
The use of raised-bed techniques in farmers' fields resulted in 25% saving in irrigation water, 30% increase in wheat yield and 74% improvement in water use efficiency (average of 4 years 2011–2014).

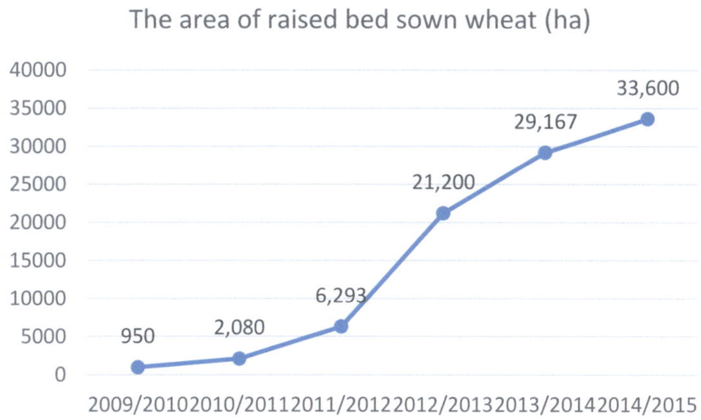

Fig. 6 The Area of raised bed sown wheat in Al-Sharkia Governorate, Egypt (ha)

In Al-Sharkia governorate, the area of raised-bed sown wheat has increased 35 times, from 950 ha in 2009/2010 to 33,600 ha in 2014/2015. In addition there was an increase in the use of certified seeds of new improved varieties. Al-Sharkia governorate's total amount of wheat delivered to the Ministry of Supply went up from 557,030 tonnes in 2009–2010 to 880,941 tonnes in 2013–2014, an increase of about 58%. Adjusted for the wheat acreage in both seasons, the additional amount of wheat produced was estimated to be 261,000 ton and has a calculated value of 52.2 million USD. In a country where 50% of its wheat demand was earlier met by imports, wheat production increased by 1.1 million MT.

A nation-wide wheat national campaign (WNC) was launched by the Egyptian government in 2011–2012, supported with a government funding of 1.7 million USD. Mass dissemination approach was adopted and 1900 demonstration fields of raised-bed system were implemented in 22 governorates of Egypt based on the same approach used in Al-Sharkia site.

At the national level, in 2015 season, in all wheat governorates the total wheat area under raised-bed was 105,000 ha. Building on the promising results of improved raised-bed technology, farmers in Egypt have now found ways of sustainable water management to better adapt to climate change challenges. Their production yield and incomes have also increased in the process. The raised bed technology is being tested in farmers' fields in other countries of the region such as Morocco and Tunisia.

Morocco

In the site of Tadla in Morocco where supplemental irrigation is practiced, the project results indicated that the deficit supplemental irrigation technology can lead to substantial saving in irrigation water which could reach an average of 644 cubic meters per ha. It is expected that at least 20% of the cereal cropped area in Tadla will be covered by the deficit supplemental irrigation technology in the coming 2 years. The resulting saving in water is expected to be in the average of 1.5 million cubic meters and can be used to irrigate an additional 400 ha using the deficit irrigation technology. Hence, and at a yield level of 7.40 t /ha, an additional production of 3000 tons of wheat is expected. At the current wheat price in Morocco the additional production of wheat is worth 1.1 Million US dollars annually.

Tunisia

In the rainfed site of Fernana in Tunisia and after 3 years of field demonstrations adoption of improved wheat varieties and agronomic practices resulted in an increase in wheat production worth of about 433,171 US dollars. Similarly in the supplemental irrigated site of Kairouan (Chebika) farmers have adopted improved wheat variety, appropriate cultural practices and irrigation management techniques. This allowed an increase in wheat production worth of 1,106,236 US dollars.

Jordan

In Jordan the dissemination of improved wheat production technologies in the two project sites had raised the yield levels from 1.66 t/ha without project intervention to 1.83 t/ha under field demonstrations. Yet the potential increase is still higher as shown by the results obtained in the wheat demonstrations (2.85 t/ha). The additional production increase in just 1 year of project interventions in a very small wheat growing area is worth 207,000 US dollars.

Reducing the Wheat Yield Gap

Under dissemination activities large scale on-farm demonstrations with active farmers' participation have been implemented using proven technology packages.

Results showed that in all countries wheat yield can be increased under all production systems by the use of improved technologies as compared to the use of own farmers' practices. An average increase of 27% can be achieved across all countries involved. The maximum yield levels achieved by leading farmers clearly shows that there exists large potential for increase in the wheat yields in the countries involved in the project and that even higher production increase could be obtained, up to 78% in case of Sudan.

Increasing Water Productivity and Reducing Water Losses at Farm Level

Affordable technologies can improve irrigation efficiencies such as application and distribution efficiencies.

In Egypt results from demonstrations in farmers' fields show for five consecutive years a clear advantage in using the raised-bed technique. The technology allowed 24% saving in irrigation water, 34% increase in wheat yield and 78% improvement in water use efficiency.

In Morocco deficit irrigation proved to be successful in saving 1/3 of irrigation water with little or no impact on wheat yield while increasing water productivity from 1.7 kg/m^3 under full (supplemental) irrigation to 2.0 kg/m^3 under deficit irrigation.

In Tunisia demonstrations on irrigation management of wheat based on water balance methods in combination with the SMS technology proved to be quite efficient. In most of the demonstration fields the water productivity was higher than the national average (0.9 kg/m^3).

With regard to drip irrigation system in wheat production, results in Morocco and Tunisia show that this system had a better water productivity compared to other systems.

Water Use Efficiency was 1.4, 1.1 and 0.9 kg/m^3 respectively for drip, basin and raised bed irrigation systems in Morocco. In Tunisia a 100 cm line spacing drip system gave higher grain yield (6.4 t/ha) compared to the 50 cm spacing (5.3 t/ha) and higher water productivity (1.22 kg/m^3 vs. 1.00 kg/m^3 for the 50 cm spacing).

Evidence on Improving Water Use Efficiency

The typical irrigation method at the field level in many countries is a surface gravity system, generally associated with low efficiency and high seepage and evaporation losses. For example, water use efficiency (the ratio of the amount of water actually utilized by the crop to the total water pumped) for irrigated agriculture in Syria is 40–60% (MunlaHasan 2007). Traditional surface irrigation methods also lead to over-irrigation and/or uneven irrigation, especially when field drainage is poor.

Supplemental irrigation (SI) is an important technology for boosting food production and food security in rainfed areas. Supplemental irrigation is understood by most farmers as the use of irrigation in addition to rainwater – regardless of the amount applied. In its true essence, SI involves the application of irrigation only when rainfall is inadequate, with the amount and timing carefully scheduled to ensure that a minimum amount of water is available during critical stages of crop growth (Oweis 1997). SI in low rainfall areas not only increases yield relative to purely rainfed production, but also substantially improves the productivities of both irrigation and rainwater (Oweis and Hachum 2004). To avoid confusion, we distinguish between improved supplemental irrigation (ISI), in which the recommended water application rates are used and traditional supplemental irrigation (TSI) where farmers use excessive irrigation rather than the recommended levels.

Several studies have been conducted to estimate the impacts of SI. For example, El-Shater (2009), Bader (2010). Shideed et al. (2005), Adary et al. (2002) and Salkini and Ansell (1992) reported almost twofold increases in wheat yield with the use of SI and related technologies relative to solely rainfed crops. Water use efficiency measurements in West Asia and North Africa revealed that wheat yield under rainfed conditions ranges from 0.35 to 1.0 kg per m^3 of water (Oweis and Hachum 2009). In comparison, ISI with appropriate management and optimal

Table 1 Profile of the sample farm households in terms of irrigation method and water use

	Zone 1	Zone 2	Total
Traditional irrigation			
Households using the practice (%)	34	70	53
Average water used m^3 per ha	2554	2744	2686
Average yield (mg/ha)	5384	4892	5040
Sprinkler irrigation			
Households using the practice (%)	66	30	47
Average water used m^3	1852	1910	1869
Average yield (kg/ha)	5840	5460	5733

Note: Some farmers who use surface irrigation methods are applying less than 1800m^3/ha while some farmers who use sprinklers apply excessive irrigation water (>1800m^3/ha)

irrigation produces an additional 2.0 to 3.5 kg of grain per m^3 of ISI, compared to solely rainfed wheat (Ilbeyi et al. 2006).

An ICARDA study (Yigezu et al. 2011b) measured output oriented technical efficiency, irrigation water efficiency, and assessed the reduction in water use inefficiency due to the adoption of modern irrigation methods (particularly sprinklers) for supplementary irrigation of spring wheat farmers in Syria. It used market and non-market valuation methods to measure the economic and environmental impacts of the shift from traditional supplemental irrigation (TSI) to improved supplemental irrigation (ISI) of wheat farms in zones 1 and 2 of the Aleppo, Deraa and Al-Hassakah provinces in Syria, which together produce 61% of the country's wheat. Table 1 shows the profile of the sample farm households. Some of the findings are described below.

At the low level of its current adoption on wheat farms in the three provinces (22.3%), ISI helps to conserve at least 120 million m^3 of water per year. The combined impact of ISI with sprinkler technologies is estimated at US$ 8.96 million per year. Introducing a water user charge of US$ 0.20 for every cubic meter applied in excess of the recommended application rate of 1800 m^3 can ensure adoption of ISI by all farmers leading to further conservation of at least 46 million m^3 of water per year (Yigezu et al. 2011a). Such a policy not only promotes groundwater conservation but also increases the productive value of water and farm profits.

Water was valued at the value marginal product. The values of marginal products at the profit maximizing levels of inputs should be equal to factor price. The marginal value product of water for sample farms is estimated at US$ 0.13. The marginal product of water varies depending on the type (TSI or ISI) and method (surface or sprinkler) of irrigation. Generally, ISI and sprinklers lead to higher marginal products (MP) of water than TSI and surface canals, respectively.

ISI consumes less irrigation water than TSI, and also saves fuel and other costs associated with pumping groundwater. Most of the pumps in Syria use diesel fuel. Adoption of ISI saves an estimated 49.8 million liters of diesel per year, worth about US$ 20 million in the three provinces.

The net effect of the shift from TSI with canals to ISI using sprinklers for the individual adopter is a US$ 235.5/ha increase in annual profit. The total profit gained by all farmers adopting ISI with sprinklers (18% of total farms) in the three provinces is estimated at US$ 4.7 million per year. The adoption of ISI at the current adoption level of 22% in the three provinces has led to a total annual increase in farm profits of US$ 9.0 in 2010. In terms of economic and social costs and benefits, even if the conserved water is used wastefully for TSI with open surface canal, the introduction of ISI in the three provinces still has a clear overall positive impact.

Despite the obvious benefits of ISI, TSI is still practiced by 78% of wheat farms with an average irrigation water application rate of 2600 m^3/ha. What can the government do to encourage adoption? One possibility is to introduce water user charges. Under the existing conditions where farmers do not pay any water user charges, the profit maximizing application rate for surface canal users is 2375 m^3/ha which is 12% lower than the current average (2686 m^3/ha). This shows that these farmers are applying unnecessarily too much irrigation. Water user charges of US$ 0.11 or higher per m^3 would force surface canal users to reduce their application rate to the recommended level (1800 m^3 of irrigation water). However, at this water user charge, the profit maximizing rate of irrigation water application for sprinkler users is 2075 m^3/ha which is still higher than the recommended level. The minimum water user charge that would force sprinkler users to reduce their application rate to the recommended level is US$ 0.20/m^3. Hence, if the government were to impose this level of water user charge for every cubic meter applied in access of the recommended 1800 m^3/ha, all farmers in Zones 1 and 2 of Syria will be compelled to apply the recommended level regardless of their irrigation technology.

Given the rapid depletion (2–6 m per year in some areas), conservation of groundwater is a priority for the Syrian government. Substantial amounts of groundwater could be conserved by introducing user charges for every cubic meter of irrigation water applied in excess of 1800 m^3/ha (the upper limit of the recommended application range). Such a policy would also increase total farm profits by US$ 16.14 million per year. The total impact of introducing the optimal water user charge of US$ 0.20/m^3 ranges between US$ 36–90 million per year. However, introduction of water user charges in Syria, and other dryland countries, could be very sensitive as nearly one-third of wheat production area in the country depends on supplemental irrigation.

Traditional surface irrigation methods, involving canals and field furrows, lead to over-irrigation and/or uneven irrigation, especially in the absence of land leveling and good drainage. Such practices have led to low levels of technical and irrigation efficiencies of scarce water resources. Technical, irrigation water, and irrigation water technical cost efficiencies were estimated using a sample of 385 wheat farms in zone 1 and zone 2 of Aleppo Governorate in Syria.

The estimated output oriented technical efficiency measures show high efficiency differential among the farms ranging from 47 to 98% with a mean value of 78% (Table 2). By more efficient utilization of inputs, with current technology, the average farmer can increase yield by 22%. Generally, irrigation water efficiency turns out to be lower than technical efficiency but has very high variability across farms. The mean irrigation water efficiency of 69% shows that everything else constant, a farmer with the average irrigation water efficiency can reduce the

Table 2 Estimates of technical, irrigation and cost efficiencies under supplemental irrigation of wheat farms in Syria, 2010

Irrigation method	Number of farms	Technical efficiency (%)	Irrigation water efficiency (%)	Irrigation water technical cost efficiency (%)
Surface SI	186	70	66	89
Improved SI	142	89	75	91
Total Farms	328	78	69.9	89.9

amount of water by 31% with no loss of yield, only by adjusting the input mix (Yigezu et al. 2011b).

The mean irrigation water technical cost efficiency is estimated at 89.9% indicating that by making more efficient utilization of irrigation water, the average farmer can reduce total cost by 10.2%. Even though the total cost reduction is not small in absolute value, it is small relative to the amount of irrigation water reduction that can be achieved (31%). This is because currently, there are no charges for pumping water from wells, so irrigation water and associated costs (including labor) account for only 35% of total cost of production.

Irrigation method is the most important variable explaining inefficiency in wheat production in the study area. Sample farms show substantial adoption rate (57%) of the sprinkler irrigation technology. After controlling for all other possible determinants of inefficiency, the efficiency differential between adopters and non-adopter can be attributed to the sprinkler technology. Farmers who use sprinklers, on the average, have 19% and 9% higher output oriented and irrigation water use efficiencies, respectively, than those using traditional surface irrigation method. However, even among the farmers who use the improved technology (sprinklers), there is still 25% gap in irrigation water efficiency which needs to be closed.

Effective Mechanized Rainwater Harvesting

Despite its low levels of rainfall (less than 200 mm/year), Badia in Jordan constitutes an important source of grazing for livestock breeders. Due to low productivity rural populations seek alternative income-earning opportunities elsewhere. This increased migration leads to the collapse of traditional land, water and vegetation management systems, causing further degradation.

In a project on Community-Based Optimization of the Management of Scarce Water Resources in Agriculture in West Asia and North Africa, which started in 2004 with funding support from Arab Fund for Economic and Social Development (AFESD), International Fund for Agricultural Development (IFAD) and OPEC Fund for International Development (OFID), ICARDA established several benchmark research sites in different agroecosystems including Jordan Badia for the implementation of integrated watershed management approach. GIS-based land suitability analysis was performed to integrate biophysical and socioeconomic factors into a comprehensive site characterization map.

The integrated approach that was used in the Badia recovery to sustain plant cover and conserve soil from degradation has resulted in community resilience to harsh environment and to the severe effects of climate change.

The use of mechanized Vallerani system has allowed the implementation of rainwater harvesting package at a large scale. The main beneficiaries are the herders and rural communities of marginal lands where sheep and goats are the main source of income for better livelihoods. This system has enabled them to grow fodder shrubs and reduce pressure on rangelands for livestock grazing in areas where rainwater is not enough to support stable and sustainable production and better livelihoods.

Beneficiary farmers are enjoying more than double the yield for barley and 1.6 times for rangeland and forage shrubs production compared to those grown without rainwater harvesting. Highly degraded lands have been rehabilitated by this improved vegetation, thus mitigating land degradation. The GPS-guided micro-catchment rainfall system has reduced cost and time required to delineate contours for the plow to follow, and tripled the construction capacity (up to 30 ha/day), improved efficiency and precision, and substantially reduced the cost of establishing micro-catchments.

In Jordan, the Vallerani rainwater harvesting package developed through scientific research has been implemented on over 3864 ha of rangeland so far, with adoption rates tripling since the start of the project. Biodiversity has increased by capturing 67–73% of available seeds in the harvested area compared to 3–6% in the control sites. Around 40–50% of precipitation is prevented from being lost by evaporation at the intervention sites.

GIS-based land suitability analysis conducted by ICARDA team suggests that a total area of 2.7 million ha in Jordan has the potential for adoption of this package, with further prospects of scaling out on an area of 300 million ha in WANA countries. IFAD is contributing to these efforts through 'Agriculture Resource Management Project (ARMP) II', a development project, where tenders for the purchase of three Vallerani machines were done in 2015. The valuation and assessment of environmental benefits associated with implementing rainwater harvesting techniques provide encouraging justification for public investment in scaling-out of these techniques in the dry areas of Jordan. The scaling-out efforts will achieve higher success if water harvesting interventions are included in plans for integrated land and water resources development taking into consideration all the necessary technical, agricultural, socioeconomic and institutional aspects and inputs.

Conservation Agriculture

Australian experience in conservation agriculture has some relevance to rainfed farming in developing countries, taking into account that small-holder farming is

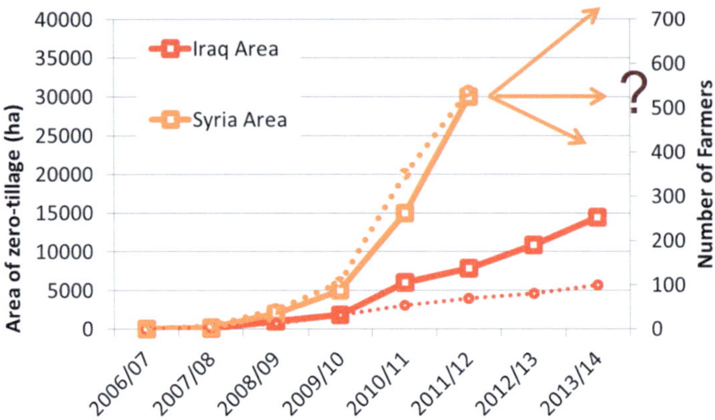

Fig. 7 Adoption growth of zero tillage in Iraq and Syria, 2006–2011 (Source: ACIAR-AusAID Iraq Project, Colin Piggin 2014)

the context for dryland developing countries whereas Australia farming is predominantly of large-scale farms. Much of productivity growth in Australian dryland farming is attributed to conservation agriculture. An Australian-funded project in West Asia, managed by ICARDA, has achieved impressive results in developing and promoting conservation cropping in Iraq and Syria, where ZT has been little known or adopted. The program helped introduce ZT in the 2006–2007 season. In the 2013–2014 season, about 100 farmers used ZT on 15,000 ha in Ninevah province in Iraq and 500 farmers on 30,000 ha in Syria, as shown in Fig. 7. A key constraint to adoption was overcome with development of locally-manufactured or modified ZT seeders, which were effective and affordable. Local farmers encouraged to try ZT were impressed with increased yields and cost savings. Farmers have been purchasing/modifying their own seeders and taking up ZT, and adoption is expected to continue to increase in both countries, and to expand to other countries in the region.

What Science Can Do to Provide Drought-Tolerant Plants to Overcome Climate Change in Semi-arid Zones (GMOs, . . . etc)

Advances in crop science to produce improved and higher-performing crops and livestock hold exciting prospects for making dryland food production systems more efficient, and more resistant to climatic pressures and new pests and diseases. More than 1000 improved cereal and legume varieties have been released by national programs in partnership with ICARDA, and adopted by farmers worldwide. Economic benefits associated with these improved varieties are estimated at more than one billion (US$) annually. Releases of plant genetic materials from ICARDA's gene banks, which host wild relatives of barley, wheat and legumes, has led to the development of crops with higher yields and greater resistance to a range of biotic stresses. Some varieties also offer large improvements in bread-making quality, nutritional value and other traits.

In rainfed farming systems, widely adapted varieties (produced through conventional breeding) are likely to be less than optimal in any particular environment. This is particularly true for grain legumes in dry areas, which is characterized by heterogeneous and variable environment. Under such environments, participatory varietal selection methods (or client-oriented breeding approach) is considered necessary for heterogeneous environments, such as rained areas, where spatial and temporal yield variability is the norm. Whereas, a centralized breeding approach is better suited to more homogeneous target environment, such as irrigated areas (Anderson, Johansen and Siddique 2015).

Modern genomics and genetic approaches coupled with advances in precise phenotyping and breeding methodologies are expected to more effectively address drought tolerance in crops. Most recent advances in plant physiology for precision phenotyping of drought include molecular dissection of drought tolerance by QTL or gene discovery through linkage and association mapping, QTL cloning, candidate gene identification, transcriptomics and functional genomics. Molecular breeding approaches such as marker-assisted backcrossing, maker-assisted recurrent selection and genome-wide selection have been suggested to be integrated in crop improvement strategies to develop drought-tolerant cultivars that will enhance food security under changing climate (Rouf Mir. et al. 2012).

To increase crop yield per unit of scarce water requires both better varieties and appropriate agronomy. The challenge is to manage the crop or improve its genetic makeup with multiple desirable traits. There is scope for developing genotypes that are able to maintain adequate floret fertility despite severe water deficit during floral development. Marker-assisted selection has helped in controlling some root diseases that limit water uptake, and in maintaining fertility in water-stressed maize (Passioura 2004).

Although marker-assisted selection is widely used in wheat, it has not contributed significantly to variety improvement for adaptation to low-yielding environments and breeding has relied largely on direct phenotypic selection for improved performance in these difficult environments. To better breed for drought tolerance, a multi-disciplinary approach, taken into account interaction among multiple stresses and plant phenology, and integrating the physiological dissection of drought-tolerance traits and the genetic and genomics tools, such as quantitative trait loci (QTL), microarrays, and transgenic crops (Fleury et al. 2010).

Wild barley represents a major source of favorable alleles for increasing the genetic variation for multiple traits including resistance to both biotic and abiotic stresses. Kalladan et al. (2013) used advanced back-cross quantitative trait locus (AB-QTL) analysis of BC3-double haploid population developed between a cultivated parent and wild accession to study the contribution of wild barley in improving various agronomic and seed quality trait under post-anthesis drought. Results indicated that wild barley contributed positively to most of the traits studied under both control and drought conditions.

Wheat genetic resources have played a significant role in wheat improvement by contributing important sources of genes for yield potential, broad adaptation, short plant height, improved grain quality, and resistance/tolerance to major abiotic and biotic stresses. In view of the threat of genetic erosion due to many factors including climate change, efforts have been made to collect and conserve wheat genetic resources in genebanks. More than 900,000 wheat accessions of wild cultivar, landraces, synthetic wheats, breeding lines, and genetic stocks are conserved in different genebanks at the global level. Management and utilization of such huge but highly valuable genetic resources is a big challenge. Application of modern tools and strategies, such as Focused Identification of Germplasm Strategy (FIGS), effective gene introgression methods, and genomics are essential in improving genetic resource utilization and improving breeding efficiency (Tadesse et al. 2015).

The narrow genetic base among cultivated chickpea accessions is limiting genetic improvement of chickpea through breeding efforts. Exploring the extent of natural variation among cultivated chickpea accessions for drought tolerance is important to develop pre-breeding and breeding strategies for chickpea. Land races from ICARDA were evaluated by Kumar, et al. (2015) for their relative Water Content (RWC) and Membrane Stability Index (MSI) which are established physiological parameters for drought tolerance. Results indicated wide variability in landraces for drought tolerance.

To combat drought in chickpea and lentil in WANA region, where rainfall takes place in winter months, the key to success of crop improvement is to maximize the water-use efficiency in water-limited environments, thus making these crops profitable to farmers and able to retain their place in the cropping systems. Results indicate that winter sowing of chickpea in low-to-medium altitude areas and lentil in high altitude areas increase productivity due to increased water-use efficiency and escape of terminal drought (Malhotra et al. 2004).

Concluding Remarks and Recommendations

Yield gap in rainfed crops remain large enough to suggest considerable scope for increasing achievable yields.

Technology Adoption There is an unsatisfactorily large gap between realized and potential yields. Technologies are available to narrow the gap but their adoption by resource-poor farming communities faces many constraints including technical, local availability of information, timely input availability, risk management, economic, social and markets (Anderson, Johansen, and Siddique 2015). Thus, there is a need to invest more on quality and rigorous adoption research to better understand how and when new technologies are used by farmers and the extent of their scalability and wide dissemination in developing countries. Adoption research is extremely important and it has been largely used in the following areas of agricultural policy (Doss 2006):

- Assessing the impact of agricultural research
- Priority setting for research
- Evaluating the distributional impacts of new technologies
- Identifying and reducing constraints to adoption

The Next Green Revolution The dramatic increases in food production during the 1st Green Revolution demanded increased resources, particularly water, fertilizers, energy and land to a lesser extent. As argued by Austin (2010), those resources are no longer available to drive another "broad-scale green revolution". Land degradation is increasingly severe, surface and groundwater resources are at or beyond sustainable utilization limits. Clearly, future food production growth must be driven by productivity growth.

Austin (2010) further argued that the next revolution must increase yields without increasing inputs of water, fertilizer, pesticide or energy. Instead, it must increase "knowledge-intensity". In contrast to the "first Green Revolution's seed-fertilizer-water package, the next must deal with sustainability and environmental friendliness- toward agro-ecology, agro-forestry, and conservation agriculture".

Historically, agricultural science has delivered component technologies aimed at increasing farm-level productivity. The future will be more complex, and individual technologies are unlikely to drive innovation. Responses must be designed within increasingly complex market and institutional arrangements, and integrated value chains. The next revolutions will entail ecological sustainable and integrated innovations such as integrated pest management, precision farming, minimum tillage, and integrated fertility management. Food security cannot be achieved without addressing underlying social inequities, poverty/hunger, vulnerability, equity and sustainability.

The inter-dependence between food and health security will become increasingly apparent, with increasing prevalence of nutritionally related diseases and other transmissible agents. The nutrition dimension of food security has evolved from an early focus on protein deficiencies, then to calorific intake and most recently to micronutrient deficiencies. The next Green Revolution must go beyond cereals, and diversify to include high value crops.

Food Prices and the Poor Higher and more volatile food prices are a major cause for concern, particularly in developing countries, where populations spend 50–80% of their income on food, compared to <15% of disposable income in developed countries (Austin 2010). One study estimated that the 2008 spike in food prices significantly reduced the purchasing power of poor people in developing countries, e.g. by up to 32% in Arab countries (Shideed 2008).

Higher prices will stimulate greater investments in food production, but smallholder farmers are often slow to respond to price incentives, because of various factors. Empirical studies in six countries showed that limited response by smallholders to the 2007–2008 increases in cereal prices, was due to water scarcity and institutional and policy factors such as unavailability of credit, inputs and information; export restrictions and price controls (Khouri et al. 2011).

There is a likelihood of more frequent commodity price spikes in the years ahead, due to increased volatility in commodity markets, more direct coupling of food and energy markets, and increased incidence of extreme weather events linked to climate change (Austin 2010). Another factor is biofuels, whose production has trebled since 2000 and is projected to double again within the next decade. Liquid biofuels represent only 0.2% of energy consumption, but account for 7% of coarse grain use and 9% of vegetable oil consumption (Austin 2010). Recent studies suggest that real prices of maize and oilseeds will increase by 26 and 18% in 2020 as a result of biofuels production; and that at least one-third of agricultural land may be converted from food production to the production of crops used in ethanol production (Austin 2010).

Water Resources and Climate Change Water is the single most binding constraint to agriculture in dry areas. Per capita water availability in West Asia and North Africa, for example, is less than world average. Large areas have annual rainfall below 100 mm per year. Extraction rates are mostly unsustainable; groundwater levels continue to fall, while salinization levels are increasing. Most countries in the region will drop below the internationally defined "water scarcity" level in the near future.

Because water is not valued, farmers and users tend to abuse and overuse this scarce resource. The second problem is that because water is not valued, people are reluctant to pay for it. As a result water is inefficiently used and thus is subject to huge losses in both domestic and agricultural sectors. Most irrigation systems are less than 50% efficient. Farmers who get water free or for little cost have no incentive to reduce their usage with water-saving technologies (Catley-Carlson 2011).

In all regions, except Europe and North America, agriculture is by far the biggest user of water, accounting for about 69% of all global withdrawals (FAO 2001). By implication, water-saving technologies for agriculture would have fundamental impacts on conservation of the global resource base.

Climate change will have significant effects on agriculture. The projection models show that much of the world's dry areas will be severely affected by climate change. Large parts of West Asia and North Africa will experience at least 20% decrease in rainfall, and up to 50% in many areas. Declines in rainfall will directly translate into reduced production, and necessitate changes in cropping patterns, and agricultural production.

Climate change will also increase risk, because extreme events (droughts, floods, temperature spikes) are predicted to increase in frequency and magnitude. Greater variability will compound the problems of dry areas, where variability is already high. Developing countries will be hard hit, particularly countries in West Asia and North Africa, South Asia and Sub-Saharan Africa with extensive dry areas. These countries are highly dependent on agriculture, not only for food security but also for employment and economic growth. The International Food Policy Research Institute has projected that as a result of climate change, global wheat production would be reduced by 47%, rice by 27%, and maize by 13% in 2050 under irrigated conditions (Rosegrant 2011).

In developing countries with large dryland areas, there is usually a close correlation between GDP growth and rainfall, because of agriculture-dependent economies with limited diversification. The persistent association between rainfall and GDP growth in Ethiopia is a striking example of the vulnerability of rainfed agriculture-based economies to climate change (The World Bank 2009).

Policy and Research Implications This chapter illustrates the huge potential of technological innovation to improve food security, but also the need for supportive policies and institutions to encourage farmers to adopt these innovations.

To ensure food and water security in dry areas, four key areas must be addressed: investment deficits in agricultural research and development, widening yield gap coupled with slowdown in productivity growth, water scarcity, and knowledge gaps. Risk is a critical factor influencing adoption and thus it needs explicit attention in the diagnosis and intervention measures (Keating et al. 2010).

New technologies may improve yield and product quality, but may not necessarily improve input use efficiency. For example, the introduction of supplemental irrigation into traditionally rainfed agriculture has increased yields and reduced yield variability. However, the benefits will be fully realized only when farmers have adequate technical know-how about the SI technology. Without this, water savings may be far less than what can be achieved.

Irrigation method is the most important variable in explaining the variation in efficiency in both technical and irrigation water efficiencies. A shift from traditional surface irrigation to modern irrigation methods, particularly sprinklers, leads to higher output oriented technical efficiency and irrigation water technical efficiency. However, even among farmers who use modern irrigation methods, there is still about 25% water use inefficiency that can be eliminated by building farmer knowledge. Similarly, ICARDA's empirical studies in Syria show that irrigation water efficiency differential between the most efficient and most inefficient farm is in excess of 60% with the average farmer lagging behind the most efficient farmer by 28%. This indicates more wasteful use of irrigation water than other inputs – and the scope for saving irrigation water without reducing wheat production. These findings illustrate the need for extension interventions – with strong policy support –to enable farmers to make best use of new technologies.

Participatory, community-led approaches are critical, particularly for land and water management technologies. This approach has been successfully used by ICARDA and its partners in different countries. For example, the IFAD-supported Badia Rangelands Development Project in Syria is a participatory rangelands management project that has introduced the concept of grazing cost recovery by the end- users (IFAD 2010).

Different strategies are required to tackle the waste in food production and consumption, estimated at 30–40% in both developed and developing countries. In developing countries public investment in transport infrastructure would reduce spoilage while better functioning markets and the availability of capital would increase the efficiency of the food chain, for example, by allowing the introduction of cold storage (Godfray et al. 2010).

Financial analyses show that the rates of return for farmers to invest in some soil-conserving technologies, in marginal dry areas, are often not high enough to trigger technology adoption. Development investments are necessary to provide farmers with incentives to stimulate adoption. Such incentives can be justified because the economic rate of return to investment is satisfactory if these costs are accounted for. In addition, there are important environmental benefits -- in the form of reducing soil erosion, maintaining soil fertility, and conserving soil moisture -- that can be generated from public investments in dry areas. Successful, sustainable natural resources management requires approaches that go beyond the conventional and biophysical aspects and that capture the holistic nature of the problem by integrating economic, environmental and social aspects (Shideed et al. 2007). Land tenure, for example, is an important factor affecting the adoption of soil-conserving technologies. The long-term environmental benefits of conservation cropping may be irrelevant to farmers whose planning horizon is limited by insecure land tenure.

Research findings are available on many of the key issues, although important knowledge gaps still exist. The studies described in this chapter show that policies create most of the conditions that lead to greater resource-use efficiency level. Examples include farm size, water allocation and costs, cropping patterns, inputs, and crop pricing. Water allocation among crops, techniques and growing seasons should consider the level of efficiency which may be attained under each option. Well designed, implemented policies are the key to efficient use of scarce resources, growth in farm income and protection of the environment.

Regardless of the high rates of return to investments in agricultural research and development (R&D), growth in public R&D spending for food and agriculture has slowed down substantially.

Tackling the twin issues of increasing food security and addressing climate change, requires developing sustainable approaches that enable smallholder farmers to grow more food and feed more people while at the same time increasing their resilience to climate change and protecting the environment. Developing such approaches requires research, scientific innovation, financing, and the active involvement of smallholder farmers themselves to ensure that solutions are demand-driven and context-specific (IFAD 2015). The EU, IFAD and CGIAR Centers, including ICARDA, research partnership has demonstrated the impact that can be achieved when science, technology, local knowledge and the readiness to innovate come together.

From ICARDA's experiences in innovative dryland agriculture, it can be concluded that for attaining water and food security in the rainfed areas in the context of climate variability, it is essential to achieve productivity growth and intensification through tested and proven scientific techniques together with enabling policy environment.

Promoting integrated systems approach is vital for the eco-efficiency criteria of sustainable food production systems. Focused attention and priority needs should be given to policy and investment in the rain-fed areas. Supporting dryland agriculture is of prime importance because it holds the key to future food security, and possibly a new Green Revolution.

References

Adary, A., Hachum, A., Oweis, T., & Pala, M. (2002). *Wheat productivity under supplemental irrigation in northern Iraq, On-farm water Husbandry Research Report Series, No. 2*. Aleppo: ICARDA.

Ahmed, M. A. M., Shideed, K., & Mazid, A. (2010). Returns to policy-oriented agricultural research: The case of barley fertilization in Syria. *World Development, 38*(10), 1462–1472.

Alexandratos, N., & Bruinsma, J. (2012). World agriculture towards 2030/2050: the 2012 revision. ESA Working paper No. 12–03. Rome: FAO.

Alwang, J. (2015, January). *Impacts and lessons learned from ICARDA's strategic plan 2007–2016*. Amman.

Alwang, J. (2016). *Major trends and prospects in dry areas: Lessons for ICARDA's 2017–2025 strategic plan*. Strategy paper, ICARDA, 2016, pp. 39+ Figures.

Anderson, W., Johansen, C., & Siddique, K. (2015). Addressing the yield gap in rainfed crops: A review. *Agronomy for Sustainable Development, 36*, 18.

Anderson, W., Johansen, C., & Siddique, K. H. M. (2016). Addressing the yield gap in rainfed crops: A review. *Agron Sustain Dev*. Review Article, 1–13. doi: 10.1007/s13593-015-0341-y.

Asian Development Bank (ADB). (2010, December). *Operational plan for sustainable food security in Asia and the Pacific*. ADB, Manila.

Austin, N. R. (2010). *The science of food security*. http://www.abare.gov.au/outlook/download/food_Austin.pdf.

Bader, A. (2010). *Economic and social assessment to the adoption of improved supplemental irrigation in wheat production system in Syria*. PhD thesis, Aleppo University, Aleppo.

Bazza, M., & Ahmed, M. (2002). *A comprehensive assessment of links between irrigation water pricing and irrigation performance in the Near East*. Paper presented at the conference on Irrigation Water Policies: Micro and Macro Considerations, Aghadir, Morocco, 15–17 June 2002.

Carberry, P., Keating, B., Bruce, S., & Walcott, J. (2010). *Technological innovation and productivity in dryland agriculture in Australia*. A joint paper prepared by ABARE-BRS and CSIRO, Canberra, July 2010. Science Review.

Catley-Carlson, M. (2011). Water, water everywhere. *Nature, 473*, 27–28.

CGIAR. (2011). *A strategy and results framework for the CGIAR*. Washington, DC: CGIAR.

CGIAR Science Council. (2006). *Impact assessment of policy-oriented research in the CGIAR: A scoping study report*. Rome: Science Council Secretariat.

Chambers, R. G., & Just, R. E. (1989). Estimating multi-output technologies. *American Journal of Agricultural Economics, 71*, 980–995.

Doss, C. R. (2006). Analyzing technology adoption using microstudies: Limitations, challenges and opportunities for improvement. *Agricultural Economics, 34*(2006), 207–219.

El-Shater, T. (2009). *Impact of alternative agricultural policies on groundwater use, food security and farmer income in stability zones Z1 and Z2 in Aleppo province*. MSc thesis, Damascus University, Syria.

ESCWA/ICARDA. (2003). *Enhancing agricultural productivity through on-farm water use efficiency. An empirical case study of wheat production in Iraq*. New York: United Nations.

FAO. (2001). *Crops and drops: Making the best use of water for agriculture*. Rome: FAO.

FAO. (2011). *Save and grow: A policymaker's guide to the sustainable intensification of smallholder crop production*. Rome: FAO.

FAO. (2013). *Food security indicators*. http://www.fao.org/economic/ess/ess-fs/ess-fadat/en/

Fleury, D., Jefferies, S., Kuchel, H., & Langridge, P. (2010). Genetic and genomic tools to improve drought tolerance in what. *Journal of Experimental Botany, 61*(1), 3211–3222.

Fraiture, C., & Perry, C. (2002). *Why is Irrigation water demand inelastic at low price ranges?* Paper presented at the Conference N irrigation water policies: Micro and macro considerations, 15–17 June 2002, Aghadir, Morocco.

Fuglie, K. (2012). Productivity growth in the global agricultural economy and the role of technology capital. In K. Fuglie, V. E. Ball, & S. L. Wang (Eds.), *Productivity growth in agriculture: An international perspective*. CAB Internaional: Oxfordshire.

Godfray, H., Charles, J., et al. (2010). *Food security: The challenge of feeding 9 billion people*. Sciencexpress Review. Sciencexpress / www.sciencexpress.org/ 28 January 2010.

Haddad, N., et al. (2011). The potential of small-scale rainfed agriculture to strengthen food security in Arab Countries. *Food Security, 3*(Supplement 1), S163–S173.

Hoekstra, A. Y., & Mekonnen, M. M. (2012, February). The water footprint of humanity. PNAS, *109*(9), 3232–3237

ICARDA. (2011). *Strategies to reduce the emerging wheat stripe rust disease*. Synthesis of a dialog between policy makers and scientists from 31 countries at the international wheat stripe rust symposium, Aleppo, Syria, April 2011. Research to Action Seties no. 1. ICARDA, Aleppo. 27 pp.

ICARDA. (2015, July 14). *Defining a rights-based approach to water for food security*. Theib Oweis. http://www.icarda.org/drylandsthinking/

IFAD. (2010). *Climate change and sustainable water resources management: IFAD's experiences in the Near East and North Africa and in Eastern and Central Europe*. Background paper for PN Side Event (Executive Summary). Governing Council, 17–18 February 2010. IFAD, Rome.

IFAD. (2015, November). *Climate change and food security: Innovations for smallholder agriculture*. EC, IFAD, CGIAR.

International assessment of agricultural knowledge, science and technology for development (IAASTD). (2009). *Central and West Asia and North Africa (CWANA)*. Wahington, DC: Island Press.

International Food Policy Research Institute (IFPRI). (2015). *2014–2015 Global food policy report*. Washington, DC: Intenational Food Policy Research Institute.

Ilbeyi, A., Ustun, H., Oweis, T., Pala, M., & Benli, B. (2006). Wheat water productivity in a cool highland environment: Effect of early sowing with supplemental irrigation. *Agricultural Water Management, 82*, 399–410.

Just, R. E., Zilberman, D., & Hochman, E. (1983). Estimation of multicrop production functions. *American Journal of Agricultural Economics, 65*, 770–780.

Just, R. E., Zilberman, D., Hochman, E., & Bar-Shira, Z. (1990). Input allocation in multicrop systems. *American Journal of Agricultural Economics, 72*, 200–209.

Kalladan, R., et al. (2013). Identification of quantitative trait loci contributing to yield and seed quality parameters under terminal drought in barley advanced backcross lines. *Mol Breeding, 32*(1), 71–90. doi:10.1007/s11032-013-9853-9.

Keating, B. A., Carberry, P., Bindraban, P. S., Senthold, A., Meinke, H., & Dixon, J. (2010). Eco-efficient agriculture: Concepts, challenges, and opportunities. *Crop Science, 50*(March–April 2010, Symposia), s1–s11.

Khouri, N., Shideed, K., & Kherallah, M. (2011). Food security: Perspectives from the Arab world. *Food Security, 3*(Supplement 1), S1–S6.

Malhotra, R., Ashutosh Sarker, & Mohan Saxena. (2004). *Drought tolerance in Chickpea and Lentil-Present status and future strategies*. Crop Science Society of America and American

Society of Agronomy, 677 S. Segpe Rd., Madison, WI 53711, USA, Challenges and Strategies for Dryland Agriculture, CSSA Special Publication no. 32, pp. 257.
Moore, M. R., Gollehan, N. R., & Carey, M. B. (1994a). Multicrop production decisions in western irrigated agriculture: The role of water price. *American Journal of Agricultural Economics, 76*, 859–874.
Moore, M. R., Gollehan, N. R., & Carey, M. B. (1994b). Alternative models of input allocation in multicrop system: Irrigation water in the Central Plains, United States. *Agricultural Economics, 11*, 143–158.
MunlaHassan, A. (2007). *Water use efficiency in Syrian agriculture*. NAPC Working Paper No. 26. Damascus, Syria.
Ncube, M., Anyanmu, J. C., & Hausken, K. (2013). *Inequality, economic growth and poverty in the Middle East and North Africa (MENA)*. Working paper series No. 195 African Development Bank, Tunis, Tunisia.
Norton, G. W., & Alwang, J. (1998, December). *Policy for plenty: Measuring the benefits of policy-oriented social science research*. Impact assessment of the IFPRI agricultural science and technology. Washington DC: IFPRI.
Oweis, T. (1997). *Supplemental irrigation: A highly efficient water use practice*. Aleppo: ICARDA.
Oweis, T., & Hachum, A. (2004). *Water harvesting and supplemental irrigation for improved water productivity of dry farming systems in West Asia and North Africa*. In: New directions for a diverse planet. Proceedings of the 4th international crop science congress, 26 Sep–1 Oct 2004, Brisbane, Australia. http://www.cropscience.org.au
Oweis, T., & Hachum, A. (2009). Optimizing supplemental irrigation: Tradeoffs between profitability and sustainability. *Agricultural Water Management, 96*, 511–516.
Passioura, John. (2004). *Increasing crop productivity when water is scarce-from breeding to field management*. In Proceedings of the 4th International crop science congress, 26 Sep–1 Oct 2004, Brisbane, Australia.
Piggin, C., Haddad, A., & Khalil, Y. (2011). *Development and promotion of zero tillage in Iraq and Syria*. Paper accepted for presentation at 5th World congress of conservation agriculture and 3rd farming systems design conference, 26–29 Sep 2011. Brisbane, Australia.
Piggin, C. (2014). Crop yields and global food security: Will yield increase continue to feed the world, by Tony Fischer, Derek Byerlee and Greg Edmeades. xxii +634pp. Canberra ACIAR Monograph No. 158- ISBN 978. *Journal of Agricultural Sciences, 152*, 1012.
Pretty, J., et al. (2010). The top 100 questions of importance to the future of global agriculture. *International Journal of Agricultural Sustainability, 8*(4), 219–236.
Reyazul, R. M., et al. (2012). Integrated genomics, physiology and breeding approaches for improving drought tolerance in crops. *Theor Appl Genet, 125*, 625–645.
Rosegrant, M.W. (2011). Impacts of climate change on food security and livelihoods. In Solh, M., & Saxena, M. C. (Eds.), Food security and climate change in dry areas: Proceedings of an International Conference, 1–4 Feb, Amman, Jordan. ICARDA, Aleppo. pp: 24–26.
Ryan, J. G. (1999). *Assessing the impact of rice policy changes in Viet Nam and the contribution of policy research*. Impact Assessment Discussion Paper No. 8. International Food Policy Research Institute (IFPRI), Washington, DC.
Ryan, J. G. (2002). Assessing the impact of food policy research: Rice trade policies in Viet Nam. *Food Policy, 27*, 1–29.
Salkini, A.B., & Ansell, D. (1992). *Agro-economic impact of supplemental irrigation in rainfed wheat production under the Mediterranean environment of Syria*. Paper presented at International conference on supplemental irrigation and drought water management, Valenzano-Bari, 27 Sep – 20 Oct 1992.
Shideed, K. (2008). Food security and its socioeconomic impact: Towards reducing the food gap in the Arab countries. *Arab Agricultural Investment Journal, 6*(2008), 31–35.
Shideed, K., Oweis, T., Gabr, M., & Osman, M. (2005). *Assessing on-farm water-use efficiency: A new approach*. Aleppo: ICARDA xiv + 86 pp.

Shideed, K., Alary, V., Laamari, A., Nefzaoui, A., & El-Mourid, M. (2007). Ex-post impact assessment of natural resource management technologies in crop-livestock systems in dry areas of Morocco and Tunisia. In H. Waibel & D. Zilberman (Eds.), *International research on natural resource management: Advances in impact assessment* (pp. 169–195). Wallingford: FAO and CABI International.

Shideed, K., Mazid, A., Ahmed, M. A., & Zahir, Z. (2008). *Policy influence and returns to policy-oriented agricultural research: The case of barley fertilization in Syria*. International Center for Agricultural Research in the Dry Areas (ICARDA) and the Syrian Ministry of Agriculture and Agrarian Reform (SMAAR). Syria, Aleppo and Damascus. 58 pp.

Solh, M. (2011). *Harnessing research and innovation for Arab food security*. 1st Arab development symposium on food and water security in the Arab world, 14–15 March 2011, Kuwait.

Tadesse, W. et al. (2015). *Wheat. A chapter in a book on: Genetic and genomic resources for grain cereals improvement*. Elsevier, ISBN: 978-0-12-802000-5

The Independent. (2015, July 12). *World entering era of global food insecurity with malnutrition and obesity side by side within countries, say leading food expert* http://www.independent.co.uk/news/world/

The World Bank. (2009, December). *Making development climate resilient: A World Bank strategy for sub-Saharan Africa*. Report No. 46947-AFR, Sustainable Development Department.

The World Bank. (2011). *Rising global interest in farmland: Can it yield sustainable and equitable benefits? Klaus Deininger and Derek Byerlee with Jonathan Lindsay, Andrew Norton, Harris Selod, and Mercedes Stickler*. Washington, DC: World Bank.

Vermeulen, S. J. (2014). *Climate change, food security and small-scale producers*. CCAFS Info Brief. CGIAR Research Program on Climate Change, Agriculture and Food Security (CCAFS). Copenhagen, Denmark. Available online at: www.ccafs.cgiar.org

World Bank. (2014). http://www.worldbank.org/en/news/press-release/2014/04/11/remittances-developingcountries-deportations-migrant-workers-wb. Accessed 29 June 2014.

World Bank. (2015). *Agriculture and rural development in MENA*. http://siteresources.worldbank.org/INTMENAREGTOPAGRI/Resources/AGRICULTURE-ENG-2008AM.pdf. Accessed 29 June 2015.

Yigezu, Y.A., Ahmed, M.A., Shideed, K.H., Aw-Hassan, A. El-Shater, T., & Al-Atwan, S. (2011a). *Economic and environmental impacts of supplemental irrigation in rained agriculture: The case of wheat in Syria*. ICARDA. Forthcoming.

Yigezu, Y.A., Ahmed, M.A., Shideed, K.H., Aw-Hassan, A. El-Shater, T., & Al-Atwan, S. (2011b). *The effect of changes in irrigation technology on technical, water use, and cost efficiencies: The case of supplemental irrigation in Syrian wheat farms*. ICARDA. Forthcoming.

Zilberman, D., & Waibel, H. (2007). Productivity enhancement and natural resource management. In H. Waibel & D. Zilberman (Eds.), *International research on natural resource management: Advances in impact assessment* (pp. 21–55). Wallingford: FAO and CABI International.

Ensuring Food Security by Improving "Freshwater Use Efficiency" or by Farming the Seas

Imad Patrick Saoud

Abstract The present chapter discusses water use and food production in a freshwater challenged world with a growing human population. Proposed approaches to ensure food security are multifaceted but simple. First, we need to improve water use efficiency in all food production systems. Second, we need to develop technology to farm dry or semi-arid areas of the world. Third, we should integrate various fields of agriculture and animal husbandry to improve water productivity. Fourth, we should learn to go for proven local technologies rather than glamourous setups that do not work in rural settings. Fifth, we should seriously start working on farming the oceans. Last but not least, we need to invest in education and training. The common thread in all these suggestions is the use of aquaculture to improve food production efficiency. These proposals are not novel ideas but most governments have failed to implement even the simplest of suggestions. However, now that climate change adaptation plans are at the forefront of international discussions, maybe more countries will implement suggestions for water use efficiency improvement summarized in this chapter in order to better ensure sustainable food production.

Keywords Arid farming • Farming efficiency • Water recycling

Introduction

Food security and livelihoods of millions of people are threatened by growing world population and associated climate change. The planet's human population is projected to exceed nine billion by the year 2050. Consequently, estimated food demand necessitates an increase in agricultural production by at least 70% (FAO 2009). This presents a challenge in light of current unsustainable food production practices, limited suitable arable land and very limited freshwater resources. Agriculture and wild harvests use excessive energy and waste freshwater resources in addition to emitting substantial amounts of greenhouse gases (GHG) (McMichael

I.P. Saoud (✉)
Department of Biology, American University of Beirut, Beirut, Lebanon
e-mail: is08@aub.edu.lb

S. Murad et al. (eds.), *Water, Energy & Food Sustainability in the Middle East*,
DOI 10.1007/978-3-319-48920-9_15

et al. 2007; Escobar et al. 2009). Moreover, climate change and global warming, associated with increased emissions of GHG also impact availability of useable freshwater (Doll and Siebert 2002; Barnett et al. 2005; Alcamo et al. 2007; Spash 2007), crop productivity (Droogers 2004; Droogers and Aerts 2005) and biodiversity (Fischer et al. 2007; Hanjra and Qureshi 2010), thus further threatening food security. Most vulnerable to these changes are semi-arid and arid regions that generally tend to be underdeveloped and poor (Rosenzweig and Parry 1994; IPCC 2007). Consequently, it is imperative to develop and integrate innovative technologies that can increase food security with minimal undesirable impacts on the environment and on livelihoods of rural communities in less developed nations. Investment in such novel ideas and technologies can potentially improve farming systems, make more efficient use of land, energy and water resources, increase food and nutritional security, enhance biodiversity and mitigate agricultural contribution to climate change. Considering the fact that freshwater resources are the limiting factor, two possible solutions are: (1) To produce more food using freshwater resources presently being exploited or, (2) To farm marine waters. In both cases, aquaculture would be a major part of the solution.

During the next 50 years, problems associated with a lack of water will affect virtually everyone on the planet (Gleick 1992). At least one-third of the world's population lives in countries experiencing medium to high water stress (Stockholm Environment Institute (SEI) 1997). In the Middle East (ME), the water situation is precarious. Middle East countries have experienced rapid population growth during the past few decades, yet limited technical and institutional development in the water sector took place (Diliman 1989; Haddadin 1989; Gleick 1992; Arlosoroff 1995; Haddad and Mizyed 1996), thus reducing per capita fresh water availability. In the Mediterranean region, agriculture consumes about 72% of available freshwater resources (Hamdy and Lacirignola 1999) and thus will be the economic sector most affected by freshwater shortages (Water and Environmental Studies Center 1995). The intuitive solution is thus to produce more food using the same freshwater resources or to use salt water. In the 1970s, aquaculture accounted for about 6% of fish available for human consumption. In 2006, the figure was 47% and some reports suggest that in 2016, more than 52% of finfish consumed around the world will have been grown in farms. The large majority of this aquaculture production is in freshwater, yet supply of adequate freshwater is one of the major problems faced by many countries.

Freshwater resources of planet earth amount to approximately 2% of all water available and the majority of this freshwater is in unusable ice in Polar Regions. The remainder marine water has traditionally been underutilized mainly because humans are terrestrial organisms and, because salt water is very corrosive to metals. However, with new technologies and materials available to us, we are now capable of producing much more food using salt waters. This can be done on coastal marginal lands in ponds and tanks or *in situ* in cages. Production can be fish, invertebrates, algae, plants or even oil seeds. However, like all agricultural practices, marine aquaculture is not totally environmentally friendly. Some of the problems will be discussed below. The remainder of the chapter will discuss the various methods in which aquaculture can supposedly improve water productivity. Some ideas are controversial, and some are not as beneficial as we would like to believe. I will limit myself to food production in freshwater stressed regions of the world because that is where the problem is most acute.

Aquaculture vs. *Terrestrial Animal Husbandry*

Environmental agriculturists today talk of a human ecological footprint of food production. The concept of an ecological footprint was first mentioned by Rees (1992) and then used by Wackernagel and Rees (1996) to estimate earth's carrying capacity of humans. Furthermore, the UNESCO-IHE, Institute for Water Education in the Netherlands developed a method of water use analysis that takes into account the quantity of water used for producing various commodities and the virtual water content of products needed for food production such as fossil fuel and land use (see Hoekstra and Chapagain 2007). By combining those methods, Steinfeld et al. (2006) showed that the footprints of livestock production are not limited to water consumption but also to water pollution by animal wastes, use of fertilizers and pesticides for the production of feeds, sediment inputs from eroded pastures and feed production, and hormones and antibiotic use. In the United States, livestock production is estimated to account for a third of nitrogen and phosphorus pollution in freshwater. In addition, degradation of ground water resources occurs through soil compaction that reduces natural infiltration, and livestock production is responsible for emission of greenhouse gases from animals and manure. Moreover, expansion of livestock production is a key factor driving deforestation and loss of biodiversity with consequent effects on water resources. Cajas-Cano and Moffitt (2008) used these calculations to compare resource use and ecological footprint of beef production *vs.* trout production. They found that if meat-eaters substituted beef with trout meat, they would significantly decrease pressure on water resources. Their findings suggest that the amount of freshwater needed for production of one kg of beef dressed and boneless averaged around 16,555 L. Alternatively, one kg of boneless trout filets requires 2700 to 8300 L of water when water needed for production of plant proteins and feed ingredients was taken into account. If water is treated before being returned to natural water bodies, the consumptive water use to produce a kg of trout meat would be even less (Engle et al. 2005).

Fish are generally more efficient animals to produce than any other domesticated terrestrial animal (Olah and Sinha 1986). As poikilotherms, fish do not use energy to heat their bodies. Because they excrete ammonia, fish use little energy in protein catabolism and nitrogenous waste excretion (Goldstein and Forster 1970). Furthermore, because they live in a dense environment that carries their weight, fish do not need heavy bones (Tucker 1969). For example channel catfish (*Ictalurus punctatus*) gain 0.85 g of weight for every gram of feed consumed, compared to 0.48 g in chickens, the most efficiently farmed warm-blooded animal. Beef cattle gain 0.13 g for every gram of food consumed on average although the number varies between free range cattle and stock fed cattle (NRC 1983; Lovell 1989). In terms of consumptive water use, fish use no more, and in many cases less, than do other animals (Brummett 1997). For plant crops, with which fish production competes both for nutrient inputs and for fresh water, the calculations are not as clear cut.

Aquaculture vs. *Terrestrial Plant Production*

"For most farming businesses, efficiency is measured in economic terms; that is, the amount of money spent on a farming activity (including costs of inputs, labor, management, opportunities on land and capital, etc.) is compared to the amount earned through the sale of produce" (Brummett 2007). In a world of 9 billion people, that calculation would no longer be sufficient to ensure sustainability of food production. It has become necessary to measure biophysical limiting factors, of which the most limiting is freshwater. Traditional farming systems also neglect the costs of environmental goods and services in their economic assessments (Berg et al. 1996; Kautsky et al. 1997). Brummett (2007) lists some of these non-tangible costs to include:

- Oxygen requirements for decomposition of organic wastes;
- Fertilizer runoff such as phosphorus and nitrogen;
- Ecological impacts of pesticides and herbicides;
- Health consequences of antibiotic use in animal feeds;
- Production of CO_2 and methane as in rice farming;
- Negative environmental impacts of introduced alien or genetically modified organisms.

Because freshwater availability and use is the major component of any modern calculation of sustainability in food production systems, especially in arid and semi-arid regions of the world where future growth in food production has to occur, I will limit the present discussion to water. In a food limited world, water cannot be considered only as an economical component of the equation. We need to quantify edible or usable dry matter produced per unit of water used; protein and essential protein produced per unit of water used; and digestible energy produced per unit of water used. When such parameters were taken into consideration in addition to economic profitability of food production systems by Brummett (2007), he concluded that in terms of edible dry matter output per unit of water, drip-irrigated cucumber is the most efficient, followed by drip-irrigated tomato and furrow-irrigated onion. In terms of crude protein production per unit of water consumed, pivot- irrigated maize ranks highest. In terms of digestible energy per unit of water, rain-fed cassava was the most efficient. However, when all factors are integrated, intensive vegetable production with drip irrigation in greenhouses is the most efficient way to use water to produce edible vegetal dry matter.

Fish production systems are different from agriculture systems in that the water necessary to fuel the system is not completely consumed. Consumption is highest in earthen ponds where seepage and evaporation can sometimes be considerable, especially in hot, dry, windy areas. Flow-through raceways must pass large quantities of water through the production unit, but the quality of water released from these systems is good and readily available for other uses, especially crop irrigation. Cages and recirculating systems consume virtually no water. The main consumptive water use of a fish is what it holds in its biomass which is about three quarters of its weight (Abdul-Rahman et al. 2011). For farms using manufactured feeds, the water requirements of the crops to produce the feeds should be added to the calculation (Piemental et al. 1997). These calculations suggest that aquaculture is of comparable efficiency to crop production only in terms of edible dry matter output per cubic meter of water used. Accordingly, aquaculture in arid areas is more efficient than agriculture only when it is highly intensive and/or integrated with other farm enterprises so that water costs are distributed across several production units. However, rural citizens living on marginal lands generally do not have the education or capital necessary to manage intensive aquaculture systems. Accordingly, our short-term goal should be to integrate aquaculture with agriculture to produce more crop per drop thus improving water use efficiency and simultaneously invest in capacity building to improve education and infrastructure in marginal areas in preparation for future intensification of aquaculture processes.

Integrated Agriculture-Aquaculture Systems

Integrated agriculture–aquaculture (IAA) allows for efficient use of water particularly in arid and semi-arid regions (Palada et al. 1999; McIntosh and Fitzsimmons 2003; Abdul-Rahman et al. 2011). It potentially reduces the cost of water and amount of fertilizer needed for crops and increases water productivity (Al-Jaloud et al. 1993; D'Silva and Maughan 1994, 1996; Azevedo 1998). Integrating agriculture with fish farming could improve fish pond water quality, reduce environmental impact of nutrient rich water discharge, reduce cost of water and amount of chemical fertilizer needed for crops, diversify farm production, increase income, and thus increase water efficiency (Billard and Servrin-Reyssac 1992; Ghate and Burtle 1993). If aquaculture is added to existing agriculture systems, it becomes a non-consumptive, productive segment that does not compete with irrigation. Moreover, the long-term performance of diversified farms is better than non-diversified enterprises for various reasons. Monoculture fish farms operate at high levels of risk from diseases, water quality and price fluctuations (Naylor et al. 2000; Pant et al. 2005), and the same idea holds for monocrop agriculture practices.

Various researchers have demonstrated the benefits of integrating agriculture with existing aquaculture facilities (Naegal 1977; Lewis et al. 1978; Watten and Buch 1984; Zweig 1986; McMurtry et al. 1990, 1993; Parker et al. 1990; Olson 1992; Al-Jaloud et al. 1993; Rakocy et al. 1993; Seawright 1993; D'Silva and

Maughan 1994, 1996; Palada et al. 1999; Cruz et al. 2000; Al-Ahmed 2004; Abdul-Rahman et al. 2011). In the Middle East, aquaculture is in its infancy but traditional animal husbandry and irrigated agriculture projects are abundant. Peoples' lives and societies were built around agriculture whilst aquaculture is not part of everyday life. I maintain that if aquaculture is introduced in addition to traditional agriculture as a method to increase water value and farm productivity, food safety in arid lands would improve while reducing energy use per unit of digestible energy produced. The argument holds true for all semi-arid populated regions of the planet.

Aquaponics

In recent years we have witnessed an explosion of interest in aquaponics. The concept is "sexy" and the projects glamorous because they are new and different. Basically, aquaponics is a process whereby farmers grow fish in a tank and pump effluent water over the roots of plants and then return the water to the fish tank. The idea is that plants would use fish metabolites as nutrients thus removing the need for fertilizers and making the crops "organic" and the water needed for plant production would be conserved thus reducing consumptive water use. Unfortunately, the reality of the matter is not so simple. First of all, the system needs continuous water pumping and aeration, thus limiting the use to areas with reliable power supplies. Secondly, plant roots are not very efficient filters and become less efficient in the presence of too many particulates in the water, thus necessitating a filter to be placed between fish tank and plants. Thirdly, fish metabolites are not a complete plant nutrient mix so some micronutrients such as iron need to be supplemented to the system in specific dosages. Fourth, since the plant water will be recycled to the fish, farmers cannot use pesticides to aid in fighting common greenhouse plant diseases and thus have to rely on natural enemies for their pests. All this requires quite a high level of education and professionalism, thus limiting the possibility of success of rural farmers. Finally, and possibly most importantly, in most regions of the world vegetables can still be produced cheaply enough using traditional agricultural practices, making aquaponically produced vegetables too expensive. I suggest that a much more suitable solution in the short term would be to produce fish in tanks and then use the water for irrigation of greenhouse produced vegetables. Aquaponics is a good idea in remote islands or very arid countries such as the UAE and Qatar where freshwater is scarce and vegetables have to be imported at high cost. However, as a method to improve food security or water productivity, it is useless.

How Do We Increase Food Production Sustainably Using Available Water?

The title above is the proverbial "million dollar question". There is no single perfect answer to the question but there are many partial answers which when put together could ensure adequate and sustainable food production in the future. Some of those answers involve technological advances, genetically modified organisms, better use of marginal lands, use of alternative or "green" energy, better storage and processing of produce and implementation of good management practices. However, this chapter is about water use and I will limit my discussion to that. There are five major underutilized categories of water that can be used for food production in addition to available freshwater. These are: 1- Marine water; 2- Fossil ground water; 3- Inland low salinity ground water; 4- Grey water; and 5- Treated sewage. Fossil ground water is not a renewable resource and will not be included in the discussion. Grey water and treated sewage water will be pooled with freshwater.

Inland Low Salinity Ground Water

A significant proportion of inland groundwater worldwide (e.g. Australia, Brazil, China, Ecuador, India, Thailand, USA) is too saline for traditional agriculture and food production practices (Saoud et al. 2003). These waters can be used to produce low-salinity tolerant aquatic organisms (Joshi and Tyagi 1994; Forsberg et al. 1996; Allan et al. 2001; Mourad et al. 2012). However, inland saline well waters (ISWW) do not have the same ionic composition as marine or brackish waters (Saoud and Davis 2005) and are not suitable for survival and growth of many aquatic species, unless chemically modulated. Saline ground waters are generally deficient in potassium (K^{+}) (Forsberg and Neill 1997; Boyd and Thunjai 2003; Saoud et al. 2007) and magnesium (Mg^{++}) (Roy et al. 2009). As the most abundant intracellular ion in all animals, potassium plays basic physiological functions including maintenance of membrane potentials and cellular volume, acid/base balance, cardiac function and nerve impulse transmission. In fish, potassium plays additional critical roles in osmoregulation (Marshall and Grosell 2006). Some robust osmoregulators such as silver perch *Bidyanus bidyanus* (Ingram et al. 2002; Doroudi et al. 2007) and rainbow trout *Oncorhynchus mykiss* (Ingram et al. 2002) grow well in non-chemically modified inland saline groundwater. However, potassium deficiency was shown to negatively impact the performance of various other high market value aquaculture species such as Pacific white shrimp *Litopenaeus vannamei* (McNevin et al. 2004; Davis et al. 2005; Roy et al. 2007; Saoud et al. 2007) Australian snapper *Pagrus auratus* (Fielder et al. 2001; Partridge and Furey 2002), barramundi *Lates calcarifer* (Partridge and Creeper 2004; Jain et al. 2006; Partridge and Lymbery 2008), western king shrimp *Penaeus latisulcatus* (Prangnell and Fotedar 2005), tiger shrimp *Penaeus monodon* (Ingram et al. 2002; Collins and

Russell 2003; Shakeeb-Ur et al. 2005), mulloway *Argyrosomus japonicas* (Doroudi et al. 2006; Partridge et al. 2006) giant prawn *Macrobrachium rosenbergii* (Fielder et al. 2008) and rabbitfish (Mourad et al. 2012). There are also other ions that affect the suitability of ISWW for aquaculture such as Mg^{++} and SO_4^{--} (Saoud et al. 2003). The knowhow to use ISWW for aquatic organism production is widely available and inland saline waters are increasingly being used for aquaculture. Two of the latest examples are European seabass being grown in Egypt and marine shrimp in inland Mexico. Governments can increase food production by facilitating technology transfer and reducing the red tape needed for investment in non-traditional agriculture.

Fresh Water

Freshwater available for human use is less than 0.1% of all water on the planet and much of that is polluted, non-sustainably utilized or inefficiently utilized. Some terrestrial areas of the world have lots of precipitation in large watersheds such as in northern Canada and in tropical rainforests but for various reasons, these areas cannot be used for large scale agriculture production. Most of the regions that are used to feed the world such as Australia, Southern Russia, Eastern Brazil, Northern Argentina, Midwest USA and California, and Western Europe, Pakistan and Ukraine tend to have periodic water problems that seem to be intensifying with climate change. These regions will continue to produce lots of food but unless there are major shifts in attitudes towards biotechnology and genetic enhancement of crops, production will not increase to feed the two billion extra people expected in the next 30 years. Terrestrial production has to start using marginal lands and often they are marginal because of water stress.

Deserts account for one third of the planet's land mass (25.5 million km^2), and are found on every inhabited continent. These harsh environments are characterized by high day temperatures and solar radiation, cold nights, little precipitation and very low relative humidity (Hochman and Brill 1994). Developing aquaculture in such environmental and physical conditions necessitates the adoption of production strategies focused on good water management which includes the use of water recycling as well as protection against solar radiation. Accordingly, the introduction of modern aquaculture technologies requiring more knowhow than traditional methods used in wet regions is necessary. Desert-suitable aquaculture systems should have a relatively small surface to volume ratio and include extremely efficient water usage technologies with biomass holding capacities of up to 50 kg fish/m^3 of water (Kolkovski et al. 2011) when possible. Additionally, all effluents from these fish farms should be reused for irrigation or other productive industries.

Many methods have been proposed to conserve or recycle water but because agriculture water is often more than 70% of a nation's freshwater utilization, water stress solutions should start with the agricultural sector. We need to start using agriculture water much more efficiently. One way to do that is to produce better nutrition using the same water (fish instead of Buffel grass) or better yet, to start producing two crops with the same water (fish and Buffel grass). We also need to start developing technologies that work in semi-arid and arid regions of the world. Accordingly, the rest of my discussion about freshwater use is going to concentrate around arid lands (regions where evapotranspiration is much greater than precipitation) with the understanding that I believe that all agriculture should be treated as if it is arid land agriculture. Many desert areas around the world have developed agriculture practices to suit their environment, and irrigate using limited groundwater supplies. Such practices were acceptable in the 1950s up till the late 1900s but no more in a world where water scarcity is increasing.

A recent project I was involved with in the arid regions of northern Mexico calculated that 800 L of water were being used to produce one kg of alfalfa and 9 kg of alfalfa were needed to produce 1 kg of beef which yielded about 300 gm of meat. Accordingly, without accounting for drinking water and polluted water, 1 kg of beef meat requires 24 tons of water just to produce feed for the animal. Similar calculations of water inefficiency are available for practically any crop planted in an arid area. The most obvious solution would be to use the water to grow fish and then use the effluent from fish farms to grow said crops. Furthermore, intensive animal husbandry in arid areas is often economically not feasible and the water used for such operations cannot then be used for irrigation. Accordingly, the importance of aquaculture in supplying protein in such areas is increasing rapidly worldwide (Crespi and Lovatelli 2011). Arid land fish farming was first suggested in 1963 and empirical research later showed that it was possible to use desert brackish waters to rear fish successfully (Fishelson and Loya 1969). The ionic profile of desert groundwater coupled with high solar radiation improves primary productivity and together create a very productive aquatic environment (Pruginin et al. 1988). By adapting existing technology to arid areas, we could increase food production significantly without competing with existing industries and without using more water.

If predictions by the World Meteorological Organization (WMO) about climate change and desertification are realized, then many areas of the world that are presently suitable for traditional agriculture will become arid. Arid lands already are home to 50 percent of the world's livestock (UNCCD 2007). In 2007, the United Nations Development Program/United Nations Office to Combat Desertification (UNDP/UNSO) reported that 13 percent of world population (approximately 313 million at the time) lived in arid zones with 92 million residing in hyper-arid areas (Smith et al. 2008). Since 2007, populations have increased as have arid areas. Consequently, a move towards using technology to improve water use efficiency in arid areas is becoming ever more important. Unfortunately, most research centers and decision makers around the world have yet to embark on this new odyssey.

Aquatic animal farming in desert ponds is not the only solution for the world's food and water problems. Greenhouse technology nowadays provides more control of parameters such as humidity, temperature, and radiation penetration thus improving production while reducing evaporation. The use of greenhouses does not necessarily require significant investments and can be used for commercial, as well as small-scale aquaculture initiatives. Coupled with vegetable farming, use of water for fish culture in greenhouses improves water use efficiency while greatly reducing the total amount of water needed for fish culture in conventional earthen ponds and for vegetables (Kolkovski et al. 2011). Many deserts have significant groundwater resources that are replenished very slowly. If technology is properly chosen so that water used for food production is highly productive and conserved, deserts could become bread baskets.

Desert aquaculture does not necessarily have to be for direct human consumption. Other commercially important and valuable organisms tolerant to salinity and temperature extremes are attractive candidates for commercial production in arid regions. Brine shrimp, *Artemia* and the unicellular green algae *Dunaliella* sp. are such examples. For example, about 60% of the world's natural ß-carotene is extracted from *Dunaliella salina* produced in large saline evaporation ponds in South and Western Australia (Benemann 2008). Huge quantities of brine shrimp are produced in the Great Salt Lake in Utah where yearly salinity and temperature variations are great. Similar environments in central Asia would be optimal for Artemia production using water that is not suitable for any other kind of agriculture.

It is easy to pontificate about the need to start using arid areas productively. However, the truth is that there are many policies that governments need to implement in order to encourage production in deserts. The most important strategies needed were suggested by Crespi and Lovatelli (2011) and are listed below:

- Promotion of aquaculture farming systems adapted to desert environments focusing on the smart use of water resources.
- Integration, as far as possible, of aquaculture activities with other existing production systems (agriculture, animal production, etc.).
- Inventory and chemical analysis of available surface and subsurface water resources to facilitate selection of suitable farm sites and species to be cultured.
- Support capacity building programs to strengthen national/local technical capacities through farmer field schools and ad hoc training initiatives.
- Provision of incentives for the establishment, upgrading and modernization of national feed processing plants.
- Support national programs on farm-made feed production to reduce dependency from expensive and often imported commercial feeds and improve the efficiency of on-farm feeding strategies particularly within more intensive farming systems.
- On-site high quality fingerlings production programs to give greater degree of independence for the farmers to obtain seed locally reducing at minimum the acquisition of the seed for small-scale aquaculture farmers living in remote areas. Small-scale aquaculture farmers would benefit from having local sources of seed available for stocking ponds/cages following a harvest. Long transport distances increase costs and reduce the viability of fingerlings stressed by high temperature and low oxygen levels.

- Promotion of national programs for the utilization of renewable energy sources (e.g. solar and wind energy) in remote areas not served by the national electricity grid.

Although these suggestions are very good, I propose that education and technology transfer coupled with a good extension service is the most important component of a successful desert production program. I have been involved in many of those projects and the only ones that I would consider a failure, or in the least not total success, are those where governments have helped in water testing, provision of feed and seed, provision of incentives and even economic help but did not really teach the farmers what to do nor assigned specific extension agents to help them.

Marine Food Production

Although the issues discussed above would improve food security if properly implemented, I believe we will not have sufficient freshwater to ensure complete nutrition for a 9 billion world. In the 1970s, aquaculture accounted for about 6% of fish available for human consumption. In 2006, the figure was 47% and some reports suggest that in 2015, more than 50% of finfish consumed around the world was grown in farms. The majority of finfish production (by quantity) is freshwater fish (reared mainly in China) but marine finfish aquaculture, including diadromous salmonids and eels, is growing at a faster rate than freshwater finfish aquaculture. In 2006, freshwater fish culture amounted to 54% of total aquaculture production whilst marine fish culture produced around 8% of total aquaculture production worldwide (the remaining production was crustaceans, mollusks and algae). However, freshwater fish production increased at an average rate of around 9% per year between 1970 and 2006 whilst saltwater fish production increased at an annual rate of approximately 16% during the same time period. The oceans were large enough to supply sufficient wild caught fish to meet world demand for marine fish, and technological constraints made freshwater fish aquaculture easier and more profitable than marine aquaculture. Two things have changed since 1970. World population has increased exponentially and technology has advanced beyond what we could imagine at the time. This means that we need to and can start producing food in the 97% of the water on the planet that is not suitable for traditional agriculture. We need to start using marine agriculture; mariculture much more than we even plan to do. In 2014, seafood provided more than 2.9 billion people with almost 20% of their intake of animal protein and researchers at the World Bank estimated that in the next 15 years, about 62% of the seafood we eat will be farm-raised. That seafood must be farmed mainly in marine waters. As Marit Solberg, the CEO of the Norwegian firm Marine Harvest said recently on the BBC, 70% of the planet is ocean but only about 2% of the world's food supply comes from the seas.

Marine aquaculture can be divided into fish culture, crustacean and molluskan aquaculture, and marine algae and plants. The present chapter is not intended as a textbook on aquaculture so I will limit the discussion to fishes grown in marine cages. Marine crustaceans and mollusks, namely shrimp and oysters, are bottom dwellers aquacultured in coastal zones and algae require sun and nutrients and labor so they are also relatively restricted to coastal zones. The open ocean is the realm of pisciculture. As suitable land becomes scarce and competition for space by other industries increases, mariculture will move to the high seas. The large majority of marine fish are already grown in cages. These cages can be simple but very effective wood and net structures as seen in South East Asia and Japan, near-shore industrial cages seen in the Mediterranean and more often than previously in Asia, advanced large net cages used mainly for salmon in Norway, Chile and Scotland, and advanced technology submersible offshore cages with special mooring devices and Kevlar or other advanced material netting that are starting to proliferate and will probably be used more and more as technology develops. These advances will greatly improve food security but marine aquaculture is not without its drawbacks. Food security means that we need to guarantee sustainability and that requires that we discuss the negatives as well as the positive aspects of the industry.

Technological constraints and cost of entry restricts open ocean cage farming to large and rich enterprises. Poor farmers might be able to afford to place cages near-shore in protected bays but legislation, environmental concerns and competition with other industries will make such prospects nearly impossible. Future growth is thus for large companies and marine finfish culture will become even more concentrated in a few large corporations than are land-based vegetal and meat agriculture today. Offshore cages require a lot of capital investment as well as operating costs, and economies of scale will only permit large companies to work. Political will and collaboration in some countries will allow fishermen cooperatives to pool resources and start offshore farms and probably some vertical integration in feed factories. Processing facilities already exist for wild-caught fish, and these will probably help develop the industry by partially paying for harvests in advance. One of the main results of such a development is an increase in fish monoculture, thus reducing drastically the number of marine finfish species on the market while increasing the tonnage sold.

The majority of freshwater fish production is constituted of a few types, namely carps, tilapias, catfishes, rainbow trout, eels and some minor species. Most of these fishes were endemic to one region of the planet but were introduced worldwide. By the time marine finfish aquaculture started to increase, various countries had instituted laws against exotic introductions. Consequently, many regions started developing technologies to rear "local" marine species. Most aquaculture research started when wild catch could not satisfy market demand. Some of the major marine finfish species aquacultured around the world are European and Japanese sea bass, gilthead sea bream, various other breams in the Asia-Pacific region, Atlantic salmon, Seriola species (jacks), red drum, groupers, and snappers. Various other species are being reared in limited amounts such as meager, rabbitfish and tuna. In the coming years, the industry will probably concentrate on a few species for the global market but small enterprises will still produce "specialty" species for local markets. Similar trends were observed for the beef, pork, chicken, catfish, tilapia, trout, salmon and shrimp industries. Even today, the number of farmed marine finfish species is low compared to the total number of marine fishes traded around the world. However, the market for those particular species is generally dominated by aquacultured fish. An example of such domination would be sea bass, sea bream, red drum, eel, salmon, etc. The result would be that although marine aquaculture will help improve food security, it will not do so at the cost of choice or diversity as terrestrial animal husbandry has by practically limiting us to poultry, pork, beef, mutton and goat in that order.

The major constraint to offshore cage culture is going to be political. Research in environmental impact of cage aquaculture is showing that many of the environmental concerns of the early days did not materialize, and the few environmental problems that did appear, have been solved or are being solved. However, offshore cages will interfere with shipping lanes, will require boats to transport feed out of harbors every day and bring in fish harvests regularly. These processes could easily be misused for smuggling, yet intensive policing would retard progress and be costly to governments. Furthermore, property rights of marine realms are not well defined in developed countries and not defined at all in most developing nations. A rush to claim good sites and speculation would be very detrimental to a nascent industry. Additional concerns could be issues like blocking transport and recreational boating, theft and protection, and control of seals, turtles and sharks. This means that for marine finfish aquaculture to keep growing rapidly, governments should be pre-emptive and begin discussing today how to legislate the industry in order to facilitate sustainable growth.

Another constraint to successful aquaculture is going to be energy usage. Energy affects aquaculture in two important ways. Traditionally, as the price of energy increased, the price of wild caught fish increased faster than that of aquacultured fish, giving aquaculture an advantage. However, when fish are reared in offshore cages, and boats need to visit them daily, the advantage may disappear and might even become a disadvantage. Furthermore, aquaculture depends on manufactured feed and the environmental effect of producing the feeds, starting with grain agriculture and irrigation, to harvesting and processing to transportation, makes the virtual energy content of manufactured feed (and greenhouse gas emissions) inhibitive. Wild fish with their "organic" perception will be difficult to compete with, thus putting even more unsustainable pressure on wild fisheries. Accordingly, marine aquaculture growth will have to be concomitant with green technologies for aquaculture. The use of wind energy and solar panels on cages is already feasible and being done albeit in its infancy stage. More needs to be implemented on sea cages as well as in agriculture, feed manufacture and transport. Moreover, as discussed with freshwater fish culture, the issues of virtual water needs of marine fish feeds should be added to the equation. In addition to using renewable energy, marine fish farmers need to develop diets using marine grown ingredients.

A good way to improve sustainability would be to start growing more omnivorous and algaevorous fishes. Digestible energy consumption in producing fish food is greater in carnivorous finfish farming than in omnivorous finfish farming. Estimates of the ratio of edible protein energy output to industrial energy inputs range from 1.4 for omnivorous fishes to more than 2.0 for carnivorous fishes. Also of importance is transport of fish. Transportation is expensive and polluting, especially airfreight. Long distance airfreight emits more than three times the amount of CO_2 per kg of fish transported than does sea freight, and more than 90 times the amount of CO_2 that is produced by local transportation of fish consumed close to source of production. These concerns would be difficult to address in pond based aquaculture because land price and policies in the countries that consume the fish would make local production prohibitive. However, offshore cages are feasible in any country in the world that has a coastline and thus fish can be produced near its markets if the political will to be truly environmentally friendly can trump improbable environmental concerns such as genetic pollution of wild stocks.

When political barriers to aquaculture growth are brought down, old technological problems will emerge as the dominant "brakes" to industry growth. Concerns such as disease, price and source of feed ingredients, availability of trained personnel, and quality of product have in the past proven to be resolvable. The main impediment to rapid growth today is seed production. Marine finfish tend to have semi-buoyant eggs and sensitive larvae with very small mouth gapes and specific nutrient requirements at first feeding. Moreover, although fish are often compared to chicken, we forget that each and every fish species has its own requirements and rearing larvae of two fish species in a similar manner could be like rearing chicken chicks and penguin chicks using similar methods. Therefore, although life cycles of various species have been closed, there are many more fish species whose larvae and juveniles are collected from the wild. This dependence on wild stocks could be both beneficial and detrimental for aquaculture as an economic enterprise. It is beneficial because aquaculturists would have access to healthy and well developed seed. It is bad for many reasons. First, the industry would depend on the whim of the environment and natural fluctuations in production. Overfishing would reduce the stock and reduce the food of other carnivorous fishes that depend on the larvae as part of their diet. Also, dependence on wild stocks would reduce selective breeding to enhance desirable traits such as increased growth or disease resistance. Finally, the growth of the aquaculture industry is partly a result of decreasing wild fisheries. By taking wild fish and rearing them, we are not truly increasing fish production but rather changing the method of production. Better to use wild stock to obtain new brood fish periodically in order to avoid inbreeding. As fish are selected for specific traits, use of wild fish will decrease but by that time enough genetic variability will exist among various hatcheries rendering inbreeding less of a risk. Feeding a future world requires that we account for climate change and the dangers of genetic inbreeding as well as political and technological impediments.

Although the rate of growth of the marine finfish farming industry will probably slow down in the near future because of constraints mentioned above, it will start growing again at a rapid rate as the problems are solved, people get richer, and investors see profits and get more confident. Governments will need to help if they intend to reduce the risks of nutritional deficiencies. They will need to develop legislation to support marine aquaculture, to invest in research and to help with crop insurance as with other branches of agriculture. Environmental groups and NGOs can help by working with farmers to solve problems instead of working against farmers to stop food production. Processors and distributors can help by reducing profit margins on seafood so that fish can become as common a food as chicken. Artificially inflated prices because of perceived rarity of product is a short sighted strategy. There is no reason why aquacultured grouper should be ten times more expensive than aquacultured catfish. The capacity to ensure food security is within our reach.

References

Abdul-Rahman, S., Ghanawi, J., Owaied, M. K., Holail, H., Farajalla, N., Haidar, M., & Saoud, I. P. (2011). Improving water use efficiency in semi-arid regions through integrated aquaculture/agriculture. *Journal of Applied Aquaculture, 23*, 212–230.

Al-Ahmed, A. A. (2004). Effluents and water quality in Tilapia farms in Kuwait. In *Aquaculture–an ecologically sustainable and profitable venture*, Book of Abstracts of the Aquaculture 2004 Conference (WAS Conference), Honolulu, Hawaii, USA.

Al-Jaloud, A. A., Hussain, G., Alsadon, A. A., Siddiqui, A. Q., & Al-Najada, A. (1993). Use of aquaculture effluent as a supplemental source of nitrogen fertilizer to wheat crop. *Arid Soil Research and Rehabilitation, 7*, 233–241.

Alcamo, J., Dronin, N., Endejan, M., Golubev, G., & Kirilenko, A. (2007). A new assessment of climate change impacts on food production shortfalls and water availability in Russia. *Global Environmental Changes, 17*(3–4), 429–444.

Allan, G. L., Banens, L. B., & Fielder, S. (2001). *Developing commercial inland saline aquaculture in Australia: Part 2*, Resource Inventory and Assessment, FRDC Project No. 98/335, NSW Fisheries Final Report Series No. 31, NSW, Australia.

Arlosoroff, S. (1995). Water resources management within regional cooperation in the Middle East. In: M. Haddad & E. Feitelson (Eds.), *The proceedings of the second workshop on joint management of shared aquifers* (pp 1–31).

Azevedo, C. M. S. B. (1998). *Nitrogen transfer using 15 N as a tracer in an integrated aquaculture and agriculture system*. Ph.D. dissertation, University of Arizona, Tucson, 105.

Barnett, T., Adam, J., & Lettenmaier, D. (2005). Potential impacts of a warming climate on water availability in snow-dominated regions. *Nature, 438*(7066), 303–309.

Benemann, J. R. (2008). Opportunities and challenges in algae biofuels production [online]. In J. R. Benemann (Ed.), *Line with Algae World 2008*, 15. [cited 14 October 2011]. ww. futureenergyevents.com/algae/whitepaper/algae_positionpaper.pdf

Berg, H., Michélsen, P., Troell, M., Folke, C., & Kautsky, N. (1996). Managing aquaculture for sustainability in tropical Lake Kariba, Zimbabwe. *Ecological Economics, 18*, 141–159.

Billard, R., & Servrin-Reyssac, J. (1992). The negative impacts of pond fish farming on the environment. In G. Barnabé & P. Kestemont (Eds.), *Production, environment and quality* (pp. 17–29). Oostende: European Aquaculture Society (EAS).

Boyd, C. E., & Thunjai, T. (2003). Concentrations of major ions in waters of inland shrimp farms in China, Ecuador, Thailand, and the United States. *Journal of the World Aquaculture Society, 34*, 524–532.

Brummett, R. E. (1997). Farming fish to save water. *Bioscience, 47*, 402.

Brummett, R. E. (2007) Comparative analysis of the environmental costs of fish farming and crop production in arid areas. In D. M. Bartley, C. Brugère, D. Soto, P. Gerber & B. Harvey (Eds.), *Comparative assessment of the environmental costs of aquaculture and other food production sectors: Methods for meaningful comparisons* (pp 221–228). FAO/WFT Expert Workshop, Vancouver, Canada, 24–28 April 2006. FAO Fisheries Proceedings No. 10, Rome.

Cajas-Canol, L., & Moffitt, C. M. (2008, September 10). Comparing footprints of trout and beef production. *Journal of the World Aquaculture Society*.

Collins, A., & Russell, B. (2003). Inland prawn farming trail in Australia. Pond Study Tests *Penaeus monodon* Performance in Low Salinity Ground Water, *Global Aquaculture Advocate*, 74–75.

Crespi, V., & Lovatelli, A. (2011). *Aquaculture in desert and arid lands. Development constraints and opportunities*. FAO Technical Workshop, Hermosillo, Mexico, 6–9 July 2010.

Cruz, E. M., Al-Ameeri, A. A., Al-Ahmed, A. K., & Ridha, M. T. (2000). Partial budget analysis of Nile Tilapia *Oreochromis niloticus* cultured within an existing agricultural farm in Kuwait. *Asian Fisheries Science, 3*, 97–305.

Davis, D. A., Boyd, C. E., Rouse, D. B., & Saoud, I. P. (2005). Effects of potassium, magnesium, and age on growth and survival of *Litopenaeus vannamei* post-larvae reared in inland low salinity well waters in West Alabama. *Journal of the World Aquaculture Society, 36*(3), 403–406.

Diliman, J. (1989). Water rights in the occupied territories. *Journal of Palestine Studies, 19*, 46–48.

D'Silva, A. M., & Maughan, O. E. (1994). Multiple use of water: Integration of fish culture and tree growing. *Agroforestry System, 26*, 1–7.

D'Silva, A. M., & Maughan, O. E. (1996). Optimum density of Red Tilapia *Oreochromis mossambicus* x *O. urolepishornorum* in a pulsed-flow culture system. *Journal of the World Aquaculture Society, 27*, 126–129.

Doll, P., & Siebert, S. (2002). Global modeling of irrigation water requirements. *Water Resources Research, 38*(4), 1037–1046.

Doroudi, M. S., Fielder, D. S., Allan, G. L., & Webster, G. K. (2006). Combined effects of salinity and potassium concentration on Juvenile Mulloway (*Argyrosomus japonicus*, Temminck and Schlegel) in inland saline groundwater. *Aquaculture Research, 37*, 1034–1039.

Doroudi, M. S., Webster, G. K., Allan, G. L., & Fielder, D. S. (2007). Survival and growth of silver perch (*Bidyanus bidyanus*), a salt-tolerant freshwater species, in inland saline groundwater from Southwestern New South Wales, Australia. *Journal of the World Aquaculture Society, 38*, 314–317.

Droogers, P. (2004). Adaptation to climate change to enhance food security and preserve environmental quality: Example for Southern Sri Lanka. *Agricultural Water Management, 66*(1), 15–33.

Droogers, P., & Aerts, J. (2005). Adaptation strategies to climate change and climate variability: A comparative study between seven contrasting river basins. *Physics and Chemistry of the Earth Parts A/B/C, 30*(6–7), 339–346.

Engle, C. R., Pomerleau, S., Fornshell, G., Hinshaw, J. M., Sloan, D., & Thompson, S. (2005). The economic impact of proposed effluent treatment options for production of trout *Oncorhynchus mykiss* in flow-through systems. *Aquacultural Engineering, 32*, 303–323.

Escobar, J., Lora, E., Venturini, O., Yáñez, E., Castillo, E., & Almazan, O. (2009). Biofuels: Environment, technology and food security. *Renewable and Sustainable Energy Reviews, 13* (6–7), 1275–1287.

FAO. (2009). *How to feed the World in 2050*. Rome: Food and Agriculture Organization.

Fielder, D. S., Bardsleya, W. J., & Allan, G. L. (2001). Survival and growth of Australian snapper, *Pagrus auratus*, in saline groundwater from inland New South Wales, Australia. *Aquaculture, 201*, 73–90.

Fielder, D.S., Raizada, S., Chadha, N. K., & Allan, G. L. (2008, August) *Development of hatchery and grow out technology for Macrobrachium rosenbergii using saline groundwater in Northern India* (pp , 3–6). Skretting Australasian Aquaculture 2008, Brisbane, Australia.

Fischer, G., Tubiello, F., van Velthuizen, H., & Wiberg, D. (2007). Climate change impacts on irrigation water requirements: Effects of mitigation, 1990–2080. *Technological Forecasting and Society Change, 74*(7), 1083–1107.

Fishelson, L., & Loya, Y. (1969). Experiments on rearing Tilapia-hybrids in brackish water pond near the dead sea. *Limnology, 17*, 602–610.

Forsberg, J. A., & Neill, W. H. (1997). Saline groundwater as an aquaculture medium: Physiological studies on the red drum, *Sciaenops ocellatus*. *Environmental Biology of Fishes, 49*, 19–128.

Forsberg, J. A., Dorsett, P. W., & Neill, W. H. (1996). Survival and growth of red drum *Sciaenops ocellatus* in saline groundwaters of West Texas, USA. *Journal of the World Aquaculture Society, 27*, 462–474.

Ghate, S. R., & Burtle, G. J. (1993). Water quality in channel catfish ponds intermittently drained for irrigation. In J. K. Wang (Ed.), *Techniques for modern aquaculture* (pp. 177–186). St. Joseph: ASAE pub..

Gleick, P. (1992). Water and conflict. In *The workshop on environmental change, water resources and international conflict*, University of Toronto, June 1991.

Goldstein, L., & Forster, R. P. (1970). Nitrogen metabolism in fishes. In J. W. Campbell (Ed.), *Comparative biochemistry of nitrogen metabolism, The vertebrates* (Vol. 2, pp. 495–518). New York: Academic Press.

Haddad, M., & Mizyed, N. (1996). Water resources in the middle east: Conflict and solutions. In T. Allan (Ed.), *The proceedings of the workshop on water peace and the middle east: Negotiating resources of the Jordan River Basin*. New York: Tauris Academic Studies.

Haddadin, M. (1989, April 4–6). Water resources in the Arab World, and its strategic importance. In *The proceedings of the conference on water resources in the Arab World, and its strategic importance*. Amman: Jordan University.

Hamdy, A., & Lacirignola, C. (1999, March 1–5) Mediterranean water resources: Major challenges towards the 21st century. In *International seminar on: Mediterranean water resources: Major challenges towards the 21st century*, Cairo (Egypt).

Hanjra, M., & Qureshi, M. (2010). Global Water Crisis and Future Food Security in an Era of Climate Change. *Food Policy, 35*(5), 365–377.

Hochman, E., & Brill, E. (1994). Fish production in a super-intensive controlled system in the Israeli Desert. In Y. C. Shang, P. S. Leung, C. S. Lee, M. S. Su & I. C. Liao (Eds.), *Socioeconomics of aquaculture*. Tungkang Marine Laboratory Conference Proceedings (Vol. 4, pp. 281–295).

Hoekstra, A. Y., & Chapagain, A. K. (2007). Water footprints of nations: Water use by people as a function of their consumption pattern. *Water Resources Management, 21*, 35–48.

Ingram, B. A., McKinnon, L. J., & Gooley, G. J. (2002). Growth and survival of selected aquatic animals in two saline groundwater evaporation basins: An Australian case study. *Aquaculture Research, 33*, 425–436.

IPCC. (2007). Climate change 2007: Synthesis report. In: R. K. Pachauri & A. Reisinger (Eds.), *Contribution of working groups I, II and III to the fourth assessment report of the intergovernmental panel on climate change*, IPCC, Geneva, Switzerland.

Jain, A. K., Kumar, G., & Mukherjee, S. C. (2006). Survival and growth of early Juveniles of Barramundi, *Lates calcarifer* (Bloch, 1790) in inland saline groundwater. *Journal of Biology Research, 5*, 93–97.

Joshi, P. K., & Tyagi, N. K. (1994). Salt affected and waterlogged soil in India: A review. In: M. Svendsen & A. Gulati (Eds.), *Strategic change in Indian irrigation* (pp. 237–252). New Delhi/Washington, DC: ICAR/IFPRI.

Kautsky, N., Berg, H., Folke, C., Larsson, J., & Troell, M. (1997). Ecological footprint assessment of resource use and development in Shrimp and Tilapia aquaculture. *Aquaculture Research, 28*, 753–766.

Kolkovski, S., Simon, Y., Hulata, G., & Ayaril, N. (2011). Desert aquaculture. In J. S. Lucas & P. C. Southgate (Eds.), *Aquaculture: Farming aquatic animals and plants* (2nd ed., pp. 107–125). Oxford: Wiley.

Lewis, W. M., Yoop, J. H., Schramm, J., & Brandesburg, H. L. (1978). Use of hydroponics to maintain quality of recirculated water in fish culture system. *Transactions of the American Fisheries Society, 107*, 92–99.

Lovell, T. (1989). *Nutrition and feeding of fish*. New York: AVI, Van Nostrand Reinhold.

Marshall, W. S., & Grosell, M. (2006). Ion transport, osmoregulation, and acid–base balance. In D. H. Evans & J. B. Claiborne (Eds.), *The physiology of fishes* (pp. 177–230). Boca Raton: CRC Press.

McIntosh, D., & Fitzsimmons, K. (2003). Characterization of effluent from an inland, low salinity shrimp farm: what contribution could this water make if used for irrigation? *Aquacultural Engineering, 27*, 147–156.

McMichael, A., Powles, J., Butler, C., & Uauy, R. (2007). Food, livestock production, energy, climate change, and health. *The Lancet, 370*(9594), 1253–1263.

McMurtry, M. R., Nelson, P. V., Sanders, D. C., & Hodges, L. (1990). Sand culture of vegetables using recirculated aquaculture effluents. *Applied Agricultural Research, 5*(4), 280–284.

McMurtry, M. R., Sanders, D. C., Nelson, P. V., & Nash, A. (1993). Mineral nutrient concentration and uptake by tomato irrigated with recirculating aquaculture water as influenced by quantity of fish waste products supplied. *Journal of Plant Nutrition, 16*, 407–419.

McNevin, A. A., Boyd, C. E., Silapajarn, O., & Silapajarn, K. (2004). Ionic supplementation of pond waters for inland culture of marine shrimp. *Journal of the World Aquaculture Society, 35*, 460–467.

Mourad, N., Kreydiyyeh, S., Ghanawi, J., & Saoud, I. P. (2012). Aquaculture of marine fish in inland low salinity well water: Potassium is not the only limiting element. *Fisheries and Aquaculture Journal, 42*, 1–12.

Naegal, L. C. A. (1977). Combined production of fish and plants in recirculating water. *Aquaculture, 10*, 17–24.

Naylor, R. L., Goldburg, R. J., Primavera, J. H., Kautsky, N., Beveridge, M. C. M., Clay, J., Folke, C., Lubchenco, J., Mooney, H., & Troell, M. (2000). Effect of aquaculture on world fish supplies. *Nature, 405*, 1017–1024.

NRC. (1983). *Nutrient requirements of warm-water fishes and shellfishes* (Rev. edn.). Washington, DC: National Research Council, National Academy Press.

Olah, J., & Sinha, V. R. P. (1986). *Energy costs in carp farming systems*. NACA/WP/86/35. Bangkok: Network of Aquaculture Centres in Asia.

Olson, G. L. (1992). The use of trout manure fertilizer for Idaho crops. In J. Blake, J. Donald & W. Magette (Eds.), *National livestock, poultry and aquaculture waste management* (pp. 198–205). St. Joseph: American Society of Agricultural Engineers, ASAE Pub. 03–92.

Palada, M., Cole, W., & Crossman, S. (1999). Influence of effluents from intensive aquaculture and sludge on growth and yield of bell peppers. *Journal of Sustainable Agriculture, 14*(4), 85–103.

Pant, J., Demaine, H., & Edwards, P. (2005). Bio-resource flow in integrated agriculture-aquaculture systems in a tropical monsoonal climate: A case study in Northeast Thailand. *Agricultural Systems, 83*(2), 203–219.

Parker, D., Anouti, A., & Dickenson, G. (1990). *Integrated fish/plant production system: Experimental results. ERL report 90-34* (pp. 12). Tucson: University of Arizona.

Partridge, G. J., & Creeper, J. (2004). Skeletal myopathy in Juvenile Barramundi, *Lates calcarifer* (Bloch), cultured in potassium-deficient saline groundwater. *Journal of Fish Diseases, 27*, 523–530.

Partridge, G. J., & Furey, A. (2002, September 16–20). Culturing snapper in Dumbleyung – A case study for determining the potential for inland saline groundwater to grow marine fish in Western Australia. In *8th national conference and workshop on the Productive Use and Rehabilitation of Saline Lands (PURSL)*, Fremantle, Western Australia, Australia, Promaco, Perth, Australia.

Partridge, G. J., & Lymbery, A. J. (2008). The effect of salinity on the requirement for potassium by Barramundi (*Lates calcarifer*) in saline groundwater. *Aquaculture, 278*, 164–170.

Partridge, G. J., Sarre, G. A., Ginbey, B. M., Kay, G. D., & Jenkins, G. I. (2006). Finfish production in a static, inland saline water body using a Semi-Intensive Floating Tank System (SIFTS). *Aquacultural Engineering, 35*, 109–121.

Piemental, D., Houser, J., Preiss, E., White, O., Fang, H., Mesnick, L., Barsky, T., Tariche, S., Schreck, J., & Alpert, S. (1997) Water resources: Agriculture, the environment and society. *Bioscience, 47*, 97–106.

Prangnell, D. I., & Fotedar, R. (2005). The effect of potassium concentration in inland saline water on the growth and survival of the Western King Shrimp, *Penaeus latisulcatus* Kishinouye, 1896. *Journal of Applied Aquaculture, 17*, 19–34.

Pruginin, Y., Fishelson, L., & Koren, A. (1988). "Intensive Tilapia farming in brackish water from an Israeli desert aquifer. In *Second international symposium on Tilapia in aquaculture* (pp. 75–81).

Rakocy, J. E., Hargreaves, J. A., & Bailey, D. S. (1993). Nutrient accumulation in a recirculating aquaculture system integrated with vegetable hydroponics. In J. K. Wang (Ed.), *Techniques for modern aquaculture* (pp. 148–158). St. Joseph: ASAE Publication 02–93.

Rees, W. E. (1992). Ecological footprints and appropriated carrying capacity: What urban economics leaves out. *Environment and Urbanization, 4*, 121–130.

Rosenzweig, C., & Parry, M. (1994). Potential impact of climate change on world food supply. *Nature, 367*, 133–138.

Roy, L. A., Davis, D. A., Saoud, I. P., & Henry, R. P. (2007). Effects of varying levels of aqueous potassium and magnesium on survival, growth, and respiration of the pacific white shrimp, *Litopenaeus vannamei*, reared in low salinity waters. *Aquaculture, 262*(2), 461–469.

Roy, L. A., Davis, D. A., Nguyen, T. N., & Saoud, I. P. (2009). Supplementation of chelated magnesium to diets of the Pacific white shrimp, *Litopenaeus vannamei*, reared in low salinity waters of west Alabama. *Journal of the World Aquaculture Society, 40*(2), 248–254.

Saoud, I. P., & Davis, D. A. (2005). Effects of betaine supplementation to feeds of shrimp reared at extreme salinities. *North American Journal of Aquaculture, 67*, 351–353.

Saoud, I. P., Davis, D. A., & Rouse, D. B. (2003). Suitability studies of inland well waters for *Litopenaeus vannamei* culture. *Aquaculture, 217*, 373–383.

Saoud, I. P., Roy, L. A., & Davis, D. A. (2007). Chelated potassium and arginine supplementation in diets of pacific white shrimp reared in low-salinity waters of West Alabama. *North American Journal of Aquaculture, 69*, 265–270.

Seawright, D. E. (1993). A method for investigating nutrient dynamics in integrated aquaculture-hydroponics system. In: J. K. Wang (Ed.), *Techniques for modern aquaculture* (pp. 137–147). St. Joseph: ASAE Publication 02–93.

Shakeeb-Ur, R., Jain, A. K., Reddy, A. K., Kuman, G., & Raju, K. D. (2005). Ionic manipulation of inland saline groundwater for enhancing survival and growth of *Penaeus monodon* (Fabricius). *Aquaculture Research, 36*, 1149–1156.

Smith, M., Veth, P., Hiscock, P., & Wallis, L. A. (2008). Global desert in perspective. In M. Smith, P. Veth & P. Hiscock (Eds.), *Desert peoples: Archaeological perspectives* (pp. 1–14). Blackwell Publishing.

Spash, C. (2007). The economics of climate change impacts à la Stern: Novel and Nuanced or rhetorically restricted? *Ecological Economics, 63*(4), 706–713.

Steinfeld, H., Gerber, P., Wassenaar, T., Castel, V., Rosales, M., & de Haan, C. (2006). *Livestock's long shadow: Environmental issues and options*. Livestock Environment and Development Initiative. Rome: Food and Agriculture Organization of the United Nations.

Stockholm Environment Institute (SEI). (1997). *Comprehensive assessment of the freshwater resources of the World.*

Tucker, V. A. (1969). The energetics of bird flight. *Scientific American, 220*(5), 70–78.

United Nations Convention to Combat Desertification (UNCCD). (2007). *Desertification, exacerbated by climate change, represents one of the greatest environmental challenges of our times* [online]. UNCCD thematic fact sheet series No. 1. Climate Change and Desertification. [cited 14 October 2011]. www.unccd.int/documents/Desertificationandclimatechange.pdf

Water and Environmental Studies Center (WESC). (1995). *Middle east regional study on water supply and demand development.* Phase I study report submitted to the German Agency for Technical Cooperation (GTZ), The An-Najah National University, Nablus.

Watten, B. J., & Buch, R. L. (1984). Tropical production of tilapia (*Sarotherodon aurea*) and tomatoes (*Lycopersicon esculentum*) in small-scale recirculating water system. *Aquaculture, 41*, 271–283.

Wackernagel, M., & Rees, W. E. (1996). *Our ecological footprint: Reducing human impact on the Earth.* Gabriola Island: New Society Publishers.

Zweig, R. D. (1986, May–June). An integrated fish culture hydroponics vegetable production system. *Aquaculture Magazine*, pp. 34–40.

Impact of Food Losses and Waste on Food Security

Wajih N. Sawaya

Abstract In the world of rising population and concerns about inequality and growing food insecurity where about 870 million people still suffer from chronic undernourishment, food waste and loss (FLW) is one of the greatest challenges of our times.

The Food and Agriculture Organization (FAO) estimates that about one-third of the edible food produced is lost or wasted each year, i.e. about 1.3 billion tons, with an economic cost estimated at about US$750 billion (excluding fish and Seafood). In addition, the environmental impact of FLW in terms of volume and cost is tremendous and represents a huge cost to the society, in terms of greenhouse gas emission, water footprint, wastage of agricultural land and biodiversity loss.

FLW occur at all stages of the Food Supply Chain (FSC) starting with the initial level of agricultural production to the consumption by people. Some causes of FLW are structural and related to infrastructure, such as shortage of cold chains, processing facilities and efficient market infrastructure as well as shortcomings in the management and implementation of best practices at different stages of the FSC. Other causes are systematic and also related to policies and regulations that are mostly driven by improper functioning systems and non-supportive policies and regulations.

The extent of the FLW is different for different commodities and also different between developing and developed countries. The FLW in developing countries mostly occur at the post-harvest level, while in developed countries, it's mostly at the retail and consumer level, mainly related to consumer behavior.

Among the major food commodities, cereals represent the largest share of FLW and comprise 53% of the total FLW. Meat represents only about 7%. However, when the impact on the environment and the economic cost are considered, reducing the meat loss and waste that has a large environmental impact, should receive as much attention as other major commodities despite its relatively small share of FLW.

Approaches/solutions to reduce the global FLW through the FSC are discussed, both at the post-harvest, handling, storage and processing levels (close to the farm)

W.N. Sawaya (✉)
Kuwait Institute of Scientific Research, Safat, Kuwait
e-mail: wsawaya@kisr.edu.kw

S. Murad et al. (eds.), *Water, Energy & Food Sustainability in the Middle East*,
DOI 10.1007/978-3-319-48920-9_16

and at the retail and consumption level (close to the fork). Many of these problems and solutions are similar to those faced by countries in the MENA region. The regional strategic framework that was developed by FAO in 2014 for the reduction of FLW in the MENA region to achieve the region's goal for the reduction of FLW by 50% in 2024 is highlighted and discussed. Actions on moving forward and the need to establish/develop the necessary pre-requisites that can catalyze the successful implementation to reduce the FLW in the MENA member countries are recommended.

Keywords Agricultural waste • Food security • Food supply chain

Introduction

Food Security is a global challenge. The Food and Agricultural Organization of the United Nations (FAO) indicated that with the global population estimated to reach over 9.0 billion people by 2050, the global demand for food will increase by 60% (FAO 2013a). Recent investments in agriculture technologies and genetic engineering, that often focus on raising productivity through higher yields, crop intensification and expanded crop acreage, have contributed to improvement in crop efficiency and will contribute to meet the food production challenge. But at the same time it imposes an enormous burden on the environment and natural resources, when it comes to limited land, limited water and limited energy, as well as on the sustainability of the Food System and Food and Nutrition Security.

Gustavsson et al. (2011) estimated that one third of the global food produced for human consumption is lost or wasted, amounting to 1.3 billion tons/year worth nearly $1.0 trillion (CTA 2012, Policy Brief, № 7). When converted into calories, the global food losses and waste (FLW) can reach about 24% of all food produced, i.e. about one out of every four calories is wasted and is unavailable for human consumption (Lipinski et al. 2013). These losses will be more than adequate to feed some 870 million people that still suffer from chronic undernourishment (World hunger and Poverty Facts and Statistics 2013). In the MENA region, the problem is even compounded where the majority of the countries suffer from "double burden of malnutrition", i.e. the co-existence of under- nutrition, with the rise in overweight, obesity and diet related chronic diseases with different scales according to the level of economic development (FAO 2015a).

FLW is an issue that has chronically been overlooked. It has now emerged as a global priority and a pillar of the UN Secretary General's Zero Hunger Challenge Initiative to cut half per capita global waste by 2030 along the food chain for all countries, particularly developing countries as well as by governments across the MENA Region (FAO 2015b).

Global efforts to reduce FLW must be holistic. When food is saved, the resources used to produce it are saved. Reducing the waste should be by not creating it in the first place. Solutions for reducing the FLW can be developed on different levels of the food system using different tools. However, it's important to ensure that technical solutions are also accessible and affordable to farming communities, noting that small farmers are the most vulnerable.

Definitions of Food Losses and Waste

Food Losses and Waste (FLW) refers to a decrease in edible food portion mass harvested or produced from plants and animals for human consumption across the food supply chain. (FSC) (Fig. 1)

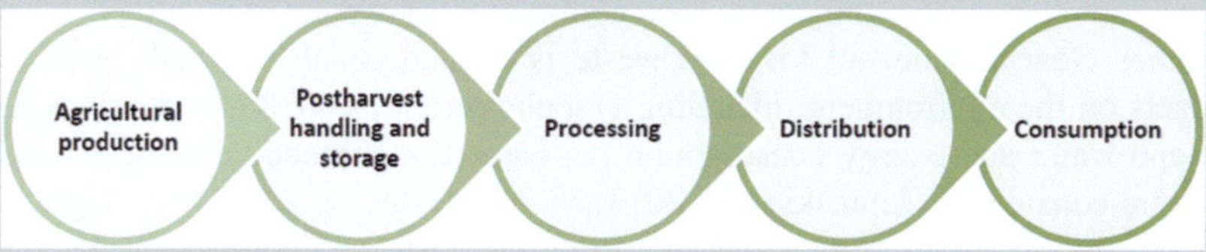

Fig. 1 Food Supply Chain (Source: FAO 2014a, Food Losses and Waste in MENA Countries)

Food Losses (FL) refers to a decrease in edible food mass intended for human consumption due to loss at all stages of the FSC prior to consumption.

Food Waste (FW) refers to decrease in food mass intended for human consumption due to food wasted at the consumer level and not eaten by people because it was discarded or left to spoil.

Food Quality Loss or Waste (FQLW) refers to degradation of the quality of food at all stages of the FSC e.g. nutrition color, taste, consistency, that translate into loss of economic value.

Extent of Global Food Losses and Waste

According to the International labor organization (ILO 2014), the global food and agriculture sector value is around US $ 8 trillion, or 10% of global GDP. It employs about one billion people, equivalent to about 1/3 of the world's work force. Food production and consumption require huge resources in terms of raw material, land, energy, water, capital and labor to grow, produce, store, transport, process, distribute, prepare and to cook about 4 billion tons for 7 billion people (Parry 2013). Reducing the FLW becomes an important and achievable approach to preserve the sustainability of the food system as well as delivering significant economic benefits.

Current data on the estimates of global FLW have until to-date relied mainly on one single source published by the FAO (Gustavsson et al. 2011). This study represents a ground breaking report on global FLW despite the authors' acknowledgement that the FLW numbers generated need to be taken with great caution.

This is due to the uncertainties and margin of error of the FLW data numbers that are generated from national and regional food balance sheets, along with complimentary information from the literature from different commodities. Efforts are underway by the UN agencies and other international organizations for the development of standardized methods "protocols" for countries and companies for measuring and monitoring FLW in the FSC, and for the improvement of the reliability and comparability of the data among users (HLPE 2014).

FAO estimates of FLW are about 32% of all food produced in the world in 2009, based on weight, i.e. one ton of grain is the same like one ton of fruit or meat. When the estimates for each of the major food commodities are converted into kcal/lost, it becomes clear that not all loss and waste is created equal, especially when the impacts on the environment, including greenhouse gases (GHGs) emissions, land use and water and energy consumption per calorie, combined with the economic cost, are considered (Lipinski et al. 2013).

Considering the environmental and economic costs, and comparing, for example, the global FLW in caloric content for cereals that comprise the largest share among the food commodities (53%) with the relatively small share of meat (7%), it becomes evident that reducing meat loss and waste should receive at least as much attention as other commodities despite its smaller share of caloric losses (Searchinger et al. 2013) (Fig. 2).

Food Losses and Waste Across the FSC

FLW along the FSC are different between developing and developed countries. They are also different among different products and between regions for the same type of products. In developing countries, about 40% of the FLW occur mostly at the production and handling and storage levels. In contrast, 40% of the FLW in developed countries occur mainly at the retail and consumer level at the end of the food chain that is mainly related to consumer behavior. This volume of FLW (222 million ton) is almost equivalent to the total food production in Sub-Saharan African (230 million tons) (Lipinski et al. 2013) (Fig. 3).

On the regional level, the total per capita production of edible parts of food for human consumption and FLW in Europe and North America are about 900 kg/capita/year and 280–300kg/capita/year, respectively, as compared to 460kg/capita/year production in Sub-Saharan Africa and South/Southeast Asia and 120–170kg/capita/year FLW, respectively. The food wasted by consumers in Europe and North America is 95–115 kg/capita/year compared to 6–11kg/capita/year for Sub-Saharan Africa and South/Southeast Asia (Fig. 4).

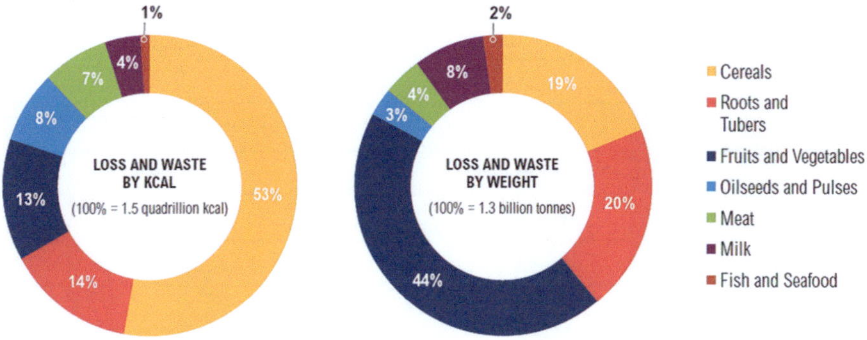

Fig. 2 Shares of Global Losses and Waste by Commodity, 2009 (Source: WRI analysis based on FAO 2011. Global food losses and food waste—extent, causes and prevention (From Lipinski et al. 2013))

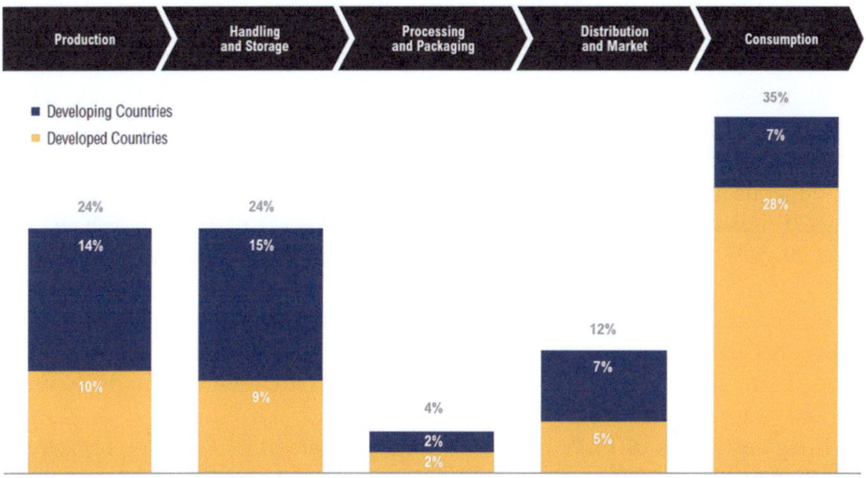

Fig. 3 Shares of Total Food Loss and Waste by Stage in the Value Chain, 2009 (Source: WRI analysis based on FAO 2011. Global food losses and food waste—extent, causes and prevention (From Lipenski et al. 2013)). Note: Number may not sum up due to rounding

Food Losses and Waste by Commodity Groups

On commodity group level, FLW follow the same trend in developed and developing countries. In developed countries, most of the FLW occur at the consumption level, while in developing countries, most of the FLW are at the post-harvest and storage levels. According to Gustavsson et al. (2011), with respect to staple crops such as wheat and rice, it's estimated that 40–50% of the wheat crop losses occur in industrialized countries at the consumption level, whereas in low-income countries where rice is the dominant staple, most of the losses are in the post-harvest and handling, as opposed to distribution and consumption levels. Fresh roots and tubers

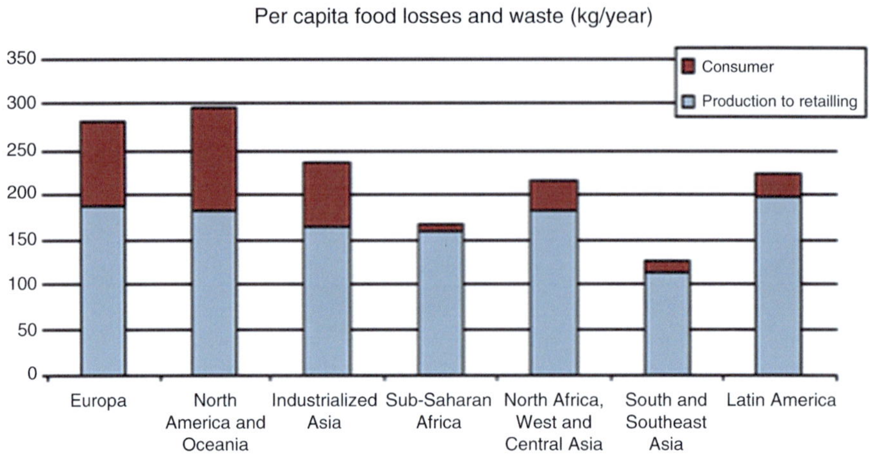

Fig. 4 Per Capita Food Losses and Waste (kg/year) (Source: Global food losses and food waste – Extent, causes and prevention, FAO 2011.)

FLW in developed countries e.g. Europe and North America and Oceania range between 50-60% versus 32–45% in developing countries such as Sub-Saharan Africa and North Africa and West and Central Asia. For other crops, the figures for oil seeds and pulses FLW are about 20% versus 20–30%, fruits and vegetables 47–52% versus 50–55%, meat products about 20–22% versus 20–28%, fish and sea food 30–50% versus 30–32% and milk and dairy 12–20% versus 20–25%, respectively (Fig. 5).

In the Arab countries, the high levels of FLW are particularly alarming, especially that the capacity to increase food production in the Arab region is very limited due to the scarcity of natural resources, thus driving heavily towards imports to meet their food needs (FAO 2015a).

Rough estimates of FLW in the MENA region are at 44% at the postharvest, handling and storage and processing and distribution levels of the FSC, and about 34% at the consumption level (HLPE 2014). The environmental burden of the FLW in the MENA region as reflected in the waste of natural resources in surface and ground water loss estimated at $42Km^3$/year, or $90m^3$ capita/year or 17% of the global foot print, despite having only 7% of world's population (FAO 2014a). In land occupation, the losses are estimated around 360 million hectares or $8000m^2$/capita versus a global average of $2000m^2$, a figure that is greater for any other region. In energy the equivalent of 2 million tons/year fertilizer losses and in GHGs emissions around 200 million tons of CO_2. Economic losses, are estimated about $60 billion/year (FAO 2014b).

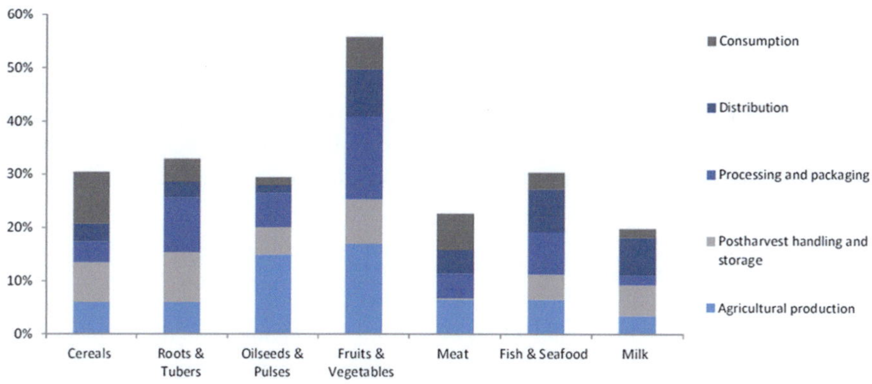

Fig. 5 Percentage of edible FLW (by weight) for each commodity group in each step of the food supply-chain for North Africa, West and Central Asia (MENA) (Source: Food security and sustainable agriculture in the Arab region, FAO 2015c.)

Causes of Food Losses and Waste

FLW occur at all stages of the FSC, starting at the initial level of agricultural production up until consumption by people. Causes are often interrelated. FLW occurring at one stage of the FSC can affect causes at the other stages across the chain. Some of these causes are stage-specific and range from biological, microbial, chemical, biochemical, mechanical, physical, physiological, technological, logistical, and organizational to psychological and behavioral causes. Other causes are linked to structural, mostly related to inadequate infrastructure and weak organization and coordination among the different actors along the FSC, and systemic, policies and regulations that are all required for the proper functioning of the food system and the development of an enabling environment to help cut down on the FLW.

Stage-Specific Causes of Food Losses and Waste Across the FSC

Pre-harvest

The pre-harvest losses comprise conditions and actions in the field, ranging from choice of the crop variety, agronomic practices (water and nutrient management, fertilization, pest/disease management, etc.) to biological and environmental factors. According to Kader et al. (2012), the choice of the right variety of produce adapted to a given location and meeting the requirement of the target market, in terms of quality and time of maturity, constitutes a vital factor at the production stage. For fruits and vegetables, poor agronomic practices such as pre-harvest pest infestation can lead to high losses (Thompson 2007). Produce that are left un-harvested because of failure to meet certain processors/retailers requirements and/or quality standards or target demand are as well contributing factors to FLW (Stuart 2009). In the U.S., it is estimated that, on average, 7% of planted fields are not harvested. Sometimes, failure to harvest could be due to economic reasons related to low market price at time of harvest or labor cost (HLPE 2014).

Harvesting and Handling

Factors that are mostly related to FLW during harvesting include poor harvest scheduling as well as rough and careless handling of the produce. Over-maturity and delayed harvesting of food grain crops such as maize, groundnuts and sorghum are reported to be major factors contributing to aflatoxin contamination (Farag 2008; Lewis et al. 2005). For fruits and vegetables, both immaturity, driven by poverty of farmers in urgent need of food and cash, as well as over-maturity that results in a short shelf-life of the product, are major causes for FLW in highly perishable produce (Sivakumar et al. 2011). Multiple handling of fruits and vegetables can also lead to increased damages (FAO 2013a). But of primary concern for handling of fruits and vegetables is the temperature management. Initial cooling of perishable foods such as fruits, vegetables, milk, meat and fish is critical for maintenance of quality. Most growers in developing countries lack on-farm cold facilities or shade, thus keeping harvested perishable produce under ambient temperature that will accelerate spoilage of the products. For milk, the initial handling conditions such as spillage, lack of appropriate milking equipment and poor sanitation during milking are determinants of the overall quality. Poor sanitation and hygienic practices constitute a major factor for losses, especially among small holder farmers, such as mastitis or water adulteration, both of which lead to rejection at the collection center of the factory (FAO 2014c).

Storage

Storage is an effective means for the delayed marketing and consumption of the product, but it does not substitute for the importance of the initial quality of the produce that dictates its shelf-life afterwards. In developing countries lack of proper storage facilities is a major cause of post-harvest losses (FAO 2011a). For example, cold storage facilities are non-existent or inaccessible to the majority of stakeholder framers in Sub-Saharan Africa. If infrastructure for initial storage is lacking, spoilage of the produce occurs within hours after harvesting (Rolle 2006; Stuart 2009). Use of low quality containers is another key factor that was identified by a WFLO in their post-harvest losses study that measured losses of 26 horticultural crops in four countries (WFLO 2010). Traditional storage of shelf-stable food such as grains by small farmers in Sub-Saharan Africa that still use grain stores made of grass, wood and mud expose the crops to rodents attack, insects, birds and fungal infections leading to big losses (Yosuf and He 2011). Lack of storage that may sometimes lead framers to sell their harvest directly after harvest may only attract low prices due to oversupply in the market.

Transport

Rolle (2006) indicated that in developing countries, lack of proper transportation vehicles, poor roads and poor/inefficient handling and management hinder proper conservation of perishable commodities during transport. Similarly, the lack of the proper logistics in developing countries impacts the quality, and consequently increases the losses of perishable products such as fish and milk due to lack of cold transportation, especially during the hot season as well as the erratic power supply to milk processors and coolers (HLPE 2014). As a result, at the distribution stage, shipments of imported perishable products are sometimes rejected and dumped/destroyed after being subjected to testing at the point of entry/exit to check adherence to phytosanitary, food safety and veterinary regulations.

Processing

Processing is an effective means for the transformation of food crops to reduce FLW and increase shelf-life. Inadequacy of processing facilities in most developing countries that often times are aggravated by seasonability of some of the processed products constitute a major cause of FLW. Technical malfunctions and inefficiencies at the processing stage can as well lead to defects to the end products such as wrong size, weight, shape, appearance or damaged packaging (HLPE 2014). Excessive trimming of certain produce such as cabbage and lettuce for example, to attain certain shape/size dictated by standards of the processor/consumer contributes to food losses, that may still be fit and safe for human consumption (HLPE 2014). Inadequate packaging is another important factor that may reduce the shelf life and increase food losses (FAO 2011a). Improper sanitization of processing equipment represents a major cause of losses of animal products. Contamination that may originate either from the processing unit or from part of the product that contaminates the whole production batch is also a cause of waste.

Retail

The quality standards of the produce to be supplied and displayed in the market are dictated by the retailers who are considered a major player in the FSC. The conditions of storage within the retail outlets (e.g. temperature and humidity) and handling practices are important determinants of the quality, shelf life and acceptability by the consumer. In the U.S. alone, Buzby et al. (2014) estimated that the in-store food losses were about 10% of the total food supply, while in Norway, food losses at the retail stage represent 18% of the FLW (HLPE 2014).

Improper practices in some open markets such as the sprinkling of water that might not be clean to minimize wilting and shriveling of fresh produce could result in unsafe food that may end up discarded. FLW at the retail stage stem out also from inappropriate stockpiling of produce on display such as fruits at different stages of maturity that are squashed or bruised and are discarded. A major cause of FLW at the retail stage is failure of adherence to the quality standards set by the processors and retailers. Products that are in terms of color, shape, size and freedom of blemishes, don't meet the set standards eventually end up in rejection at delivery. The same applies to cut fruits and vegetables that have become popular products in modern supermarkets and retail stores. Cut fruits and vegetables are prone to more spoilage, if not purchased within the short shelf-life of these products, they are then discarded if they remain unsold.

Consumption

At the consumption level, FLW represent a major problem in developed countries. The demographic changes during the last few decades and the income growth have impacted the food consumption patterns. Overconsumption and emergence of obesity as a world-wide problem have also lead to more waste by the consumer that increases with household wealth. Assessment studies on consumer waste are mostly limited to the U.S. and Europe. These studies address the social and cultural role as drivers in food attitudes and consumption, and may not be extended literary to developing countries. Segre' (2013), reported on types of causes of consumer attitudes leading to waste. These causes are mainly linked to food preference and food consumption habits.

At the household level, fresh products make up most of the wasted foods. The U.S. Department of Agriculture reports that a typical American throws out 40% of fresh fish, 23% of eggs and 20% of milk. Citrus fruits and cherries top the list of fruits, and sweet potatoes, onion and greens are commonly wasted vegetables (Muth et al. 2011). Much of the household waste is due to over purchasing, food spoilage and plate waste. The Waste and Resources Action Program (WRAP 2009) conducted a household survey of food waste in the U.K. They found that 41% of the food is wasted, because too much of it was cooked or served, and 54% of the waste is because the food was not consumed in time. In another study conducted by (WRAP 2013), it was reported that food waste of households with children produced 41% more food waste than similarly-sized households without children. Evans (2011a, b) showed in his studies on ethnographic analysis of everyday food practices in English households, that household provisioning routines and time management, accounting for family tastes and food safety concerns, can drive day-to-day waste, in-spite of awareness of it. Buying habits represent another element that plays an important role in food waste increase. Buying less often and in greater quantities with the possibility of products loosing quality by time contributes to increase in waste.

At the Food services level, food discards in school canteens and restaurants as well as fixed price buffet (eat-as-much-as-you-can) and supersized portions and refills of soft drinks are all factors that promote obesity and waste (Lipinski et al. 2013). Stuart (2009) estimated that 24 to 35% of school lunches in the U.K. end up in waste. Cohen et al. (2013) demonstrated that in the middle school students, Boston, U.S.A, that averages measured in calories, students discard about 19% of their entrees, 47% of their fruits, 25% of their milk and 73% of their vegetables. In Finland, Silvennoinen et al. (2012) reported that an average of 20% of food handled in restaurants and catering is wasted, with more waste in self-serving buffet restaurants. Williams's et al. (2012), in Sweden, reported that packaging contributes 20-25% of food waste. Portioning and pack size that entice consumers to buy large packs and bulk offers to maximize value for money, are another factor that increase food waste (Fusions 2014).

Structural, Systemic and Policies and Regulations Causes of Food Losses and Waste

The interdependence of the FLW at different stages of the FSC, necessitates the proper functioning of all activities along the chain. This requires the collective support of and coordination by all actors along the FSC, as well as proper functioning systems and supporting policies and regulations.

Structural causes of FLW, from the producer to the consumer, can be linked to the following:

- Lack of adequate infrastructure including storage, cold chains, processing facilities and efficient market physical infrastructure, considered to be crucial to limit FLW.
- Lack of investment and implementation of food practices, especially for small farmers/enterprises, depriving them from adoption of technologies to reduce FLW, as well as lack of implementation of good practices at different stages of the food chain, including training, organization, extension services and low-paid unskilled workers.
- Lack of a well-functioning and integrated food chain management, especially in low-income countries, that require an enhanced efficiency and proper management by all stakeholders at the farmers level, transporters, store keepers, the food processing industry, shop keepers and supermarkets, among others.
- Confusion around food date labeling. This represents an indirect cause of FLW at retail and consumer levels. It stems from the belief that expiry dates are linked to food safety, beyond which they need to be discarded, when in reality they are more related to food quality. In Europe for example a product with a "use by" date can't be sold after the date.

Systemic and Policies and Regulations causes of FLW, mostly driven by improper functioning systems and non-supportive policies and regulations including:

- Agricultural policies for investment in food production with front and back-linkages to increase food production, but that are not coupled with equally important measures for investment in related infrastructures across the FSC for storage, transport, logistics, marketing, etc. as well are R&D, capacity building and extension services, all of which will affect the extent of FLW.
- Surrounding policies and regulatory framework – Such as those related to bans on redistribution of oversupply food or its use in animal feed, lack of or partial implementation of food safety regulations, food hygiene regulation, food labelling and packaging regulations and waste disposal regulations, can all adversely affect FLW.
- Systemic causes that impact the overall extent of the global FLW and that are linked to the functioning of the food chain based on the country conditions. In low-income countries, systemic causes are mostly linked to financial, managerial and technical limitations across all stages of the FSC. In medium/high income countries, FLW causes can happen at the same stages but for different reasons that are mainly due to consumer behavior and wastage at the consumption level.
- Other systemic causes, such as Standardization of produce whether in size, color or even degree of freshness, represent a major cause of food losses. In developed countries, a greater part of food losses is related to urban lifestyle reflected in consumer behavior, as consumers can afford the luxury of wasting food and rejecting buying them for aesthetic or other reasons that do not meet the standards.
- Urbanization and globalization introduce new trends in food consumption that risk FLW. The trends towards freshness, being fruits and vegetables or other animal product perishables, such as livestock products and fish, along with the growing fluxes of the seasonal fresh products from all over the world, require a very efficient system as well as all required conditions at the different stages of the food chain for its proper functioning. This can be particularly challenging for transitioning countries confronted with rapid change of food demand, while they still lack the basic infrastructure.

Implication/Potential Benefits of Reducing Food Losses and Waste

With the increase in population and cost of food, and concerns about inequality and growing food insecurity, FLW is one of the greatest challenges of our time. (**CSACC** 2011)

The economic impact of FLW in terms of the volume of Waste and cost is tremendous. Roughly, one third of the food produced globally end up in waste based on weight (1.3 billion tons) (Gurtavsson et al. 2011). (FAO 2013a) estimated that the value of food wasted is around $750 billion to 1.0 trillion, based on production price versus retail. By 2030, it is estimated that the middle class will be more than double, to almost 5 billion people (Kharas 2010). The three billion extra middle class people globally could generate an additional 280 million tons of food waste per year, which could take the cost of consumer food waste to more than $600 billion (WRAP 2015)

Environmental impacts of FLW along the food chain also represent huge cost to society. The environmental footprint of food wastage (food loss and waste) is assessed through four different model components, carbon footprints, water footprints, land and occupation of food wastage and biodiversity impact complemented by economic quantification. Without accounting for GHGs emissions from land use change, estimates of the carbon foot print of food produced and not consumed is about 3.3 billion tons of CO_2 equivalent. If the food waste was a country, it would be the third biggest emitter of GHGs globally after China and the U.S.A. (FAO 2013a). Major commodity contributors to the carbon footprint are cereals (34%), meat (21%) and Vegetables (21%). While the carbon footprint of animal product origin accounts altogether for about 33% of total carbon footprint, the wastage volume is only 15%.

The global blue water footprint, that includes surface and ground water resources, for the agriculture production of food wastage in 2007 is estimated at about 250Km^3. In Terms of volume, this volume of water represents almost three times the lake of Geneva. The major commodity contributors to the blue water footprint of food wastage are cereals (52%) and fruits (18%), whereas their contributions to food wastage are 26% and 16%, respectively.

Land occupation includes surface, cropland and grassland, occupied by food produce but uneaten because of wastage. The total global amount of food wastage in 2007 occupied almost 1.4 billion hectares of land. This represents close to 30% of the world agricultural land area (FAO 2013a). The major contributors to land occupation food wastage are meat and milk, representing 78% of the total surface, whereas their total food wastage is only 11%.

Impact of food wastage on biodiversity is difficult to estimate, however the negative impact of agriculture expansion into wild areas creates biodiversity loss, including mammals, birds, fish and amphibians (FAO 2013a).

As global demand for food, fodder and bioenergy grows, many of the agriculture systems are depleting soil fertility, biodiversity and water resources (CSACC 2011). The problem is exacerbated where every year, an estimated 12 million hectares of agricultural land, which would potentially produce 20 million tons of grains, are lost by degradation (Bai et al. 2008). The levels of GHGs already in the atmosphere that will continue to grow, have also impacted climate change, where extreme weathers such as high temperatures, draughts and floods have already become more frequent and severe and have negative social, economic and ecological consequences. In the coming decades, the global climate change will have an adverse effect on the agriculture production system, particularly in areas that are currently suffering from food insecurity and are expected to experience disproportionately negative effects (CSACC 2011).

With such global food waste and related environmental pressure on natural resources, it becomes clear how the reduction of this wastage at the global, regional, and national scales would have a substantial positive effect on national and societal resources. The reduction of food waste would not only reduce the burden of degradation and depletion of natural resources, but will also reduce the burden on the food system to raise food production by 60% to face the additional pressure of the global population growth to over 9.0 billion in 2050.

Approaches/Solutions to Reduce Food Losses and Waste

According to the IMF Regional Economic Outlook (2014), an estimated 34% of food supplies suitable for human consumption in the MENA region are being lost or wasted. The FLW are estimated to be 31% in cereals, 33% roots and tubers, 29% oil seeds and pulses, 56% in fruits and vegetables, 23% in meats, 30% in fish and seafood and 20% in milk, amounting up to 210kg/capita/year, equivalent to 594 Kcal/capita/day (Gustavsson et al. 2011). In view of the region's resources scarcity, low productivity and rapid population growth, and the major challenges in closing the widening import gap to meet its food needs, such levels of food waste are not only uneconomical but also detrimental to the environment and food security.

The MENA region countries economics are diverse in terms of economic structure, living standards and food security. This diversity is also reflected on the FLW. In low and middle- income countries with more diversified economic structures, including an agriculture sector, most of the FLW occur at the beginning of the food chain, i.e. post-harvest, handling and storage. In high income countries, such as the GCC, the FLW occur mostly at the end of the food chain, i.e. at consumption level, similar to those of the developed countries. Over the last decade, high incomes, rapid population growth, urbanization and changing consumption patterns with the proliferation of super markets, even in low-income countries, have exacerbated all forms of malnutrition, particularly overweight and obesity, thus creating the phenomenon of double –burden malnutrition, i.e. the coexistence of under-and over-nutrition.

FLW exists throughout the entire FSC for any given commodity. Approaches/solutions to reduce the FLW across the supply chain will be discussed both at the post-harvest, handling and storage and processing levels **(close to the farm)** and the retail and consumption level **(close to the fork).**

Many of the problems and issues being faced in the MENA region related to FLW, as well as solutions at different phases of the FSC are similar to those being encountered in other regions of the world. Hence, most of the suggested approaches/solutions to reduce FLW discussed here apply also to the MENA region.

Approaches/Solutions Close to the Farm

Improved Post-Harvest Handling and Storage Methods

Preventing food losses due to premature harvesting by small farmers in need for cash or due to food deficiency could be addressed through organizing small resource – poor farmers into groups to produce a variety of cash crops or animals. This will help facilitate receiving credit from financial institutions or advanced payment from buyers of the produce (Gustavsson et al. 2011).

At the harvesting, handling and storage phase, methods used differ for different foods. **For agronomic crops**, part of the initial production of cereals lost or wasted in the MENA Region is estimated at about 20% and for oil seeds and pulses at about 27% (Gustavsson et al. 2011). For crops such as grains and legumes, simple low-cost storage methods can drastically cut grain losses, especially for small-scale farmers, who frequently loose a substantial volume of the crop due to damage from pests and spoilage. Among those methods is hermetic storage technology in air tight containers, such as metal silos or hermetic plastic bags, which reduces the oxygen level thus reducing pest damage. It also avoids the application of pesticides that are more costly and more hazardous to health. The FAO has built more than 45,000 small, metal storage silos (250–1000 kg), just big enough for use by a single farmer in 16 different countries. These have cut down food loss during the storage phase to almost zero (Lipinski et al. 2013). Researchers at Purdue University, U.S.A. have developed a simple reusable plastic bag "Purdue Improved Cowpea Storage" (PICS) in which grain is stored in three interlocking plastic bags that locks out pests and keep grain fresh for months. The bags may also be useful for other crops, such as "Maize" (Lipinski et al. 2013).

For horticultural perishables, poor handling practices of food produce are detrimental, particularly the use of improper containers for harvest and transportation. Using plastic returnable containers instead of plastic sacks during transport has significantly cut down food losses during handling and storage, particularly among fruits and vegetables (Kader et al. 2012). In developing countries, 19% of fruit and vegetables loss occurs in the handling and storage phase due mostly to bruising and squashing (Lipinski et al. 2013). Minimizing losses at this stage will reduce further losses later in the value chain (Adegbola et al. 2011).

For the preservation of quality of harvested perishable foods, including animal products, the most important factor is temperature control. The lack of sufficient and efficient cold chain infrastructure is a major contributor to FLW. ***The importance of cold chain can't be overemphasized.*** The FAO (2011b) "Developing the cold chain for Agriculture "in the near East and North Africa (MENA), considered the cold chain development in MENA region (pre-cooling, refrigerated transport, refrigerated commercial storage, refrigerated display during marketing and refrigerated maintenance at home) an integral element for improving the efficiency of the food system, reducing FLW and insuring delivery of safe and nutritious food. It emphasized the necessity to develop strategies that should be adapted to specific commodity groups and geographic and socio-economic condition as well as good management, including maintenance.

Investment in cooling and maintaining a cold environment is expensive, especially in hot temperature climates, and is dependent on the availability of electricity sources which may be out of reach for small-holder farmers. Under these circumstances, the choice of technology, taking into consideration decisions that affect energy efficiency and materials used, become critical for the cold chain in MENA. For example, if refrigerated vehicles are not available, covered or insulated vans could be appropriate in certain cases to help prevent warming. Solar power is another alternative, especially in the absence of regular energy source. Evaporative cooling, especially in dry and low humidity air conditions, can reduce temperature by 10–15^{0}C and maintain high humidity, which is beneficial to the quality of horticultural produce (HLP 2014). But at the end, an efficient cold chain can't improve the quality and safety of food commodities, but rather preserve them. Such can only be insured by applying good production practices (GPP), good harvesting practices (GHP), good manufacturing practices (GMP) and good hygienic practices (GHP) at different phases of the FSC.

The vital importance of collaboration among multiple stakeholders involved, as well as government intervention to provide the key services such as infrastructure, legislation, capacity building and awareness, are essential and required (FAO 2013b). Several examples exist of investments by companies in the MENA region that have successfully adapted cold chain technology, however these investments have largely been made to meet the food import requirements of the export market (the EU in particular). Given the gap between export-oriented food chains and domestic traditional food handling, there should be more focus on technologies that are also economically accessible to the vast number of small- scale food producers and food chain SMEs. Developing and enforcing quality and safety standards can also contribute to bridge the large gap between domestic and export-oriented supply chains. Adoption of good harvesting practices, development of proper transport infrastructure and conditions as well as sorting, grading, packing, packaging and storage infrastructure, are all essential to support the proper functioning of the FSC.

Food Processing

Food processing is an effective way to conserve perishable products. Technologies employed in food processing such as pasteurization, sterilization, canning and packaging; contribute to extension of the products shelf-life, thus reducing losses and waste in the food chain (Langelaan et al. 2013). In general, developing food processing necessitates developing the infrastructure and appropriate processing technologies based on insight into the market demands and forecasts in-light of the FSC structure.

Within the food processing sector, substantial parts of the raw materials that enter the factory end up as by-products. Technical innovations can enable to volarize these co-products as food and feed. For example, the production of fruits and vegetables involve numerous phases: pre-cooling, washing and disinfection, peeling, trimming, deseeding, cutting, sorting, dipping, drying, storage, packaging, labelling and distribution (James and Ngarmsak 2011). Opportunities can be created through industrialization of these stages for better volarization of by-product production of juices and jams (Verghese et al. 2013), as feed and for bioenergy production and/or compost. Additional benefits ensue when these operations are localized in rural areas; in such cases, it can also reduce costs of transport, reduce urban waste as well as reduce the environmental impact of FLW (HLPE 2014).

In developing countries, the lack of processing facilities causes substantial food losses. Capacity of the processing industry to process and preserve fresh farm produce to be able to meet the demands constitutes a major problem. Part of the problem stems from the seasonability of production and the cost of investment in processing facilities that will not be fully utilized all the year. Prevention approaches that may provide solutions to this problem is to develop contract framing linkages between processors and farmers, where governments should create a better "enabling environment" and investment climate to stimulate investment by the private sector in the food industry and work closely with the farmers to address supply issues (Gustavsson et al. 2011). On the other hand, the introduction of small scale, low-cost food technology processing methods, complemented with capacity building of small farmers and supported by extension services, will contribute to the proliferation of an artisan industry in rural areas. This will also provide work opportunities to women and contribute to the improvement of their livelihoods.

Approaches/Solutions Close to the Fork

Food waste close to the fork arises from within retail and households and hospitality and food services. They are mostly related to consumer behaviors.

Retail and Household Food Waste

Consumer studies have shown that consumers are often time unaware of the amount of FLW and its cost. Consumer behavior is identified as being complex, interlinked and self-reinforcing where self-awareness can be a powerful trigger to promote consumer behavioral change (Bond et al. 2013). According to Quested et al. (2013), there are two main ways of reducing the amount of FLW at home, by influencing people's actions or by making changes in the food sold, such as changing packaging or extending shelf-life.

Raising awareness on the extent of consumers can be an effective way to influence people's actions motivate change. Awareness raising campaigns help overcome this lack of information, such as when individuals record their wasted food and are provided with practical advice on how to achieve savings once waste has been identified. Several materials are used in different countries to raise awareness of the scale and value of food waste. In developed countries, food losses are closely linked to cultural attitudes and standardization of the product offered to consumers that are mostly related to appearance and quality standards set by retailers for fresh products. Some of the produce do not even leave the farm, because of rejection by retailers/supermarkets due to the imposed rigorous quality standards regarding weight, size and appearance of crops. As a result, large portions of crops never leave the farm, even though some of these rejected crops are diverted to animal feed that originally were aimed for human consumption (Stuart 2009). Engaging the consumers in decision making through surveys to explore their willingness to buy heterogeneous produce, as long as the taste is not affected, may provide alternatives that will reduce losses (Stuart 2009). Consumers can influence quality standards, especially if they are offered a broader quality range products in retail stores at reduced prices. Sales closer to consumer without having to pass the strict quality standards set by retailers could be another approach to reduce the amount of rejected crops. This could be achieved through, e.g. farmers markets and farm shops (Stuart 2009).

On the other hand, much better use of packaging that has been designed to protect food could dramatically reduce the amount of food thrown away, but may come at an environmental and financial cost. The WRAP report (2015) indicated that the significant advances achieved in packaging material and related technologies over recent decades, with intelligent/smart materials designed to keep specific products at best for longer, have significantly contributed to the reduction of FLW at the retail and household levels. Novel packs for fresh meat, for example, that enable consumers to divide the packaging without exposing a portion of the meat, had significantly reduced the amounts of fresh meat thrown away as a result of not being used in time (WRAP 2015). Studies conducted by Quested et al. (2013) recommended detailed measures that consumers can implement on their own to reduce FLW. These include:

- Better planning of purchases to avoid buying more than needed
- Avoid impulsive or advance purchasing of food that is not required immediately
- Better understanding of the distinction between "best before" and "use by" dates
- Better storage practices and stock management at home
- Better evaluation of the portions that need to be prepared
- Better food preparation techniques, to avoid food being not eaten, and FLW due to the preparation method
- Making full use of fruits and vegetables to extract all the nutritional benefits
- Better knowledge on how to use the leftovers on other recipes instead of discarding

Hospitality and Food Services Food Waste

The hospitality and Food services sector (restaurants, canteens, catering, hotels, etc.) are considered key players for the reduction of FLW at the end of the food chain, i.e. consumption level. They can play a double role by developing strategies to reduce their own FLW and at the same time raise awareness of consumers. To establish a waste reduction strategy in food services, a first step is to measure and track the amount, type of and the reason of FLW, which is usually different for different business. For instance, a guide to reduce FLW in hospitality sector in Spain – Catalonia (Alicia/UAB 2012), has been developed that provides detailed measures from stock management to menu design. Also, different tools have been developed to assess FLW and provide options to reduce them. The U.S. Environmental Protection Agency, for example, developed a FLW assessment tool to compute costs and benefits of waste reduction options, such as source reduction, reuse of leftovers, use as animal feed, composting, as well as savings in GHGs. Other sophisticated systems can help operators to monitor weighing the quantities discarded based on type of food and calculate their cash values (HLPE 2014). A more plausible approach to reduce FLW is to use a combination of business and consumer oriented strategies. One straight forward approach is to reduce the portion sizes and offer smaller sizes at a lower price. This approach will give the smaller-appetite customer the choice to order smaller meals, leave less and save more.

In Brazil and Portugal restaurants "a kilo", consumer pay for what they take as opposed in a buffet style food service "all-you-can-eat". This is a good example for the consumer to get to know the economic value of food and cost of waste (Generalitat de Catalunya 2011). Replacing the "pay-by-weight" and adjust the size of meal to the real needs would save not only in meals cost and waste reduction, but it can also save the retailer money, as less food would need to be prepared. Another example is the introduction of tray-less system in cafeteria-style food service, where customers rather carry the food they purchase on plates. One study of dining halls in 25 Universities in the U.S. found that going tray-less reduced food waste by 25-30% (Aramark 2008).

Food Date Label

Simplifying date label packaging of food and drinks such as "use by", "sell by" and "best before" helps to reduce consumer confusion. One study in the U.K., for instance, found that a fifth of food thrown away by households in the U.K. is thrown due to food being perceived as "out of date" due to labeling, where in fact some of the food was still fit for human consumption (WRAP 2011). Date labelling that is a requirement by regulatory agencies, for example Food And Drug Administration in the U.S.A., refer to food quality when labels indicate "best before" or "display until" which advise staff and customers on quality, while "use by" dates relate to food safety. However, for the consumers, all these dates are perceived as being a measure of food safety which may lead them to throw away edible food that is still entirely safe to eat. To reduce this confusion, governments can provide guidance on what dates manufacturers and retailers should print on their packaging. In 2013, TESCO a main retailer in the U.K. piloted the use of a single date code in its U.K. stores on meat "use by", fruits and vegetables "best before". Results were encouraging that TESCO dropped out the "display until" and sticked to a single code "use by" for pre-packed meat, that has been rolled out in more than 3000 stores in the U.K (Lipinski et al. 2013). The simplified date coding system "best before' for fruits and vegetables was rolled out by the end of 2013. TESCO also developed awareness material to the consumer with the information needed to understand these dates. WRAP (2015), after working with governments, regulators and industry in the U.K. to update official guidance on food labelling, indicated that most retailers in the U.K. already moved to a single date label (removal of "sell by" or "display until" dates). They are switching from "use by" to "best before" dates where possible (for example on fresh produce, hard cheese and pasteurized fruit juice), and that "use by" should only be used where there is a food safety issue.

Surplus Food Redistribution

Non-marketable Products at Discount Prices

As earlier indicated, standardization of products offered to consumers is a major cause of FLW in modern retailing systems. In these systems, products are rather defined as marketable or not. The concept of "not marketable" refers here mainly to the quality aspect of the products, which is often not linked to their edibility. Such products are still fit for human consumption pending its safety, and can find its way for the consumer avoiding the FLW stream. Studies conducted by the FAO in Kenya on several products reported different selling prices reflecting different levels of FQLW (FAO 2014c). In the U.S., some retailers sell such products at discounted prices (NDRC 2013). In Spain, consumer surveys revealed that more than half the consumers would be ready to buy such products (MAGRAMA 2013).

Food Banks

Food banks provide a means to give edible food value, that would have been lost, and serve as one of the solutions that contributes to people in need. Food banking has emerged in different countries as non-government initiatives in the form of non-government organizations (NGOs) collecting food to distribute to people and/or utilize them in kitchens to feed the needy. Food bank models vary greatly among countries. In traditional models, all actors in the FSC can be engaged to donate food that has commercial value. Producers can donate products that are ripe and ready for harvesting but whose market prices do not cover production cost. Processors can donate food that faced problems in packaging, for example, or from cancellation of orders from buyers, but that are still fit for human consumption. Distributors and retailers can donate foods that are too close to the expiration date or fruits and vegetables that are bruised and rendered not marketable. The proper functioning of the food banks relies on voluntary involvement of NGOs with the help of specialists that have a secondary role in the food chain, such as logistics, information technology and legal services. However, the role of the governments to guarantee an institutional environment that favors donations, especially with respect to tax incentives, wherever it exists, and legal immunity is essential (Aiello et al. 2014).

Corporate Social Responsibility

During the last decade, there has been a rapid growth in activities associated with Corporate Social Responsibility (CSR). More companies now include in their annual business corporate reports a section dedicated to making positive social or environmental impact on society that could also lead to more sustainable systems and less FLW. In FLW, businesses can play an effective role in its reduction. Reporting on monitoring and reduction of FLW in business activities and supporting activities that lead to reduction of FLW with their suppliers, at consumer level or elsewhere, businesses can give FLW an effective awareness role. In addition, this can also bolster their images as role models towards reduction of FLW. TESCO, in its commitment to transparency published figures about FLW within its own operations and within the total supply chain, amounting to 28.500 tons at its stores and distribution centers for the first six months of 2012 alone. Since 2011, the largest retailer in the Netherland, Ahold publishes data in its CSR report on their FLW, that represented 1–2% of total food sales, with fresh food waste at 2–3% (HLPE 2014). In Argentina, as reported by HLPE (2014), the CANALE group is supporting a training program in 90 schools, where 2000 school children and 100 teachers have been trained on sustainable food consumption. Also in Argentina, the firm “Mondelez” with the Argentinian Red Cross succeeded in distributing over 3.6 million tons of fresh fruits and vegetables from 2009 to 2012 to more than 230,000 people that were fit for consumption, but with cosmetic imperfection that makes them unmarketable.

Consumer Awareness Campaigns

In-spite of the fact that changing consumer food consumption and waste pattern is not easy, yet consumer awareness campaigns can help influence consumer behavior. Food retailers can play a key role in reducing food waste at the consumption level due to their close interaction with their customers. For example, in the U.K., the Co-operative Group that owns over 2800 grocery stores took the initiative to inform customers for improving food storage and lengthening of shelf-life of fruits and vegetables by printing tips onto the plastic produce bag. This was done in an effort to assist customers understand better best storage practices for increasing the shelf-life of their purchases. The Co-operative Group has also shifted away from "buy-one-get-one-free" promotion for perishable foods to using price reductions (Ferguson 2009).

HPLE (2014) reported on a number of campaigns to reduce food waste. In China "Empty Plate" campaign that targeted public food consumption and reception and banquets in 2013 mobilizing public media and a number of TV stations at provincial level that resulted in significant reduction in restaurant food wastage. In the Republic of Korea, "Half-bowl" campaign and a "New Container" that encourages people to order half bowl of rice to reduce food restaurant waste by 20%. In Denmark, the "Stop Wasting Food" campaign empowers consumers to take action and individual initiatives such as cooking left overs, shopping more wisely and distributing surplus food. In the Netherland, the campaign "Feed Battle" to track food waste in households through keeping a diary on how much food is wasted over a three week period combined with practical tips and specific interventions. It resulted in wasting 20% less food in a period of three weeks (Bos-Brouwers et al. 2013).

Enabling Environment and Supportive Policies and Institutions

Implementation of change along the FSC to reduce FLW is largely affected by many socio-economic and cultural factors. Specifics vary by country and by the type of markets, value products and others. As a first step, one needs to characterize the local food sector development needs on case by case basis (Kader et al. 2012). Development of strategies to reduce FLW in any country requires a mix of policies and institutions. It also requires a conducive environment that facilitates the coordination and collaboration among the different actors and stakeholders, strengthens organization of the FSC, fosters investment in infrastructure, mobilizes financial and other service sectors, promotes best practices at each step of the FSC and improves its overall functioning efficiency.

Successful implementation of FLW strategies do not only require access to key pre-requisites of technology, financing, markets, etc. but also government policies that promote stability of raw material supplies, favor cost efficiency, empower farmers, promote the role of SMEs, and facilitate the participation of the private sector. Additional requirements include innovative business models, strong institutions, and adequate support services such as research and development, extension services, capacity building, safety and quality standards and information systems to name a few (Da Silva and Mhlanga 2011).

Integration of FLW concerns in policies, being it through existing relevant policies such as agriculture, the food chain and consumption, where numerous public policies interact, or by devising specific FLW reduction policies to address the interdependence of actions that cumulatively end up creating FLW, is essential and required (HLPE 2014). In addition, the creation of alternative business models and approaches that support the proper functioning of the FSC based on the socio-economic status and needs of every country, represents vital tools for the development of the food system sustainability. A variety of business models and approaches have been in use, both in developed and developing countries. These include associations, business networks, agro-based clusters, food parks, export processing zones, public-private partnerships (PPP), etc. Several of these can be operated feasibly at the small and medium scale and address the challenges faced by SMEs in accessing advanced technologies, food safety and quality standards, markets, etc.

Way Forward for the MENA Region

The FAO (2014a) developed through a series of consultation studies and workshops and in alignment with SAVE FOOD "Global Initiative on Food Losses and Waste Reduction" a regional strategic framework for the reduction of FLW in the MENA region: This strategic framework targets national actions for achieving the region's goal for the reduction of FLW by 50% in 2024. (Fig. 6)

Some countries have begun by adopting this regional framework that will feed into their national action and implementation plans (FAO 2014a). Successful implementation depends on each of the member countries of the MENA. In parallel, concerted efforts should be expended to establish/develop basic and necessary pre-requisites that can catalyze the successful implementation of the national FLW reduction plans:

- Developing mechanisms to foster effective collaboration among the academic, industry, public and private sectors, and promoting better linkages and communications among the different actors across the FSC to help reduce duplication of efforts and encourage adoption of cost-effective food loss prevention measures.
- Investing more in Research and Development with a focus on how it can help reduce FLW at all stages of the FSC and improve the food safety and quality systems.
- Establishing different instruments and policies directed at different forms of organizations and models, e.g. association, networks, clusters, PPP etc. to empower small scale farmers and SMEs and enhance their ability to organize production and procurement, adopt modern and most appropriate technologies

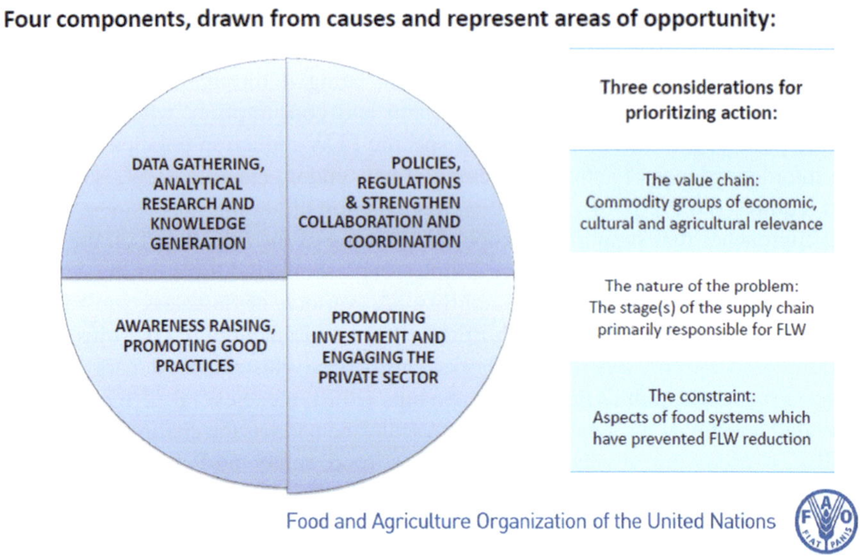

Fig. 6 Four components, drawn from cause and represent areas of opportunity (Source: Food Losses and Waste in the MENA countries (Akdeniz University – Turkey), FAO 2014.)

and best practices, mobilize financial and other support services and build competitiveness capacity in both domestic and global market.

- Investing more in capacity building at all levels for the reduction of FLW along the FSC, including manpower build-up and training on practices and procedures related to good agriculture production practices, good post-harvest practices, good hygienic practices and good manufacturing practices.
- Developing platforms for knowledge sharing and dissemination, including market information, among public and private sector actors, and empowering farmers and SMEs to make informed choices about the most appropriate inputs for their needs.
- Providing the resources to improve the infrastructure needed to support the efforts for the reduction of FLW across the FSC, particularly those related to proper storage including cold chain, transportation, and distribution including wholesale and retail markets, as well as food safety and quality systems.
- Mobilizing all actors and consumers for awareness and action to reduce FLW at the different stages of the FSC, with the overall aim to address the challenge of sustainability of the food system.

At the same time, MENA countries can capitalize and benefit from UN organizations' wide experience and expertise in the area of FLW to facilitate and help in the planning and implementation of national FLW reduction plans. In this respect:

- U.N. organizations should be encouraged to continue its engagement with member countries to assist in the developing the agro-production and agro-

industry and reducing food losses and waste, through promoting continuing dialogue and advising on policies, institutions and services to create the enabling environment that support achieving the goals towards reduction of the FLW in the MENA region.
- U.N. agencies programs should be reinforced, especially those related to capacity building and human build-up, in areas related to FLW across the FSC, including food standards, safety and compliance systems and innovation and technology diffusion,
- U.N. agencies should enhance its involvement in supporting country members in MENA to address problems of both-under and over-nutrition and promote healthy eating and lifestyle.

References

Adegbola, A., Bamishaiye, E. I., & Olayemi, F. (2011). The factors affecting the adoption of the re-usable plastic vegetable crate in three local government areas of Kano State, Nigeria. *Asia Journal of Agricultural Science, 3*(4), 281–285.

Aiello, G., Enea, M., & Muriana, C. (2014). Economic benefits from food recovery at the retail stage: An application to Italian food chains. *Waste Management, 34*(7), 1306–1316.

Alícia/UAB (Fundació Alícia/Universitat Autonoma de Barcelona). (2012). Aprofitem el menjar!, Una guia per a la reducció del malbaratament alimentari en el sector de l'hostaleria, la restaració i el càtering http://www.alicia.cat/uploads/all/guia_malbaratament.pdf

Aramark Higher Education. (2008). *The business and cultural acceptance case for trayless dining*. Accessible at: http://www.aramarkhiered.com/assets/docs/whotepaper/ARAMARK%20Trayles%2Dinning%20July202008%20Final.PDF

Bai, Z. G., Dent, O. L., & Schaepman, M. E. (2008). *Global assessment of land degradation and improvement identification by remote sensing*. International Soil Reference and Information Center (SRIC): Wageningen.

Bond, M., Meacham, T., Bhunnoo, R., & Benton, T.G. (2013). Food waste within global food systems. A Global Food Security Report. www.foodsecurity.ac.uk.

Bos-Brouwers, H. E. J., Scheer, F. P., Nijenhuis, M. A., Kleijn, F., & Westerhoff, M. (2013). FoodBattle: Reductie milieudruk voedselverspilling op het snijvlak van supermarkt & consument. Wageningen http://www.rwsleefomgeving.nl/onderwerpen/duurzaam_produceren/ketenaanpak/downloads/foodbattle-reductie/.

Buzby, J., Wells, H. F., & Hyman, J. (2014). *The estimated amount, value, and calories of postharvest food losses at the retail and consumer levels in the United States* (Economic Information Bulletin No. (EIB-121)); http://www.ers.usda.gov/publicaitons/eib-economic-informaion-bulletin/eib121.aspx

Cohen, J. F., Richardson, S., Austin, S. B., Economus, C. D., & Rimm, E. B. (2013). School lunch waste among middle school students: Nutrients consumed and costs. *American Journal of Preventive Medicine, 44*(2), 114–121. doi:10.1016/j.amepre.2012.09.060.

CSACC. (2011, November). Commission on sustainable agriculture and climate change – "Achieving Food Security in the Face of Climate Change". Summary for policy makers from the Commision on Sustainable Agriculture and Climate Change.

CTA (Technical Center for Agriculture and Rural Cooperation). (2012, September). *Going to waste – Missing opportunities in the battle of improved food security* (CTA policy brief, № 7).

da Silva, C., & Mhlanga, N. (Eds.). (2011). *Innovative policies and institutions to support agro-industries development*. Rome: FAO.

Evans, D. (2011a). Beyond the throwaway society: Ordinary domestic practice and a sociological approach to household food waste. *Sociology*.
Evans, D. (2011b). Blaming the consumer – Once again: The social and material contexts of everyday food waste practices in some English households. *Critical Public Health, 21*(4), 429–440.
FAO. (2011a). Global food losses and food waste – Extent, causes and prevention, by J. Gustavsson, C. Cederberg., U. Sonesson, R. van Otterdijk & A. Meybeck. Rome (http://www.fao.org/docrep/014/mb060e/mb060e00.pdf)
FAO. (2011b). The regional office of the near East and North Africa – Agro-industries and infrastructure – Policy Brief: Developing the cold chain for Agriculture in the Near East and North Africa (MENA)
FAO. (2013a). Food wastage footprint: Impacts on natural resources. "Summary Report"
FAO. (2013b). *Report of the expert consultation meeting on food losses and waste reduction in the near east region: Towards a regional comprehensive strategy*, 18–19 December 2012. Rome: Sharm El Sheikh.
FAO. (2014a, September 25–26). Food losses and waste in the MENA countries: Status, prospects and strategies for Reduction. In Workshop on agricultural trade and food security in Euro-Med area, Turkey.
FAO. (2014b). Food wastage footprint, Full cost accounting.
FAO. (2014c). Food loss assessments: Causes and solutions case studies in small-scale agriculture and fisheries subsectors. Kenya: Banana, maize, milk, fish. Global Initiative on Food Loss and Waste Reduction – Save Food. Rome (http://www.fao.org/fileadmin/user_upload/save-food/PDF/Kenya_Food_Loss_Studies.pdf).
FAO. (2015a). Regional coordination mechanism (RCM) issue brief for the "Arab Sustainable Development Report". Food Security and Sustainable Agriculture in the Arab Region, FAO.
FAO. (2015b). Regional overview of food insecurity: Near East and North Africa "Strengthening Regional Collaboration to build resilience for food security and nutrition". Rome: Food and Agricultural Organization of the United Nation.
FAO. (2015c). Regional coordination mechanism (RCM). Food Security and Sustainable Agriculture in the Arab Region.
Farag, D. (M. Diaa El-Din Hamed El-Sayed Farag). (2008, February) Aflatoxins: Awareness and control. In Third Dubai International Food Safety Conference, organized by Dubai Municipality, United Arab Emirates (pp. 23–27). http://www.foodsafetydubai.com/prevconf/files/3FSC05.pdf.
Ferguson, I. (2009). Food waste – Influencing consumer behavior. Presentation.
FUSIONS. (2014). Drivers of current causes of food waste, threats of future increase and opportunities for reduction. FUSIONS. http://www.eu-fusions.org
Generalitat de Catalunya. (2011). Un consum responsable dels aliments, propostes per a prevenir i evitar el malbaratament alimentari. Barcelona: Departament de Territori i Sostenibilitat. http://www20.gencat.cat/docs/arc/Home/LAgencia/Publicacions/Centre%20catala%20del%20reciclatge%20(CCR)/guia_consum_responsablebr.20.11.12.pdf.
Gustavsson, C. C., & Sonesson, U. (2011). Global food losses and food waste. Rome: FAO
High Level Panel of Experts (HLPE) on Food Security and Nutrition. (2014, June). Food losses and waste in the context of sustainable food systems, HLPE Report # 8.
IMF. (2014). Regional economic outlook, update.
International Labor Organization (ILO). 2014. Global Employment Trends.
James, J. B., & Ngarmsak, T. (2011). *Processing of fresh-cut tropical fruits and vegetables: A technical guide*. FAO: Bangkok.
Kader, A., Kitinoja, L., Hussein, A., Abdin, O., Jabarin, A., & Sidahmed, A. (2012). *Role of agro-industry in educing food losses in the middle east and North African region*. Rome: FAO.
Kharas, H. (2010). *The emerging middle class in developing countries* (OECD Development Center Working Paper number 285); http://www.oecd.org/social/poverty/44457738.pdf

Langelaan, H. C., Pereira da Silva, F., Thoden van Velzen, U., Broeze, J., Matser, A. M., Vollebregt, M., & Schroën, K. (2013). Technology options for feeding 10 billion people. Options for sustainable food processing. State of the art report. Science and Technology Options Assessment. Brussels: European Parliament. http://www.europarl.europa.eu/RegData/etudes/etudes/join/2013/513533/IPOL-JOIN_ET(2013)513533_EN.pdf.

Lewis, L., Onsongo, M., Njapau, H., Schurz-Rogers, H., Luber, G., Nyamongo, S.J., Baker, L., Dahiye, A.M., Misore, A. & Kevin, D.R. (2005). Aflatoxin contamination of commercial maize products during an outbreak of acute aflatoxicosis in Eastern and Central Kenya. *Environmental Health Perspective,* 113(12), 1763–1767.

Lipinski, B., Hanson, C., Lomax, J., Kitinoja, L., Waite, R., & Searchinger, T. (2013, May). Reducing food loss and waste. World Resources Institute, working paper.

MAGRAMA (Ministerio de Agricultura Alientación y Medio Ambiente). (2013). Estrategia "más alimento, menos desperdicio".

Muth, M. K., Karns, S. A., Nielsen S. J., Bubzy, J. C, & Wells, H. F. (2011). Consumer-level food loss estimates and their use in the ERS loss-adjusted food availability data. Technical Bulletin No. (TB-1927), USDA.

NRDC (Natural Resources Defense Council). (2013). The dating game: How confusing food date labels lead to food waste in America (htpp://www.nrdc.org/food/files/dating-game-report.pdf).

Parry, D. L. (2013). Analyzing food waste management methods. *BioCycle, 54*(6), 36.

Quested, T. E., Marsh, E., Stunell, D., & Parry, A. D. (2013). Spaghetti soup: The complex world of food waste behaviour. *Resources, Conservation and Recycling, 79*, 43–51.

Rolle, R.S. (2006). Improving postharvest management and marketing in the Asia-Pacific region: Issues and challenges trends. In R. S. Rolle, ed. *Postharvest management of fruit and vegetables in the Asia-Pacific region* (pp. 23–31). Tokyo: Asian Productivity Organization. ISBN: 92-833-7051-1.

Searchinger, T., Hanson, C., Ranganathan, J., Lipinski, B., Waite, R., Winterbottom, R., Dihshaw, A., & Heimlich, R.. (2013). *The Great Balancing Act.* Working paper, Installment 1 of Creating A Sustainable Food Future. Washington, DC: World Resources Institute. Accessible at: http://www.worldresourcesreport.org

Segrè, A. (2013). *Vivere a spreco zero, una rivoluzione alla portata di tutti.* Venice: Marsilio Editori. ISBN 978-88-317-1583.

Silvennoinen, K., Katajajuuri, J. M., Hartikainen, H., Jalkanen, L., Koivupuro, H. K., & Reinikainen, A. (2012, November 12–15). Food waste volume and composition in the Finnish supply chain: Special focus on food service sector. In Proceedings Venice 2012, Fourth International Symposium on Energy from Biomass and Waste, Cini Foundation, Venice, Italy https://portal.mtt.fi/portal/page/portal/mtt/hankkeet/foodspill/Food%20Waste%20Volume%20and%20Composition%20Focus%20on%20Food%20Service%20Sector.pdf.

Sivakumar, D., Jiang, Y., & Yahia, E. M. (2011). Maintaining mango (Mangifera indica L.) fruit quality during the export chain. *Food Research International, 44*(5), 1254–1263.

Stuart, T. (2009). *Waste: Uncovering the global food scandal.* London: W.W. Norton Co..

Thompson, A. K. (2007). Pre-harvest factors on post-harvest life. In A. K. Thompson (Ed.), *Fruit and vegetables: Harvesting, handling and storage.* Oxford: Blackwell Publishing Ltd.

Verghese, K., Lewis, H., Lockrey, S., & Williams, H. (2013). *Final report: The role of packaging in minimising food waste in the supply chain of the future.* Melbourne: RMIT University.

Waste and Resources Action Program (WRAP). (2015, February). Strategies to achieve economic and environmental gains by reducing food waste.

WFLO. (2010). http://ucce.ucdavis.edu/files/datastore/234-2143.pdf

Williams, H., Wikström, F., Otterbring, T., Lófgren, M., & Gustafsson, A. (2012). Reasons for household food waste with special attention to packaging. *Journal of Cleaner Production, 24*, 141–148.

World Hunger and Poverty Facts and Statistics. (2013). http://www.worldhunger.org

WRAP. (2009). Household food and drink waste in UK; http://www.wrap.org.uk

WRAP. (2011). Consumer insight: Date labels and storage guidance. Assessable at: http://www.wrap.org.uk/sites/files/wrap/technical_report_dates.pdf

WRAP. (2013). Household food and drink waste in the UK 2012, http://www.wrap.org.uk/content/household-food-and-drink-waste-uk-2012

WRAP. (2015). Strategies to achieve economic and environmental gains by reducing food waste

Yusuf, B. L., & He, Y. (2011). Design, development and techniques for controlling grains post-harvest losses with metal silo for small and medium scale farmers. *African Journal of Biotechnology, 10*(65), 14552–14561.

Foodborne Disease in the Middle East

Ewen C.D. Todd

Abstract Food safety is a concern worldwide and according to the World Health Organization, developing countries are probably more at risk of foodborne illness because many of these, including those in the Middle East, have limited disease surveillance and prevention and control strategies. Specifically, the Middle East and North Africa (MENA) region has the third highest estimated burden of foodborne diseases per population, after the African and South-East Asia regions. However, it is difficult to determine what the burden is since little is published in peer-reviewed journals or government reports for public access. This chapter reviews 16 autonomous nations, namely, Afghanistan, Bahrain, Egypt, Iran, Iraq, Israel, Palestine, Kuwait, Lebanon, Oman, Pakistan, Qatar, Saudi Arabia (KSA), Syrian Arab Republic (Syria), United Arab Emirates (UAE) and Yemen. Countries range in size from Bahrain with 1.8 million inhabitants to Pakistan with a population of 184 million. Agriculture and local food production is much influenced by water availability for irrigation. Water shortages are most severe in the Gulf countries which rely on aquifers, desalination, and recycled waste water for most of their water supplies. This means that most food is imported which is expensive if not subsidized through petrodollars. This impacts food security which is a particular concern in countries under conflict, particularly, Syria, Yemen and Iraq. Gastrointestinal infections are frequent in this region from *Salmonella* Typhi and other *Salmonella* spp., *Shigella* spp., *Campylobacter jejuni* and *C. coli*, rotavirus, hepatitis A virus, parasites, and more rarely from *Aeromonas, Yersinia enterocolitica, Brucella* spp., and Middle East Respiratory Syndrome coronavirus (MERS-CoV). Reports indicate that children are the most susceptible and that many isolates are multidrug resistant. Chemical contamination of water supplies and crops are probably more of a concern than published reports indicate, because of widespread indiscriminate use of fertilizers, antibiotics, and pesticides, coupled with increased industrial pollution affecting the water supplies. Like many other parts of the developing world, foodborne disease surveillance is limited and outbreaks are most often reported through the Press but with insufficient detail to determine the etiological agents and the factors contributing to the outbreaks, leading to

E.C.D. Todd (✉)
Ewen Todd Consulting, Okemos, MI 48864, USA
e-mail: todde@msu.edu

S. Murad et al. (eds.), *Water, Energy & Food Sustainability in the Middle East*,
DOI 10.1007/978-3-319-48920-9_17

speculation to the cause by those interested or responsible for food prevention and control. However, there are some well investigated outbreaks in the region that have those details, and reveal where the shortcomings of both the establishments and the inspection systems have been. Where the causative agents are known, the kinds of pathogens are generally similar to those found in the West, e.g., *Salmonella*, but many outbreaks seem to have short incubation periods that point to a toxin of some kind of chemical or biological origin, but these are almost never identified. Because of sectarian warfare, residents and refugees have been given food that has made them sick and solders? have been deliberately poisoned. Research has been focused on microbial contamination of locally-sold foodstuffs and manager and employee knowledge of food safety and hygienic conditions in food preparation establishments. An innovative pilot project in Qatar is to use seawater and sunlight for raising crops through the Sahara Forest Project. All countries have some kind of food establishment inspection system, but they tend to be punitive if faults are found in management or employees on the premises rather than being used for their education for improving food safety. Restaurants may be closed down and owners and employees fined for often unspecified infringements. However, some food control agents are moving towards employee training through seminars and courses before problems occur, which is a good disease prevention strategy. Unfortunately, many of the food handlers are from Asian countries with languages other than Arabic and English, which makes effective food safety communication and training difficult. Tourists visiting popular resorts in Turkey and Egypt have suffered from foodborne illnesses, usually of unknown origin but poor hygienic conditions are blamed with law suits following, and the adverse publicity affects the long-term viability of some of these resorts. Food exports, important for local economies, have occasionally been contaminated resulting in recalls and sometimes illnesses and deaths, notably fenugreek seeds from Egypt (*E. coli* O104:H4), pomegranate arils from Turkey (hepatitis A virus), and tahini from Lebanon (*Salmonella*). Overall, in recent decades, the Middle East has made strides towards improving food safety for both residents and foreign visitors or ex-pat workers. However, within the countries there are large discrepancies in the extent of effective public health oversight including food safety and food security. Currently, almost all of the countries are involved to a greater or lesser extent in the civil wars in Syria and Yemen, or are affected through political tensions and strife in Egypt, Iraq, Iran, Israel, Palestine, Lebanon and Turkey. In addition, the current overproduction of oil on a world-wide scale has led to a rapid decrease in revenues to most Gulf states. All this points to a severe setback, and an uncertain foreseeable future for improvements in obtaining both sufficient and safe food for residents in this region.

Keywords Food borne diseases • Food safety • Prevention strategies

Introduction

The World Health Organization (WHO) Eastern Mediterranean region, comprising countries in the Middle East and North Africa (MENA), has the third highest estimated burden of foodborne diseases per population, after the African and South-East Asia regions. According to the WHO (2015a), more than 100 million people living in this region are estimated to become ill with a foodborne disease every year and 32 million of those affected are children under 5 years. Diarrheal diseases caused by *E. coli*, norovirus, *Campylobacter* and nontyphoidal *Salmonella* account for 70% of the burden of foodborne disease. An estimated 3000 people die each year from unsafe food, caused primarily by diarrheal diseases, typhoid fever, hepatitis A, and brucellosis. Both typhoid fever and hepatitis A are contracted from food contaminated by the feces of an infected person and the source of brucellosis is typically unpasteurized milk or cheese from infected goats or sheep. Half of the global cases of brucellosis are in people living in this region, with more than 195,000 people infected every year, causing fever, muscle pain or more severe arthritis, chronic fatigue, neurologic symptoms and depression. Cholera, which after a short incubation period of 2–5 days causing severe diarrhea and dehydration, is returning to those countries with limited public health infrastructure caused by conflict, such as Iraq (Agence France-Presse 2015).

The list of countries covered by this chapter is similar to that of WHO but leaving out North African countries except Egypt (which has territory in eastern Asia) and adding Turkey which is not always considered in the region because it is not Arabic, but has interesting food safety data. Therefore, the countries under review are Afghanistan, Bahrain, Egypt, Iran, Iraq, Israel and Palestine, Kuwait, Lebanon, Oman, Pakistan, Qatar, Saudi Arabia (KSA), Syrian Arab Republic (Syria), United Arab Emirates (UAE) and Yemen. Gulf countries Bahrain, Kuwait, Oman, Qatar, KSA and the UAE have similar social, political, economic, culture, religion, language and ancestry with several similarities in their food control systems and food safety programs (Al-Kandari and Jukes 2009).

A Food and Agriculture Organization (FAO) report covering international investments in agriculture in the Near East (not identical to the countries chosen for this chapter, but many of the findings apply) states that this region is characterized by a mix of very different countries' resources and incomes (Tanyeri-Abur and Elamin 2011). The wealth in the richer countries of the region is primarily dependent on oil revenues and the past economic growth has been closely linked to the oil market; about 43% of regional Gross Domestic Product (GDP) is concentrated in the high income countries (Qatar, Kuwait, UAE, Saudi Arabia, and Bahrain) which are home to only11.8% of the population in the region, and many of these are expatriates working in these countries. The report indicates that food insecurity varies sharply in the region but overall the percent of the undernourished population does not exceed 57% in most countries of the region, except for Sudan, Mauritania, Djibouti and Yemen where the proportion of undernourished exceeds 25%; however, in 2016 these percentages will be totally out of date for countries like Syria and Iraq and in neighboring countries where refugees have reached because civil war and jihadi terrorist groups have put considerable stress on public health facilities and food availability. The countries in the region however, are largely similar when it comes to the challenges in achieving sustainable agriculture and food security. For most of these countries, the overwhelming concern is to secure adequate and stable supplies of food at the national level, making food security a concern for both rich and poor countries of the region (Tanyeri-Abur and Elamin 2011). The three major problems affecting most of the countries are (i) limited water availability; (ii) population growth; and (iii) heavy dependence on food imports. Water scarcity in particular, is the most critical development problem in the region and the single most important factor in limiting agricultural growth, and water availability has been declining steadily since the late 1950s. The region as a whole has 70% less availability of renewable water per person in 2003–2007 than in 1958–1962. Lack of water for irrigating crops but also for potable water supplies affects many of the countries, particularly in the Gulf region. It is important to note that the wealthiest countries are also those with the highest water depletion record, namely, the UAE and Qatar. The unprecedented growth in investment in agriculture is in large part a result of the food crisis of 2007, which brought about a rethinking of agricultural support policies, mostly in countries of the Gulf and particularly Saudi Arabia, which has invested heavily in the last 30 years in large-scale agricultural production using up valuable water resources. Saudi Arabia announced in January 2008 that it would phase out wheat and agricultural production in the course of the next 8 years. In July 2008, Qatar and UAE took similar policy decisions (Tanyeri-Abur and Elamin 2011).

Crops grown in the region may serve as fresh food sources for the population, but much of the food is imported with limited locally processed products, and if the policies of KSA, Qatar and UAE expand to other countries, more will be imported in the future (Tanyeri-Abur and Elamin 2011). Thus, the main foodborne disease issues are with homemade, restaurant and street food, where isolated claims of illness are followed up by inspections and possible punitive action by public health agencies responsible for food safety. Those countries that rely on tourism for their main source of GDP have sometimes been damaged by adverse publicity, e.g., Egypt, and to a lesser extent, Turkey and Lebanon. According to the Food and Agriculture Organization, less than 4% of the world agricultural trade is conducted in the region. Even though by tradition many of these countries relied on growing their own food, today some of these countries import almost 90% of their food; 2.3% of the food in the world alone was imported to Saudi Arabia and United Arab Emirates in 2007, and the food trade balance in food in Middle East is negative, estimated at over 50 US $ billion dollars (Tajkarimi et al. 2013). There are specific restrictions prevalent in the Arab-speaking countries related to Islam and Judaism with the prohibition of eating pork and blood, the drinking of alcohol, and mixing dairy foods and meat under Halal and Kosher food laws. Therefore, parasites related to pigs, e.g., *Trichinella* and *Taenia* spp., are unlikely to be prevalent in these populations. However, there are many Muslim and Jewish feast occasions with large gatherings such as Eid linked to Ramadan and particularly the Muslim

Hajj, which put a strain on food preparation, distribution and storage. Good health conditions for travelers to Saudi Arabia for the pilgrimage to Mecca (Hajj) are critical and any incident that occurs has to be quickly contained to prevent extensive infectious disease outbreaks (Memish and Al Rabeeah 2011).

Traditional Middle Eastern foods are mainly related to legumes, leafy greens, fruit, dairy products and meat on special occasions; details can be found in Brittin (2010). In urban areas today, grocery stores and supermarkets can supply most of the food requirements of a family but imported foods tend to be expensive. Also, some fruit and vegetable items are seasonal and are only available once or twice a year such as local plums, almonds and bananas, which tend to be cheaper than imported varieties. Quality of raw produce in stores varies but they often have short shelf lives and can spoil quickly because of harvesting ripe products, bruising, and high storage temperatures. Traditional rural foods include aromatic stews, stuffed vegetables, wild leaves, pulses and cracked wheat, and occasional goat or lamb meat. A typical Middle Eastern meal starts with a variety of cold and hot mezze (appetizers), salads and pastries, especially in Greece, Turkey and Lebanon. Many contain herbs, cheese, pickles, nuts, seeds, and parsley and lettuce are widely eaten in salads or traditional mezzes. Most mezzes are vegetarian and fresh fruits and vegetables are an integral and important part of the cuisine when they are in season. Tabbouleh, a salad where parsley is a major ingredient with small pieces of tomato, and some bulgur (ground wheat) in it, is often served in leaves of romaine lettuce or raw cabbage. Almost as popular is fattoush, a mixed bowl of lettuce, tomatoes, cucumbers, and fried or toasted pita chips, typically seasoned with a dusting of sumac and pomegranate molasses. Since leafy greens do not have a final decontamination step, they are at risk from environmental fecal contamination as reported in Lebanon by Faour-Klingbeil et al. (2016). Hummus, a smooth chickpea paste made with tahini/tehineh, lemon juice or citric acid, garlic and salt, and often served with olive oil, is the most ubiquitous mezze. Since tahini and hummus are major exported products from the region, particularly Lebanon, they are prone to *Salmonella* contamination, and are sometimes recalled from other countries, which is damaging to the local economies. Dairy products are also served regularly at meals and these are locally made or imported. Labneh, strained yogurt, very similar to Greek yogurts, is widely used as a base for mezze which might have olive oil, pine nuts or za'atar (a mixture of thyme, sumac, and sesame seeds) added. Cheeses including the popular haloumi are frequently served in restaurants. Shawarma/shwarma is frozen or refrigerated raw or marinated meat (lamb, beef or chicken) cooked on a vertical rotisserie popular throughout MENA countries and now frequently seen in western nations. Higher fish consumption tends to be close to where these are locally caught, either sea or river netted. One example from Iraq is masquf (split large fish cooked on stakes over a fire, and eaten outdoors by a river, served with slices of tomato and onion and Arab bread. Crustaceans are less frequently eaten but can be obtained from imports. Cosmopolitan foods are widely available in the larger cities, as are multinational fast-food chains.

Foodborne illnesses have been sporadically reported throughout the region over the past decades and global assessments of the kinds of problems encountered reviewed, e.g., Todd (2001) and Al-Mazrou (2004) and more recently by Tajkarimi et al. (2013). These last authors indicate that reporting foodborne disease is functioning well in Jordan, Kuwait, Oman, Saudi Arabia and UAE, compared to other countries in the region. However, the foodborne outbreak surveillance systems in Middle Eastern developing countries are still limited with reporting of less than 1% of the actual outbreaks; one reason is that many foodborne illnesses occur in homes and those ill may not visit medical care facilities. In addition, available laboratory analytical support for public health agencies is often minimal or lacking, even though some research institutions may have up-to-date equipment and technical expertize. Change is gradually coming and a food and drug authority has been established in both Saudi Arabia and Jordan (Al-Kandari and Jukes 2009). Also, new food legislation has been initiated by Egypt, Lebanon and Syria (Tajkarimi et al. 2013), but is currently stalled in last two countries. Improvements in inspection service, hand held computers, customized software and improved surveillance systems are some examples of developments in food safety systems in the region. Jordan, Saudi Arabia and Bahrain have been developing unified food safety activities from farm to fork (Al-Kandari and Jukes 2009). However, there is a need for substantive food safety education for all foodservice staff. Increasing quality and quantity of the food safety training and human resources in governmental agencies in the region will improve the public health infrastructure. For example, the Municipality of Dubai has established an international annual food safety conference to improve the food safety education system of those in the region, now in its 10th year (2016).

The following sections of the chapter focus on five aspects: gastrointestinal infections; foodborne disease outbreaks in specific countries; food safety related research and surveys; issues relating to tourism and exported food; and government oversight of the food industry, with specific examples from countries in the region.

Gastrointestinal Infections

Gastrointestinal diseases are frequently encountered in the Middle East and many etiological agents have been identified where specific studies have been carried out to look for bacterial, viral and parasitic pathogens. The average annual incidence of culture-proven shigellosis in Israel was 97/100,000 from 1998 to 2012, but each reported case was considered to represent 25 cases indicating the high burden of the disease in the country (Cohen et al. 2014). Orthodox Jewish communities, living in highly crowded conditions and with a high number of children aged <5 years were the epicenter of country-wide biennial propagated epidemics of *S. sonnei* shigellosis. *S. flexneri* was the leading *Shigella* serogroup in Israeli Arabs. Isolates showed high rates of resistance to ampicillin and trimethoprim/sulfamethoxazole, but very low rates to quinolones and third-generation cephalosporins. There is no indication if foods or water were vehicles of these shigellosis cases. Also, in Israel a study of pregnancy-related listeriosis cases from 1998 to 2007, identified 166 cases, resulting in a yearly incidence of 5–25 cases per 100,000 births (Elinav et al. 2014). There were 29 fetal deaths, two neonate deaths and one maternal mortality. The incidence of Israeli pregnancy-associated listeriosis has a high yearly variability and is one of the highest worldwide. The geographical distribution varied greatly between years and had a different epidemiological pattern compared with non-pregnancy-related listeriosis. The sources of the infections were not studied but all listeriosis cases have a foodborne link. This has to be further researched as to diet, and the unawareness of the Israeli public of the risk for certain food products contributing to the extremely high incidence in Israel, in both general and pregnancy-associated listeriosis, as occurs in other countries. A total of 132 stool samples were collected from Palestinian patients with acute diarrhea from which 12 (9.1%) yielded enteropathogenic bacteria. *Salmonella, Campylobacter coli/ jejuni*, and *Aeromonas hydrophilia* were isolated in equal numbers from samples 3/12 (25% each), *Shigella boydii* 2/12 (16.7%), *Yersinia enterocolytica* 1/12 (8.3%) (Abdelateef 2011). Many strains were antibiotic-resistant. Children younger than 5 years old were more susceptible to infectious diarrhea; in addition, diarrhea was more frequent in those living in crowded houses, and in houses rearing poultry, including pigeons. *Salmonella enterica* serovar Typhi continues to be an important public health problem in Kuwait. Analysis of the isolates from 163 patients, collected between 1995 and 2003, showed that the majority were from patients from the Indian sub-continent, and many strains were drug resistant (Dashti et al. 2008). Typhoid fever in Kuwait is predominantly associated with those who have traveled from endemic areas to work in Kuwait.

The circulation of enteric viruses among the population of Cairo, Egypt, between March 2006 and February 2007 was studied by Kamel et al. (2009). At least one type of virus was detected in 50% of fecal samples, 57.4% of which were positive for rotavirus, 26% for norovirus, 10.4% for adenovirus, and 1.7% for astrovirus. Over 10% of infections were mixed infections. Among the noroviruses, half belonged to the predominant GGII.4 cluster which were similar to those circulating elsewhere, but there were also new GGII.4 variants that were not associated with any previously known GGII.4 isolate. Although norovirus is rarely implicated in foodborne outbreaks compared with the US and other western countries, it is clearly present in Egypt. Further studies are required to assess the disease burden of enteric viruses in Egypt and the impact of atypical strains. The disease burden of hepatitis A and E in Egypt is one of the heaviest worldwide, based on serological analysis, with HAV infections occurring very early in life, with almost 100% seropositivity after the first years of life (Kamel et al. 2011). To determine the actual contamination levels in the environment, these authors conducted a survey of HAV and hepatitis E virus (HEV) in sewage in Cairo. HAV was detected by RT-PCR in 11 of 76 (15%) sewage samples. In addition, all the HAV-positive samples were also positive for enteroviruses. That only one stool sample was HEV-positive might be explained by the lower level of excretion of the virus in stools, the fragility of the virion in the environment, and technical difficulties in concentrating and amplifying the virus with standard methods. Bacterial etiology was found in 15.2% of cases of childhood diarrhea in Dhahira, Oman, mostly *Shigella sonnei* and to a lesser extent *Salmonella* (Patel et al. 2008). Antibiotics were prescribed in 36.2% of cases and the resistance to the common antibiotics tested was low. One reason for the low pathogen isolation rate could be that many cases had viral etiology. Rotavirus was detected in stool specimens from 1712 (49%) of 3470 children, who were admitted to 11 regional public hospitals in Oman for a median of 3 days with severe diarrhea (Al Awaidy et al. 2009). A diverse rotavirus strain pattern in Oman was identified with G2 (37%), G1 (38%), and G9 (11%) accounting for most of typeable strains. The authors estimated the burden for the Omani government at US$791,817 and US $1.8 million annually to treat rotavirus-associated diarrhea in the outpatient and hospital settings, respectively. They recommended a rotavirus vaccination program that would substantially reduce the burden of severe diarrhea among children in the country.

Unlike the above countries where the health care system functions for most residents, though not always to Western standards, the same cannot be said for Pakistan, particularly in rural areas. Poor nutrition combined with diarrheal and other foodborne diseases puts the population at risk for serious illness and death, especially among infant and children in Pakistan (Akhtar 2015). Cholera, campylobacteriosis, *E. coli* gastroenteritis, salmonellosis, shigellosis, typhoid, and brucellosis have been demonstrated to be the major foodborne illnesses in the country as well as infectious diseases caused by viral and parasitic agents. Many fatalities have been associated with food poisoning but the actual agent has rarely been determined. Many health experts believe that rapid spread of gastrointestinal diseases cannot be controlled if the public has no awareness of prevention and control measures against cholera and other forms of gastroenteritis, and that in most parts of the country, sewage is continuously contaminating streams, lakes, springs, wells, and other drinking water sources (Qasim 2009). In May 2011, an epidemic of diarrhea and gastroenteritis occurred in Kamalia, Toba Tek Singh, with over 30 children and others being admitted to hospitals which had few medical supplies. Apart from lack of potable drinking water, the main reason given for the rise in cases was the heat of summer when there were frequent power cuts so that food "rots" or becomes "stale" (Islam 2011). In remote areas of Pakistan, cholera has been responsible for many outbreaks. Two examples in July and August of 2013, both in areas of conflict near Afghanistan, give an idea of local but severe outbreaks. In one case authorities seemed not to want to be involved and in the other vaccinations are carried out. Although water is the primary vehicle of the *Vibrio cholerae* pathogen, it can easily contaminate prepared foods through poor hygienic practices. In July 2013, five deaths from cholera occurred in Pashtoon Kot area, Balochistan region of Pakistan (Federally Administered Tribal Areas) along the Afghan border (Staff 2013), some 70 km from Quetta, in the absence of any emergency medical aid. The condition of an additional 20 people suffering from the disease was said to be critical. A local tribal elder expressed the fear that outbreak of cholera might cause loss of life at large scale. He complained that the doctor and paramedics deployed at the basic health center in Panjpai live in Quetta and are rarely seen at the center. Officials of the provincial health department appeared to be unaware about the cholera outbreak and loss of lives (or ignored these), as they sent no medical teams to the affected area. In fact, Pakistani government rebuffed international media's claims, and did not respond to requests to dispatch healthcare professionals to the Balochistan area. It was assumed the outbreak would continue without medical aid. In 2012, cholera outbreaks killed hundreds of people, mostly children, in flood-hit districts of Nasirabad, Jaffarabad and Jhal Magsi where waterborne diseases were reported at a large scale because of consumption of contaminated water by local people. In August 2013, two people died and 320 others had fallen ill, following a cholera outbreak in Kurram Tribal Agency near Afghanistan (Hussain 2013). Dhand and Kudiad Khel were the worst-hit areas but vaccinations were carried out amid tight security, and tribesmen were instructed not to drink water directly from the well and boil it first instead since the wells had been contaminated from the rain water. Around 100 people were shifted to Parachinar Headquarters Hospital, while others were discharged after medical aid.

Sometimes diseases kept at bay by functioning public health systems come back when these break down as is occurring in a few of the countries embroiled in internal strife and outside attacks. For instance, in Iraq in October, 2015, >1800 cases and 6 deaths of cholera occurred which started along the Euphrates valley in September with the governorates of Baghdad and Babil, south of the capital, being the worst affected with more than 500 cases each. The epidemic then spread to the northern autonomous Kurdish region, which hosts hundreds of thousands of people displaced by conflict from other parts of Iraq (Agence France-Presse 2015). A previous outbreak killed four people in the Kurdistan region in 2012. The United Nations says the number of people displaced by conflict in Iraq since the start of 2014 has topped 3.2 million which would exacerbate the spread of the disease. Authorities blamed the cholera outbreak mostly on the poor quality of water caused by the low level of the Euphrates. Limited vaccination programs are in place in areas of conflict. In October, 2015, two persons arriving in Kuwait from Iraq tested positive for cholera and both were provided proper treatment and recovered. The Ministry of Health recognized that further cases could be discovered among people arriving from Iraq, but because Kuwait has a well-structured health infrastructure with water and sewers grids, and a supply of healthy and safe food, the disease should not spread into the Kuwaiti population (Anonymous 2015a).

Probably there are some cases in Yemen and Syria, countries also with limited public health infrastructures, but have yet to be identified. In Saudi Arabia, a country with a well-maintained health system, the main infectious disease concern today are the infections and deaths arising from exposure to the Middle East respiratory syndrome corona virus (MERS-CoV), which has reservoirs in camels and bats (Todd and Greig 2015). A potential food source for this virus and other pathogens is from unpasteurized camel milk, as camel farmers drink the milk as well as being exposed through other aspects of camel contact. This brief review indicates that diarrheal diseases, caused by cholera, dysentery, hepatitis A, salmonellosis, shigellosis, typhoid fever, and other enteric diseases through water and food are major contributors to ill health in the region in agreement with the WHO (2015b) report on global estimates of foodborne diseases.

Foodborne Disease in Specific Countries

In the region, not very many outbreaks of foodborne disease tend to be investigated, or at least reported publically, and those that are tend to have fatalities or are very large. For instance, in June, 2009, two children and one adult were brought to a hospital in Dubai, UAE, with suspected food poisoning (vomiting) after they ate take-away food (the father was out of town). Although the mother eventually recovered, the two young children (5 and 7 years old) died, one on arrival and the other the next day. The cause was not determined (Saberi and Scott 2009). It is not known if the family or restaurant was primarily responsible for the deadly gastro-intestinal attack as bacteria can multiply quickly in the hot summer months, and the public had been recently warned to minimize eating out at this time of year, especially at smaller eateries where hygiene levels are often of lower standard. A toxin was likely involved to cause fatalities so rapidly, but it could have been an accidental contamination of the food with a chemical such as a pesticide, as much as it could have been with an enterotoxin produced by *Staphylococcus aureus* or *Bacillus cereus* through careless ambient temperature storage. Unfortunately, this was one episode in a string of incidents, most of them with fatalities, in the county. In April, 2007, a 3-year-old died of suspected food poisoning in Sharjah, and in August, a 10-year-old girl died of food poisoning in Abu Dhabi. In March, 2008, six people fell ill after eating buffet food at a restaurant in the large Ibn Battuta Mall, Dubai; in November of the same year, 14 employees at a cement factory were hospitalized after consuming what was considered rotten food prepared at the factory kitchen in another Emirate, Ras Al Khaimah. In May, 2009, a 4-year-old girl died of suspected food poisoning in Sharjah. The Indian family of four rushed to the hospital after series of vomiting but were too late to save the girl. Dubai has been reporting foodborne outbreaks and cases through its foodborne disease investigation and surveillance system since 2011; in that year there were 1663 cases reported in the first nine months (Saseendran 2014). In 2013, 1123 suspected cases of foodborne illnesses were reported but only 518 cases were confirmed. No deaths were reported since the surveillance system was in place.

Egypt has had a particular problem with foodborne illnesses in universities and schools, mostly without a confirmed etiology, which seem to be related to poor food quality. Food poisoning is not uncommon in Egyptian university dormitories, where basic hygiene standards are often not observed, but the following outbreak was one of the largest. On April 2, 2013 hundreds of Egyptian students angered by a mass outbreak of food poisoning at a Cairo university stormed the offices of the country's top Muslim cleric and university president, Ahmed el-Tayeb, because of the 479 students who were hospitalized after a meal served at the university dormitories in the Nasr City district of Cairo (Associated Press 2013). The university is affiliated with Al-Azhar mosque, the world's foremost seat of Sunni Muslim learning, and awards degrees in sciences and humanities, as well as in religious studies. In the protest, thousands of Al-Azhar students blocked roads, broke into el-Tayeb's offices by the main campus, and chanted slogans against the university's management. The causative agent was unknown, and only with the incubation period, types of symptoms and their duration would it be possible to consider the potential etiologies of this illness. Because of their poor quality, campus meals were not very popular before they were being blamed for the current food poisoning outbreak. Although investigators were not able to find a specific cause, the university suspended its food services director and some other staff members. Within a few weeks food poisoning affected 161 students on April 29, 2013 at the same university, Al-Azhar (Masriya 2013). Investigations were initiated within the University and by the Ministry of Health, and apparently "bad tuna" had been served at the campus cafeteria; no further details were given. If tuna was the vehicle of the outbreak, scombroid poisoning was the likely cause of the illnesses. The allergic-like symptoms generally begin 15–60 minutes after ingestion and usually resolve in a few hours. Scombroid fish poisoning occurs after fish, most frequently tuna, with high levels of accumulated histamine or other biogenic amines, is eaten. But "bad tuna' could equally be contaminated with bacterial or viral enteric pathogens with a longer incubation period. A month later there was another outbreak. Because at least three outbreaks of food poisoning occurred at Al-Azhar University between April and May 2013 with over 700 cases of food poisoning detected in the university's male dorms, the dorm's director, the university's kitchen manager and eight chefs were sentenced in November, 2013, to 5 years in prison with a financial bail. In a similar situation, Egypt's top prosecutor ordered a swift investigation into the 178 cases of food poisoning reported in two primary schools in October, 2014, in Suez (Masriya 2014). An official of the Ministry of Education indicated that the poisoning was caused by the consumption of milk provided by the schools. The distribution of milk to all schools in the governorate was halted until the milk's validity was ensured. If milk was responsible, the etiological agent could be *Bacillus cereus* enterotoxin if the onset time was short, or less likely an infectious disease pathogen such as *Salmonella* or *E. coli* O157:H7. On January 1, 2015, 150 female students were diagnosed with food poisoning at Al-Azhar University in Upper Egypt's Assiut/Asyut Governorate, by the banks of the Nile, and were briefly hospitalized in an Assiut city (Anonymous 2015b). This follows a similar incident which occurred in April 2014 when 29 students, also in the girls' dormitories, contracted food poisoning on the university campus in Luxor. This report also flags two major poisoning incidents involving at least 500 students ill consecutively at its campuses in Cairo in 2013 (probably the ones already discussed). The reason given for these repeated mass foodborne illnesses among university students is the quality of the food served them. Apparently cheap, subsidized food is poorly stored, cooked and distributed to the poorer university students. In most cases the attorney general would open a criminal investigation that would be closed without knowing the microbiological cause of these outbreaks. The ProMED-MENA Editor speculated that enterotoxins of *Staphylococcus aureus* were the most probable cause of such communal food poisoning, as a toxic dose of less than 1.0 microgram in contaminated food is sufficient to produce symptoms of staphylococcal intoxication. This toxin level is reached when *S. aureus* populations exceed 100,000/g, a condition likely to be present in these university kitchens because of intense pressure on them to feed a huge number of students in a short time, taking into consideration that most of these kitchens lack basic hygienic measures with regard to safe food handling. The Editor also considered *Shigella*, with its low infective dose (10–200 depending on the species) as another possible agent. However, the incubation period and symptoms of *S. aureus* intoxication and shigellosis or dysentery are quite different.

Pakistan is similar to Egypt in that much of the country is rural but with very large cities with high populations (total population is 89 million in Egypt and 184 million in Pakistan, the most populous of all Middle Eastern countries). In September, 2010, more than 250 of the 1400 flood victims at a relief camp in Bengali Boys Sindhi Section School in Ibrahim Hyderi vomited after eating cooked food and then fell unconscious; 59 of them had to be taken to a nearby hospital (Aligi 2010). A local philanthropist had been providing cooked food to the flood victims but by the time the food arrived at the relief camp, the cooked rice had turned "stale". Since the rice did not show any sign of spoilage, it was served to the flood victims. A similar incident had taken place 3 days earlier at another town where more than 80 flood affectees had fallen unconscious after consuming "stale" food and 10 were hospitalized. None was seriously affected. During the investigation, it was noticed that the sanitary situation in and around the relief camps was very poor. Even though the reason for the illness was not determined, the police took action against the donor and two caterers. In fact, based on the information of the vehicle and the symptoms, *Bacillus cereus* enterotoxin which is known to be produced in boiled rice, was the most likely agent.

In the following two outbreaks yoghurt is blamed for the serious illness and deaths though details of the symptoms are not given. Rapid onset of symptoms indicates the presence of a toxin of some kind, although yoghurt is not a food known to be frequently contaminated with pathogens because of its high acidity. Either the yoghurt was made under very unhygienic condition with the source of the milk perhaps being spoiled (possibly containing *Bacillus cereus* enterotoxin), or a chemical had been added accidently such as a pesticide, or deliberately and illegally to enhance the flavor. However, it is possible other foods were involved and yoghurt was not the contaminated vehicle. In January 2011, in Lahore, a hospital employee died and two other employees became critically ill after eating contaminated yoghurt. The three employees ate rice with yoghurt at a local restaurant (Ians 2011). Action was taken against the restaurant owner and manager. No further details are known. In early April, 2014, a Rawalpindi family of ten became seriously ill after eating a home-prepared evening meal where yoghurt was suspected to have been the contaminated food, and they were taken to a hospital, where a teenage boy and 7-year-old girl died (Asghar 2014). The surviving family members remained in critical condition for some time but eventually recovered; the cause of the illnesses was not discovered, although it was postulated by a relative who had eaten the yogurt with the meal that it was possibly poisonous or, strangely he thought a lizard might have fallen into it.

In February 2012, at least four people died and another seven were hospitalized in a critical state after eating home-cooked biryani (a dish made with spices, rice and meat or vegetables) in a suburb of Karachi (Mahmood 2012). The owner of a grocery shop, who provided the ingredients, was arrested, and a sample taken for analysis. It is not known if any toxin was found. A month later in March, in Faisalabad, more than 60 children and women were ill after eating contaminated aalo-chanay (potatoes, chick peas, onions, tomatoes and spices) purchased from an unidentified vender (Anonymous 2012c). As soon as the children ate the aalo-chaney, they felt ill and started vomiting. Although they were immediately rushed to a rural health center, one boy died. A medical opinion was given that the eaters suffered from "diarrhea and cholera". However, the onset was too rapid for anything but a toxin of some kind, most likely heat-resistant since the aalo-chanay was cooked. Also, in March 2012, as many as 47 student nurses and eight staff nurses were hospitalized with acute food poisoning at a hospital in Rawalpindi after eating food at the nursing hostel, but none was critically ill (Anonymous 2012d). The nurses residing in the hostel started reporting complaints of vomiting and diarrhea along with high-grade fever at an undisclosed time after a meal. The hospital administration was criticized for failing to provide safe food and drinking water to its employees and demanded immediate inquiry into the case, but none was reported on. The illnesses are consistent with an enteric infection such as *Salmonella* or norovirus.

In April 2015, at least 20 constables suffered from diarrhea and were admitted to hospitals when they ate food during the Sehat Ka Insaf program, which is a blanket method of administering the polio vaccine along with eight other vaccines, hygiene kits and vitamin A drops in order to circumvent polio-specific terrorist attacks in Pakistan. Local administration purchased packed food, including piece of chicken and juices from a local supporter (Mayar 2015). No further details are given but the chicken could have been undercooked or cross-contaminated with enteric pathogens such as *Salmonella* and *Campylobacter*; if the packs had been left at ambient temperatures for some time, these pathogens could have multiplied on the chicken to large numbers. Over thirty children in Faisalabad were hospitalized over 4 days because of diarrhea and gastroenteritis, three seriously, and other children were expected to be ill. Undetermined contaminated food was postulated as the cause, more than usual because of the extreme seasonal heat combined with frequent power outages to allow rapid bacterial growth in contaminated food. The unavailability of clean drinking water was mentioned as a contributing factor to the increasing number of gastrointestinal disease cases. Hospital administrators complained that vaccines and medications were required but were not forthcoming from the Health Department. Probably many family meals were contaminated because of the lack of potable water and any unspecified enteric bacterial pathogens present could grow rapidly in the heat. Children are more vulnerable than healthy adults to infections which might explain the high proportion of sick children seeking medical help.

Botulism outbreaks occur periodically in Iran. In a study of stool and serum specimens of 115 patients with clinical symptoms of botulism, who were at inpatient and outpatient medical centers in Tehran and other areas of Iran, between April 1984 to August 1994, specimens of 73 patients showed the toxin and spores of *C. botulinum* (Modarres 1997). Type E was the most common causative agent found in this study, being responsible for 71.2% in all specimens; other etiologic types, in order of frequency were types A (16.4%) and B (12.3%). Type E strains are typically associated with fish and freshwater and marine sediments. The results of this study indicate that the cases had consumed salted fish, smoked fish and canned fish, along with cans of green beans and cucumbers. A similar result over a decade later confirms that *C. botulinum* type E is a major pathogen in Iran. In Gilan Province, of 146 fish samples collected in 2008, 11% of processed fish and 7.5% of non-processed fish contained *Clostridium botulinum*, mainly type E (Tavakoli and Imani Fooladi 2011). The processing is insufficient to kill the spores or reduce much of toxin produced because the fish tend to be partly cooked with the intestines kept intact. A total of 131 traditional food product samples (57 cheese, 11 kashk [a type of dried yoghurt or thick cream], and 63 salted fish) were examined using a bioassay method for detection of *Clostridium botulinum* toxin (Hosseini et al. 2010). Standard monovalent antitoxins were used to determine the toxin types. *C. botulinum* toxins were detected in 4.6% of examined samples (3.5% of cheese samples and 6.4% of salted fish samples). None was found in kashk samples. *C. botulinum* types A and E were dominant in cheese and salted fish samples, respectively. Consumption of these traditional foods either raw or processed may contribute to foodborne toxicity in Iranian populations. In May 2014, a quick-thinking mother immediately brought her 7-month old boy to an Israeli hospital when she saw he was suffering from vomiting, difficulty in breathing, listlessness, glassy-eyed, apathetic, and an inability to nurse or eat (Bender 2014). A doctor at the hospital diagnosed the child as suffering from infant botulism. He decided to treat the baby with the antitoxin stored in the emergency stocks, even before they got back the lab test results. The hospital like all Israeli medical facilities keep ample supplies of biological and chemical warfare antidotes on hand in case of war or terrorist attacks, and staffers are regularly drilled in dealing with the symptoms of various chemical, neural and blister agents. The infant started recovering soon after the administration of the antidote. In the rare disease of infant botulism, spores of *Clostridium botulinum* are ingested and the infant's flora is not mature enough to prevent germination and slow growth of the toxigenic pathogen. It is entirely possible that infant botulism occurs more frequently in the region but is not diagnosed.

Foodborne disease surveillance depends on an infrastructure of reporting and diagnosis in hospitals, epidemiologists, and food testing laboratories. Lebanon is an example of a country where modernization in public health seems to occur at a glacial pace. However, diseases including those of foodborne and waterborne origin, are documented and published. The Law of December 31, 1957 regarding communicable diseases in Lebanon mandates all physicians, from private or public sectors, in hospitals or ambulatory services, to declare to the epidemiologic surveillance unit of the MOPH all diseases considered a risk to public health. The data available at the Ministry of Public Health (MOPH) are compiled from different sources, and the declaration of cases remains irregular and insufficient (MOPH 2012). In 2011, foodborne and waterborne diseases were the most frequently reported in Lebanon at a rate of 0.4 ‰ (total of 1468 cases), with the highest rate in the Bekaa (0.7 ‰) and the lowest in the South (0.1 ‰). The most common infection was viral hepatitis A, which represented 30.5% of the total food and waterborne diseases with 448 cases. There were also 362 cases of typhoid (24.7%), 311 cases of food poisoning (unspecified, 21.2%), 186 cases of dysentery (12.7%), 34 cases of brucellosis (9.1%, 15 cases of parasitic worms (1.0%) and 12 cases of hydiatic cyst (0.8%). No cases of cholera and trichinosis were declared. Hydiatic cyst (cystic echinococcosisis) caused by *Echinococcosis* (typically *E. granulosis*) is acquired by contact with animal feces contaminated with tapeworm eggs. Sources include contaminated food (meat), water, and animal fur. Cysts containing tapeworm larvae may grow in the body for years before symptoms appear. When cysts become large, they may cause nausea, weakness, coughing, and belly or chest pain.

Occasionally, well-investigated outbreaks are published; the following two examples are from Lebanon and neighboring Jordan. In May 2004, 32 employees suffered from diarrhea, fever, and abdominal pains 5.5–22.5 h (mean, 10.3 h) after eating chicken noodles au gratin at a catered lunch served at a bank cafeteria (Hanna et al. 2009). A few cases had systemic infections. *Salmonella* Enteritidis (SE) was confirmed in stool and blood cultures within 48–72 h after hospital admission of the first cases, and also in leftovers of the suspect food. The same dish had been served at the bank in the past with no apparent health problems. Preparation normally started in the evening prior to the day the dish is served. However, in this instance, some of the constituents had been prepared 3 days ahead, because the dish was to be served on a Monday, immediately after the week-end closure. No *Salmonella* was found in rectal and nasal mucosal swabs taken from all 18 kitchen workers, or in the tanker water supply (although it had high fecal coliform counts), but SE was found in a frozen batch of the same raw chicken breast consignment that had been used for the chicken noodles. The batch of chicken came from a large producer of poultry and eggs in Lebanon, who was advised of its potential involvement in a major foodborne outbreak. However, the investigators were refused access to the poultry-producing facility. It is highly likely that contaminated chicken carcasses had been, and would continue to be, shipped to many parts of Lebanon. That the same SE strain occurred in the patients, the raw chicken, and the leftover food was confirmed through random amplified polymorphic DNA polymerase chain reaction (RAPD-PCR). It would appear the 3-day delay in the chicken noodle preparation was significant in allowing the *Salmonella* present in the ingredients not only to survive but probably to grow; undercooking, cross-contamination, inadequate storage and reheating all may have played a role in the outbreak, but no more information was available to determine which of these were the key factors in the outbreak. The bank management decided to sue the caterer and because they were aware of apparently inefficient way that public authorities were conducting the procedure, they took the initiative to call upon an independent investigative team to obtain solid evidence to win any court action. The caterers, concerned that they would be the only party blamed for the *Salmonella* outbreak, had succeeded in concealing some raw and cooked items from destruction by the public health authorities, which was their normal practice after a complaint. These items were central to establishing contamination upstream from the caterer's kitchen. No action seems to have been taken against the poultry producer who was the source of the SE, a pathogen that is invasive of flocks and difficult to eradicate. The authors complained about the obsolete Lebanese laws dating back to the 1930s that still governed what should be done following a report of "food poisoning". Public health officers are mandated to stop the spread by destroying allegedly contaminated food items and closing down incriminated facilities. Hanna et al. (2009) stated that this kind of action is generally lauded by the public but does not help determine the cause to develop appropriate prevention and control strategies. They also complained that because no investigation is typically done, many non-implicated foods and ingredients are wastefully discarded.

The Jordanian example is over two decades old, but is worth noting in detail. In September 1989, a 183-case outbreak of salmonellosis occurred in a university hospital in Amman after employees, patients and visitors ate in the cafeteria. The incubation period ranged from 16 to 72 h. Symptoms included diarrhea (88%), fever (71%), abdominal pain (74%), dehydration (34%), and bloody stool (5%); 84 were hospitalized (Khuri-Bulos et al. 1994). Cultures of eight food items were negative, but stool culture on 90 of 180 patients and 11 of 61 kitchen employees yielded *Salmonella* Enteritidis (SE) group D1. A cohort study revealed a food-specific attack rate of 72% for the steak and potato meal and 18% for the rice and meat meal. Stratified analysis of the steak and potato meal revealed that the potatoes were implicated most strongly. Cultures were obtained from all kitchen employees, who showed no symptoms of illness, but 11 of 61 grew SE group D1. One asymptomatic, culture-positive employee had prepared the mashed potatoes on September 23, 16 h before the first case presented at the hospital emergency with severe gastroenteritis symptoms. All of the food workers had negative stool cultures 3 months earlier. The potatoes were mashed by machine, but peeled after boiling and mixed with milk by hand, using a ladle but no gloves. Two different batches, the first of which was served exclusively to hospitalized patients and the second to a few remaining patients and employees, were prepared and served within 30 to 60 minutes of preparation. From the epidemiological data it can be assumed that the infected handler fecally contaminated only the second batch of potatoes, thus sparing most of the highly susceptible inpatients from exposure. Furthermore, while potatoes clearly were implicated, individuals who ate steak only had an elevated risk of being attacked. This probably was due to surface contamination of foods being served on the same plate. Kitchen employees harboring *Salmonella* were excluded from work until they had three negative stool cultures taken 1 week apart; it took 6 weeks for them to return to work. Stool surveillance that was routinely carried out in the hospital was ineffective in detecting infected employees to prevent this outbreak and the investigators recommended that employees adhere to proper hygienic practices including thorough washing of hands, especially when preparing food.

Today, *Salmonella* is only one of many of the pathogens that can be encountered in foodborne illness. One of the newer pathogens, well established in the West is norovirus (NoV), which causes more cases of foodborne disease in the U.S. than any other agent (Scallan et al. 2011). In May 2009, a significant increase in acute gastroenteritis (AGE) cases was noted in the American health clinic at Incirlik Air Base (IAB) in Adana, Turkey. This increased rate of AGE led to discussions with local Turkish military public health authorities, which confirmed that the Turkish military community and the residents of Adana were also experiencing an anecdotal increase in AGE illnesses (Ahmed et al. 2012). An epidemiologic investigation was launched to attempt to identify the cause and possible source of this AGE outbreak at IAB from May to June with the peak incidence of cases during the week of May 31–June 6, with a total of 71 patients seeking medical care at the clinic. Of the total 187 infected persons, 82 patients completed the case survey, 79% reported diarrhea, 46% reported vomiting, and 29% reported fever. The median number of days between symptom onset and clinic visit was 2 days. During the 7 days prior to symptoms, 73% of respondents reported travelling off base, 56% reported eating off base, and 24% reported using an outdoor pool. This outbreak had a significant negative operational impact, degrading mission readiness with nearly 20% of the American population in a 1-month period affected. Initiation of a clinic case-based investigation yielded 37 stool specimens in which NoV was detected in 43%, with 81% of the positive NoV specimens identified without a copathogen. DNA sequencing data demonstrated that several relatively rare genotypes of NoV contributed to this outbreak; four different genotypes were isolated from 16 positive specimens. Two of the NoV strains were previously reported in Iraq and only from deployed troops, while the other two genotypes were reported in South Africa and in the US. In Turkey, little systematic data on circulating NoV genotypes exist. However, GIIb/GII.4 strains have been frequently identified in Turkish children with gastroenteritis; strains belonging to this genotype have been found in Europe and mainly in children. Previous reports from British troops deployed to Iraq indicated that two NoV strains isolated were responsible for cases of gastroenteritis there. Similar mixed NoV outbreaks have been previously observed and are often attributed to systematic failure of cooking/cleaning/drinking water supplies (Ahmed et al. 2012). One limitation of this investigation was that the survey was not used to capture data from a control group, those without recent AGE, preventing carrying out a risk factor analysis. Another limitation was the lack of environmental samples that could be tested for NoV in order to track the source of outbreak. From anecdotal information, it is likely many in the local population and the Turkish military base were ill, but a formal outbreak investigation in the Turkish population was never performed. From the multiple genetic types involved, one specific contaminated food or water source seems unlikely. The largest Turkish NoV outbreak was in Keçiborlu province of Isparta county between April 5 and

17, 2010, with 1428 patients seeking medical help from the healthcare centers, after suffering from nausea, vomiting and abdominal pain (more frequent than diarrhea) (Sözen et al. 2014). Because of underreporting, the number of affected people was estimated to be higher. Municipal water was the suspected source but no samples tested positive. As a cautionary note, the authors suggest that NoV may not be the only causative agent of gastroenteritis outbreaks, especially from an undetermined fecal source, and bacterial, viral and parasitic agents should be examined together with the NoV.

In Saudi Arabia, a national policy for reporting, notifying, and recording incidents of bacterial food poisoning was established in 1984 (Al-Joudy et al. 2010). Since then *Salmonella* food poisoning outbreaks have been reported from different regions of KSA, exhibiting seasonal and regional variations, with chicken, meat, and rice being commonly incriminated food items, and frequently reported in the Saudi Epidemiological Bulletin. Al-Mazrou (2004) reviewed the history of foodborne outbreaks in KSA and saw an increase over the last few decades, especially those caused by *Salmonella*, with the main food vehicles being chicken, meat and eggs, and *S.* Enteritidis being the most frequent *Salmonella* serovar responsible. According to ProMED editorials, restaurants and communal feasts and institutional feeding (such as in school cafeterias, hospitals, nursing homes, prisons, etc.) where large quantities of food are prepared several hours before serving are the most common settings in which foodborne illness incidents occur (http://www.promedmail.org). For instance, in 2005, a hospital in the Jizan Region received 19 suspected food poisoning cases that were ill after taking meals from a restaurant, including a woman who suffered from severe diarrhea, abdominal pain, vomiting and dizziness (Fagbo 2005). The restaurant was closed down and three of its workers were detained pending the results of laboratory tests. The report of an investigative committee could not find a specific cause, but noted that the restaurant had earlier been responsible for some hygienic violations. In 2015, 80 cases suspected of foodborne illness after eating a meal at a restaurant were admitted to various hospitals in the Najran region (AlHayat 2015). Most of the cases were not seriously ill. No report was given on the samples that were taken from the suspected restaurant, which was closed temporarily. There is an interesting observation related to variant Creutzfeldt-Jakob Disease (vCJD); four cases have occurred in the US since the disease was first diagnosed in the United Kingdom in 1996 linked to consumption of cow meat suffering from bovine spongiform encephalopathy (BSE); two of these were associated with the United Kingdom (where BSE was first reported), but one came from Saudi Arabia and the most recent case in 2014 had extensive travel to the Middle East and Europe (CDC 2014). This may indicate some source of vCJD in the Middle East including Saudi Arabia.

One of the big concerns for KSA is the annual Hajj with millions of Muslims from around the world converging on Mecca, in Saudi Arabia, each year. No other mass gathering can compare with the Hajj, either in scale or in regularity, and various communicable disease outbreaks of various infectious diseases have been reported repeatedly, during and following the Hajj (Memish 2010). In 2006, an outbreak during the Hajj occurred where all the cases came from one tent occupied by 50 soldiers located in a government camp in Mina, Makkah province, near Mecca (Al-Joudi 2007). The camp was served by a catering company that prepared and distributed three meals daily (breakfast, lunch, and dinner). A case was defined as any individual who developed diarrhea with or without abdominal pain after eating at the camp in Mina in January, 2006. Of the soldiers who were interviewed, 16 (39%) had developed gastroenteritis, most commonly manifested by diarrhea (100%), and abdominal pains (87.5%). The mean incubation period was 12.6 $\pm$ 4.9 h and the epidemic curve suggested a common point source outbreak. Out of three served meals, lunch with a rice dish was found to have a statistically significant association with illness. Unfortunately, no food remnants were found for sampling, and the results of stool cultures of all diarrhea patients, and rectal swabs from all food handlers were inconclusive. Temperature abuse was cited as a contributory factor in this outbreak. Based on the incubation period and symptomatology, *Bacillus cereus* would be the most likely etiological agent. Another example of a foodborne illnesses associated with the Hajj occurred in 2011 when 81 Bangladeshi pilgrims were taken to hospitals in Madina (Medina) after eating a meal prepared by an unlicensed caterer (ProMED-MENA 2011). They suffered from abdominal pains associated with diarrhea and vomiting. The pilgrims were all treated and discharged, except for one who remained hospitalized. Samples of the food they had eaten were sent for analysis but the results are not known. Considering the mass of people converging on this small part of the Middle East, it is surprising there are not more foodborne disease outbreaks. This may mean excellent food control by the authorities or some illnesses are simply not recognized and reported.

At least 750 Bahrainis suffered from food poisoning after eating catered sandwiches served during a wedding celebration, the biggest mass poisoning outbreak in the country's history (ProMED-MENA 2002). The wedding took place in the Safala village, near the eastern island of Sitra. All eventually recovered after treatment but one man who had sickle cell disease, died. Teams were formed to investigate the outbreak, and blood specimens from all workers at the bakery who prepared the egg, cheese, and mayonnaise sandwiches along with leftover sandwiches and their ingredients on the caterer's premises were sent for bacteriological analysis. The bakery which supplied the sandwiches was closed by the Public Health Directorate at the ministry pending the investigation's results. Unfortunately, no final report was released to the public. The ProMED editor considered the etiological agent could be *Salmonella* or *Staphylococcus aureus* enterotoxin, depending on the length of the unstated incubation period.

The region experiences some unusual type of illnesses relating to on-going hostilities. For instance, the United Nations has been sending aid to reach besieged towns in Syria, close to the Lebanese border, but in October, 2015, it sent hundreds of boxes of "moldy" high-energy biscuits past their 'sell-by' date in September (320 of the 650 boxes transported) to Zabadani and Madaya, apparently causing food poisoning (Afanasieva et al. 2015; Muhkalalati and Kieke 2015). Officials stated these could be the only cause of an outbreak of food poisoning among almost 200 residents who came to makeshift hospitals, mainly children who had vomiting, diarrhea and abdominal swelling almost immediately after eating the biscuits. The biscuits were described as "moldy and rotten and had been poorly stored". Apparently, when the last aid order that was sent was filled, there was a shortage of food. The Red Crescent, who was filling the order, took some of the expired goods to complete it. However, these biscuits had only just expired and normally would not have posed any health risks to those eating them. Nevertheless, the words poorly stored suggest that moisture may have encouraged microbial growth (visible mold more likely than bacteria because fungi can grow aerobically in the presence of the presumably elevated sugar content in the high energy biscuits). Also, contributing to the symptoms, the residents of Zabadani and Madaya had been blockaded for 120 consecutive days, and their immune systems were extremely weak.

Refugees are also at risk of gastrointestinal diseases from contaminated water or food. Up to two million Syrian migrants fleeing Syria due to the civil war were living in Turkey, and supplying them with safe and secure food supplies is a challenge for any host country. One incident, no doubt, one among many indicates the risk of contaminated food. In April, 2015, five security forces were injured after Syrian migrants in a tent city in Turkey's southeastern province of Mardin reportedly attacked guards over allegedly being poisoned from the lunch at the camp (Anadolu Agency 2015). Some 17 Syrian migrants were detained after the incident; 66 Syrian migrants out of the 5230 currently residing in the Temporary Sheltering Center in Mardin's Derik district applied to the center's hospital with symptoms of food poisoning, dizziness, and vomiting. After treatment they were discharged, none of them in a critical condition. Although an investigation was conducted and samples from the lunch sent to the lab for analysis, no further information was available on the outbreak. ProMED speculated that if the lunch food was the vehicle, it would be a short incubation illness likely caused by *Staphylococcus aureus, Bacillus cereus, Clostridium perfringens*, or possibly a non-biological toxin. These illnesses may cause vomiting, diarrhea, or both, and are usually short in duration (less than 24 h), and are not associated with prominent fever.

In Iraq, no recent foodborne disease outbreaks have been published, but no doubt many have occurred in the last decades with so much public health infrastructure dismantled. Only the most newsworthy of outbreaks are being covered by the press today. Iraq and other Middle Eastern countries are in sectarian turmoil and on two occasions Islamic State (ISIS/IS/ISIL) fighters (jihadis) were likely poisoned by cooks who infiltrated their camps. In November, 2014, a group of defected Syrian soldiers (Free Syrian Army men) who posed as cooks reportedly poisoned ISIS militants after they ate a contaminated lunch at the Fath El-Sahel camp, where 1200 of them were based (Gee 2014). Apparently about a dozen of the jihadis were killed and 15 taken to nearby field hospitals. The 'cooks' immediately fled, along with their families, with the help of fellow revolutionaries. Seven months later, in July 2015, 45 jihadis died after ingesting an Iftar meal eaten by 145 ISIS militants (Akbar 2015; Variyar 2015). It remains unclear whether the jihadis, who were breaking their Ramadan fast in Mosul, Iraq, died of accidental food poisoning or intentional poisoning, but it is likely a repeat attack of the earlier incident described above. The nature of the poison or details of the illnesses in either episode are not known. However, in both episodes, onset and severity of the attack were rapid, probably caused by a relatively tasteless chemical in lethal doses added to one or more foods. Targeting the military by any means including poisoning food has always been a strategy of opposing forces. In February 2012, a deliberate attack was foiled when Afghan border police detected a significant amount of bleach in fruit and coffee stored at their main border checkpoint between Afghanistan and Pakistan, a likely attempt to poison the Afghan security forces (Tucker 2012). The police decided that although none of this food had been consumed, the level of contamination was high enough to cause serious injury, and it must have been done intentionally. There had been previous incidents of intentional food poisoning aimed at Afghanistan's civil defense forces, including an episode in Kabul in 2011 when several people were sickened. In 2012, in southern Helmand province militants killed four Afghan policemen and two civilians inside a police checkpoint by poisoning their yoghurt coordinated with an attack (Anonymous 2012a). There had been several recent poisoning incidents involving members of the Afghan National Police, as part of attempts by the Taliban to infiltrate the security forces; three police officers were reported missing, along with their weapons and a police vehicle, following that attack in Helmand province. Taliban militants had first poisoned the police officers' yoghurt before launching a full scale attack on the checkpoint. Similar tactics had been used by insurgents in Helmand before. The same thing happened again in January 2016 when a rogue policeman collaborating with insurgents in southern province of Uruzgan shot dead 10 colleagues after first poisoning their food, but no further details are given (Reuters 2016).

Food Safety Research and Surveys

Turkey does have food laws that are supposed to limit food contamination and resultant foodborne illnesses. The 2000 Turkish Food Code stipulates that all Turkish food businesses have to provide food hygiene training commensurate with the work activities of their staff. To see what progress had been made in this area Baş et al. (2006) evaluated knowledge, attitudes, and practices concerning food safety issues among food handlers in Ankara, conducting face to face interviews and administrating questionnaires. The majority of the 764 food handlers who responded (47.8%) had not taken a basic food safety training (and probably most of non-respondents had not either). The mean food safety knowledge score was 43.4 ± 16.3 of 100 possible points. The self-reported hygienic practices showed that only 9.6% of those who were involved in touching or distributing unwrapped foods always used protective gloves during their working activity. Of those food handlers who used gloves, only 8.1% and 3.8% always washed their hands before putting them on and after removing them, respectively. In addition, there was a difference handlers' scores depending on where they worked. Scores were higher for food handlers in catering establishments (50.4 ± 9.4), school food services (52.5 ± 9.2) and hospital food services (50.9 ± 9.2) than restaurants (47.3 ± 8.6), hotels

(47.4 ± 8.3), takeaways (44.1 ± 2.5) and kebab houses (37.9 ± 6.7). These scores may also be biased upwards since they were self-reported and not observed practices. The study demonstrated that food handlers in Turkish food businesses often have lack of knowledge regarding the basic food hygiene, e.g., critical temperatures of hot or cold ready-to-eat foods, acceptable refrigerator temperature ranges, and cross-contamination. Those who were trained scored better, and the authors stated there was an immediate need for education and increasing awareness among food handlers regarding safe food handling practices.

In Istanbul from 2005/2006, thermophilic *Campylobacter* was isolated from 11.1%, 21.6%, and 50.4% of beef, mutton, and chicken samples tested, respectively (Bostan et al. 2009). There was no significant seasonal variation in the prevalence of the pathogen. *C. jejuni* was the species most commonly isolated from chicken meat, while *C. coli* was the most common in beef (63.3%) and mutton (63.9%) carcasses. *Campylobacter* isolates were most often resistant to tetracycline (69.1%), followed by trimethoprim-sulfamethoxazole (64.2%), nalidixic acid (58.1%), erythromycin (56.9%), enrofloxacin (48.8%), ciprofloxacin (42.7%), chloramphenicol (36.2%), and gentamicin (26.0%). The results of this study suggest that a high proportion of meat samples, particularly chicken carcasses, are contaminated by campylobacters, most of which are antimicrobial-resistant strains. In Yemen, the prevalence of *Salmonella* in food was determined in Sana'a city from April 2009 to April 2010 by Ahmed (2013). Of the 362 different food samples collected from local markets, *Salmonella* spp. were isolated from 26 (7.2%). The highest prevalences were in red meat (14.7%), chicken (12.1%), eggs (11.8%), cooked foods (9.5%), raw milk and milk products (5%), juices (4.8%), vegetables (4.4%), sandwiches (3%), and pastries (2.6%). Serogroups identified were B, C1, C2-C3, D1, E1, and E4, and some foods contained more than one isolate with different serogroups, especially red meat.

Because handlers in foodservice facilities play a major role in transmission of foodborne diseases (Greig et al. 2007), studies have been carried out to demonstrate their knowledge of practices related to food safety. In Jordan, Osaili et al. (2013) measured food safety knowledge of food handlers working in fast food restaurants in the cities of Amman and Irbid. A total of 1084 food handlers in 297 fast food restaurants participated in this question survey study. The overall knowledge of food handlers on food safety concepts was considered to be fair (69.4%). The food safety aspect with the highest percentage of correct answers was "knowledge of symptoms of foodborne illnesses" (81.7%) and "personal hygiene" (79.9%), while the lowest percentage of correct answers was for "safe storage, thawing, cooking and reheating of the foods" (52.4%), critical practices to prevent the survival and growth of pathogens. The mean knowledge score of "personal hygiene" reported in the study was much higher than 51.5% and 31.8% reported by Martins et al. (2012) and Baş et al. (2006), for the food handlers in Portugal and Turkey, respectively. Also, only 31.9% of respondents considered the duration of hand washing to be ≥20 s. When they were asked how they check that the poultry is sufficiently cooked, only 31% knew "when the meat has the correct thermometer reading", although 85% of the respondents had thermometers in their restaurants. About 25%

of them answered that poultry is cooked "when it has been cooked for the stated time" (24%) and "when it looks cooked" (27%). About 50% of them would store leftovers on the steam table (40%) and in the refrigerator (53%) while about 27% of the correspondents would store leftovers at room temperature in kitchen or in the oven. A low percentage of the respondents (20%) reheated leftovers to the appropriate temperature (73 °C). About 40% and 20% of the respondents had heard about *Salmonella* and hepatitis A virus, respectively, but $\leq$10% of the respondents knew about *Listeria monocytogenes, Staphylococcus aureus, Bacillus cereus, Escherichia coli* O157:H7, *Clostridium perfringens, Campylobacter jejuni*, or *Shigella*. Food workers who had enrolled in a food safety training course had significantly higher total food safety knowledge score than those who did not take any training. There was no association between the experience or any other characteristic of food workers and total food safety knowledge score. This study suggests adopting proper food safety education training courses to food handlers, periodic evaluation of food handlers' knowledge and food safety training course materials. Also, the authors considered that better pay for food handlers would improve the food safety status in foodservice institutions. Similar concerns over practices that could lead to food contamination and foodborne illnesses were demonstrated in Lebanon. A survey was conducted in Beirut to evaluate the knowledge, attitudes and practices related to food safety issues of food handlers (n = 80) in foodservice establishments (n = 50), and to assess the influence of management type on enactment of safe practices on food premises (Faour-Klingbeil et al. 2015). The data suggest that while respondents do have some knowledge of food safety aspects, substantial gaps in their knowledge and self-reported practices associated with critical temperature of foods and cross contamination remain, therefore posing health risks to consumer health. Food handlers in corporate managed food outlets showed a significantly higher awareness on food safety practices. It is concluded that the management type is an integral element of the Theory of Planned Behavior that influence food handlers' practices and substantiate the need for more research work on safe food handling in the context of food safety culture framework in food businesses. As in many other MENA countries, there is a critical need for food safety education interventions and technical guidance fostered by synergistic participation of the private and public sector to support food handlers in SMEs (small and medium sized enterprises).

Parasites are not often looked for in Middle Eastern countries but they are frequent, and one of the ones of most concern for pregnant women is *Toxoplasma gondii* which is transmitted through undercooked meat and cat feces. Since stray cats are common in some localities, of 240 fecal samples of stray cats examined in Kuwait, 22 (9.2%) were found to be infected with oocysts of coccidian protozoa (Abdou et al. 2013). *Toxoplasma gondii* was found in 2.1%, and cats <6 months old had higher infection rate with oocyst of enteric protozoa than older cats. A sero-survey of the 240 stray cats revealed that 19.6% were positive to *T. gondii* IgG. *Toxoplasma* sero-positivity was observed in a higher number of adult cats compared to younger ones suggesting that with age the risk of exposure to *T. gondii* increases. Thus, pregnant women handling cats and particularly kittens or cleaning out sand boxes have a chance of infecting their fetuses and eating raw meat.

In Pakistan, enteric pathogens are present not only in water but also foods contaminated from the environment or through human actions. Mishandling of foods allows these pathogens to contaminate and multiply in them. For example, street-vended fruit salads, locally called fruit chats, offered for sale at high ambient temperatures without coverings, and khoya and burfi, two indigenous sweet dairy products, and locally produced ice cream are often heavily contaminated with *Enterobacter, E. coli, Klebsiella, Salmonella* and *S. aureus* (Akhtar 2015). These contamination scenarios have led to outbreaks with cases severe enough to be hospitalized. Bus and train stations where pulses (edible seeds of various crops as peas, beans, or lentils), ground meat dishes, and chickpeas are sold to passengers, and are also heavily contaminated with bacteria including *Clostridium perfringens*. Sweet dishes and home-prepared foods in small communities are commonly contaminated with *S. aureus, C. perfringens*, and *Bacillus cereus* leading to rapid intoxications. One study confirmed campylobacters to be present in 48% of 1636 tested samples of milk and meats and 40.9% of vegetables in three major cities of Pakistan (Akhtar 2015). A wide array of vegetables is routinely consumed in this country and serve as a rich source of vitamins, minerals, bioactive compounds, and fiber but these can be sources of enteric infections if they are consumed contaminated. *Shigella* spp. has been shown to develop resistance and is generally thought to be a major cause of foodborne illnesses, especially among the poor where health care facilities are minimal; shigellosis is associated with poor sanitary conditions and unsafe water for drinking and preparing foods. Possible etiologies can be postulated in the following outbreaks.

Unfortunately, it is not only pathogens that give rise to food-associated disease. Soomro et al. (2008) highlighted the indiscreet use of pesticides in agriculture and its impact on environmental pollution. Despite the increased production cost associated with extensive use of pesticides, their use is common in developing countries. Numerous studies have demonstrated substantial levels of pesticide residues in various foodstuffs in Pakistan, and the groundwater has been observed to be considerably polluted in many parts of Punjab and Sindh provinces of Pakistan (Akhtar 2015). Commonly used open rural wells in the Punjab were polluted with six pesticides: bifenthrin, λ-cyhalothrin, carbofuran, endosulfan, methyl parathion, and monocrotophos. In the Hyderabad region 61% of the tested samples of eight vegetables (cauliflower, green chili, eggplant, tomato, peas, bitter gourd, spinach, and apple gourd) were found to be contaminated with pesticide residues exceeding maximum recommended limits (MRLs) (Tariq et al. 2003; Anwar et al. 2011). Heavy metals such as cadmium (Cd), copper (Cu), lead (Pb), and zinc (Zn) arising from increased industrialization can contaminate agricultural soils and these can be found in fruits (including widely-consumed mangoes), fruit juices, vegetables directly from soil uptake or from the processing and packaging (Akhtar 2015). For instance, spinach, coriander, and peppermint, grown in Sindh province contained 0.90–1.20 mg/kg of arsenic resulting in a total ingestion of arsenic 9.7–12.2 μg/kg body weight/day in diet (Arain et al. 2009; Khan et al. 2010).

Aluminum concentration in branded and nonbranded biscuit samples from Hyderabad were found to range 7.4–84.3 and 34.5–70.2 mg/kg, respectively (Jalbani et al. 2007). Similarly, Javed et al. (2009) detected higher concentrations of Cd, Cr, Ni, and Pb residues (mg/L) in bovine and goat milk. Pakistani foods are more prone to aflatoxin contamination because of the warm and humid climate, and the situation is exacerbated by malpractices during handling and storage of edible commodities (Mobeen et al. 2011). Samples of broken rice, wheat, maize, barley, and sorghum ranged 15–45% with the highest aflatoxin concentration (15.5 μg/kg), in wheat samples (Akhtar 2015). Chilies are widely eaten and exported, but aflatoxin levels can be eightfold higher than the EU permissible limits to pose a potential health risk to Pakistani consumers; concentrations can be reduced by more appropriate care and handling of the chilies at pre- and postharvest stages. Nuts and dried fruits in Pakistan are cultivated and processed in the northern areas and have been shown to have aflatoxin levels above the EU limit of 4 μg/kg in up to 70% of samples (Ahmad et al. 1989; Luttfullah and Hussain 2011). Aflatoxin M1 in milk and milk products requires regular monitoring in Pakistan since 3% of the total tested samples of milk were found to exceed the US tolerance limit of 0.5 μg/L (Hussain and Anwar 2008; Hussain et al. 2010), and buffalo milk had higher levels of aflatoxin compared with cow's milk. Intentional deception of consumers by blending low cost and inferior quality ingredients to make more profit of food intended for sale is prevalent in Pakistan, where families are exposed toxic dyes, sawdust, soapstone, and harmful chemicals in beverages, oil or ghee, bakery products, spices, tea, sweets, bottled water, and especially milk and milk products where more than 50% of samples tested have had adulterants added (Akhtar 2015).

One of the more innovative research projects to provide more home-grown food is in Qatar. The Sahara Forest Pilot (SFP) pilot study demonstrated that there are significant comparative advantages using saltwater for the integration of food production, revegetation and renewable processes: (1) seawater cooling system for greenhouses supports production of high-quality vegetables throughout the Qatari summer, and reduces freshwater usage to less than half that of comparable greenhouses in the region; (2) solar and desalination technologies were successfully integrated as designed into the SFP system, such as the greenhouse and evaporative hedges providing wet-cooling efficiencies without cooling towers; (3) the external evaporative hedges provide cooling of up to 10°C for agricultural crops and desert revegetation with vegetable and grain crops growing outdoors throughout the year; (4) commercially interesting algae showed good tolerance to heat and high evaporation rates in the leftover salty water (miss22 2013; Clery 2013). The concentrated solar power plant uses mirrors in the shape of a parabolic trough to heat a fluid flowing through a pipe at its focus. The heated fluid then boils water, and the steam drives a turbine to generate power. Hence, the plant has electricity to run its control systems and pumps, and can use any excess to desalinate water for irrigating the plants. In summary, SFP allows food production in all 12 months of the year (3 crops) with half the fresh water usage than in comparable greenhouses. On the basis of the pilot success, SFP is now engaged in studies aimed at building a

20-hectare test facility near Aqaba in Jordan, large enough from the 1-hectare operation in Qatar to demonstrate a commercial enterprise.

Issues Involving Tourism and Food Exports

Tourism is popular in several Middle Eastern countries, particularly beach and coastal resorts in Egypt and Turkey. Tourism has been the major economy in Egypt for many years but can be threatened not only by civil unrest and terrorism but also by foodborne illness (Costa 2008). Tourists might not stop coming to Egypt due to a few reports of diarrhea; however, widespread reporting of severe cases, and lawsuits, will make tour operators much more selective, and bring pressure on the Egyptian hospitality industry to improve its hygienic standards. The greater challenge is for Egypt to ensure that it has the capacity to sustain a safe food supply for its own people. In doing so, it provides safe food for those who want to explore its rich history and seaside resort areas. Multiple reports of illness have been reported from Nile River cruises and a resort town on the coast. From September to November, 2008, 34 cases of hepatitis A imported from Egypt were reported to the German public health authorities (Bernard and Frank 2009). Investigations pointed to a continuing common source of infection, most likely linked to Nile river cruises. In addition, eight cases from France had been travelling on a Nile cruise and one on a Red Sea diving safari (Couturier et al. 2009). One specific cruise ship was mentioned by six of ten Belgian cases (Robesyn et al. 2009). Those who took a Nile cruise had typically done this in combination with a hotel stay. At least three different ships and three different hotel accommodations were mentioned in the travel histories of the French cases. The patients affected had not been vaccinated, which emphasized the need for more effective travel advice before trips to hepatitis A endemic countries (Sane et al. 2015). Possible sources of infection might have been contaminated food obtained from a common food catering company consumed onboard, contaminated tap water supplies for the ships' bunkers, or a common exposure on shore (e.g., a restaurant where tourist groups from various ships were taken during day trips). As all of these ships continuously traveled up and down a short stretch of the river (Aswan to Luxor and back) with standard must-see stops along the way, the cases possibly shared an exposure on land. Both the long incubation period of hepatitis A (15–50 days) and long delays in collecting information on the individual cases precluded any rapid intervention on location. No specific food source was identified but it could have been juices as recognized in an earlier major outbreak. In 2004, tourists returning from Egypt included 351 hepatitis A case-patients from 9 European countries who were infected with a single HAV strain (genotype 1b) (Frank et al. 2007). The case-control study identified orange juice most likely contaminated during the manufacturing process, e.g., by an infected worker with inadequate hand hygiene or by contact of fruit or machinery with sewage-contaminated water. Citrus fruit and citrus juices have occasionally been implicated as vehicles of HAV and *Salmonella* infections, with contamination typically occurring during production, or preparation just before consumption. As HAV is resistant to acid, it likely can survive for prolonged periods in orange juice. It is also possible that leafy greens could contribute to foodborne illness in Egypt. An international study of contamination of 562 leafy green lettuce and spinach samples taken between 2011 and 2013 from 23 open-field farms in Belgium, Brazil, Egypt, Norway, and Spain showed that the Egyptian samples were the most contaminated at 55.6% (Liu et al. 2016). These authors claimed that temperature had a stronger influence than did management practices on *E. coli* presence and concentration. Region was a variable that masked many management variables, including rainwater, surface water, manure, inorganic fertilizer, and spray irrigation. Temperature, irrigation water type, fertilizer type, and irrigation method should be systematically considered in future studies of fresh produce safety.

Also in the spring of 2008, a young couple was ill with vomiting and abdominal cramps after their first meal at a Sharm El Sheikh 4-star hotel in the Egyptian coastal resort area, and they remained there in their bedrooms for the rest of their week (This is Staffordshire 2008). Both continued to have ongoing issues 6 months later, with one of them suffering from reactive arthritis. Other guests also complained about diarrhea. They stated that the food was disgusting; the meat was undercooked, the buffet was left out for long periods of time, with new food being piled on top of the old food, and there were flies landing on food items. In August 2014, a family stayed at a resort hotel, also in Sharm El Sheikh, and all suffered severe symptoms including diarrhea, stomach cramps, and vomiting. They were put into the hotel clinic given antibiotics and intravenous drips but had not completely recovered after they returned home (Galley 2014). At the time other guests were also ill. They noticed that the food including chicken and beef, appeared to be undercooked a couple of times, and that one of the chefs touched raw meat and then touched cooked meat without changing gloves. The booking company confirmed that "a very small number" of guests staying at the resort in 2014 reported that they had been unwell, "with symptoms similar to a virus". The company said that guests were offered the appropriate support and advice by their overseas holiday advisors. It claimed that all of its hotels were subject to stringent monitoring and audits and this hotel achieved an extremely high score in its audit carried out in the summer of 2014. However, high audit scores do not necessarily correlate with day-to-day safe hygienic practices (Powell et al. 2013).

The popular beach resort of Sarigerme, Turkey, on the Aegean Sea also has had a reputation for gastroenteritis, with repeat problems of foodborne illness with British tourists on vacations organized by tour companies, although the actual hotels were different. In 2009, an outbreak of gastric illness at this resort led to £1.7 m paid out in compensation, with 595 people suffering from infections including *Salmonella, Cryptosporidium, Campylobacter* and *E. coli* (Hutchison 2015). In September, 2012, hundreds of British holidaymakers suffered from salmonellosis after returning from a hotel complex in Sarigerme (Disley 2012). Final figures may have been close to 1000, and several were hospitalized. In October 2014, the Swannell family had booked a week's stay at the First Choice Holiday Village resort in Sarigerme, when Mark Swannell, 46, fell seriously ill a few days into the break with diarrhea, abdominal pain, nausea and lethargy (Hutchison 2015). He said that some of the food he was served at the hotel had been undercooked, with some chicken bloody in the middle, food was not served at the correct temperature, food was left uncovered for prolonged periods of time, and the same food had been served more than once. The family stated that cutlery, crockery and table linen used in the restaurant was not up to standard, and they saw cats in the public areas of the hotel and in the restaurant. Legal action was taken.

In addition to ill tourists in Middle Eastern countries, contaminated exported food can affect those abroad, as illustrated in the following U.S. outbreak. From March to August 2013, of 165 patients identified with hepatitis A in ten states, 69 (42%) were admitted to hospital, two developed fulminant hepatitis, and one needed a liver transplant, but none died (Collier et al. 2014). Almost all cases reported consuming pomegranate arils (seeds) from one retail chain. Hepatitis A virus genotype IB, uncommon in the Americas, was recovered from specimens from 117 people with hepatitis A virus illness. Pomegranate frozen arils imported from Turkey were identified as the vehicle early in the investigation by combining epidemiology, genetic analysis of patient samples, and product tracing. The product was then removed from store shelves, the public warned not to eat the seeds, recalls took place, and post-exposure prophylaxis with both hepatitis A virus vaccine and immunoglobulin was provided. This investigation showed that modern public health actions can help rapidly detect and control hepatitis A virus illness caused by imported food. Egyptian trade has also been adversely affected by exports. In 2013, there were three outbreaks of hepatitis A sickening 400 persons in 16 - European countries. In the first report in April, 106 persons in four Scandinavian countries were infected with hepatitis A (Andrews 2014). Epidemiological investigations traced those cases to frozen strawberries grown in Egypt and Morocco, though no strawberries were found to be positive for HAV. The second outbreak in April was larger in extent with 107 ill in 14 countries, all having recently visited Egypt, and the outbreak strain of the virus had the same subgenotype as the first outbreak associated with strawberries. An epidemiological investigation into the second outbreak suggested the likely source was strawberries or another fruit distributed to hotels in Egypt. The third outbreak was reported in Germany in May, after nine Germans were infected with hepatitis A after traveling to Italy. This third outbreak infected about 200 Italian residents, as well as nine Germans, one Dutch traveler and five Polish travelers; 21 Irish residents with no travel history to Italy were infected by the same strain of the virus. Separate investigations in Italy and Ireland both implicated imported frozen mixed berries as the source, with most of those berries coming from Eastern Europe. It is not known if these berries came from other regions, such as Egypt, or were local to Eastern Europe. Contributing factors to the larger number ill was lack of vaccination. Because HAV infections were declining in Europe over the last few decades, fewer people had developed antibodies to repel the virus. Couple that with the fact that hepatitis A was not on the vaccination schedule for citizens of many of the countries affected, and the result was a highly susceptible population. Also, most of the European travelers to Egypt were not advised to get hepatitis A vaccinations when staying in all-inclusive resorts, which were attracting an increasing number of Europeans traveling to Egypt. Further, the investigators believe contamination of the berries occurred early in the food production chain. Investigators suspect that irrigation water contaminated with sewage water likely contaminated the strawberries in the two outbreaks connected to Egypt. But the contamination might have also been caused by infected workers in the field or the processing facility, or by contaminated water sprayed on the berries sometime before distribution. The outbreaks indicate that fresh and frozen berries are efficient vehicles of HAV infection, as previously demonstrated in the US and elsewhere (Palumbo et al. 2013). European authorities agreed that "the 2013 experience demonstrated the absolute necessity for extensive collaboration between countries and between the public health and food sectors to identify as quickly as possible the vehicle of infection and, ideally, to control the outbreak in a timely fashion."

A more serious outbreak damaged Egypt's food export trade. In July 2011, the European Union (EU) banned the import of certain Egyptian seeds and beans till at least October following an official report that a single batch of Egyptian fenugreek seeds probably caused two European outbreaks of *E. coli* infections responsible for 4211 ill persons and at least 50 deaths. A task force of health officials set up by the European Food Safety Authority (EFSA) reported that one lot of fenugreek seeds imported from Egypt was the most likely common link between the two outbreaks in northern Germany and in Bordeaux, France (Anderson 2011). Both were traced back a year and a half to a shipment of 33,000 pounds (15,000 kg) of fenugreek seeds, that was loaded onto a ship at the Egyptian port of Damietta on November 24, 2009. On the ship's arrival at Antwerp, Belgium, the seeds were barged to Rotterdam to clear customs. The sealed container was trucked into Germany to an unidentified importer, who resold most of the lot. An unidentified German company then resold about 150 pounds of the seeds to the German sprouter, which is believed to be the source of the sprouts that caused the extensive German outbreak. The German importer also sold about 800 pounds of sprout seed to the English company Thompson & Morgan, which repackaged the seeds into 1.75-ounce (50 grams) packages. Those packages were shipped to a French distributor, who resold the seeds to about 200 garden centers around France. Investigators believe that one of those packets was the source of the second European outbreak with 16 cases in the Bordeaux area. Because the seeds were likely contaminated with *E. coli* O104:H4 at some point before leaving the importer, and more contaminated seeds could be in circulation, it was deemed appropriate to consider all lots of fenugreek from the Egyptian exporter as suspect. Soil contact or animal or human fecal contamination of the seeds likely occurred during their production or distribution in Egypt. Even a negative laboratory test of those seeds could not be interpreted as proof that a batch was not contaminated. Trace-forward findings indicate the German importer sold seeds from the suspected lot to 70 companies, and the shelf life of the seed can be up to 5 years.

By mid-October, 2011, the European Commission (EC) lifted import restrictions on fresh and chilled podded peas and green beans and other fresh produce from Egypt, but the ban on Egyptian seeds and sprouts, scheduled to expire on October

31, was to be extended until the end of March, 2012, following an "unsatisfactory audit" of seed producers in Egypt (News Desk 2011). The extended ban involved arugula sprouts, leguminous vegetable sprouts (fresh or chilled), soy bean sprouts, dried (shelled) leguminous vegetables, fenugreek seeds, soy beans and mustard seeds. The EC audit showed that measures taken by the Egyptian authorities to address shortcomings in the production of seeds that may be sprouted for human consumption were not sufficient "to tackle the identified risks." Those shortcomings were not seen in the growing and processing sites for fresh peas and beans, and therefore those vegetables were no longer considered a food safety risk.

There is no need for actual illnesses to occur to affect trade. Recalls, seizures, and bans can be employed by importing countries if standards are not met, and force exporting countries like Egypt to take action. For instance, in 1999 the EC suspended the import of peanuts from Egypt due to the presence of aflatoxin in concentrations in excess of maximum levels specified in EU regulations (Technical Cooperation Department 2003). Egypt is a major peanut exporting country and the European markets then accounted for 68% of its peanut exports. This decision was repealed on 1 December 1999 and was replaced by another Decision, which imposed a requirement for certification to accompany every consignment and required systematic analysis of consignments and documentation by the importing member state. Under this system only 18 Egyptian exporters were allowed to ship to the EU. In August 2003, the 1999 Decision was replaced by another Decision that required the competent authorities in EU Member States to undertake random sampling and analysis of 20% only of peanut consignments from Egypt for aflatoxin B1 and total aflatoxins. This improvement came as a result of the efforts that the Egyptian Government put in complying with the requirements of the EU. To this end, the Egyptian Ministries of Agriculture and Land Reclamation (MALR) and Ministry of Foreign Trade (MOFT) issued Ministerial Decree No. 2/2000, which covered all stages of production, processing, sampling and exporting of peanuts. The main provisions of the decree were:

exported peanuts must be produced, inspected and prepared according to set scientific procedures; and exporters who violate the rules would be suspended for 1 year;

The decree also established the legal limit for aflatoxin in peanuts in both the domestic and EU export markets. In the Egyptian domestic market, the legal limit was 5 mg/kg aflatoxin B1 and 10 mg/kg total aflatoxin content. For the EU market, the legal limits were 2 mg/kg aflatoxin B1 and 4 mg/kg total aflatoxin content. In addition, the decree specified the sampling procedures that must be followed for export certification. In September of 2001 the Food and Veterinary Office sent a mission to Egypt to assess Egypt's compliance with its certification system requirements. A number of recommendations on steps Egypt should take to improve the control system of foodstuffs intended for export to the EU were made. In response, the Egyptian authorities declared that they were taking actions to address the Mission's recommendation. But to achieve that there was a need to coordinate among a number of Egyptian agencies involved in the production and export of peanuts and aflatoxin control: MALR, The Central Administration for Plant Quarantine (CAPQ), The Agricultural Research Center (ARC), The Ministry of Foreign Trade (MOFT), and The Customs Service. Also a laboratory capable of testing for mycotoxins was necessary. Alongside this; Egypt had technical assistance from international organizations in order to build human and physical capacities necessary for achieving compliance. The action by the EU forces Egypt to improve the safety of its peanut production which would be beneficial both to Europeans and to all who eat products made from Egyptian peanuts, including the domestic consumers.

Lebanon used to be a tourist haven but is less today because of a seemingly dysfunctional government following a civil war. The country produces food for both the domestic and overseas markets. Unfortunately, some exported food has caused illnesses and recalls. Twenty-three cases of *Salmonella* Bovismorbificans in eight states and in the District of Columbia (Washington, D. C.) from August to November, were linked epidemiologically to hummus eaten at three Mediterranean-style restaurants in the D. C. area, all owned by the same individual (Goetz 2012). Although samples collected from all ingredients used to make the hummus tested negative for any *Salmonella*, the hummus was recalled and the outbreak ceased. During its investigation of the restaurants, the D.C. Department of Health discovered multiple food safety violations at the establishments, including inadequate food temperature control, insufficient hand washing, and the presence of pests and insects, which had to be corrected. It is not clear if any abusive temperature conditions could have allowed growth of the *Salmonella* in the hummus. The public was not notified because by the time the hummus had been withdrawn from the market, there were no further cases. However, the contaminated ingredient in the hummus was not discovered until May, 2012, when a traceback by the U.S. Food and Drug Administration (FDA) revealed that the tahini used to make the hummus in one of the restaurants had recently been associated with recalls in Canada for contamination with S. Cubana (September 2011) and *S.* Senftenberg (February 2012). All tahini linked to these outbreaks had been imported from the same company in Lebanon. The FDA then mandated that all tahini products coming from this Lebanese company be tested for *Salmonella* before entering the U.S. and has recommended that U.S. and Canadian officials partner to inspect the tahini manufacturing plant. This was the first time *S.* Bovismorbificans had been implicated in a tahini outbreak in the U.S. As a result of this outbreak, the author stated it is important for public health officials and consumers to be informed that products made with imported sesame paste have been shown to be associated with *Salmonella* outbreaks and that they should be considered as possible sources for foodborne illness in the future. In fact, contaminated sesame seed paste was in the news a few days before a CDC report on the outbreak was made public, after a supply of contaminated tahini was stolen from a California importer's warehouse, where it was being stored because a sample had tested positive for *Salmonella*. The tahini, which had also been imported from Lebanon but from a different manufacturer, was awaiting destruction, and the FDA warned the public that the stolen, potentially contaminated tahini may be on the market. Lebanese tahini has been implicated in several outbreaks in the past and subject to recalls (Harris et al. 2015).

Government Oversight of the Food Industry

Government oversight of the food industry is variable across the region with many regulations stemming back to colonial days, but modernization changes are gradually being considered or implemented. Unfortunately, where some Middle Eastern countries are slowly moving forward to improve food safety, others are slipping back in their oversight because of conflict and lower public health priorities. There are relatively few large food processing operations except those managed by multinational companies, and most of the government oversight is on SMEs particularly small foodservice outlets. The states in the Gulf Cooperation Council (GCC), each have an aggressive food safety policy but do not always follow identical approaches, some of which are well-established and some of which are innovative. The KSA has had a food inspection system in place for many years with reports of outbreaks published regularly, though no doubt it could be improved with more cooperation between the Ministry of Health, the municipalities and the Saudi Food and Drug Authority (SFDA). The SFDA was established under the Council of Ministers resolution no (1) dated January 1, 2003, as an independent body that directly reports to the Prime Minister (El Sheikha 2015). The SFDA is responsible to regulate, oversee, and control food, drug, medical devices, as well as set mandatory standard specifications thereof, whether they are imported or locally manufactured. The control and/or testing activities can be conducted at the SFDA or any other agency's laboratories. Moreover, the SFDA is in charge of consumers' awareness on all matters related to food, drug and medical devices and associated other products and supplies. The SFDA has to negotiate with the MOH their mutual responsibilities following specific foodborne disease instances or consumer complaints.

Bahrain claims to have one of the more advanced food control systems in the region. In July 2008, as ambient temperatures heated up, the Ministry of Heath urged people to make sure the food they consume is properly stored during the summer months to avoid microbial growth and risk of food poisoning, e.g., keeping meat and fish at $\leq 4^{\circ}C$ and to cook food thoroughly (Haider 2008). The Ministry was aware that both visitors and locals want to eat safe food, especially as Bahrain is moving towards more tourism with people are eating out more often. The Ministry ordered shops to provide appropriate storage facilities, e.g., coolers and refrigerators, for food as part of its efforts to protect the public's health. Inspectors were checking food stalls, ice-cream parlors and vegetable shops to ensure that customers were not being sold contaminated or rotten products. The Ministry claimed to thoroughly investigate any complaints it receives, and to facilitate this a new hotline number was launched by the Ministry for general public to report food contamination complaints against supermarkets, restaurants, coffee shops and hotels. Specific advice for consumers included: being careful when buying salads; fruits and vegetables should be washed thoroughly before they are consumed; and dairy products such as milk, cheese and eggs, should always be refrigerated, since microorganisms grow faster in these products. The Ministry claimed that Bahrain has one of the best food control methods and food safety records in the region, and could even act in the future as a consultant in this field for other countries, including other GCC states. By 2015, government oversight had stepped up. In April, the Ministry of Health warned people against buying food advertised on social media or sold on the street by unlicensed retailers in Bahrain, either made in people's homes or by street hawkers (Anonymous 2015c). The Ministry stated that control of these home operations is difficult if someone suffers from food poisoning since inspectors are not allowed to go into homes. Many homes sell food without a license and some would-be entrepreneurs even have barns where they slaughter livestock and market the meat illegally. There were 54,968 inspection visits conducted in 2014 by 25 inspectors from the Food Safety and Licenses group, which closed 41 of around 7000 registered outlets. Inspections cover imported food from ports right up to where it reaches restaurants and food outlets; 71,886 visits revealed around 883,584 tonnes of imported food were permitted for consumption, but 1873 tonnes were considered as non-consumable (rejected), during the same period. One of the more recent important programs is the Smart Inspection Project launched in April 2012. Inspectors, many with Masters and PhD degrees, visit restaurants and coffee shops to take food samples, as well as explain to staff how to store food and ensure its safety (Anonymous 2015c). It includes awarding food outlets that achieve a 100% food safety standard a blue sticker, while those meeting 80% of standards get a green sticker. Outlets that fail to achieve basic standards are warned with a red sticker. The total number of outlets assessed between August 2013 and February 2015 was 241; 17 were presented with blue stickers, 174 with green stickers and 50 with red stickers. This project features daily inspections and is focused on small food outlets, some of which have caused food poisoning in the past. Inspection visits depend on the hygiene of each outlet and the complaints received about them; some require two or more visits annually. High-level restaurants already have certified inspectors for evaluation and most of them require only one visit per year. The Ministry's ultimate goal through this project is to decrease cases of foodborne disease, particularly important as Bahrain is increasing its tourism efforts and, thus, ensuring food safety is essential. To support the Ministry's initiatives, live demonstrations on food safety practices were promoted in kitchens in hypermarkets. However, if red sticker facilities fail to take advantage of educational material, they may be punished for neglecting food safety standards and guidelines though public prosecution.

In a bid to improve standards of hygiene in restaurants, Qatar's Supreme Council of Health (SCH) increased the number of spot checks on food outlets and has launched a hotline for residents to report food poisoning (Walker 2014). The council is responsible for monitoring food establishments and implementing Qatar's food laws along with the Ministry of Municipality and Urban Planning (MMUP/Baladiya). The SCH embarked on an intensive inspection campaign, collecting food samples from all restaurants and food outlets in the country including suppliers. The inspection teams, which include specialized doctors from the SCH's communicable diseases department and the environmental health inspection department, also medically check workers responsible for preparing food to ensure they are not carrying infections. Those found to be handling food in an unhygienic way would be immediately dismissed. Following a hotline complaint call, a report is filed, a team from the SCH visits the affected people, then inspects the related food outlet and collects samples for laboratory examination. The latest crackdown was in response to the illness of a family of four which suffered food poisoning after eating chicken, rice and salad at a popular Turkish restaurant which was closed down because a medical report prepared by the SCH's environmental health section confirmed that the outlet served contaminated food and violated health regulations. Tests conducted in the Central Food Laboratory at SCH found three types of bacteria causing diseases in food served by the restaurant. Medical tests on the victims also showed that they were infected by the same bacteria, as well as one of the restaurant workers. Another popular Turkish restaurant was closed for 2 months after it was found that several customers were treated in the hospital for food poisoning symptoms including intense nausea, vomiting and diarrhea. As part of the SCH's new campaign, experts would undertake community awareness drives, and organize seminars and training sessions about food contamination to improve understanding among owners and workers in food establishments. Other closures occurred because of serving food with moldy ingredients, rotten vegetables in the kitchen, insects in pasta, and generally violating the provisions of the food law. The MMUP increased the number of spot-checks and naming and shaming erring establishments on its website in Arabic. The 2014 amendments to the food law gave greater powers to authorities to fine and close down venues that break the law including temporarily closing down establishments if it has violated food safety and hygiene regulations, and also has the power to recommend severe penalties. A follow up to one of these closed Doha Turkish restaurants was after a trial when five staff were each been handed fines, jail sentences and deportation orders after they were found guilty of causing food poisoning to approximately 20 customers ill with vomiting, nausea and diarrhea (Santacruz 2015). The restaurant was accused of serving spoiled and unsafe food on October, 2014. An affected pregnant woman gave birth to her baby 2 months prematurely. The manager of the restaurant was fined approximately $2500 and sentenced to spend 3 months in jail while three other staff members were each fined approximately $2000 and sentenced to 1 month in jail. During an inspection it was found that another staff member did not hold the necessary health certificate and was subsequently fined approximately $1800 and also sentenced to 1 month in jail. As well as the staff members being sentenced to jail and fined, the Court of Environmental Misdemeanours also found that the restaurant itself was guilty of causing the food poisoning outbreak, and issued the restaurant with approximately $8100 in fines and ordered it closed for a further 3 months. In other parts of the world these penalties would seem unduly harsh, as it would be difficult for this restaurant ever to recover financially.

Coupled with education, there has been recent enforcement blitzes on food establishments such as hotels, restaurants and bakeries by Oman municipalities, and a leading bakery in Muscat was closed down because of rats in the premises in late December, 2015 (Staff 2015). This led food safety experts and the public to call for stricter rules and heftier fines to be imposed after surprise checks conducted by the Muscat Municipality, especially when it was disclosed that nearly half the restaurants in the Bausher area were not following food safety standards. Surprise inspections by the Muscat Municipality at 125 restaurants in Bausher found that around 53 restaurants did not meet food safety standards and were violating rules formulated by the municipality. Also, in the same time frame, Ibri Municipality officials were forced to shut down 42 commercial shops and they destroyed more than 3000 km of outdated food in 2015. According to the Municipality's officials, 698 health violation letters were issued throughout the year, as well as 541 warnings were issued to different institutions operating in the wilayat of Ibri.

There are no easily-accessible reports on government oversight in Pakistan and inspection actions are more likely to be released to the public through the Press. In 2015, the Islamabad Capital Territory (ICT), Administration conducted a drive against adulterated food items with unannounced inspections of food outlets in different markets and imposed fines amounting to Rs110,000 (about US$1000) on owners for unhygienic conditions at their premises including restaurants, cafes, bakers, candy (sweet) stores, and a hotel was sealed (APP 2015). Cleanliness conditions at the outlets' kitchens were found unsatisfactory and unhygienic while workers had not been vaccinated against viral diseases. Some business owners were also paying less to their workers in contravention of the minimum wages act. Business owners were directed to improve cleanliness conditions and ensure food safety standards failing which strict action would be taken against them. A cattle market was also ordered to "beef up" its security. Punjab, Pakistan's most populous province, has a population that is more than double that of California, and Lahore, the provincial capital, has a vast array of food outlets. From the available Press reports, the Punjab Food Authority (PFA) has a mixed record of oversight of food operations. A PFA team visited the Polo Ground restaurant at the Race Ground Park and found expired food, blocked sinks and unhygienic conditions in the kitchen and food storage area in contrast to the claimed high quality standards by the management of the supposedly high-class restaurant (Raza 2015). The team faced resistance from the management but it managed to enter the kitchen for inspection. PFA officials said the kitchen condition was similar to that of an ordinary road-side eatery, dispelling general perception that restaurants serving the elite follow higher standards of hygiene and food safety. However, the PFA in Lahore had received a complaint that an assistant food safety officer had received Rs50,000 (about US$475) bribe from the restaurant owner so he could keep his restaurant open (Anonymous 2015d). Another restaurant on Peco Road sealed by the PFA for poor hygiene and unsanitary conditions of its workers in the second week of March, was opened for business the very next day. Typically, according to the PFA's standard operating procedure (SOP), a restaurant sealed for the first time may resume business after a week. At the end of the week, the proprietor has to submit an affidavit assuring the authority that all problems pointed out by the food safety officer had been taken care of prior to reopening it for business. The PFA Director General (DG) had constituted a three-member committee to probe the complaint of bribery but it was later shelved. Similar situations occurred when restaurants that had reopened before the stipulated period for closure had expired. In the first week of 2015, a restaurant was fined Rs25,000 (about US$240) for unhygienic conditions and lack of soaps in the workers' washrooms, instead of following the PFA SOPs of sealing the premises. The SOPs regarding duration of closure and required permission from the PFA DG were stated to be flouted openly. However, a PFA spokesperson denied any wrongdoing, and the SOP was being observed to the letter. She said a written permission from the DG used to be mandatory in order to de-seal restaurants, but now an operations deputy director can also issue permission for it. She also stated that the restaurant on Peco Road had not reopened on orders of the PFA; its owner had de-sealed it illegally. These reports indicate that there may be some illegal activities including bribery by inspectors but miscommunication on how much leeway inspection staff have on prevention and control practices may be more of the issue.

In mid-2015 Ayesha Mumtaz became the new operations director of the PFA, tasked with ensuring food in Punjab is unadulterated and safe (Reeves 2015). Her self-declared war on unhygienic food generated so much publicity in the last 6 months that she became a household name in Pakistan. Mumtaz says many food producers know nothing about hygiene but are willing to learn. There's also a hardened mafia who are only interested in profit, she says. Everyone in the street seems to know about Mumtaz. Storekeepers begin shooing away customers, hauling down the shutters, and heading into the shadows in the hope that Mumtaz's scrutinizing eye will not fall on them. These traders would sooner lose business than risk a visit from a woman whose campaign to clean up the kitchens and food factories of Pakistan has made her a national celebrity. She declared that the PFA cannot allow them to get away with their "perverse" activities and to "play havoc" with the lives of the people. Consumers are unaware that the cakes and sweets that they buy over the counter are produced amid unhygienic conditions. She has found spoons encrusted with filth, fly-blown cans of gooey liquid lying around haphazardly, dirty containers, grimy rags and rusty tin cans, moldy scraps of cake, all involved in making cakes and sweets to be sold to the public. Civil servants in Pakistan are often accused of being lazy and corrupt. Mumtaz is being feted as a rare example of a government official who actually champions the public's rights. She and her inspectors have so far raided more than 13,000 businesses, and Pakistanis seem to approve. Her fans call Mumtaz the Fearless One. Hundreds of thousands have clicked like on the PFA's Facebook page in appreciation of her work. There was a very famous hotel in the heart of Lahore that she inspected and found the chiller where they keep all the foods together (vegetables with chicken, meat), but also a big rat; this became big news for the public. However, there are complaints that she does her raids with police and cameras to be broadcast nationally even before the owners are convicted, according to the Lahore Restaurant Association.

In 2011, the Abu Dhabi Food Control Authority (ADFCA) planned to check all food handlers by 2012. The Authority's emirate food safety training (EFST) program, started in 2008, provides basic training in food hygiene and safety to those who work in food outlets (Olarte 2011). According to the ADFCA, small catering businesses in most countries have the lowest standards of food safety, and most workers in Abu Dhabi's 2500 small restaurants are illiterate and do not speak fluent Arabic or English, making it a challenge for them to understand and follow safety guidelines and regulations; 69% of managers and 73% food handlers in the capital speak South Asian languages such as Urdu, Hindi and Malayalam (Pennington 2014). The training is now offered in four languages — English, Arabic, Urdu and Malayalam — which the majority of food service personnel speak, and covers basic food hygiene issues including staff hygiene, food temperature, cross-contamination, cleaning and sterilization. To help them understand and follow food-safety rules, the ADFCA is using photographs to teach employees how to handle food safely according to international standards. The scheme is an extension of a pilot involving 600 small restaurants carried out in 2012–2014. As part of the efforts to ensure retention of their learning, the ADFCA conducted spot checks at 94 food outlets in Marina and Khalidiya Malls, and gave guidance and advice to staff for those with violations, rather than just penalizing them, the normal practice in most Middle Eastern countries. The field operations manager at the ADFCA noted that the differing cultures, education and languages are the barriers that sometimes hinder food handlers from carrying out what they are trained to do. He recommends that supervisors should quiz them on hygienic and safety issues so that they know how to properly prepare and serve food. Those who have learning difficulty or are illiterate are given assistance through illustrations, in order to make it through the lessons and pass the examination. One of the critical elements of food safety that the ADFCA has to monitor and ensure, is that food handlers are aware of cold ready-to-eat food being kept at 4°C, while hot food should be kept and served very hot $>$ 60°C. The ADFCA categorizes the food premises and carries out inspections based on their risk factors — high, medium and low. Restaurants and hypermarkets belong to the high-risk group; warehouses to the medium risk; while groceries, honey shops and vegetable and fruit outlets are considered low risk.

Recently, the establishment of the Egyptian Food Safety Authority was initiated by the Minister of Trade and Industry, with the support of the Ministry of Health and the Ministry of Agriculture. It would be responsible for food safety and consumer protection through the provision of sound data and guidance to deal with processed or genetically modified food in accordance with food safety standards (Anonymous 2012b). The strategic plan for the new draft law includes a revision of all Egyptian laws and legislation that deal with food safety since 1893, including around 2446 other legislations. The authority would need to apply food safety standards on imported food the same way it does for locally produced foodstuffs. Adopting the draft law would in effect cancel all existing laws and create one food safety law for the country. The Food Safety Authority plans to monitor the foods consumed by Egyptians of different age groups as a basis for where to put resources. Another issue to be faced is that studies in Egypt based on

US statistics have revealed that the cost of food spoilage costs the country 144 million Egyptian pounds annually. The Chamber of Food Industries indicated that a unified body for food safety to apply international quality specifications and unite regulators was lacking. This reduced the competitiveness of local products, especially since most foreign countries do not recognize Egyptian regulations. It was hoped that investors in food industries would bring in new investments to the sector in the upcoming period if a Food Safety Authority were to be established, as per a ministerial decision issued in 2011. The Food Safety Authority has received several approvals from governments that ruled during the 4-year period following the revolution, but apparently nothing has been yet finalized until recently (Mefreh and Saeed 2015).

In a similar way to Egypt, the Lebanese government has been debating a new law on food safety for many years but unlike Egypt, it has yet to make much progress. Lack of agreement at the parliamentary level has resulted in different ministries (Health, Agriculture, Industry, Environment, Tourism) taking action as they see fit. The latest was in November 2014, when the Minister of Health conducted an extensive campaign of inspections in Lebanese establishments and naming of facilities that did not meet the Ministry's expectations (Naylor 2014). The minister personally revealed that numerous supermarkets, bakeries, butchers and restaurants had been violating food safety and sanitation standards. They shut down slaughterhouses, restaurants, supermarkets and other retailers selling contaminated food. For instance, changes needed to be made for the slaughterhouse to conform to health standards; the report said livestock must be hanged during slaughter and not laid on the ground and that the abattoir should also be equipped with refrigerators and storage units for separate types of meat and their cuts. However, discord among ministries is apparent with the tourism minister trying play down the publicity of the health minister's food safety blitzes by saying "We are in favor of full transparency, but we feel like we were 'deceived' because the food safety situation in Lebanon is good and better than other countries. We apologize to tourists, but more importantly, any of the Ministry of Health staff is ready to apologize to the Lebanese citizens for the public sector's failures throughout the years?" (Yaliban 2014). Foodborne disease surveillance is limited in Lebanon and cannot be used to indicate the actual level of foodborne illnesses in the country. Lebanese food exports are also being required to conform to international standards. Tahini made from sesame seed paste is a major food export to the West, but recalls of tahini manufactured in Lebanon because of *Salmonella* contamination are more frequent than they should be; one recent example was a Health Hazard Alert for Certain Clic, Al Nakhil and Al Koura brand Tahina products that may have contained Salmonella, Recall/advisory dated August 16, 2013 posted from Canadian Food Inspection Agency [also see tahini/hummus linked illnesses under **Foodborne disease in specific countries**]. Under the new US Food and Drug Administration Food Safety Modernization Act, foreign companies importing foods to the US must demonstrate that they have the operational plans and facilities sufficient to produce safe food before they can ship any product to the US (FDA 2015), which is causing some concern among Lebanese tahini manufacturers and government agencies. Thus, although there is knowledge about foodborne disease and other food safety issues within government, industry and academia, the political inertia means that many foodborne illnesses will continue to occur but not be properly reported or know what factors were present to cause the outbreaks. Industry currently is taking the lead; apart from companies promoting food safety like Boecker and GWR Food Safety, MENA Food Safety Associates (MEFOSA) (http://www.mefosa.com/), based in Beirut, assists MENA companies hone their competitive edge by establishing and verifying procedures and practices that ensure quality, wholesome and safe products through consulting, auditing and training services in HACCP, GMPs, and hygienic practices.

However, Lebanon's lack of a coordinated system of government oversight of the food industry pales into insignificance compared to that in Syria. Prior to the war, Syria's healthcare system had hospital and doctor levels equivalent to other middle-income countries such as Brazil, Turkey and China, with life expectancy of 76 years, and most of the disease burden being similar to that in the West with non-communicable diseases, but four years of violence have changed all of that. Child vaccination levels dropped from 90% pre-conflict to 50% in March 2014 (Templeton 2015). As a result, outbreaks of diseases that had long been under control have spread across the land and into neighboring countries: hepatitis, measles, leishmaniasis, multi-drug-resistant tuberculosis, typhoid and even polio, which had not been seen in the Middle East for 20 years. Life expectancy has dropped by two decades. Medical personnel are clearly targeted because they are seen as potential enemies helping the opposite side. The majority of Syria's doctors have been killed or fled the country (>700 medical workers have been killed since 2011). The situation has been called the worst humanitarian catastrophe this century, and the worst concerted attack on healthcare in living memory. At least 300,000 Syrians have been killed and more than 11 million others have been forced from their homes since the conflict began on March 15, 2011, with over four million people in areas that are hard to reach for humanitarian aid, and 4·5 million have fled mostly to neighboring Turkey, Lebanon, Jordan, and northern Iraq, while others have sought safety in Europe, provoking a political crisis in the 28-member bloc (Devi 2016).

Another Middle Eastern country under stress but with less publicity is Yemen. Currently there is little government oversight into food as there is little to be had. The situation in Yemen is characterized by large-scale displacement, civil conflict, food insecurity, high food prices, endemic poverty, diminishing resources, and movement of refugees and migrants (WFP 2016). The UN World Food Programme (WFP) has been in Yemen since 1967. In 2014, WFP conducted a Comprehensive Food Security Survey which found that 41% of the people (10.6 million) were food insecure, of which some five million were severely food insecure, meaning they were unable to buy or produce the food they need to survive. The organization's protracted relief and recovery operation (PRRO), aims to reach six million people between mid-2014 and mid-2016 with 366,734 metric tons of food and US$74.5 million in cash and vouchers at an overall cost of US$491 million. If the conflict continues, this goal is unlikely to be met in time since both the airport and shipping port are areas being fought over. The WFP has been attempting to bring in relief supplies but cannot do so under fire, which means that only small amounts are occasionally delivered to the country (Mukhashaf and Miles 2015). One example of this occurred in Aden on July 20, 2015 when a ship docked after waiting a month to unload enough U.N. food aid to feed 180,000 people for a month. Previous repeated attempts to send ships to Aden were been blocked due to severe fighting in the port area. The PRRO is aligning WFP's activities with moves to increase the Government's capacity to respond to the crisis and will promote recovery and resilience to enable food insecure households and communities to better withstand and recover from the effects of conflict and shocks.

Discussion and Conclusions

There are many similarities as well as substantial differences in the descriptions of issues concerning food safety and foodborne disease of each country in the region. Gastrointestinal diseases are frequent throughout the Middle East with some countries identifying their etiologies, such as Egypt, Kuwait, Israel, Pakistan, Turkey, Yemen. These include bacteria and parasites, e.g., *Salmonella, Shigella, Campylobacter*, enterotoxigenic *E. coli* (ETEC), *Giardia, Entamoeba*, and occasionally enteric viruses such as HAV and norovirus. However, none of the countries has a well-functioning foodborne disease surveillance system, but a few report on a regular basis like KSA, and starting recently, Lebanon with PulseNet. Mostly it seems that only large outbreaks or ones with fatalities that are reported on, and mainly through the Press. These outbreaks are often related to point sources which are in most cases communal foods prepared for a large number of individuals as in feasts, student hostels, schools, campuses, or military camps. However, the actual etiological agents and the factors contributing to outbreaks are only rarely determined. One example is a very large outbreak in Bahrain in 2002 with at least 750 people suffering from foodborne illness after eating contaminated egg-and-mayonnaise sandwiches served at a wedding party, but the etiology was not determined, even though clinical specimens and food samples were analyzed, at least in a publically-released report (ProMED-MENA 2002). Based on the type of preparation including the length of time taken for preparation of the implicated food and the time from consumption to the appearance of symptoms of foodborne illness, the types of symptoms, and what has already occurred historically in foodborne disease outbreaks, possible agents can be surmised, such as *Bacillus cereus* and staphylococcal enterotoxins, and *Salmonella, Shigella*, or norovirus infections, but ProMED is continually asking for more information once an outbreak is announced, and hardly ever receiving it (ProMED-MENA 2014). All this indicates that even if clinical specimens or food samples are taken and analyzed, laboratories are only rarely able to determine an etiologic agent, or at least report on their results. Most agents described with the little information available seem similar in all the MENA countries and to those encountered in the West. However, a few pathogens are more likely to be restricted to a few nations, such polio in Pakistan, cholera in Iraq, MERS-CoV in KSA, and botulism in Egypt and Iran where river fish are often eaten (one case of infant botulism was diagnosed in Israel but it is a rare disease anywhere); the first two are more likely transmitted though water or poor hygienic conditions, the third by camels, and only botulism exclusively through food. Brucellosis is widespread in the Middle East but only a few country studies indicate its link to meat or dairy products. Much of the Middle East is in the throes of conflict which results in unique situations in specific countries to exacerbate foodborne disease or food poisonings; these include relief agencies supplying "stale" food to those trapped and starving by the Syrian civil war, almost lack of food at all in Yemen, deliberate poisonings of enemies in Afghanistan, Syria and Iraq, accidental pesticide poisonings in Iran, preventing unsafe food being sold to those on the Hajj in KSA, improperly prepared catered food for foreign troops in bases in Afghanistan, Iraq, Kuwait, KSA, and Turkey. Countries where tourism is a major source of income can be adversely affected by bad publicity over complaints over food served in resorts, such as in Egypt and Turkey. Also, Gulf countries tend to employ workers from India and other surrounding territories, and these are typically housed in camps or separate communities from citizens and visitors, and are transported to work sites and back; conditions are not always conducive to safe food, and outbreaks are occasionally reported either from their work sites or their overnight residences where meals are prepared or catered. Most food to many of these countries is imported, especially those with limited agricultural land and adequate water supplies; fruits and fresh vegetables, tend to be grown in rural or peri-urban settings for local consumption and these can be contaminated at source through polluted river or well water, such as in the Bekaa Valley of Lebanon and mountain communities in Pakistan, and the Nile, Tigris and Euphrates fluvial plains. On one occasion, Iranian watermelons were recalled and future sales banned in KSA, Qatar, and UAE because they were suspected of being poisoned or were injected with pesticides (nobody claimed to be ill after eating the melons), because holes were found in a few of them. However, the rationale of Iranian farmers deliberately losing money seems to counter this argument, and it is more likely a sectarian economic barrier (Abdullah 2015). In fact, with the temporary ban the price of watermelons went up in the countries that had banned them. Random tests carried out on the fruit confirmed they were free from any chemical substances, insecticides or other pollutants. The holes were most likely caused by emerging insect pupae. Countries outside the Gulf region reported no problems with the imported Iranian melons.

Where some processed foods are exported, there is a risk of the importing countries recalling these if they cause foodborne illnesses or contaminants are found in them. This has happened in Egypt with hepatitis A virus in strawberries and *E. coli* O104:H4 in fenugreek seeds causing serious illnesses in Europe and restricting further trade for an extended period. The same issue affected Turkish pomegranate arils and Lebanese tahini (made from imported ground sesame seeds), both containing *Salmonella*, exported to the US. Large to medium operations for broiler chickens and egg layers in KSA, Kuwait, Lebanon and other countries try and meet national standards or international guidelines for *Salmonella* but are not always achieved, resulting in recalls and fines. Governments are also aware of increasing concern over *Campylobacter* in chickens, as widely-eaten poultry is a major source of this pathogen, but campylobacteriosis is not often cited as causing foodborne disease. Raw milk (cow, sheep and camel) and raw milk cheese are still widely consumed in the Middle East at the local level, though not usually obtained through supermarkets, and the risk of infections is high, as it is in other parts of the world, but with the added concern of *Brucella* spp. and MERS CoV (the latter in the Gulf countries where camels are bred and milked), both serious pathogens. Yoghurt, surprisingly since it is acidic and is a source of gut beneficial lactobacilli, apparently was the foodborne vehicle to cause illnesses and deaths in Afghanistan, Israel, and Pakistan. No agent was found in any of the samples. In the Afghani example, the yoghurt was claimed to be deliberately poisoned; in the Israeli one, it was apparently "stale" given to Palestinian prisoners; there were two episodes in Pakistan, one was from a home-prepared meal and the other from a restaurant which served rice and yoghurt.

For prevention and controls strategies, most countries seem to rely on local authorities (municipalities) to do inspection of food facilities, more typically restaurants than processing plants as there are far more of them. Illegal sales for unapproved products by local entrepreneurs are sometimes an issue, e.g., home-slaughtered meat in Bahrain, and Palestinians shipping food to Israel. These illegal operations probably occur more often in porous borders within the region, and are only recognized when authorities decide to become vigilant in this area. Some countries have conducted research and surveys much more than others based on the publication record, e.g., Egypt, Israel, Palestine, KSA, Turkey, and to a lesser extent, Iran, Lebanon, Pakistan, UAE, and Yemen, but some research may occur without formal publication in recognized journals, making it difficult to have a true picture of how food safety problems are recognized and controlled. A few surveys have shown that home makers and food employees have limited knowledge of food safety, as in other regions. Thus, some agencies or industry associations, sometimes in collaboration with outside organizations like FAO or WHO, have attempted to train food employees in basic HACCP principles, including best hand hygiene practices, and speakers give the latest food safety issues at the annual Dubai International Food Safety Conference, now in its 10th year.

A few governments have established food safety agencies that have broad powers to inspect and control without overlapping responsibilities; these include Jordan and KSA with food and drug administrations, UAE with Abu Dhabi and Dubai food control authorities, Oman with its National Food Quality and Safety Centre, and Pakistan with a Punjab Food Authority. Egypt and Lebanon are initiating food safety authorities. Israel, Palestine and Jordan have a cross-border agreement to collaborate on food safety issues. Typical of many food control agencies in developing countries, periodic campaigns are launched to "crack down" on foodservice operations and sometimes processing plants. These are usually stimulated by complaints of the public, or the need for the responsible ministry to be seen doing something to justify its existence in compliance with regulations (if they exist). This has occurred recently in Lebanon, Qatar and Pakistan. One issue is that poorly constructed or out-of-date regulations may be interpreted in different ways by the owners and the agencies (Kullab 2014). If a violation is found, the facility may be fined and/or temporarily closed down until it has satisfied the inspectors at the next visit. In one extreme instance in Qatar, the owners and employees, were fined, imprisoned and deported. Unfortunately, although the names of those at fault are often publicized by the media, their specific violations and how they relate to the regulations are not usually documented or at least publically released. Another issue is that whether illnesses are suspected or not following a complaint, inspectors often insist that all food be discarded as soon as a sufficient violation, which may be unrelated to the complaint, has been determined; this prevents any samples being taken for outbreak investigations (Hanna et al. 2009), as well as using the outbreak for a teaching tool for the owner and other similar operations.

In conclusion, some progress has been made in the surveillance of foodborne disease in the Middle East, but the disease's health and economic burden is barely being considered in many countries for future decision-making policies, an issue that is being tackled at the global level (WHO 2015b). Food control agencies seem to be trying to stop apparent abuses but have limited resources to do much more. This region, in particular, is severely strained because of sectarian distrust, on-going civil wars, and terrorist attacks, with refugees from Iraq seeking shelter toward Europe but stalled in Turkey and Lebanon for long periods of time. The crisis in Syria is considered the greatest humanitarian disaster of the twenty first century, or even since World War II, and it looks like the on-going fighting including outside armed forces will make food insecurity in the affected countries even worse in the foreseeable future. Less public attention has been directed to Yemen where food insecurity is a major concern. This coupled with Gulf countries losing their wealth over low oil prices and a resultant stagnant global economy means a focus on food safety will likely become lower in priority for many of these countries. Since secure food has to be safe, as illustrated by "stale" food being issued to besieged Syrian residents and prisoners, it is important that relief agencies and countries themselves be aware of the risk of foodborne diseases associated with immunocompromised persons, particularly children. However, even in countries where the food supply is acceptable, inadequate hygienic practices put the local and tourist population at risk of illness and exported foods jeopardize industry profits and a poor reputation for future trade. As demonstrated by KSA, Jordan and UAE, single agencies or multiple agencies with clear-cut roles responsible for food safety, should be pursued by governments in consultation with industry and academia. Duplication creates ambiguities for enforcement and education strategies as well as being unnecessarily costly. Water supplies are also critical and some governments are weaning away farmers from depleted groundwater aquifers, and making irrigation more efficient where there are sustainable supplies. Water for irrigation and processing has to be both free of pathogens and unacceptable levels of chemicals, and effectively treated waste water can substitute for groundwater. The Sahara Forest Project in Qatar is one example of a very dry country using seawater resources effectively; an even larger project is being considered from the 1-hectare in Qatar to a 20-hectare test facility in Jordan (Clery 2013). All these issues are being compounded by climate change and expected higher temperatures in already arid lands, which will make the region all the more dependent on more expensive imported foods. Gulf counties have enough petro-dollars to afford these, but other countries are struggling to be self-sufficient for the near future even if the fighting ceases. The repair to destroyed infrastructure will be immense, coupled with the lack of trained personnel to create a restored food system at all levels from primary production through food processing, foodservice, and retail to the home.

References

Abdelateef, N. (2011). *Antimicrobial resistance for enteric pathogens isolated from acute gastro-enteritis patients in Gaza strip, Palestine*. Thesis. The Islamic University – Gaza. Available at: http://library.iugaza.edu.ps/thesis/95584.pdf. Accessed 14 Dec 2015.

Abdou, N. E., Al-Batel, M. K., El-Azazy, O. M., Sami, A. M., & Majeed, Q. A. (2013). Enteric protozoan parasites in stray cats in Kuwait with special references to toxoplasmosis and risk factors affecting its occurrence. *Journal of the Egyptian Society of Parasitology, 43*(2), 303–314.

Abdullah, A. (2015, May 4). Sale of watermelons with holes stopped in UAE. *Khaleej Times*. Available at: http://www.khaleejtimes.com/nation/general/sale-of-watermelons-with-holes-stopped-in-uae. Accessed 12 Mar 2016.

Afanasieva, D., Miles, T., & Karouny, M. (2015, October 30). U.N. causes food-poisoning with deliveries of old, 'moldy' biscuits to Syria, says rights body. Reuters. Available at: http://www.reuters.com/article/us-mideast-crisis-syria-aid-idUSKCN0SO24L20151030. Accessed 20 Jan 2016.

Agence France-Presse. (2015, October 20). Iraq cholera cases grow, spread to Kurdish region. Relief Web. Available at: http://reliefweb.int/report/iraq/iraq-cholera-cases-grow-spread-kurdish-region. Accessed 30 Dec 2015.

Ahmad, M. A., Khan, B. A., Shamsuddin, Z. A., & Anwarullah, M. (1989). Presence of Aflatoxin B1 in the shelled peanuts in Karachi. *Pakistan Journal of Scientific and Industrial Research, 32*, 526–527.

Ahmed, M. (2013.) Incidence and distribution of *Salmonella* serogroups in some local food in Sana'a – Yemen. Al-Nasser University, Yemen. Available at: http://journals.sfu.ca/alnasseru/index.php/alnasseru/article/view/2. Accessed 5 Jan 2016.

Ahmed, S. F., Klena, J. D., Mostafa, M., Dogantemur, J., Middleton, T., Hanson, J., & Sebeny, P. J. (2012, May 11). Viral gastroenteritis associated with Genogroup II norovirus among U.S. military personnel in Turkey, 2009. PLoS One.DOI: 10.1371/journal.pone.0035791. Accessed 21 Dec 2015.

Akbar, J. (2015, July 09). 45 ISIS fighters killed by poisoned Ramadan meal. Daily Mail Online. Available at: http://www.dailymail.co.uk/news/article-3153194/45-ISIS-fighters-die-eating-poisoned-Ramadan-meal-Iraq.html. Accessed 30 Dec 2015.

Akhtar, S. (2015). Food safety challenges—a Pakistan's perspective. *Critical Reviews in Food Science and Nutrition, 55*(2), 219–226.

Al Awaidy, S. A., Bawikar, S., Al Busaidy, S., Baqiani, S., Al Abedani, I.,Varghese, R., Abdoan, H. S., Al Abdoon, H., Bhatnagar, S., Al Hasini, K. S., Mohan, P., Shah, S., Elamir, E., Klena, J., Ahmed, S. F., Teleb, N., Parashar, U., & Patel, M. M. 2009. Considerations for introduction of a rotavirus vaccine in Oman: Rotavirus disease and economic burden. *Journal of Infectious Diseases, 200*(Suppl. 1), S248–S253.

AlHayat. (2015, March 25). Foodborne illness – Saudi Arabia: (Najran) restaurant. Archive Number: 20150330.3265029. ProMED-MENA . Available at: http://www.promedmail.org/post/20150330.3265029. Accessed 25 Jan 2016.

Aligi, I. (2010, September 07). 250 flood victims suffer food poisoning. *The Daily Times*. Available at: http://archives.dailytimes.com.pk/karachi/07-Sep-2010/250-flood-victims-suffer-food-poisoning. Accessed on 24 Nov 2015.

Al-Joudi, A. S. 2007. An outbreak of foodborne diarrheal illness among soldiers in mina during Hajj: The role of consumer food handling behaviors. Journal of Family and Community Medicine.14(1): 29–33.

Aljoudi, A. S., Al-Mazam, A., & Choudhry, A. J. (2010). Outbreak of food borne Salmonella among guests of a wedding ceremony: The role of cultural factors. *Journal of Family and Community Medicine, 17*(1), 29–34.

Al-Kandari, D., & Jukes, D. J. (2009). A situation analysis of the food control systems in Arab Gulf Cooperation Council (GCC) countries. *Food Control, 20*, 1112–1118.

Al-Mazrou, Y. Y. (2004). Food poisoning in Saudi Arabia. Potential for prevention? *Saudi Medical Journal, 25*(1), 11–14.

Anadolu Agency. (2015, May 01). Syrian migrants attack gendarmerie in southeastern Turkish tent city, five injured. *Hurriyet Daily News*. Available at: http://www.hurriyetdailynews.com/Default.aspx?pageID=238&nid=81827&NewsCatID=341. Accessed 29 Dec 2015.

Anderson, R. (2011, July 6). One Egyptian seed shipment: Two outbreaks. *Food Safety News*. Available at: http://www.foodsafetynews.com/2011/07/one-egyptian-seed-shipment-two-outbreaks/#.VldO3narSM8. Accessed 27 Nov 2015.

Andrews, J. (2014, November 05). What can we learn? Food Safety News. Available at: http://www.foodsafetynews.com/2014/11/europes-3-hep-a-outbreaks-from-2013-what-can-we-learn/#.VldPLXarSM8. Accessed 28 Nov 2015.

Anonymous. (2012a, April 03). Afghanistan: Militants 'kill police by poisoning food'. *BBC News* . Available at: http://www.bbc.com/news/world-asia-17597797. Accessed 4 Feb 2016.

Anonymous. (2012b). New food safety law to come up in Egypt. Clean. Available at: http://www.cleanmiddleeast.ae/articles/18/new-food-safety-law-to-come-up-in-egypt.html. Accessed 13 Jan 2016.

Anonymous. (2012c, March 09). Pakistan-Minor dies of food poisoning. *The News International*. Available at: http://www.thenews.com.pk/article-38916-Minor-dies-of-food-poisoning. Accessed 24 Nov 2015.

Anonymous. (2012d). Pakistan – Over 50 nurses hospitalized after food poisoning. *The News*. Available at: http://www.thenews.com.pk/Todays-News-Over-50-nurses-hospitalised-after-food-poisoning-6-95733. Accessed 15 Mar 2016.

Anonymous. (2015a, October 04). MoH on cholera alert; two patients detected. IndiansinKuwait.com. Available at: http://www.indiansinkuwait.com/ShowArticle.aspx?ID=36788&SECTION=0. Accessed 2 Feb 2016.

Anonymous, (2015b, January 01). 150 'food poisoning' cases at Al-Azhar dorms in Egypt's Assiut. Ahramonline. Available at: http://english.ahram.org.eg/NewsContent/1/64/119286/Egypt/Politics-/-food-poisoning-cases-at-AlAzhar-dorms-in-Egypts-A.aspx. Accessed 27 Nov 2015.

Anonymous. (2015c, April 07). New alarm over 'unhealthy' food in Bahrain. Trade Arabia. Available at: http://www.tradearabia.com/news/HEAL_279164.html. Accessed 11 Mar 2016.

Anonymous, (2015d, April 25). New guidelines: Punjab Food Authority giving 'unsanitary' restaurants an easy time of it. Tribune Express. Available at: http://tribune.com.pk/story/875525/new-guidelines-punjab-food-authority-giving-unsanitary-restaurants-an-easy-time-of-it/. Accessed 25 Nov 2015.

Anwar, T., Ahmad, I., & Tahir, S. (2011). Determination of pesticide residues in fruits of Nawabshah district, Sindh, Pakistan. *Pakistan Journal of Botany, 43*, 1133–1139.

APP. (2015, January 13). Surprise inspections: Pakistan officials fine food outlets, seal hotel. *Express Tribune*. Available at: http://tribune.com.pk/story/820907/surprise-inspections-ict-offi cials-fine-food-outlets-seal-hotel/. Accessed 25 Nov 2015.

Arain, M. B., Kazi, T. G., Baig, J. A., Jamali, M. K., Afridi, H. I., Shah, A. Q., Jalbani, N. & Sarfraz, R. A. (2009). Determination of arsenic levels in lake water, sediment, and foodstuff from selected area of Sindh, Pakistan: Estimation of daily dietary intake. *Food and Chemical Toxicology, 47*, 242–248.

Asghar, M. (2014, April 03). Teenage boy, sister die of food poisoning. *Dawn*. Available at: https://www.dawn.com/news/1097361/teenage-boy-sister-die-of-food-poisoning. Accessed 25 Nov 2015.

Associated Press. (2013, April 03). Egypt students storm office of top al-Azhar cleric. *The Record*. Available at: http://www.therecord.com/news/world/article/911866--egyptian-students-angered-by-food-poisoning-outbreak-storm-cleric-s-office. Accessed 27 Nov 2015.

Baş, M., Ersun, A. Ş., & Kıvanç, G. (2006). The evaluation of food hygiene knowledge, attitudes, and practices of food handlers in food businesses in Turkey. *Food Control, 17*, 317–322.

Bender, D. (2014, May 28). Israeli Doctor's bio-warfare serum saves infant from botulism death. Thealgemeiner. Available at: http://www.algemeiner.com/2014/05/28/israeli-doctors-bio-war fare-serum-saves-infant-from-botulism-death/#. Accessed 12 Dec 2015.

Bernard, H. & Frank, C. (2009). Cluster of hepatitis A cases among travellers returning from Egypt, Germany, September through November 2008. Euro Surveillance. 14(3).

Bostan, K., Aydın, A., & Ang, M. K. (2009). Prevalence and antibiotic susceptibility of thermophilic Campylobacter species on beef, mutton, and chicken carcasses in Istanbul, Turkey. *Microbial Drug Resistance, 15*(2), 143–149.

Brittin, H. C. (2010). *The food and agriculture around the world handbook*. New Jersey: Pearson Education.

CDC (US Centers for Disease Control and Prevention). (2014). Confirmed variant Creutzfeldt-Jakob Disease (variant CJD) case in Texas. Available at: http://www.cdc.gov/ncidod/dvrd/vcjd/other/confirmed-case-in-texas.htm. Accessed 20 Nov 2015.

Clery, D. (2013, November 7). Desert farming experiment yields first results. *Science Magazine*. Available at: http://news.sciencemag.org/asiapacific/2013/11/desert-farming-experiment-yields-first-results. Accessed 9 Feb 2016.

Cohen, D., Bassal, R., Goren, S., Rouach, T., Taran, D., Schemberg, B., Peled, N., Keness, Y., Ken-Dror, S., Vasilev, V., Nissan, I., Agmon, V., & Shohat, T. (2014). Recent trends in the epidemiology of shigellosis in Israel. *Epidemiology and Infection, 142*(12), 2583–2594.

Cohen, D., Gargouri, N., Ramlawi, A., Abdeen, Z., Belbesi, A., Al Hijawi, B., Haddadin, A., Sheikh Ali S., Al Shuaibi, N., Bassal, R., Yishai, R., Green, M. S. & Leventhal, A. (2010). A middle east subregional laboratory-based surveillance network on foodborne diseases established by Jordan, Israel, and the Palestinian Authority. *Epidemiology and Infection, 138* (10), 1443–1448.

Collier, M.G., Khudyakov, Y. E., Selvage, D., Adams-Cameron, M., Epson, E., Cronquist, A., Jervis, R. H., Lamba, K., Kimura, A. C., Sowadsky, R., Hassan, R., Park, S. Y., Garza, E., Elliott, A. J., Rotstein, D. S., Beal, J., Kuntz, T., Lance, S. E., Dreisch, R., Wise, M. E., Nelson, N. P., Suryaprasad, A., Drobeniuc, J., Holmberg, S. D., & Xu, F., for the Hepatitis A Outbreak Investigation Team. (2014). Outbreak of hepatitis A in the USA associated with frozen pomegranate arils imported from Turkey: An epidemiological case study. *Lancet Infectious Diseases, 14*(10), 976–981.

Costa, R. (2008). Food-borne Illness and its Effect on Tourism in Egypt. Food Safety and Environmental Health Blog. Available at: http://www.safefoodsblog.com/2008/11/articles/international-food-safety/foodborne-illness-and-its-effect-on-tourism-in-egypt/. Accessed 27 Nov 2015.

Couturier, E., Roque-Afonso, A. M., Letort, M. J., Dussaix, E., Vaillant, V., & de Valk. H. (2009). Cluster of cases of hepatitis A with a travel history to Egypt, September–November 2008,

France. Euro Surveillance, 14(3): pii=19094. Available online: http://www.eurosurveillance.org/ViewArticle.aspx?ArticleId=19094. Accessed 15 Mar 2016.

Dashti, A. A., Jadaon, M. M., Habeeb, F., West, P. W., Panigrahi, D., & Amyes, S. G. (2008). Salmonella enterica Serotype typhi in Kuwait and its reduced susceptibility to ciprofloxacin. *Journal of Chemotherapy, 20*(3), 297–302.

Devi, S. (2016, March 12). Syria's health crisis: 5 years on. *Lancet, 387*, 1042–1043. www.thelancet.com http://www.thelancet.com/pdfs/journals/lancet/PIIS0140-6736(16)00690-5.pdf. Accessed 11 Mar 2016.

Disley, J. (2012, September 09). British holidaymakers in Turkey hit by salmonella. Mirror. Available at: http://www.mirror.co.uk/news/uk-news/british-holidaymakers-in-turkey-hit-by-salmonella-335726. Accessed 27 Dec 2015.

El Sheikha, A. F. (2015). Food safety issues in Saudi Arabia. *Nutrition and Food Technology, 1*(1): doi http://dx.doi.org/10.16966/nftoa.103. Accessed 15 Mar 2016.

Elinav, H., Hershko-Klement, A., Valinsky, L., Jaffe, J., Wiseman, A., Shimon, H., Braun, E., Paitan, Y., Block, C., Sorek, R., & Nir-Paz, R. for the Israeli Listeria Study Group. 2014. Pregnancy-associated listeriosis: clinical characteristics and geospatial analysis of a 10-Year period in Israel. Clinical Infectious Diseases 59 (7): 953–961.

Fagbo, S. (2005, February 04). Food poisoning – Saudi Arabia (Jizan): Request for information. Archive Number: 20050204.0380. ProMED-MENA. Available at: http://promedmail.org/post/20050204.0380. Accessed 25 Jan 2016.

Faour-Klingbeil, D., Kuri, V., & Todd, E. C. D. (2015). Investigating a link of two different types of food business management to the food safety knowledge, attitudes and practices of food handlers in Beirut, Lebanon. *Food Control, 55*, 166–175.

Faour-Klingbeil, D., Murtada, M., Kuri, V., & Todd, E. C. D. (2016). Understanding the routes of contamination of ready-to-eat vegetables in the Middle East. *Food Control*, 62, 125–133. doi: http://dx.doi.org/10.1016/j.foodcont.2015.10.024. Accessed 15 Mar 2016.

FDA (US Food and Drug Administration). (2015). Making certain imported foods meet U.S. standards under FDA Food Safety Modernization Act. U.S. Food and Drug Administration, Washington, D. C. Available at: http://www.fda.gov/Food/GuidanceRegulation/FSMA/ucm257980.htm. Accessed 28 Jan 2016.

Frank, C., Walter, J., Muehlen, M., Jansen, A., van Treeck, U., Hauri A. M., Zoellner, I., Rakha, M., Hoehne, M., Hamouda, O., Schreier, E., & Stark, K. 2007. Major outbreak of hepatitis A associated with orange juice among tourists, Egypt, 2004. Emerging Infectious Diseases 13(1), 156–158. Available from http://wwwnc.cdc.gov/eid/article/13/1/06-0487. Accessed 15 Mar 2016.

Galley, J. (2014, November 10). Dream five-star holiday 'ruined' after family struck down by vomiting bug. *Mirror*. Available at: http://www.mirror.co.uk/news/world-news/dream-five-star-holiday-ruined-after-4604283. Accessed 27 Nov 2015.

Gee, C. (2014). A dozen ISIS fighters killed after chefs infiltrate camp and POISON terrorists' lunch. *Daily Mirror*, 08 November. Available at: http://www.mirror.co.uk/news/world-news/dozen-isis-fighters-killed-after-4590573. Accessed 20 Jan 2016.

Goetz, G. (2012, November 29). CDC: Salmonella from tahini sickened 23 last year. *Food Safety News*. Available at: http://www.foodsafetynews.com/2012/11/government-reports-2011-sal monella-outbreak-linked-to-tahini/#.VqqcjvkrKM8. Accessed 28 Jan 2016.

Greig, J. D., Todd, E. C. D., Bartleson, C. A., & Michaels, B. (2007). Outbreaks where food workers have been implicated in the spread of foodborne disease. Part 1. Description of the problem, methods and agents involved. *Journal of Food Protection, 70*, 1752–1761.

Haider, A. (2008, July 19). Poisoning alert over improper food storage. *Gulf Digital News (GDN Online)*. Available at: http://archives.gdnonline.com/NewsDetails.aspx?date=04/07/2015&storyid=223513. Accessed 6 Jan 2016.

Hanna, N. M., Adib, S. M., & Daoud, Z. (2009). Food-borne salmonella outbreak at a bank cafeteria: an investigation in an Arab country in transition. *La Revue de Santé de la Mé diterranée orientale, 15*(2), 470–474.

Harris, L. J., Palumbo, M., Beuchat, L. R., & Danyluk, M. D. (2015). Outbreaks of foodborne illness associated with the consumption of tree nuts, peanuts, and sesame seeds. In Outbreaks from tree nuts, peanuts, and sesame seeds. Available at: http://ucfoodsafety.ucdavis.edu/Nuts_and_Nut_Pastes. Accessed 15 Mar 2016.

Hosseini, H., Reza, H., Tavakoli, H. R., Meshgi, M. A., Khaksar, R., Hosseini, M., & Khakpour, M. (2010). Survey of *Clostridium botulinum* toxins in Iranian traditional food products. *Comparative Clinical Pathology, 19*(3), 247–250.

Hussain, M. (2013, August 16). Deadly disease: two dead, over 300 affected in Kurram's cholera outbreak. *Express Tribune*. Available at: http://tribune.com.pk/story/590683/deadly-disease-two-dead-over-300-affected-in-kurrams-cholera-outbreak/ Accessed 25 Nov 2015.

Hussain, I., & Anwar, J. (2008). A study on contamination of aflatoxin M1 in raw milk in the Punjab province of Pakistan. *Food Control, 19*, 393–395.

Hussain, I., Anwar, J., Asi, M. R., Munawar, M. A., & Kashif, M. (2010). Aflatoxin M1 contamination in milk from five dairy species in Pakistan. *Journal of Food Protection, 21*, 122–124.

Hutchison, J. (2015, January 14). 'Blood in the chicken and cats around food': Father-of-three hospitalised with severe gastric illness at SAME Turkish holiday resort where 595 fell ill in 2009. *Online DailyMail*. Available at: http://www.dailymail.co.uk/travel/travel_news/article-2909698/Blood-chicken-cats-food-Father-hospitalised-severe-gastric-illness-family-holiday-Turkish-resort-595-fell-ill-2009.html. Accessed on 30 Dec 2015.

Ians. (2011, January 28). Man dies after eating contaminated yoghurt. *Gaea Times*. Available at: http://health.gaeatimes.com/2011/01/28/man-dies-after-eating-contaminated-yoghurt-30264/. Accessed 24 Nov 2015.

Islam, S. (2011, May 23). Gastro epidemic: Rising temperatures turn stomachs. *Express Tribune*. Available at: http://tribune.com.pk/story/173943/gastro-epidemic-rising-temperatures-turn-stomachs/ Accessed 10 Feb 2016.

Jalbani, N., Kazi, T. G., Jamali, M. K., Arain, B. M., Afridi, H., & Baloch, A. (2007). Evaluation of aluminum contents in different bakery foods by electrothermal atomic absorption spectrometer. *Journal of Food Composition and Analysis, 20*(3–4), 226–231.

Javed, I., Jan, I. U., Muhammad, F., Rahman, Z. U., Khan, M. Z., Aslam, B., & Sultan, J. I. (2009). Heavy metal residues in the milk of cattle and goats during winter season. *Bulletin of Environmental Contamination and Toxicology, 82*, 616–620.

Kamel, A. H., Ali, M. A., El-Nady, H. G., de Rougemont, A., Pothier, P., & Belliot, G. (2009). Predominance and circulation of enteric viruses in the region of Greater Cairo, Egypt. *Journal of Clinical Microbiology, 47*, 1037–1045.

Kamel, A. H., Ali, M. A., El-Nady, H. G., Deraz, A., Aho, S., Pothier, P., & Belliot, G. (2011). Presence of enteric hepatitis viruses in the sewage and population of Greater Cairo. *Clinical Microbiology and Infection, 17*, 1182–1185.

Khan, S., Rehman, S., Khan, A. Z., Khan, M. A., & Shah, M. T. (2010). Soil and vegetables enrichment with heavy metals from geological sources in Gilgit, northern Pakistan. *Ecotoxicology and Environmental Safety, 73*, 1820–1827.

Khuri-Bulos, N. A., Khalaf, M. A., Shehabi, A., & Shami, K. (1994). Foodhandler-associated *Salmonella* outbreak in a university hospital despite routine surveillance cultures of kitchen employees. *Infection Control and Hospital Epidemiology, 15*, 311–314.

Kullab, S. (2014, April 19). Experts emphasize need for food safety measures after scandals. *Daily Star*. Available at: http://www.dailystar.com.lb/Article.aspx?id=253840&link=News/Lebanon-News/2014/Apr-19/253840-experts-emphasize-need-for-food-safety-measures-after-scandals.ashx. Accessed 27 Jan 2016.

Liu, C., Hofstra, N., & Franz, E. (2016). Impacts of climate and management variables on the contamination of preharvest leafy greens with *Escherichia coli*. *Journal of Food Protection, 79* (1), 17–29.

Luttfullah, G., & Hussain, A. (2011). Studies on contamination level of aflatoxins in some dried fruits and nuts of Pakistan. *Food Control, 22*, 426–429.

Mahmood, N. (2012, February 07). Pakistan- Food poisoning claims four lives. Pakistan Observer . Available at: http://pakobserver.net/detailnews.asp?id=139008, Accessed Nov 2015.

Martins, R. B., Hogg, T., & Otero, J. G. (2012). Food handlers' knowledge on food hygiene: the case of a catering company in Portugal. *Food Control, 23*, 184–190.

Masriya, A. (2013, April 30). Egypt: Azhar's food poisoning cases rise to 161 – MOH. *AllAfrica*. Available at: http://allafrica.com/stories/201305010321.html. Accessed 27 Nov 2015.

Masriya, A. (2014, October 22). Egypt: 178 cases of food poisoning in two Suez primary schools. *AllAfrica*. Available at: http://allafrica.com/stories/201410230249.html Accessed 27 Nov 2015.

Mayar, M. R. (2015, April 8). 20 cops suffered from diarrhea during Sehat Ka Insaaf drive. *The Nation*. Available at: http://nation.com.pk/national/08-Apr-2014/20-cops-suffered-from-diarrhea-during-sehat-ka-insaaf-drive. Accessed 25 Nov 2015.

Mefreh, M, & Saeed, S. (2015, October 08). Establishment of Food Safety Authority will allow investments in food industries. *Daily News Egypt*. Available at: http://www.dailynewsegypt.com/2015/10/08/establishment-of-food-safety-authority-will-allow-investments-in-food-industries/. Accessed 13 Jan 2016.

Memish, Z. A. (2010). The Hajj: communicable and non-communicable health hazards and current guidance for pilgrims. *Eurosurveillance* 15 (39). Available at: http://www.eurosurveillance.org/images/dynamic/EE/V15N39/art19671.pdf. Accessed 25 Jan 2016.

Memish, Z. A., & Al Rabeeah, A. A. (2011). Global perspectives for prevention of infectious diseases associated with mass gatherings. *Lancet Infectious Diseases, 12*, 66–74.

miss22. (2013, June 22). Qatar pursues water and food security. Global Risk Insights. Available at: http://globalriskinsights.com/2013/06/qatar-pursues-water-and-food-security/. Accessed 12 Jan 2016.

Mobeen, A. K., Aftab, A., Asif, A., & Zuzzer, A. S. (2011). Aflatoxins B1 and B2 contamination of peanut and peanut products and subsequent microwave detoxification. *Journal of Pharmacy and Nutrition Science, 1*, 1–31.

Modarres, S. (1997). A survey of *Clostridium botulinum* in food poisoning in Iran. Tehran University Medical Journal. 55(6):30–34. Available at: http://tumj.tums.ac.ir/browse.php?a_code=A-10-25-1610&slc_lang=en&sid=1. Accessed 10 Feb 2016.

MOPH (Ministry of Public Health). (2012). National Health Statistics Report in Lebanon. Beirut: Ministry of Public Health, Government of Lebanon. Available at: file:///C:/Users/Ewen/AppData/Local/Microsoft/Windows/INetCache/Content.Outlook/SRQ0M881/20120000_MoPH_National_Health_Statistics_Report_in_Lebanon.pdf. Accessed 15 Feb 2015.

Mughal, R. (2010, December 6) Hepatitis A spreading because of lack of awareness. *Tribune*. Available at: http://tribune.com.pk/story/86427/hepatitis-a-spreading-because-of-lack-of-awareness/ Accessed 25 Nov 2015.

Muhkalalati, G., & Kieke, S. (2015, October 26). Aid delivers food poisoning as residents' immunity 'extremely weak'. *Syria:Direct*. Available at: http://syriadirect.org/news/aid-delivers-food-poisoning-as-residents%E2%80%99-immunity-%E2%80%98extremely-weak%E2%80%99/. Accessed 20 Jan 2016.

Mukhashaf, M., & Miles, T. (2015, July 21). U.N. ship brings food aid to Yemen's Aden as fighting rages. *BBC News*. Available at: http://www.bbc.com/news/world-middle-east-29319423. Accessed 14 Feb 2016.

Naylor, H. (2014, November 19). In food-crazed Lebanon, a war over tainted chicken and messed-up meze. *The Washington Post*. Available at: https://www.washingtonpost.com/world/middle_east/in-food-crazed-lebanon-a-war-over-tainted-chicken-and-messed-up-meze/2014/11/18/76a14136-6e7b-11e4-a2c2-478179fd0489_story.html. Accessed 15 Mar 2016.

News Desk. (2011, October 21). EU ban on Egyptian fenugreek seeds extended. *Food Safety News*, . Available at: http://www.foodsafetynews.com/2011/10/ban-on-egyptian-fenugreek-seeds-extended/#.Vli8T3arSM8. Accessed 27 Nov 2015.

Olarte, O. (2011, March 22). 86% of food handlers complete hygiene and safety training. *Khaleej Times*. Available at: http://www.khaleejtimes.com/article/20110322/ARTICLE/303229813/1002, Accessed 4 Dec 2015.
Osaili, T. M., Jamous, D. O., Jamous, A., Obeidat, B. A., Bawadi, H. A., Tayyem, R. F., & Subih, H. S. (2013). Food safety knowledge among food workers in restaurants in Jordan. *Food Control, 31*(1), 145–150.
Palumbo, M., Harris, L. J., & Danyluk, M. D. (2013). Survival of foodborne pathogens on berries. FSHN13–12, Food Science and Human Nutrition Department, UF/IFAS Extension, Gainesville, Florida. Available at: https://edis.ifas.ufl.edu/fs236. Accessed 15 Mar 2016.
Patel, P. K., Mercy, J., Shenoy, J., & Ashwini, B. (2008). Factors associated with acute diarrhoea in children in Dhahira, Oman: a hospital-based study. *Eastern Mediterranean Health Journal, 14*(3), 571–578.
Pennington, R. (2014). *Abu Dhabi food authority provides clearer picture of safety rules in restaurants*. The National UAE. Available at: http://www.thenational.ae/uae/abu-dhabi-food-authority-provides-clearer-picture-of-safetyrules-in-restaurants. Accessed 24 Dec 2016.
Powell, D. A., Erdozain, S., Dodd, S., Costa, R., Morley, K., & Chapman, B. J. (2013). Audits and inspections are never enough: A critique to enhance food safety. *Food Control, 30*(2), 686–691.
ProMED-MENA. (2002, July 2). Mass poisoning in Bahrain. Available at: http://promedmail.org/post/20020703.4663. Accessed 15 Mar 2016.
ProMED-MENA. (2011, November 02). Foodborne illness – Saudi Arabia: (Al Madinah), Hajj pilgrims. Archive Number: 20111102.3259. Available at: http://promedmail.org/post/20111102.3259. Accessed 25 Jan 2016.
ProMED-MENA. (2014). Foodborne illness – Egypt: (Suez) school children, milk susp, RFI. Editorial comments. Archive Number: 20141023.2891144. Available at: http://promedmail.org/post/20141023.2891144. Accessed 11 Feb 2016.
Qasim, M. (2009, July 20). NIH confirms cholera case in capital. *News International*. Available at: http://www.thenews.com.pk/TodaysPrintDetail.aspx?ID=188937&Cat=6&dt=7/20/2009. Accessed 25 Nov 2015.
Raza, A. (2015, April 3). Polo Ground eatery fined for expired food, lack of hygiene. *The News International*. Available at: http://www.thenews.com.pk/Todays-News-5-310414-Polo-Ground-eatery-fined-for-expired-food-lack-of-hygiene. Accessed 25 Nov 2015.
Reeves, P. (2015). Pakistan's food safety czar declares 'war' on unhygienic food. *NPR Salt Blog*. Posted on December 29. Available at: http://www.npr.org/templates/transcript/transcript.php?storyId=459835645. Accessed 29 Dec 2015.
Reuters. (2016, January 26). After poisoning, rogue cop shoots dead 10 policemen in Afghanistan. *Hindustani Times*. Available at: http://www.hindustantimes.com/world/after-poisoning-rogue-cop-shoots-dead-10-policemen-in-afghanistan/story-rdRLBpczKFhRjJJerNoBxJ.html. Accessed 5 Feb 2016.
Robesyn, E., Micalessi, M. I., Quoilin, S., Naranjo, M., & Thomas, I. (2009). Cluster of hepatitis A cases among travellers returning from Egypt, Belgium, September through November 2008. *Euro Surveillance*, 14(3): pii=19095. Available online: http://www.eurosurveillance.org/ViewArticle.aspx?ArticleId=19095. Accessed 15 Mar 2016.
Saberi, M. & Scott, K. (2009, June 14). Dubai siblings die of suspected food poisoning. *Gulf News*. Available at: http://gulfnews.com/news/uae/general/dubai-siblings-die-of-suspected-food-poisoning-1.28846. Accessed 11 Mar 2016.
Sane, J., MacDonald, E., Vold. L., Gossner, C., Severi, E., on behalf of the International Outbreak Investigation Team. (2015). Multistate foodborne hepatitis A outbreak among European tourists returning from Egypt– need for reinforced vaccination recommendations, November 2012 to April 2013. *Euro Surveillance*, 20(4): pii=21018. Article DOI: http://dx.doi.org/10.2807/1560-7917.ES2015.20.4.21018. Accessed 15 Mar 2016.

Santacruz, S. (2015, May 14). Jail sentences and deportation for Qatar food poisoning. *Australian Food Safety News*. Available at: https://www.foodsafety.com.au/2015/05/jail-sentences-and-deportation-for-qatar-food-poisoning/. Accessed 12 Jan 2016.

Saseendran, S. (2014, January 31). 518 cases of food borne diseases recorded in Dubai. *Khaleej Times*. Available at: http://www.khaleejtimes.com/lifestyle/health-fitness/518-cases-of-food-borne-diseases-recorded-in-dubai. Accessed 11 Mar 2016.

Scallan, E., Hoekstra, R. M., Angulo, F. J., Tauxe, R. V., Widdowson, M.-A., Roy, S. L., Jones, J. L., & Griffin, P. M. 2011. Foodborne illness acquired in the United States—Major pathogens. *Emerging Infectious Diseases, 17*(1), 7–15.

Soomro, A. M., Seehar, G. M., Bhangar, M. I., & Channa, N. A. (2008). Insecticides in the blood samples of spray-workers at agriculture environment: The toxicological evaluation. *Pakistan Journal of Analytical & Environmental Chemistry, 9*, 32–37.

Sözen, H., Gönen, I., & Beydilli, H. (2014). An outbreak of norovirus gastroenteritis in a county in Turkey. *Journal of Microbiology and Infectious Diseases, 4*(1), 26–29.

Staff. (2013, July 16). Cholera claims five lives. *Dawn*. Available at: http://www.dawn.com/news/1029567/cholera-claims-five-lives. Accessed 25 Nov 2015.

Staff. (2015, December 30) Muscat municipality shuts down leading bakery in Ruwi. *Times of Oman*. Available at: http://foodsafety.einnews.com/article/304062229/7UuJiJ5FyCzvXi4j. Accessed 11 Jan 2016.

Tajkarimi, M., Ibrahim, S. A., & Fraser, A. M. (2013). Food safety challenges associated with traditional foods in Arabic speaking countries in the Middle East. *Trends in Food Science and Technology, 29*, 116–123.

Tanyeri-Abur, T., & Elamin, N. (2011). An Overview of International Investments in Agriculture in the Near East. Available at: http://www.fao.org/fileadmin/templates/est/INTERNATIONAL-TRADE/FDIs/Egy_Mor_Sud.pdf. Accessed 7 Feb 2016.

Tariq, M., Shahzad, A., & Ishtiaq, H. (2003). Pesticides in shallow groundwater of Bahawalnagar, Muzafargarh, D.G. Khan and Rajan Pur Districts of Punjab, Pakistan. *Environment International, 30*, 471–479.

Tavakoli, H. R., & Imani Fooladi, A. A. (2011). Determination of contamination with Clostridium botulinum in two species of processed and non-processed fish. [Article in Persian]. *Journal of Gorgan University of Medical Sciences, 13*(1), 79–87.

Technical Cooperation Department. (2003). "Regional Programmes for Food Security in the Near East: Towards Sustainable Food Security and Poverty Alleviation", High-Level Technical Workshop, Executive Summary, Jeddah, 8–9 October 2003. Food and Agriculture Organization, Rome. Available at: http://www.fao.org/docrep/MEETING/007/AD388E/ad388e00.htm#TopOfPage. Accessed 4 Dec 2015.

Templeton, T. (2015, March 17). Syria's shameful healthcare quagmire. *New Internationalist Blog*. Available at: http://newint.org/blog/2015/03/17/syria-healthcare/. Accessed 20 Jan 2016.

This is Staffordshire. (2008, November 8). Couple may take action on holiday nightmare sickness. *The Sentinel*. Available at: http://www.stokesentinel.co.uk/Couple-action-holiday-nightmare-sickness/story-12530636-detail/story.html. Accessed 27 Nov 2015.

Todd, E. C. D. (2001). Foodborne and waterborne disease in developing countries – Africa and the middle east. *Dairy Food and Environmental Sanitation, 21*(2), 110–122.

Todd, E. C. D., & Greig, J. D. (2015). Viruses of foodborne origin: a review. *Virus Adaptation and Treatment, 7*, 27–45.

Tucker, R, (2012, February 27). Afghan cops; food poisoning at border post. *CBS News*. Available at: http://www.cbsnews.com/news/afghan-cops-food-poisoning-at-border-post/. Accessed 4 Feb 2016.

Variyar, M. (2015, July 8). 45 Isis fighters die in 'Iftar Poisoning'; 100 more ill after eating Ramadan meals in Mosul. *International Business Times*, India Edition. Available at: http://www.ibtimes.co.in/45-isis-fighters-die-iftar-poisoning-100-more-ill-after-eating-ramadan-meals-mosul-638488. Accessed on 20 Nov 2015.

Walker, L. (2014, December 1). Amid growing complaints, SCH launches food poisoning hotline in Qatar. *Doha News*. Available at: http://dohanews.co/amid-growing-complaints-sch-launches-food-poisoning-hotline-qatar/. Accessed 12 Jan 2016.

WFP (World Food Programme). (2016). *Yemen: current issues and what the World Food Programme is doing*. World Food Programme, Rome, Italy. Available at: https://www.wfp.org/countries/yemen. Accessed 14 Feb 2016.

WHO (World Health Organization). (2015a). *WHO's first ever global estimates of foodborne diseases find children under 5 account for almost one third of deaths*. World Health Organization, Geneva. Available at: http://www.who.int/mediacentre/news/releases/2015/foodborne-disease-estimates/en/. Accessed 7 Feb 2016.

WHO (World Health Organization). (2015b).*WHO estimates of the global burden of foodborne diseases. Foodborne diseases burden epidemiology reference group 2007–2015*. World Health Organization, Geneva. Available at: http://apps.who.int/iris/bitstream/10665/199350/1/9789241565165_eng.pdf?ua=1. Accessed 9 Feb 2016.

Yaliban. (2014, November 18). Food scandal: Food poisoning rate in Lebanon lowest, says tourism minister. Yaliban Blog. Available at: http://yalibnan.com/2014/11/18/food-scandal-food-poisoning-rate-in-lebanon-lowest-says-tourism-minister/. Accessed 15 Mar 2016.

MIX
Papier aus verantwortungsvollen Quellen
Paper from responsible sources
FSC® C105338

If you have any concerns about our products,
you can contact us on
ProductSafety@springernature.com

In case Publisher is established outside the EU,
the EU authorized representative is:
Springer Nature Customer Service Center GmbH
Europaplatz 3, 69115 Heidelberg, Germany

Printed by Libri Plureos GmbH
in Hamburg, Germany